Stair Builders Handbook

By T. W. LOVE

Craftsman Book Company
6058 Corte del Cedro P.O. Box 6500 Carlsbad, CA. 92008

Library of Congress Cataloging in Publication Data

Love, T W
Stair builders handbook.

1. Stair building -- Handbooks, manuals, etc.

I. Title.
TH5667.L68 694'.68 74-4298

ISBN 0-910460-7-8 Thirteenth printing 1990

Chapter 1

STAIRWAY DESIGN AND CONSTRUCTION

This book has all the information necessary to design and construct the right frame staircase for any job. The tables in this handbook give precise dimensions for treads and risers for every rise between 3 feet and 12 feet and for every riser height from 6 inches to 8¼ inches. These tables will be invaluable in selecting a riser height and tread width that meet job requirements for total rise and total run, angle of incline, head room and length of carriage. The craftsman will also find here complete information on quantity of materials and settings for the steel square. Every stairway in this handbook uses the most widely accepted relation between riser and tread: Riser height plus tread width equals 17½ inches.

Chapter 2 explains the basic principles of stair design and construction and Chapter 3 shows how stairs are actually constructed. Chapter 4 covers layout and fitting of newel posts and hand rails. Chapter 5 defines the terms used in the tables and shows how the tables are used in practice.

Chapter 2

STAIR FUNDAMENTALS

There are many different kinds of stairs but all have two main parts: treads people walk on and the stringers, carriage or horse which supports the treads. A very simple type of stairway consisting only of stringers and treads is shown in figure 1. Treads in the type shown here are called plank treads, and this simple type of stairway is called a "cleat" stairway because of the cleat attached to the stringers to support the treads. A more finished type of stairway has treads mounted on two or more sawtooth-edged stringers and includes risers as shown in figure 2. The stringers shown here are cut out of solid pieces of lumber (usually 2" x 12") and are therefore called "cutout" or "sawed" stringers.

Stairways may be straight, curved, "L" shape, "U" shape or a combination of several shapes. Straight stairs, that is stairs that rise from one floor level to the next without changing direction, are most common. Straight stairs, however, may not be practical where horizontal space is limited. The chance of a harmful fall is greatest on long straight stairways. Any long straight run should include a landing to break falls and give the climber a place to rest while ascending.

Figure 3 shows three views of the common "L" shaped stairway with one landing. Notice the 90 degree change of direction in the stairway. Where space is limited "winders" or "pie shaped" treads may be substituted for the landing in the "L" shaped stairway. Figure 4 illustrates a typical layout of the tread where a "winder" is used in place of a landing. Note that the width of the tread 18 inches from the narrow end of each tread should be not less than the tread

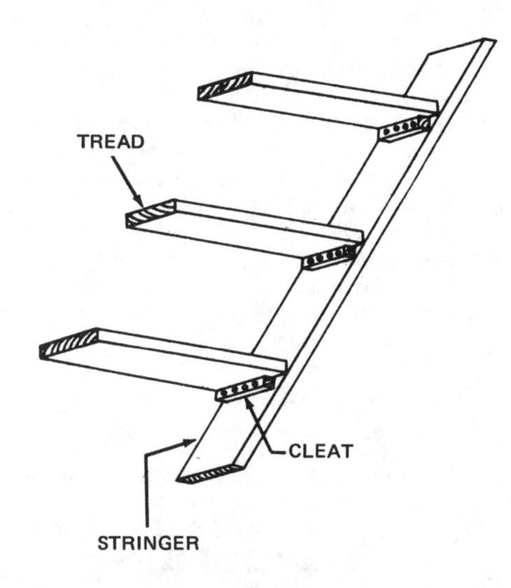

Cleat Stairway
Figure 1

width of the straight run. Many building codes require a full 9 or 10 inch run to each stair 12 inches from the narrow end of the stair. Some codes also require that the run of the stair be at least 6 inches at the narrow end of the stair. "Winders" should be avoided when possible because of the danger of falls on the narrow portions of the treads. Space limitations may make spiral or "U" shaped stairways more advisable. These designs are used when maximum rise is necessary in limited horizontal distance.

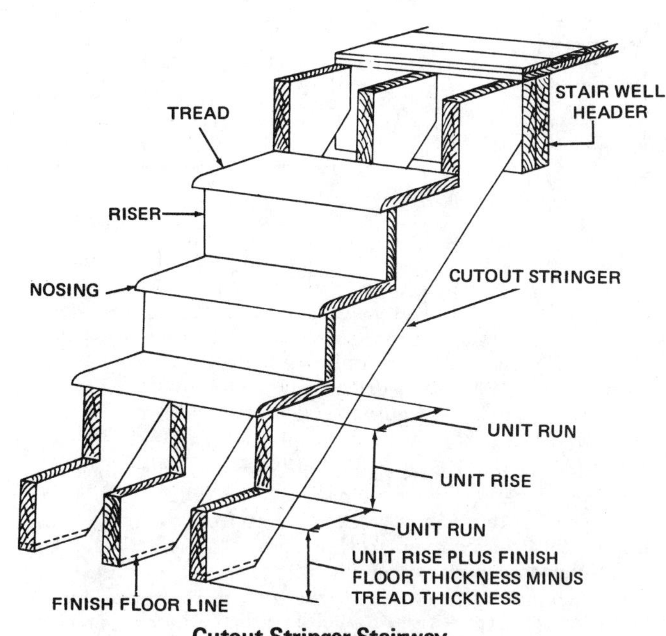

Cutout Stringer Stairway
Figure 2

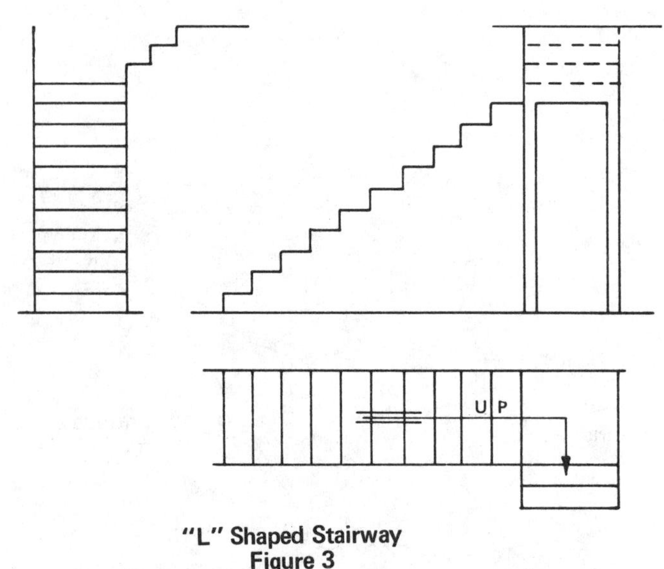

"L" Shaped Stairway
Figure 3

Since ancient times, architects and builders have recognized a natural porportion of rise to run of each stair. The rise of a stair is the vertical distance from the top of one horizontal level or tread to the top of the next tread. For most purposes, a rise of about 7 inches per stair is best. A rise much more than 7 inches per stair seems to tire the climber unnecessarily. A rise of much less than 7 inches per stair makes the stairway longer than necessary. Just as important as the rise of each stair is the width of each tread. Treads too wide or too narrow don't have the right "feel" to the climber and seem awkward. There is a length of rise and run which seems most comfortable to the largest number of adults. Usually this length is thought of as some riser dimension added to some tread dimension which equals 17½ inches. For example, where the riser is 7 inches the tread would have to be 10½ inches. By this "Rule of 17½" most any combination of rise and run will have a comfortable "feel" so long as the rise is not less than 6 inches or more than 8 inches per stair. A 6 inch rise (with an 11½ inch tread width) will result in an angle of climb of about 27 degrees. A rise of this type would be best suited for persons

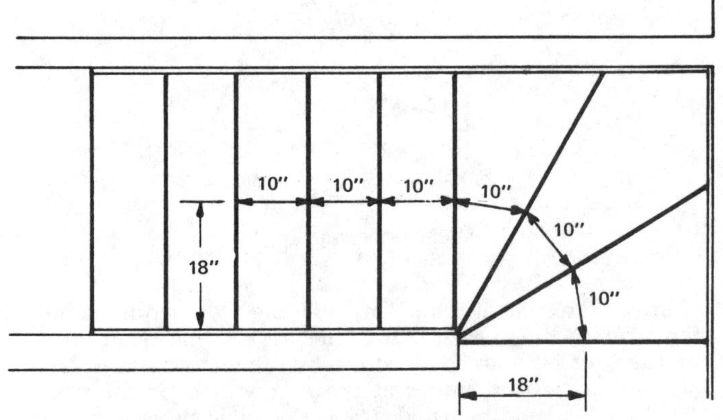

Winders Used In An "L" Shaped Stairway
NOTE: Some building codes require a minimum run of 6 inches at the narrow end of each stair.
Figure 4

of restricted physical ability or for stairways of monumental character such as an entrance to a public building. A rise of 8 inches (with a tread width of 9½ inches) produces an incline of about 40 degrees. Many basement stairs or stairs built in restricted areas may have up to a 40 degree angle of incline. Most interior stairs are designed with an incline of between 33 and 37 degrees as this produces a safe, economical stairway with the natural "feel" most adults find comfortable. Where enough horizontal space and head room are available, the designer should select a rise and run combination which yields an incline of between 33 and 37 degrees. Most building codes require a rise of no more than 7½ inches and a run of no less than 10 inches in other than private residential construction. In private homes most codes permit a rise of as much as 8 inches and a run of as little as 9 inches.

Several additional fundamentals of stair construction are widely recognized. First, when building a staircase where a door opens at the top of a stairway, such as in a basement, a landing is always provided at the top of the stairs. This landing should be long enough so that the door, when fully open, does not extend over the first step. The door should be hung so that it does not reduce the width of the landing by more than 3½ inches. All landings should be at least as long as the stairway is wide but should not exceed 4 feet in length if there is no change in direction of the stairway. See figure 5. Also, landings should be used to break any stairway which rises 12 feet or more. Landings are placed half way between the top and bottom of the staircase when possible.

Adequate head room must be maintained while ascending the stairs and many building codes prescribe minimum

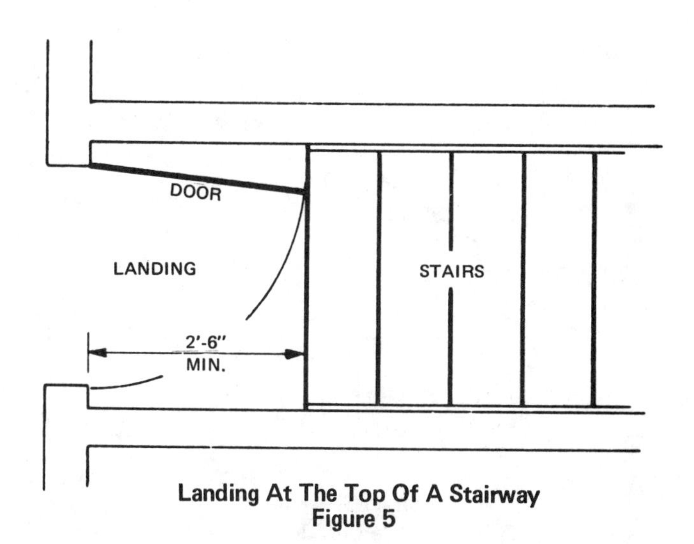

Landing At The Top Of A Stairway
Figure 5

head room requirements. Generally, head room of 6 feet 8 inches will be enough for main stairs and 6 feet 6 inches will be enough for basement stairs. See figure 6. The angle of incline of the stairway, which is determined by the rise and run of each stair, will dictate the well opening required to maintain adequate head room. Figures 7 and 8 illustrate typical framing details for well openings. When the length of a well opening is restricted, as in figure 8, the stairs should be designed with the well opening in mind.

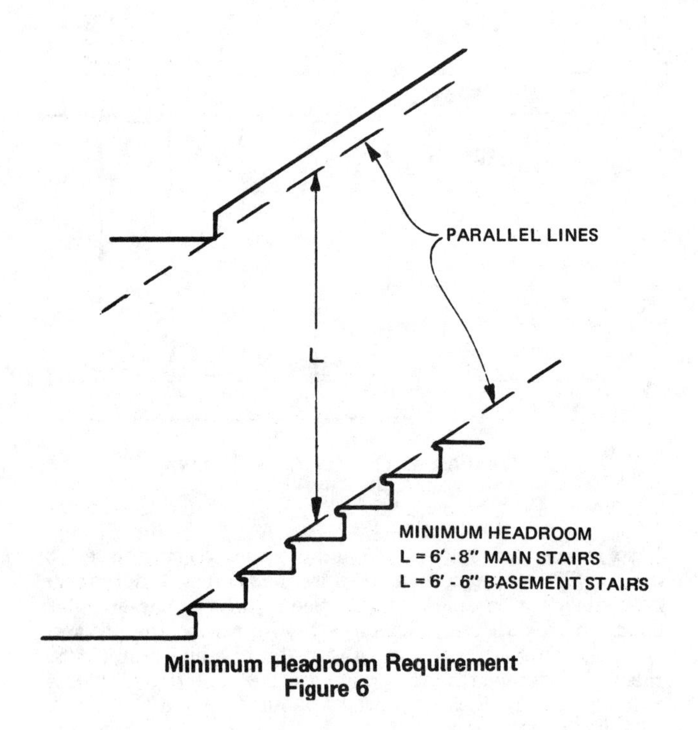

Minimum Headroom Requirement
Figure 6

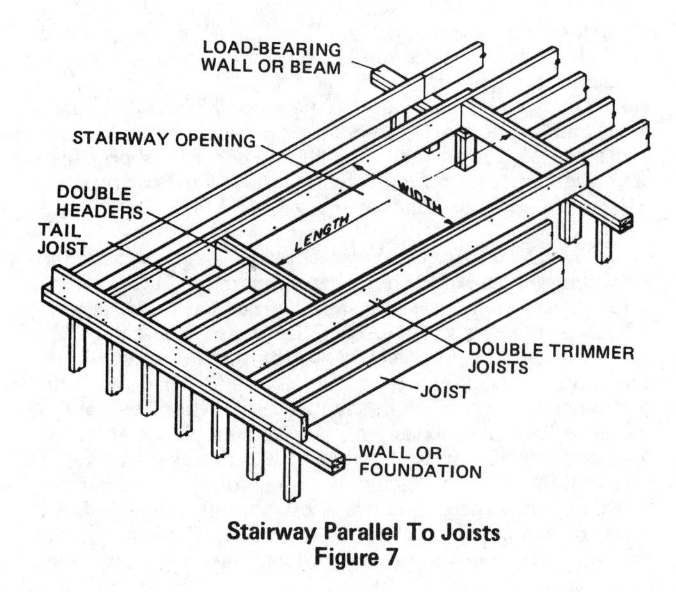

Stairway Parallel To Joists
Figure 7

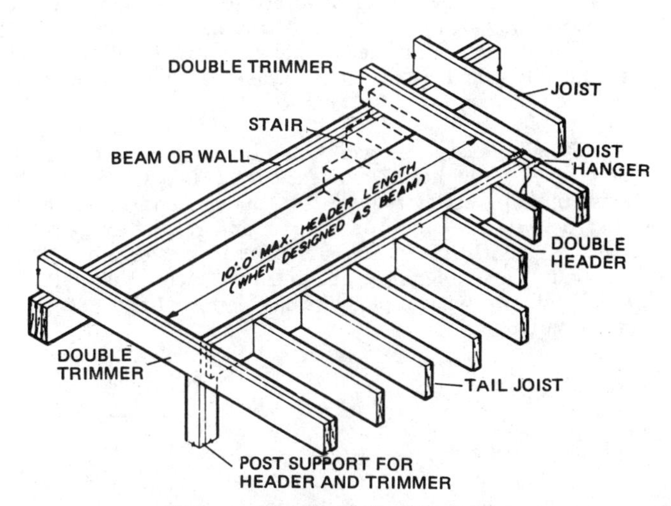

Stairway Perpendicular To Joists
Figure 8

Most stairs are built with "nosing" or a protruding edge on the front of each tread. The projection of one tread over the tread below is usually about 1¼ inches and is designed to give the climber a wider base of support on each stair. See figure 9. The nosing is not considered when calculating the run (horizontal distance) of each tread but must be considered when ordering materials.

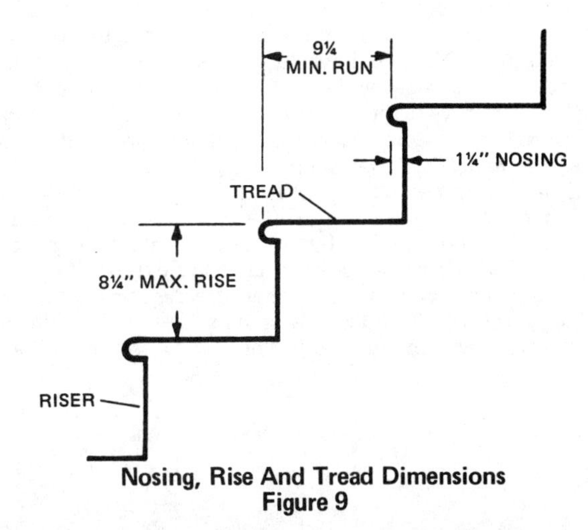

Nosing, Rise And Tread Dimensions
Figure 9

The width of main stairs should be not less than 2 feet 8 inches clear of the handrail. Many stairs are designed with a distance of 3 feet 6 inches between the enclosing side walls. This will result in a stairway with a width of about three feet. Split level entrance stairs are even wider. For basement stairs, a clear width of 2 feet 6 inches may be adequate. Most codes provide widths of at least 36 inches for other than private residential construction and at least 30 inches for private residential stairways. The handrails can project up to 3½ inches into the prescribed width.

A continuous handrail should be used for at least one

side of the stairway when there are more than four risers. Most building codes require at least two handrails on stairways in public buildings regardless of how few risers there are in the stairway. In public structures, stairways more than 88 inches wide should have an intermediate handrail approximately half way between the two side handrails. When stairs are open on two sides, a protective railing should be provided on each side even in private residential construction. The top of the handrail should be not less than 30 inches and not more than 34 inches above the nosing of each stair. The handrail must extend the full length of the stairs and, in other than private residential construction, one of the handrails should extend 6 inches beyond the top and bottom riser. The end of each handrail should terminate in a newel post or safety terminal. On landings or horizontal areas, the height of the handrail should be 2 feet 10 inches. Handrails which project from the wall should allow at least 1½ inch clearance between the wall and the handrail.

Finally, every stairway should be designed so that each riser is equal in height and each tread is equal in width. Most building codes require that the rise and run of each stair be within ¼ inch of the rise and run of every other stair in that flight. Often stairways in public use are required to meet tolerances of less than ¼ inch. The tables in this handbook will simplify the calculation of rise and tread dimensions.

Chapter 3

STAIR LAYOUT AND CONSTRUCTION

The construction of stairs is most often left to the most experienced carpenter on the job. Yet, most journeymen carpenters would agree that more material and time have been wasted tearing down poorly designed or poorly built stairways than in any other framing job. No matter how experienced the craftsman, each stairway presents its own design and construction problems. Even in highly repetitive jobs where many similar stairways are constructed on one site, each stairway to be built may be somewhat unique. The floor to floor rise of each stairway built must be measured with accuracy and the craftsman must select the right tread and riser combination so that every rise and run is within ¼ inch of every other rise and run. Many times the plans are inadequate or the actual floor to floor dimension does not correspond with the floor to floor dimension on the plans. Consequently, the craftsman who actually builds the stairway is required to design the stairway before he builds it. More and more stairways are being designed since the increasing value of land recommends multiple story structures.

Stair design and construction is not complex and can be mastered by anyone with patience and an understanding of elementary carpentry. Let's assume that the first and second floors have been framed out and an opening has been left for the stairway. The long dimension of the stairway opening may be either parallel or at right angles to the joists. However, it is much easier to frame a stairway opening when its length is parallel to the joists.

For basement stairways the rough opening may be about 9 feet 6 inches long by 32 inches wide (two joist spaces). Openings in the second floor for main stairways are usually a minimum of 10 feet long. Widths may be 3 feet or more. Depending on the short header required for one or both ends, the opening is usually framed as shown in figure 7 when joists parallel the length of the opening. Nailing should conform to that shown in figure 10.

When the length of the stair opening is perpendicular to the length of the joists, a long doubled header is required as in figure 8. A header under these conditions without a

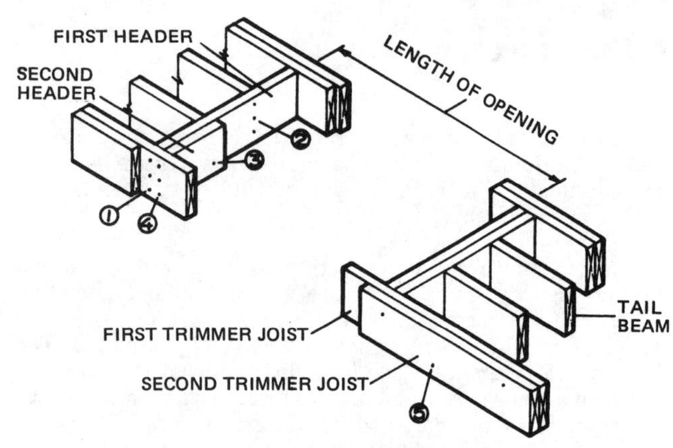

Framing for floor openings: (1) Nailing trimmer to first header; (2) nailing header to tail beams; (3) nailing header together; (4) nailing trimmer to second header; (5) nailing trimmers together

Figure 10

supporting wall beneath is usually limited to a 10-foot length. A load-bearing wall under all or part of this opening simplifies the framing immensely, as the joists will then bear on the top plate of the wall rather than be supported at the header by joist hangers or other means.

The framing for an "L"-shaped stairway is usually supported in the basement by a post at the corner of the opening or by a load-bearing wall beneath. When a similar stair leads from the first to the second floor, the landing can be framed-out. See figure 11. The platform frame is nailed into the enclosing stud walls and provides a nailing area for the subfloor as well as a support for the stair carriages.

Once the rough opening is framed, measure the exact distance from the first floor to the second floor. If an intermediate landing is planned, measure the vertical distance between the first floor and the landing. Estimate how much the finish flooring material on both the upper and lower floors will increase or reduce this distance and adjust the distance you measured accordingly. Find the finished *Floor to Floor Rise* on the top line of the tables which follow Chapter 5 in this handbook. Each *Floor to Floor Rise* dimension has several possible riser and tread combinations. Select the riser and tread combination that best suits the well opening, total run available and materials on hand. Remember, when possible, select a stair rise and run combination which yields an angle of incline between 33 degrees and 37 degrees and has the smallest riser "overage" or "underage". The last two figures in the column you have selected for your stairway give you the settings for your steel square. You will use these figures to locate the risers and treads on the stair carriage.

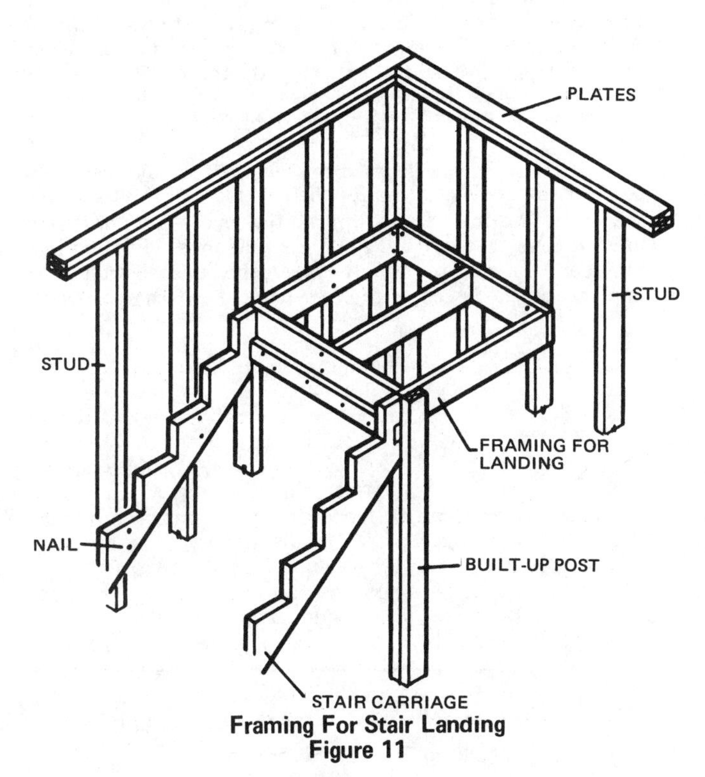

Framing For Stair Landing
Figure 11

The carriage supports the load on the stairs and can be made in either of several ways. No matter whether you are building a "cleat", "sawed" or "housed" stairway, you must cut the carriage to fit and locate the position of treads and risers on the carriage. Let's assume you are building a cut-out or sawed stair carriage. These stair carriages are made from 2 x 12 planks. The effective depth below the tread and riser notches must be at least 3½ inches (See figure 12). Such carriages are usually placed only at each side of the stairs. However, an intermediate carriage is required at the center of the stairs when the treads are 1 1/16 inch thick and the stairs are wider than 2 feet 6 inches. Three carriages are also required when treads are 1 5/8 inch thick and stairs are wider than 3 feet. The carriages are fastened to the joist header at the top of the stairway or rest on a supporting ledger nailed to the header.

Select a sound 2 x 12, 18 to 20 inches longer than the *Length Of Carriage* dimension you find in the appropriate table following Chapter 5 and begin at one end of the plank. First you must cut the floor line which will rest squarely on the lower floor. Lay the body or long end of the framing square across one corner of the plank. Move the square so that the tread distance intersects the top of the plank on the body and the rise distance intersects the tongue on the top of the plank. See figure 13A.

Sound craftsmanship requires that these dimensions be marked and cut accurately. A set of stair gauges or some device to hold the correct dimensions will be useful here. Mark the outline of both the rise and tread dimensions on the plank. Extend the tread dimension in a straight line to the bottom edge of the material and cut along this line.

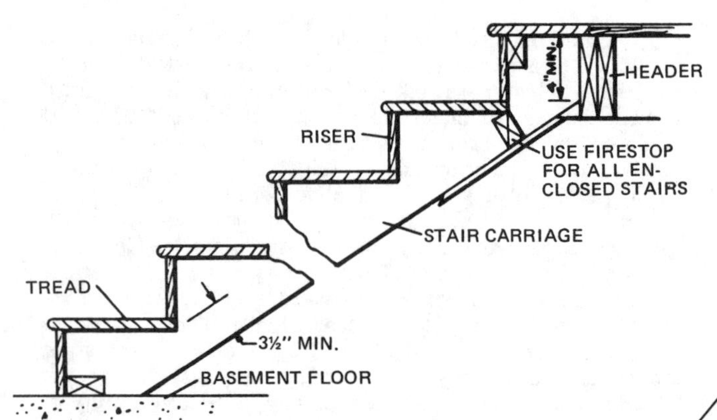

Carriage Detail
Figure 12

This is the floor cut. If this is to be a cutout carriage, cut along the line marked for the first rise (See figure 13B).

Next, place the tread dimension on the square body so that it meets the top edge of the first rise and the rise dimension on the tongue so that it meets the top edge of the carriage. See figure 14.

Repeat this process marking and cutting tread and rise dimensions for as many steps as there are in the stairway. When the last rise is reached, extend the line of the last rise to the bottom edge of the carriage. See figure 15. Cut along this extended line if the carriage is to meet the end of the wellhole as illustrated in figure 16C. If the carriage is to meet the wellhole as in 16A or 16B, allow the appropriate additional material beyond the last rise and cut parallel to the last rise. For additional strength, the carriage may be supported below the header or extend beyond the end of the well opening below the upper floor.

After the carriages have been cut, they should be "dropped" or lowered to allow for the thickness of the tread and the finish flooring. You will recall that every rise dimension must be the same within a ¼ inch tolerance. The carriage has been designed to reach between the finished first floor and the finished second floor or landing. If the carriage is

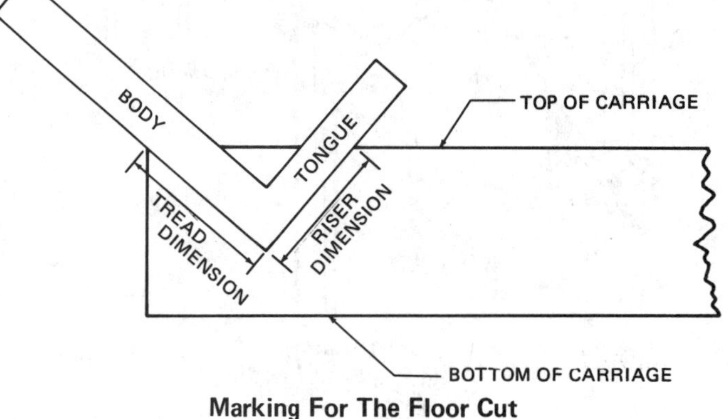

Marking For The Floor Cut
Figure 13A

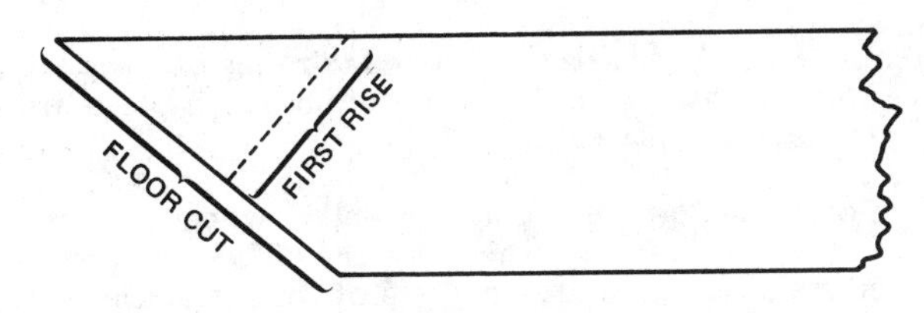

Floor Cut Made And Marked For First Rise
Figure 13B

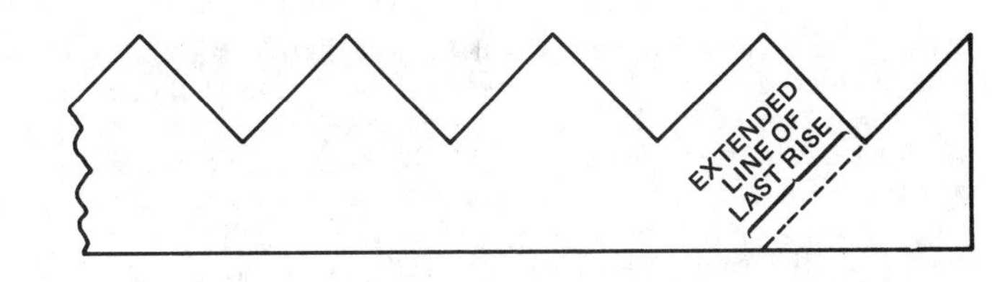

Cutting Top End Of Carriage
Figure 15

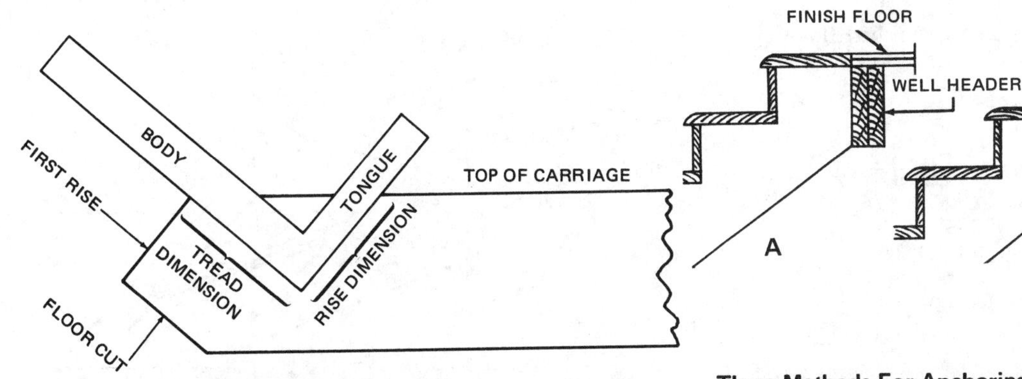

Floor And First Rise Cuts Made And Measuring For First Tread And Second Rise
Figure 14

Three Methods For Anchoring The Upper End Of A Stairway

Figure 16A Full Tread Width Extension
Figure 16B Partial Tread Width Extension
Figure 16C Top Riser Flush Against Header

installed on the subfloor, it will be lower than the height you planned for by the thickness of the finished flooring material. However, the tread material will raise the level of the tread surface above the top of each cutout on the carriage. Usually the tread thickness will be greater than the depth of the finish flooring material. When this is the case, this difference must be cut from the bottom end of each carriage parallel to the floor cut. In figure 17, 5/16" has been marked to be cut off the lower end of the carriage. The height of the finish flooring material, ¾", has been subtracted from the thickness of the tread material, 1 1/16 inch. When the stairway is completed, the rise of each step, including the first step and last step, will be 7½ inches. When the carriage rests on the finished floor level, the carriage should be "dropped" the full thickness of the treads. Once the carriage is cut and "dropped" it should be checked for fit. Each tread cutout should be level when the carriage is in place.

Stair supports are usually installed when the carriage cannot be anchored to adjacent walls. These supports are called stair bridges and usually are of 2 x 4 material. They are framed similar to any partition except that the top plate is parallel to the stair carriage.

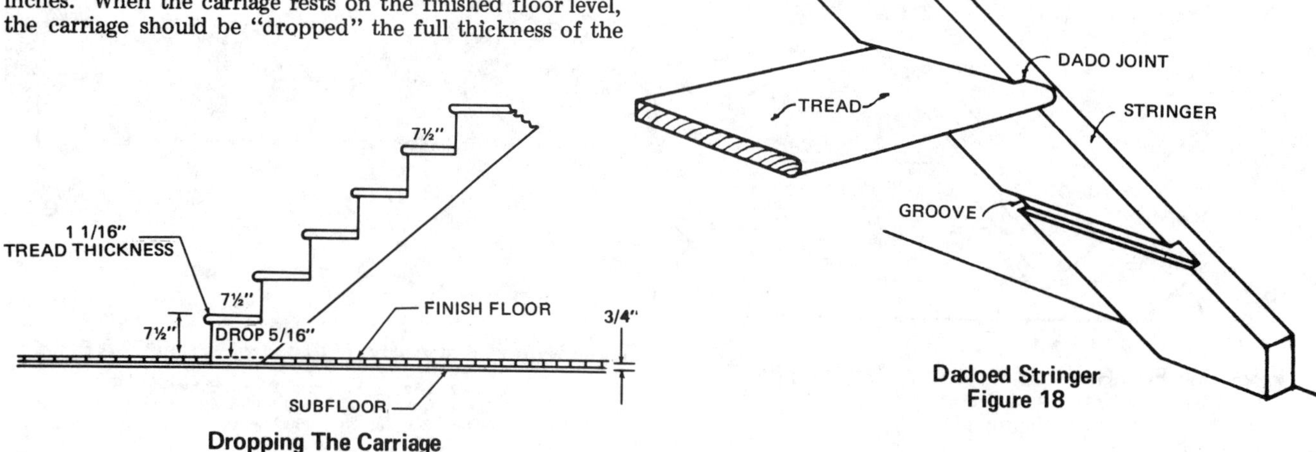

Dropping The Carriage
Figure 17

Dadoed Stringer
Figure 18

Cleat stairways (consisting only of a carriage and tread as shown in figure 1) should use 2" x 10" stock for the carriage. This design is not as desirable as the cutout carriage style where a wall is adjacent to one side of the stairway or where appearance of the finished stairway is important. The stringers should be at least 1 5/8 inch thick and wide enough to give a full width bearing for the tread. If cleats are used, they should be at least 25/32 inch thick, 3 inches wide and as long as the width of the tread. The treads can be only 1 1/16 inch thick unless the stairs are more than 3 feet wide.

A similar open stairway uses dado joints instead of cleats to support the treads. See figure 18. This type of construction is quite common in steep stairs or ladders for attic or scuttle openings. If dado joints are used, they should be only one third as deep as the stringer is thick.

How To Build Cleated Or Dadoed Stairs

1. Select the required stock. Use clear dressed stock free from defects.

2. Lay out the stringers in the same way as in laying out carriages.

   NOTE: If the treads are to butt against the stringers and are to be supported by cleats, the tread marks on the stringers represent the tops of the finish treads and the top of the cleat should be the thickness of the tread below this line. See figure 19. If the treads are to be dadoed into the stringers, assume that the tread marks on the stringers represent the tops of the dado cuts.

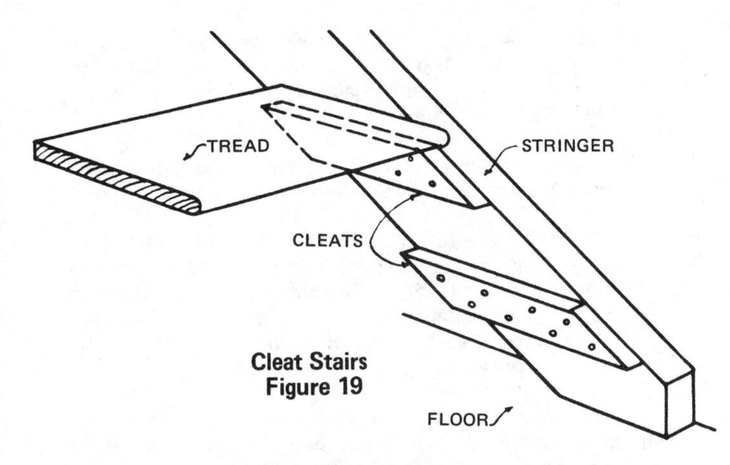

Cleat Stairs
Figure 19

3. Determine the length of the treads and cut the required number to this length.

4. Cut the required number of cleats and chamfer the edges that will show.

5. Nail the cleats to the proper marks below the tread marks on the stringers. Use nails long enough to reach within ½ inch of the combined thickness of the stringers and cleats.

6. Assemble the treads in place on the cleats or in the

dadoes and nail through the stringers into the ends of the treads. Use 16d common nails if the stringers are 1 5/8 inch thick and cleats are used, or 10d casing nails if dado joints are used.

7. Square the assembled stairs and fasten them in place in about the same way as stair carriages are fastened.

   NOTE: If the lower floor and side wall is of masonry, some means should be used to fasten the stringers firmly to these surfaces. Expansive shields and lag screws or wood blocks inserted into the masonry may be used for this purpose.

A somewhat more finished staircase for a fully enclosed stairway combines the rough notched carriage with a finish stringer along each side (figure 20). The finish stringer is fastened to the wall before carriages are fastened. Treads and risers are cut to fit snugly between the stringers and fastened to the rough carriage with finishing nails (figure 20). This may be varied somewhat by nailing the rough carriage directly to the wall and notching the finished stringer to fit (figure 21). The stringers are laid out with the same rise and run as the stair carriages, but they are cut out in reverse as shown in figure 21.

The risers are butted and nailed to the riser cuts of the wall stringers and the assembled stringers and risers are laid over the carriage. The assembly is then adjusted and nailed so that the tread cuts of the stringers fit against the tread cuts of the carriages. The treads are then nailed on the

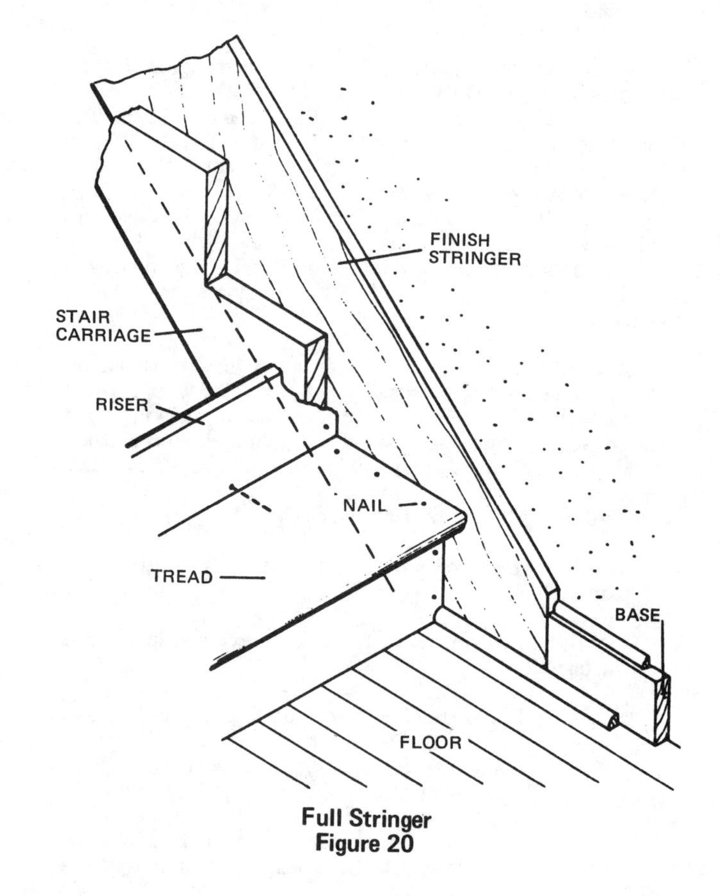

Full Stringer
Figure 20

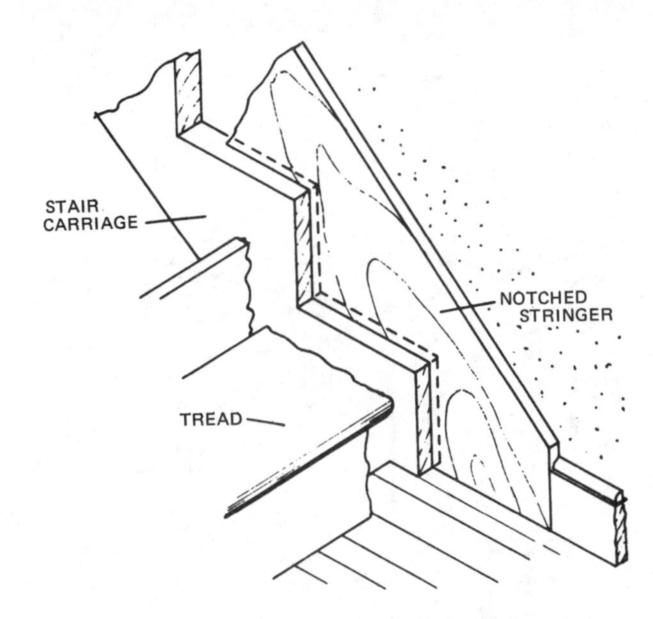

Cut Out Or Notched Stringer
Figure 21

tread cuts of the carriage and butted to the stringers the same as in figure 20. The finish stringer may be 25/32 inch or 1 1/16 inch thick and wide enough to cover the intersection of the tread and riser cut in the carriage and to reach about 2 inches beyond the tread nosing. This width is usually 12 inches to 14 inches. If the stringer is not cut out at the riser and tread marks, it is laid out in about the same way as a rough carriage. Only the level cut for the floor, the plumb cut for the header, and the plumb cuts at the top and bottom of the stringer for the baseboards are made. This is perhaps the best type of construction to use when the treads and risers are to be nailed to the carriages. This method saves time and labor in installing stairs but it is difficult to prevent squeaks when the stairs are stepped on. Another disadvantage is that any shrinkage of the frame of the building causes the joints to open up and to present an unsightly appearance. The notched stringer method has the advantage of having the two stringers tied together by the nailing of the risers to them. This prevents the two side stringers from spreading and showing open joints at the ends of the treads. Since the risers are nailed to the stringers, the face nailing of the risers to the carriages as in figure 20 is eliminated. This type of stringer is used where the carriages were fitted permanently to the bridgework at the time when the building was framed.

Sometimes the treads are allowed to run underneath the tread cut of the stringer. This makes it necessary to notch the tread at the nosing to fit around the stringer.

How To Build Stairs With A Full Stringer

NOTE: It is assumed that the temporary carriages have been removed and that the finish lumber to be used in the stair is clear stock and sanded.

1. Lay out and cut the stair carriages as outlined above for cut out stringer carriages. Use 1 5/8 inch stock.

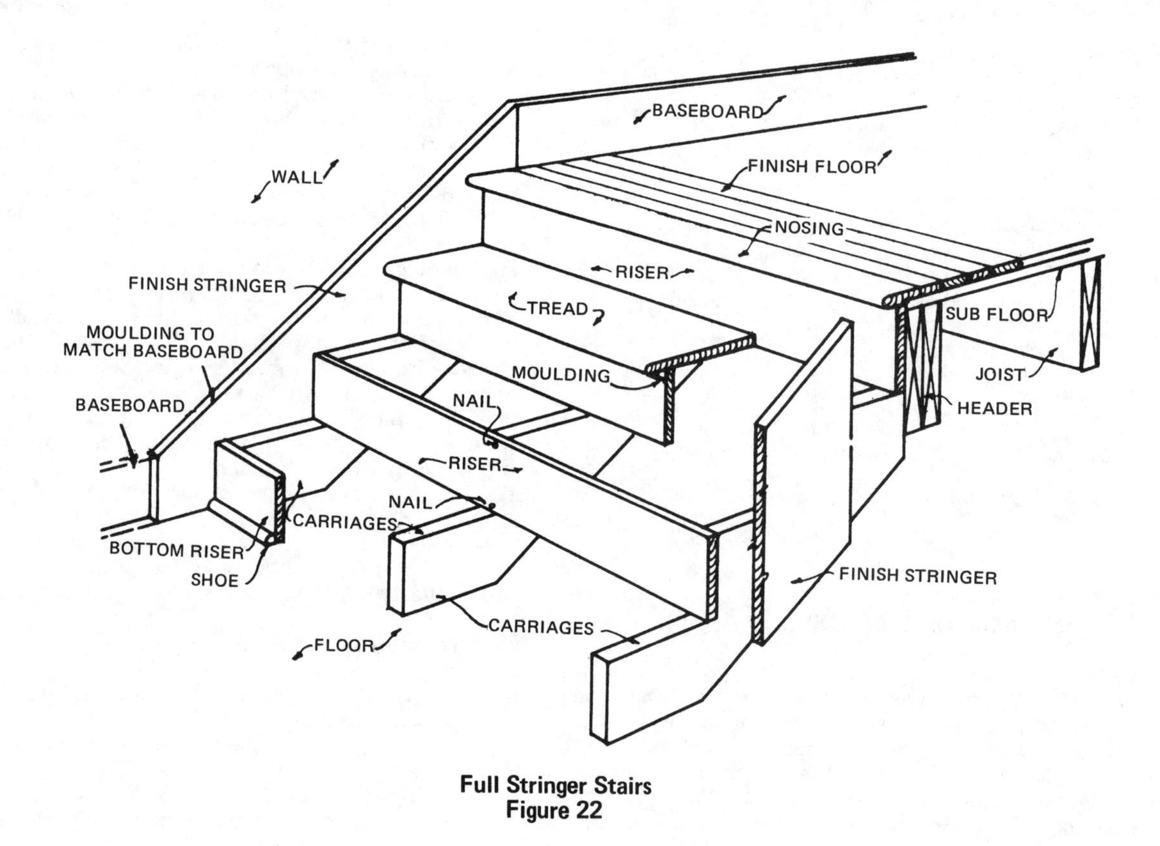

Full Stringer Stairs
Figure 22

2. Select the stair stringers of the proper thickness, width and length and lay out the exact length and the bottom and top cuts. Make the line of the bottom cut level to meet the floor line and the line of the baseboard cuts plumb to meet the baseboard at the top and bottom of the stair. See figure 22.

3. Lay out a right and a left hand stringer and cut along the bottom and top marks on the stringer.

4. Nail the right hand stringer to the stair carriage, keeping the bottom level cut of both pieces even and the top edge of the stringers at least 4 inches above the cut outs of the carriage. See figure 22.

   NOTE: Be sure the bottom edge of the carriage is parallel with that of the stringer.

5. Nail the left hand stringer to a carriage in the same manner.

6. Place the built-up stringers in their proper places against the header and side walls of the stair opening. Nail them to the headers and side walls.

7. Select the required number of riser boards. These are generally 25/32 inch thick, 7½ inches wide and as long as the distance between the inside faces of the two stringers.

8. Rip a riser board for the first riser to a width equal to the height of the first tread cut of the carriage above the floor. See the bottom riser of figure 22. Square the ends of this riser to length so that it fits tightly against the finish stringer on each side of the stairs.

9. Face nail the riser board to the riser cuts of the three carriages with two 8d finishing nails in each carriage. Keep the nails about 1 inch from the top and bottom edges of the board. See figure 22.

   NOTE: The nails at the top will be covered by the moulding that is to be placed underneath the tread nosing and the nails at the bottom of the riser will be covered by the floor shoe, or on the other risers, by the thickness of the tread.

10. Cut, fit and nail the remaining riser boards in a similar manner.

11. Cut the treads to the same length as the riser boards and fit them in place. Face nail them with three 8d finishing nails at each tread cut of each carriage.

    NOTE: After each tread is face nailed, drive 8d common nails through the back of each riser board into the back edge of each tread. Space the nails every 8 inches between the carriages.

12. Cut a piece of rabbeted nosing stock to the same length as a tread and face nail it to the top edge of the top riser and to the subfloor. Set all nails that will

show and sand the surfaces where necessary.

13. Fit and nail cove moulding under the nose of each tread.

How To Build Stairs With Cut Out Stringers (Figure 21)

NOTE: When cut-out stringers are used, the same general procedure is followed except that the finish stringers are cut to fit against the tread cuts of the carriage and the face of the riser boards. The risers extend the full width of the stair opening. It is assumed that the carriages are permanently placed so the finish stringers, risers and treads will fit them.

1. Lay out the finish right and left hand stringers using the same figures as on the carriages. Mark the cuts at the top and bottom of the stringers for the floor and baseboard cuts. These cuts should be the same as shown in figure 22.

2. Lay out the riser and tread cuts.

3. Cut along these marks with a crosscut saw. Be careful not to break the wood where the riser and tread cuts meet.

4. Temporarily nail a stringer to the wall on the left hand side of the stair opening. Keep the riser cuts of the stringer approximately 1¼ inch from the riser cuts of the carriages and the tread cuts of the stringer on top of the tread cuts of the carriage.

5. Measure the distance between the finished walls at the top and bottom of the stairs to find the lengths of the riser boards.

6. Cut riser boards ½ inch shorter than these lengths.

7. Place the top riser board between the riser cut of the left hand stringer and the riser cut of the carriage.

8. Place the right hand stringer in the proper position on the right hand wall of the opening.

9. Mark the face of the riser board along the inside surface of both stringers. Be sure that there is a space of ¼ inch on each side between the outside of the stringers and the wall.

10. Follow the same procedure for the bottom and intermediate riser boards.

11. Remove the stringers and risers. Face nail the risers to the riser cuts of the stringer, keeping the tops of the risers tight against the tread cuts of the stringers and the face of the stringer in line with the marks on the faces of the risers. Nail the riser boards to both stringer in the same way.

12. Replace the assembled stringers on the carriages and adjust them so that the riser boards are tight against the riser cuts of the carriage and the tread cuts of the

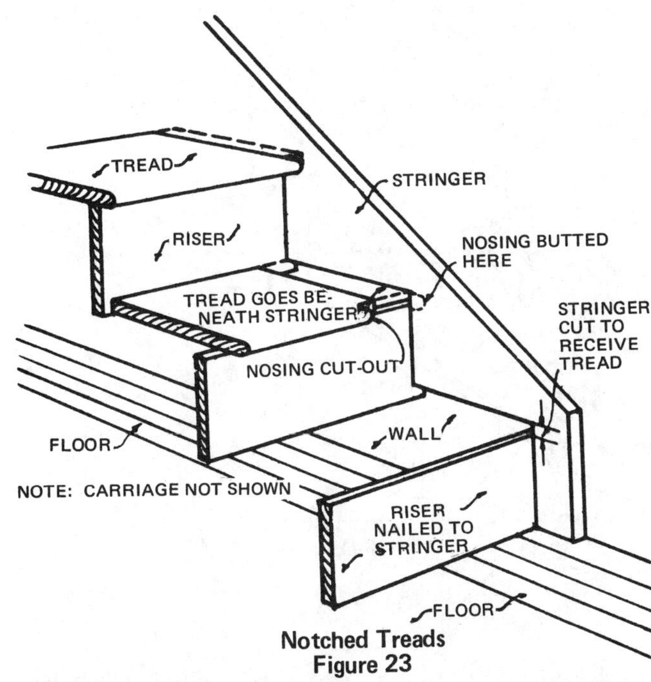

Notched Treads
Figure 23

stringers are tight against the tread cuts of the carriages.

13. Nail the stringers to the walls in this location.

14. Cut, fit and nail the treads in the same way as for the built-up stringer.

NOTE: Some carpenters prefer to allow the treads to run under the stringers the same as the risers. See figure 23. The stairs in this case are built in practically the same way except that in laying out the cut-out stringer no deduction is made at the bottom riser mark for the thickness of the tread. The treads are notched to fit underneath the stringers at the nosings.

The Housed Stringer Stairway

Housed stair stringers are frequently considered a mill job but these stringers may be housed by the carpenter even if modern power woodworking tool are not available. The methods used in laying out the stringers, cutting the risers and treads and assembling the stairs are quite similar to these processes in other stairwork.

A closed type of housed stringer staircase is enclosed by walls on both sides of the staircase. The stringers are housed out to receive the ends of the treads and risers. This type is similar to the cut-out stringer stairway in which the treads and risers extend through the thickness of the stringer. In this case they only extend approximately 3/8 inch into the stringers. The treads and riser boards are then wedged into the dado joints of the stringers.

Figure 24 shows a housed stringer in various stages of construction. The lay out of the treads and risers is similar to that of the cut-out type of stair stringer except that before the stringer is laid out, a mark is gauged about 1½ inches from the bottom edge on the face of each stringer. See line X-Y, figure 24. This line acts as a measuring line the same as in laying out roof rafters. The purpose of using this line instead of the edge of the stringer is to provide room for the riser and tread boards to be supported by the wedges. The steel square at A, figure 24, shows the position in which the square is used on the gauge line in laying out the stringer.

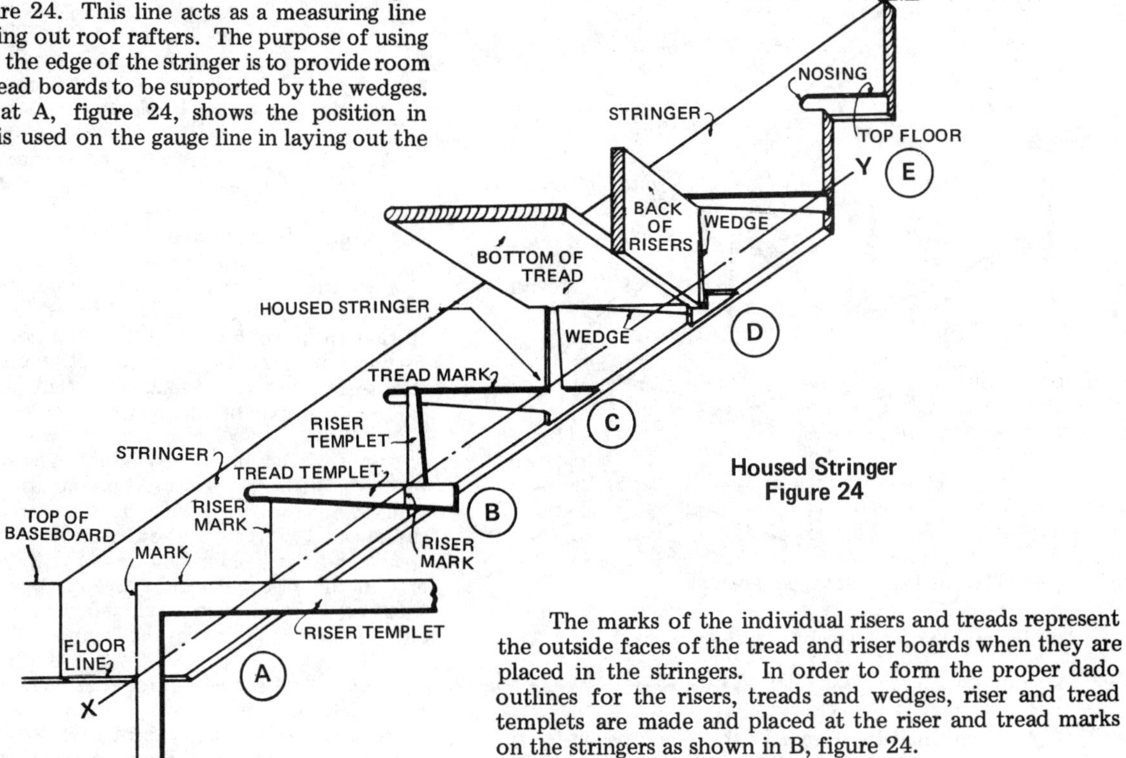

Housed Stringer
Figure 24

The marks of the individual risers and treads represent the outside faces of the tread and riser boards when they are placed in the stringers. In order to form the proper dado outlines for the risers, treads and wedges, riser and tread templets are made and placed at the riser and tread marks on the stringers as shown in B, figure 24.

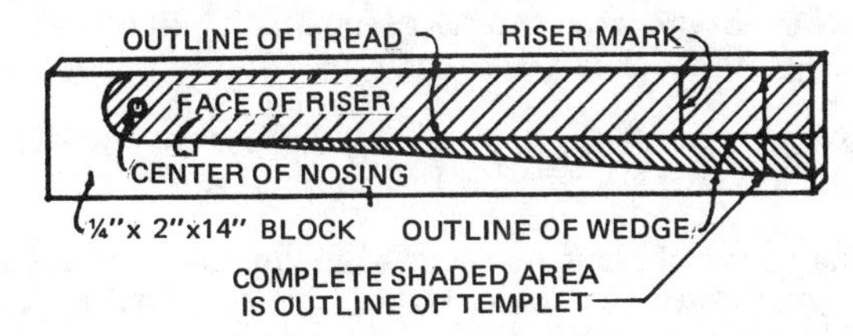

Tread Templet
Figure 25

The tread templet is laid out by drawing a straight line on the face of a board about ¼ inch by 2 inches by 14 inches. On this line is drawn the exact end section of the tread stock to be used and also the outline of the wedges to be used to tighten the tread boards in the dado joints of the stringer. See figure 25. The exact width of the tread stock including the nosing is measured and marked on the templet as shown by the riser mark in figure 25. This mark is also placed on the other face of the templet so that the templet may be used for both the right and left hand stringers. The center of the nosing profile is also located as shown in figure 25. This mark is transferred to the other side by drilling a small hole through the templet.

The riser templet is laid out in the same way by using the end section of a riser board and the wedge outline on a thin piece of board for the tread templet. See figure 26. This templet is placed with its straight edge along the riser marks on the stringer. The outline of the templet is then marked along the tapered side of the templet on the stringer face. See B, figure 24.

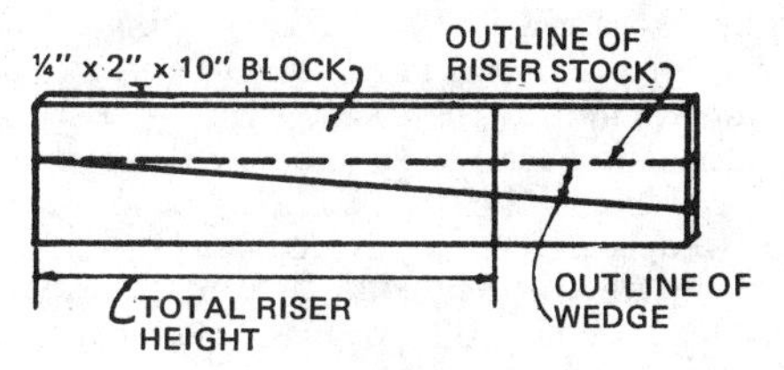

Riser Templet
Figure 26

The tread templet is placed in a similar manner with its straight edge on the tread mark of the stringer and the riser mark (figure 25) in line with the riser mark on the stringer. See B, figure 24. The location of the hole in the templet is marked on the stringer with a scratch awl. This point shows the center of the hole that is to be bored in the stringer to form the round end of the dado to fit the nosing of the tread.

The layout at the top end of the stringer where the baseboard and the stringer meet is shown in E, figure 24. The nosing at the floor line is laid out with the tread templet to show the same nosing projection from the face of the riser as on the other treads. The nosing should be housed out in the same manner as the other treads. The riser cut directly below the nosing is cut completely through the stringer and the top riser is nailed to this surface. The depth of the tread and riser dado cuts is shown at C, figure 24.

The length of the treads is determined and they are inserted into the tread dadoes. They are then wedged, glued, and nailed from the outside of the stringers and into the ends of the treads. All risers, except the top one, are cut to the same length as the treads. The top riser is about 1¼ inch longer as it does not fit into the dado cuts but extends to the outside of the stringers. The other riser boards are inserted, wedged, glued and nailed into the riser dadoes the same as the treads. Moulding is sometimes fitted between the stringers under the nosing of the treads. Figure 24 at D shows one riser and tread in place in the housed dadoes and the wedges glued to the undersides of the treads and risers.

Some stairs are built with a rabbeted joint at the back of the tread and also at the top of the riser. However, if the treads and risers are properly jointed, driven up tight, wedged and nailed in this position, the butt joint is satisfactory and saves much labor.

Stock for the various parts of stairs is generally obtained from a mill in partially finished form. Treads and risers may be obtained completely machined and sanded but somewhat oversize. Rabbeted nosing stock for the edge of a landing or for the top step is usually obtainable in rough lengths. Standard wedges are also available. If the mill is furnished with the exact dimensions of the stair well, the parts can be completely machined and then assembled on the job.

How To Build Housed Stringers

The first step in building a housed stringer stairway is the making of the tread and riser templet.. It will be assumed that the tread stock is 1 1/16 inch thick and is nosed.

1. Select a straight piece of stock approximately ¼ inch thick, 2 inches wide and 14 inches long.

2. Lay out and plane one edge of the templet straight and square and taper the opposite edge as shown in figure 25.

   NOTE: Be sure the width of the templet at the nosing is exactly 1 inch.

3. Select a similar piece of wood and lay out and cut the templet for the risers. See figure 26.

How To Lay Out And House The Stringers

1. Select the stringer stock. This should be at least 1 1/16 inch thick and from 10½ to 14 inches wide. The length depends on the length of the stairs. Allow about one foot at each end for the top and bottom cuts of the stringer.

2. Sometimes regular tread stock is used for stringers. If so, plane the nosed surface flat so that moulding may be fitted to this surface after the stairs are in place.

3. Plane the bottom edge of the stringer straight.

4. Set a marking gauge to about 1½ inches and gauge a line from this edge. See line X-Y, figure 24.

5. Lay out the tread and riser marks with a steel square. Use a scratch awl or a fine hard lead pencil in marking the stringers.

   NOTE: It is well to make a pitch board or to use a "fence" on the steel square. Both side stringers must be cut to exactly the same dimensions so that the stairs are square and sound.

6. Lay out the right and left hand stringers and check them for accuracy and length before doing any templet or housing work. Figure 27 shows how to check the stringers for length and accuracy of layout.

7. Place the tread templet with its straight edge exactly over a tread mark on the stringer. Adjust it so that the riser mark on the templet is also exactly in line with the riser mark on the stringer. Make a point on the stringer by putting the scratch awl through the hole in the templet to locate the center of the nosing. Also mark along both sides of the templet on the stringer. See B, figure 24.

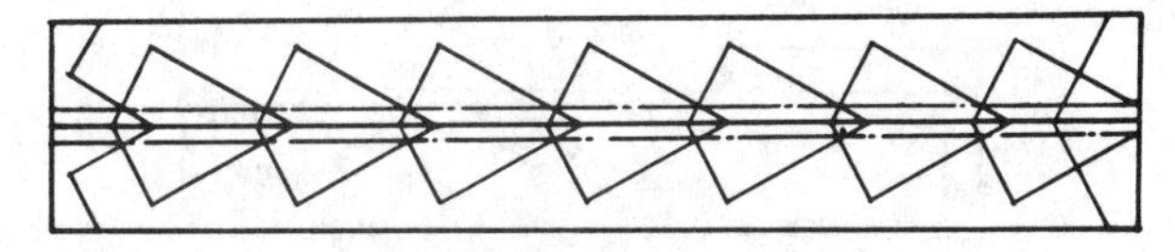

Checking Stair Layout
Figure 27

8. Bore a hole with a Forstner or center bit using the point marked through the tread templet as a center. Bore the hole approximately 3/8 inch deep. Locate the center of a second hole so it will overlap the first and be within the top and bottom lines of the tread. See figure 28.

9. Use the back of a 1½ inch butt chisel to chisel along the tread marks between the holes (figure 28).

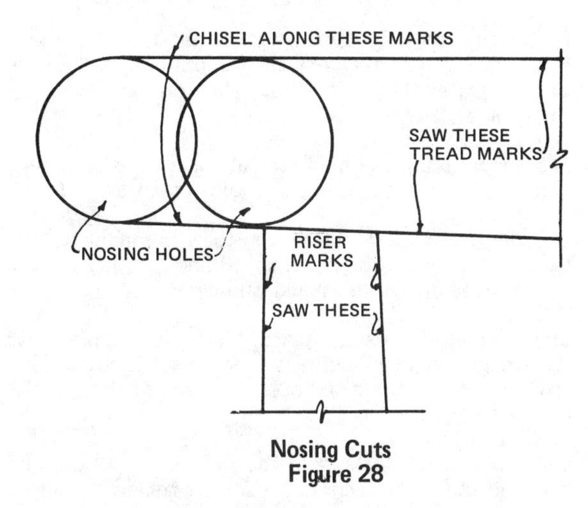

Nosing Cuts
Figure 28

10. Cut 3/8 inch deep along the tread marks with a saw. Start with the tip of the saw at the holes and continue

back the length of the tread mark.

NOTE: In using the back saw to cut the housed joints, tip the saw a trifle so that the edges of the joint will be undercut. This will allow a tighter joint between the top of the tread and the stringer cut and will help hold the wedge in between the bottom of the tread and the edge of the housing.

11. Chisel out the stock between the cuts. Take this stock out carefully and to a depth of about ¼ inch. Leave the remaining 1/8 inch to be taken out smoothly by the router plane.

12. Set the router plane blade to take a cut 3/8 inch deep and use it to bottom out the joint to an even depth.

13. Cut the other tread housings in the same manner. Naturally, a power router will speed and simplify steps 8 to 13 above and steps 15 and 16 below.

14. Mark the riser cuts by placing the riser templet with its straight edge exactly over the riser mark on the stringer. Mark along the opposite edge of the templet on the stringer.

15. Cut along these lines with the back saw. Chisel and rout out the stock the same as for the treads.

16. Finish cutting and routing for all the treads and risers including the nosings at the top of both stringers.

17. Make the top and bottom cuts of both stringers with a crosscut saw.

How To Assemble The Stairs

1. Select the tread stock. Use only clear stock free from imperfections and sanded to a finish on all surfaces that will show in the assembled stairs.

2. Square the pieces and cut them to length.

   NOTE: Assume that the distance between the two walls of the stair well is 3 feet 6 inches.

3. Check the width of the stair well at the top and bottom and several intermediate points to see that no distance is less than 3 feet 6 inches so that the assembled stairs will easily fit between the walls.

4. Deduct from this assumed distance of 3 feet 6 inches twice the thickness of a stringer from the bottom of the housed joint to the outside face of the stringer (figure 29). From this figure subtract 1 inch.

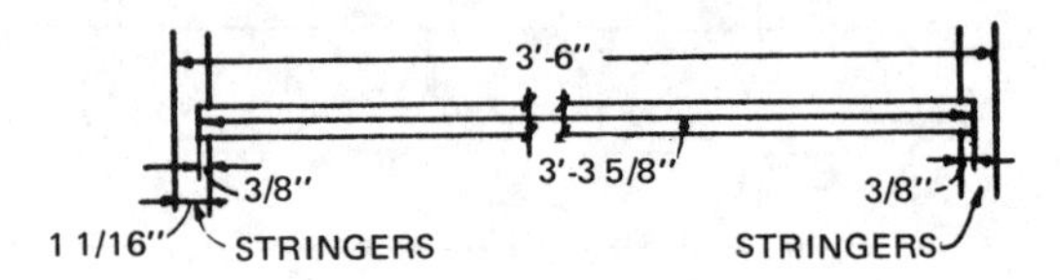

Length Of Treads
Figure 29

NOTE: Assuming that the stringers are 1 1/16 inch thick, the distance from the bottom of the housed joint to the outside face of the stringer would be 11/16 inch. Adding 11/16 inch for the other stringer would give 22/16 or 1 3/8 inch. 3 feet 6 inches minus 1 3/8 inches = 3 feet 4 5/8 inches. Subtracting 1 inch more would give 3 feet 3 5/8 inches, the length of the treads. The 1 inch is an allowance for fitting the assembled stairs in the well hole. This allowance is also made because the stairs have a tendency to spread when being assembled and the ends of all of the joints may not come up tight. The space between the stringer and wall will later be covered up with moulding.

5. Square and cut the treads and the nosed piece for the top step to length.

6. Rip the required number of wedges for both risers and treads or obtain them already cut.

7. Place the stringers on saw horses which are toenailed to the subfloor and spaced far enough apart to properly support the length of the stringers. See figure 30.

8. Apply glue to the housed joint in which the tread is to be inserted.

9. Insert a tread in the top housing of one stringer and tap it so the nosed section fits into the curved part of the housed joint.

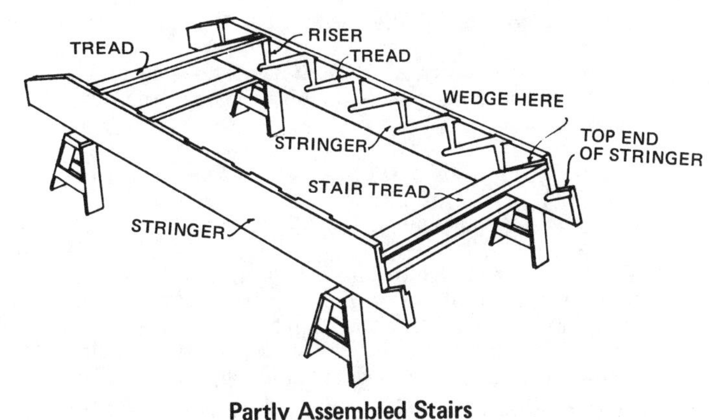

Partly Assembled Stairs
Figure 30

10. Glue the wedge and drive it between the bottom side of the tread and the edge of the housing. Drive the tread and the wedge alternately until the tread nosing and the top surface of the tread fit perfectly against the edges of the housed joint that will show in the assembled stairs. The back edge of the tread must also be in line with the riser cut of the riser housing.

11. Drive an 8d common nail through the stringer into the nosing to pull the stringer up tightly against the tread end. Drive at least two more nails into the tread but use more if it is necessary to bring the tread up tightly

against the bottom of the housed joint.

12. Insert the bottom tread in the bottom housing of the same stringer and fasten it to the stringer in the same manner.

13. Insert the opposite ends of these treads in the top and bottom housings of the opposite stringer and fasten them in the same manner.

14. Toenail the top edge of one stringer with 8d finishing nails to the tops of both saw horses. Be sure the stringer is straight. Place the steel square between the back edge of the top tread and the surface of the stringer. Bring the stairs into a square position at this point and toenail the loose stringer to the saw horses. Check the diagonally opposite corners of the stairs for squareness.

15. Insert the remaining treads and fasten them into the housings in the same manner.

    NOTE: Be sure the back edge of each tread is perfectly flush with the front cut of each riser housing. If it does not reach this point, chisel off the riser cut until it is even with the back edge of the tread. If the tread projects beyond this point, plane or chisel off the back edge of the tread very carefully to a straight line.

16. Cut the risers to the same length as the treads. The top riser will be about 1 3/8 inch longer than the rest as it must be face nailed to the stringers at the top cut.

17. Nail the top riser to the stringers and to the back edge of the top tread. Be sure the top of this riser is even with the bottom of the housed joint of the nosing.

18. Rip the bottom riser to width and insert it into the bottom riser housings in the same general manner as the treads.

19. Install the remaining risers and fasten them the same as the treads.

20. Nail the back of the risers to the back edges of the treads with 8d common nails. Space the nails about 8 inches apart.

21. Cut angle blocks from a 2 x 4 and glue and nail them in place with shingle nails. Put one block in the middle of the stair width at the intersection of the back surface of each riser and tread. See figure 31.

22. Loosen the stringers from the horses, turn the stairs over and fit mouldings underneath the nosing. Nail them to both the riser and tread surfaces with 1¼ inch brads.

23. Nail through the top surfaces of the treads into the risers with 8d finishing nails. Space the nails about 8 inches apart and set these nails.

Finally, place the stairs in the well hole.

1. Place the stairs in the well hole with the top riser

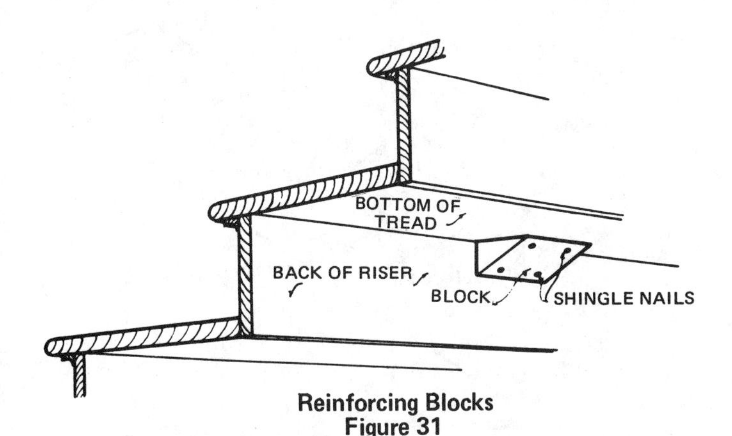

Reinforcing Blocks
Figure 31

against the header. Adjust the top edge of the housed joint for the nosing so it is level with the top of the finished floor. To do this, it may be necessary to shim the back of the top riser out from the face of the header on one side of the well.

2. Center the stairs between the two side walls of the well and nail the riser securely to the header.

   NOTE: If the finish floor has not been laid, be sure to use blocks of finish floor stock under the bottom ends of the stringers.

3. Locate the studs in the side walls and nail through the stringers into them with 10d or 12d finishing nails.

4. Insert the rabbeted nosed piece into the top nosing housed joints and fur it solidly to the top of the subfloor or header over its entire length. Nail it to the header so that it will be forced tightly into the housed joints.

5. Set the nails and cover the stairs with building paper and wood cleats to protect the nosings and other surfaces.

6. Cut and fit mouldings on top of the stringers at the side walls of the well hole and under the top nosing of the stairs.

How To Build Housed And Open Stringers

Often housed stringer stairways are built along a single wall with an open stringer at the open side of the stairs.

1. Select the material for the open stringer. This is generally 1 1/16 inch x 11½ inches by the length of the stairs. Stock 25/32 inch thick is sometimes used but it is hardly strong enough, especially for stringers over 3 feet long.

2. Lay out the open stringer in the same manner as the notched stringer illustrated in figure 21.

   NOTE: Figure 32 shows how to lay out the stringer and how to miter the riser cuts.

3. Make the tread cuts of the stringer first. These cuts should be square with the face of the stringer.

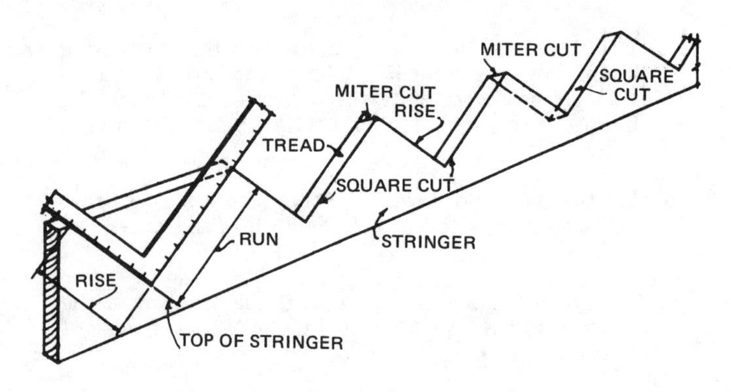

Layout Of Cut And Mitered Stringer
Figure 32

4. Make the top and bottom cuts of the stringer. These cuts should also be square with the face.

5. Make the miter cuts along the riser marks. Use the riser marks as the long point of the miter.

6. Lay out and cut a carriage to fit behind the mitered stringer (figure 33).

7. Adjust this carriage to the inside face of the stringer. Keep the tread cuts of the carriage even with the tread cuts of the stringer and the riser cuts of the carriage in line with the short ends of the miter cuts of the stringer. Nail the stringer and the carriage together temporarily.

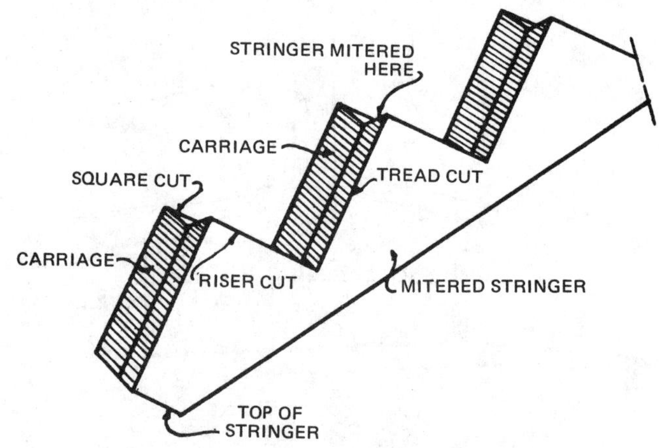

Assembly Of Carriage And Stringer
Figure 33

8. Lay out and house the wall stringers as described previously for housed stringer stairways.

9. Determine the length of the risers.

NOTE: To find the length of the risers, lay out on the subfloor the position of the wall stringer and open stringer as shown in figure 34. The outside stringer usually projects about 1 inch over the outside wall of the well hole. Measure the length of the riser as shown in figure

34. This will give the length to the long point of the miter cut.

10. Cut a miter on one end of each riser and cut the opposite end square.

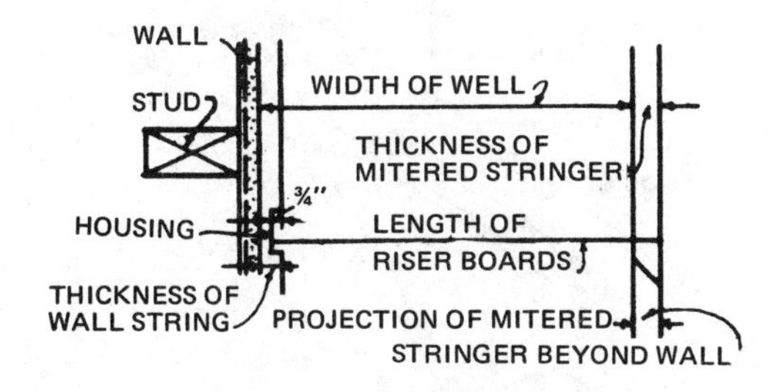

Length Of Mitered Risers
Figure 34

NOTE: Generally the top and bottom tread boards are assembled in the housings first and the stair frame is then fastened to the saw horses and squared as in assembling the housed stairs. The treads and risers may also be installed into the housed stringer first and then the open stringer assembled to the opposite ends of the treads and risers. This procedure sometimes makes it difficult to square the stairs and should be avoided, especially in long stairs. In short stairs it is satisfactory.

11. Place the square ends of the risers into the riser housings of the wall stringer. Nail, glue and wedge them as in the housed type of stairs.

12. Assemble the mitered ends of the risers to the mitered riser cuts of the open stringer. Keep the top edges of the riser boards even with the tread cuts of the open stringer. Nail the miters with finishing nails to the stair carriage riser cuts and to the miter of the open stringer.

13. Permanently nail the carriage to the open stringer.

14. Find the length of each tread by measuring from the bottom of the tread housing to the outside edge of the open stringer and adding on the amount the end of the tread will project over the open stringer. See A, figure 35. This projection should be the same distance as that of the nosing over the riser on the front of the step.

15. Lay out and make the miter and straight cuts on the ends of the treads as shown in figure 35.

16. Cut, glue and nail the mitered pieces to the ends of the treads.

    NOTE: If balusters are to be tenoned into the treads, leave the end pieces loose until the hand rail has been erected.

17. Insert and fasten the treads the same as in the housed type. Face nail the treads to the tread cuts of the open stringer and to the risers.

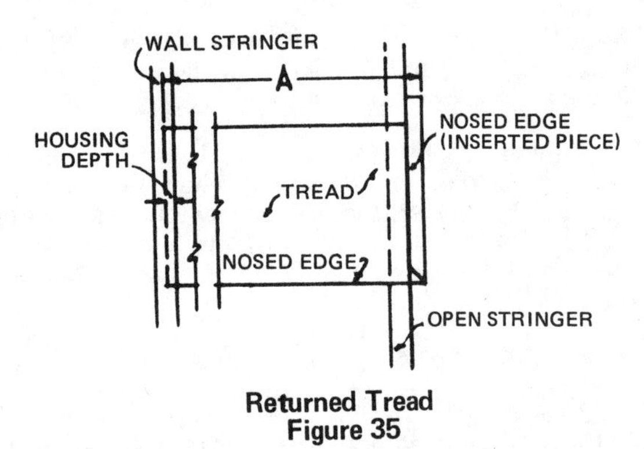

Returned Tread
Figure 35

18. Cut, fit and nail the moulding under the tread nosing at the face of the risers and under the return of the tread nosing at the stringer. Return the moulding on itself at the back edge of the tread. See figure 36.

19. Erect the stairs against the header in the same way as in the housed stairs but be sure the outside stringer is parallel to the wall.

20. Nail the top of the stairs to the header and nail the wall stringer to the studs.

21. Cut and fit a moulding along the bottom edge of the open stringer where it meets the wall. Nail this moulding temporarily until the newel posts have been fitted to the stairs.

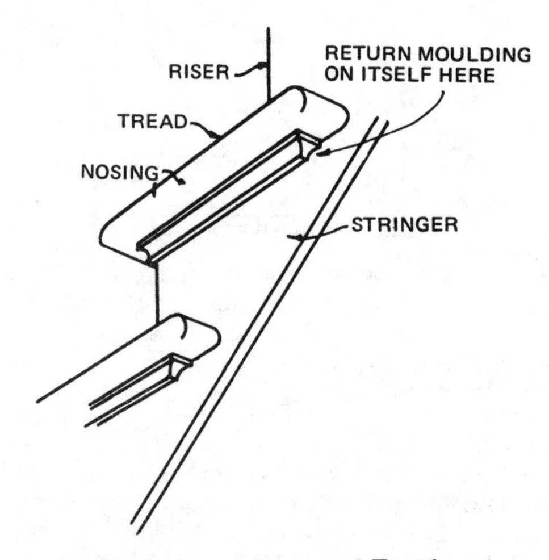

Moulding At Returned Treads
Figure 36

Chapter 4

NEWEL POSTS AND HANDRAILS

Newel posts and handrails are made in many shapes and sizes. However, there are basic rules for placing them on the stairs. Only square, solid newel posts and straight handrails will be described here because the principles involved in layout and fitting newel posts and handrails are the same regardless of the type of posts and rails called for.

Newel posts are built on the stringer, risers and treads of stairs to form a support for the balusters and handrails. In straight run staircases they are generally placed at the starting step and the top step. These are called the starting and landing newels. In "L" shaped stairs there is a third post at the platform. This is called a platform or angle newel. Newel posts may be of the hollow square built-up type or the turned and square solid type. They may be finished plain or capped and paneled depending on their size and the style of the staircase.

Figure 37 shows the different lengths of the newels. These are of the solid square type in a colonial style.

Figure 38 shows the plan view and elevation views of an "L" shaped staircase. The locations of the starting, angle and landing newels show how they are fitted to the stringers of the stairs and why the newels are of different lengths.

If the newel posts are 4½ inches or more wide, the side of the post may be cut out to allow the nosing of the tread to enter the post, thus giving support to the nosing at this point. If the posts are less than 4½ inches wide, it may be necessary to butt the nosing against the side of the newel so that the post will not be weakened by cutting the post

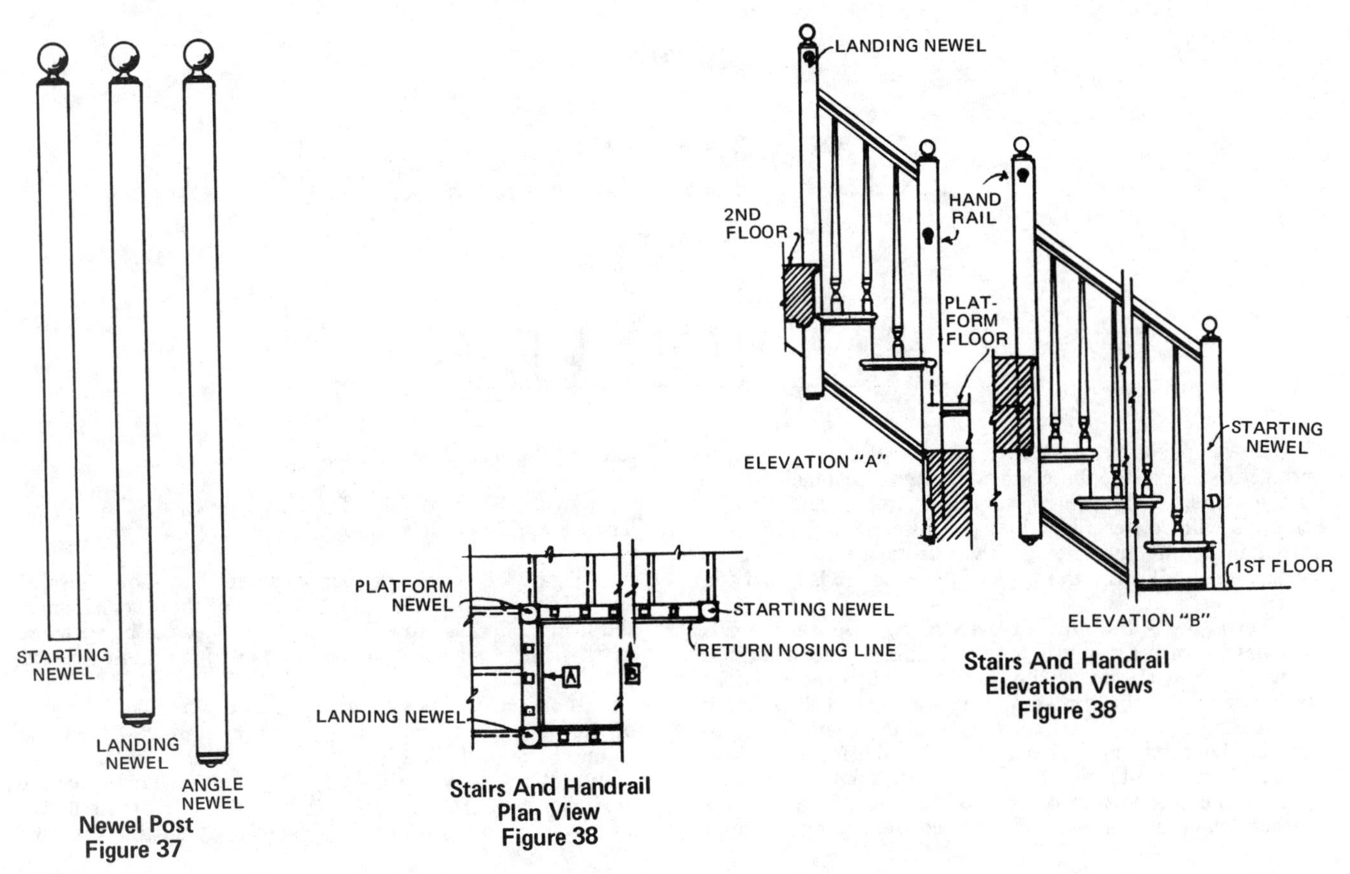

Newel Post
Figure 37

Stairs And Handrail
Plan View
Figure 38

Stairs And Handrail
Elevation Views
Figure 38

for the nosing of the tread. In the drawings, narrow newels are shown and the nosings are butted to the posts as shown by dotted lines. When hollow square posts are used the nosings are generally allowed to enter the post.

How To Lay Out And Fit The Starting Newel

NOTE: It is assumed that the proper newel post is on hand and that the stairs are permanently placed for the fitting of the newels.

1. Lay out the lengths, position of the top of the handrail, and the outlines for the stringer, nosing and riser.

   Figure 39 shows the plan of the starting newel and how it is placed in relation to the outside stringer and the first tread and riser. In locating the exact position of this newel on the stringer and tread, a center line must be drawn on the surface of the newel which is to be fitted to the stringer. See side A, figure 40. This center line must line up with the center line of the thickness of the stringer (figure 39). Another center line should line up with the center line of the riser B, figure 39.

   Figure 40 shows an elevation of how the surfaces A and B of the newel post are laid out to fit the stringer, tread and riser. The surface A in the plan of figure 39 represents the surface A in the elevation of figure 40. The sections marked X show the material to be cut out of the post so the stringer will fit into the post and so the post will rest on the top of the first tread. The sides of the post opposite sides A and B will not

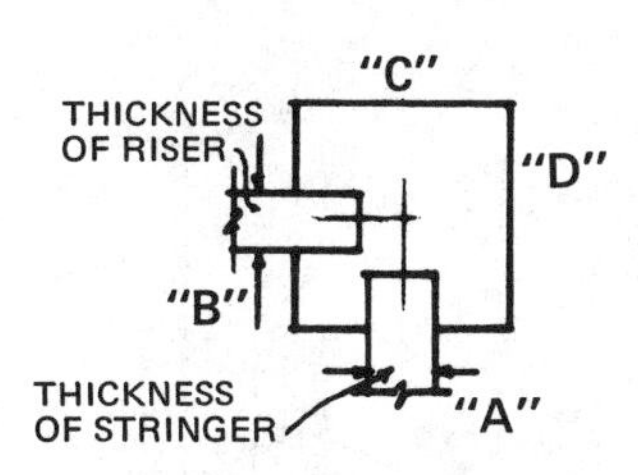

Starting Newel Diagram
Figure 39

have to be cut out as they will face the front and the outside of the staircase and extend to the floor. See figure 38. The elevation B shows the surface D of the newel shown in figures 39 and 40. In laying out the starting newel, the height of the handrail above the tread must first be considered. This distance is from 2'-6" to 2'-8" from the top member of the handrail to the tread directly below the handrail. After this location has been laid out on the post, the bottom cut of the post is measured from it. This would be the height of one riser from the top of the finish floor to the top of the tread. The cuts for the stringer, tread and riser are located from the bottom cut of the post. The depth of the cuts into the posts should be only deep enough to allow the post to be centered to the stringer and riser. When laying out the newel it is advisable to check the surfaces of the stair stringer, riser and tread to see that they are perfectly straight so that the straight lines may be used on the layout of the newel.

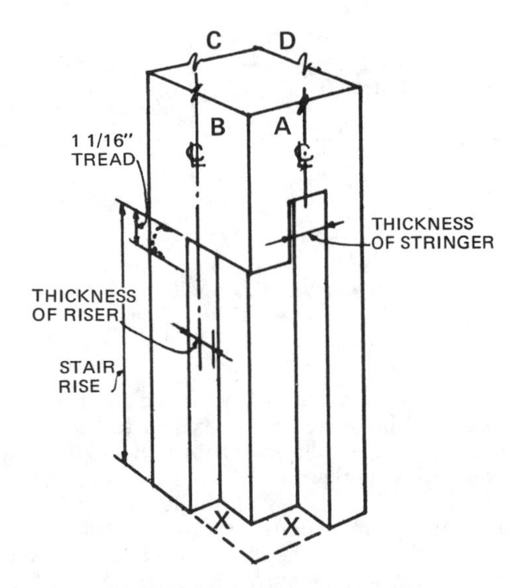

Layout Of Starting Newel
Figure 40

If narrow newel posts are used, the cuts in the newels may be left square at the riser line and the tread nosings may be butted to the side of the post. If the size of the newels is such as to provide enough projection over the side of the stringer to allow the treads to run into the newels, cut the nosings of the treads into the newels as follows.

2. Use a brace and bit to make the cut for the nosing of the tread. The bit may also be used to start the cut for the stringer by boring a series of overlapping holes along the layout lines on the newel. Power routing equipment will speed and simplify these cuts.

3. Carefully cut out the stock marked X in figure 40. Keep ½" inside the layout marks until the surplus stock has been taken out. Then carefully pare to these lines.

4. Cut out the section deep enough to allow the post to fit over the stringer and riser so that the center lines of these members line up with the center lines of the newel.

5. Square and cut the bottom edge of the post and erect it in position on the stairs. Check the fit, and if it is satisfactory, temporarily nail and brace the newel plumb both ways to the floor, tread, stringer, and riser surfaces.

How To Lay Out And Fit A Platform Or Angle Newel

The angle newel is somewhat longer than the starting or landing newel. This is to allow the handrail of the first flight to meet the angle newel 2'-6" above the top tread of the first flight and also to allow the handrail of the second flight to meet the adjacent side of this newel 2'-6" above the first tread of the second flight. See elevations A and B, figure 38. In other words, one rail would be the height of one riser above the other handrail at the angle newel. The

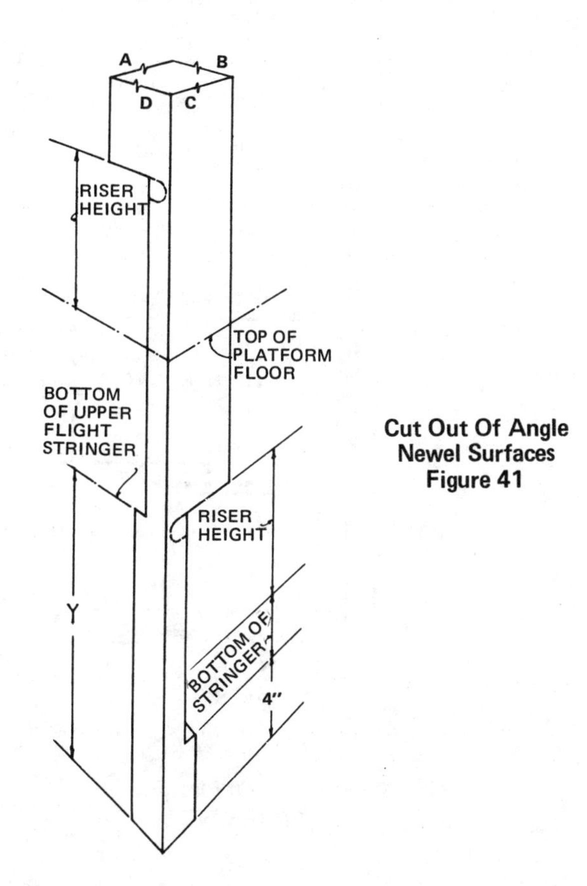

Cut Out Of Angle Newel Surfaces
Figure 41

center lines of this newel should be in line with the center lines of the stringers of the upper and lower flights. Four side surfaces of this newel will have to be cut to fit the surfaces of the upper and lower flights. See figure 38. The only tread nosings that will show on the two sides of the post that face the platform will be the top nosing of the lower flight and the bottom nosing of the upper flight. See figure 41.

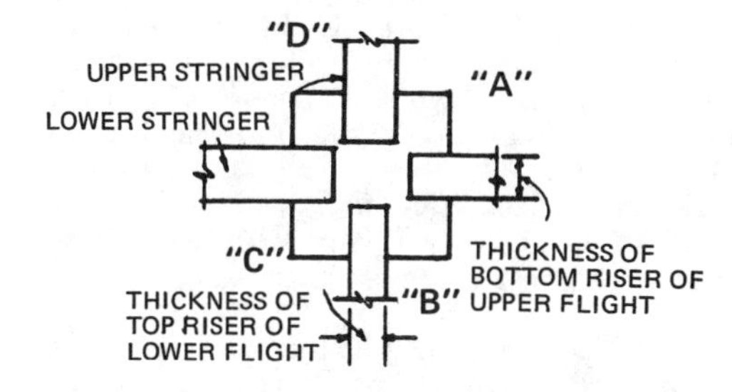

Diagram Of Angle Newel
Figure 42

Figure 42 shows a diagram of the angle newel and the surfaces into which the stringers of the first and second flights would enter. It also shows how the top riser of the first flight and the bottom riser of the second flight would enter the sides of this post. In order to start the layout of the surface D as shown in the diagram in figure 42 and in the elevation in figure 43, it is necessary to know how far the bottom of this newel is to be located below the bottom edge of the stringer of the first flight. See A, figure 43. This distance is generally about 4 inches to allow a moulding to fit on the bottom edge of the stringer and to butt against the side of the newel. See figure 41. This point may be marked on the wall surface where the post is to be fitted. From this line measure the vertical distance to the top tread of the lower flight. See Y on figures 41 and 43. This point locates the top of the tread return shown on side C of the newel in figure 41. From this point measure the distance of one riser from the top tread to the platform finish floor line as shown at R, figure 43. This point is transferred to the side C of the post and is shown as the floor line in figure 41. From this point measure the height of the first riser of the upper flight and transfer this distance to side D of the newel. This point locates the top of the first tread nosing return of the upper flight.

The two adjacent faces of this newel would show the tops of the platform front nosing and the tread immediately above the platform floor line. See R, figure 44. The center lines of the post should be kept in line with the center lines of the risers.

To lay out these sides, transfer the platform floor line of figure 41 to side A of figure 44. Locate the thickness of the bottom riser of the upper flight on the center line of side A. Also locate A, figure 44. Lay out for the tread of the lower flight from the platform floor line and the center line on side B. The height of the handrail is measured from the top of the first tread of the second flight. See H, figure 44.

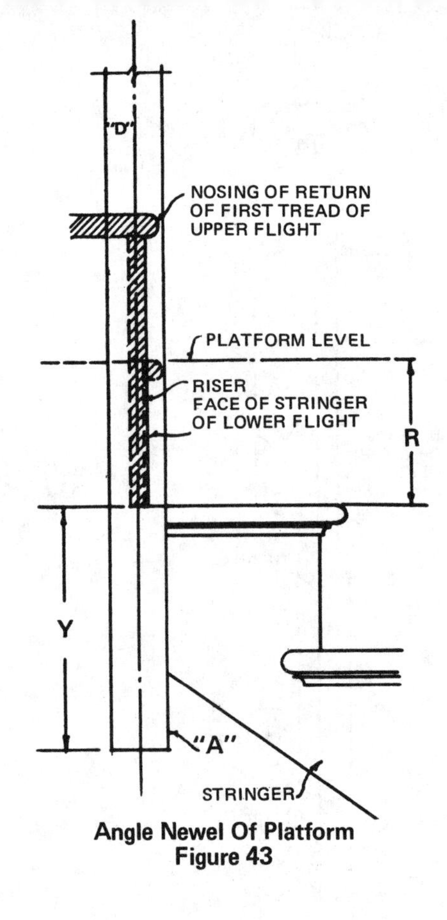

Angle Newel Of Platform
Figure 43

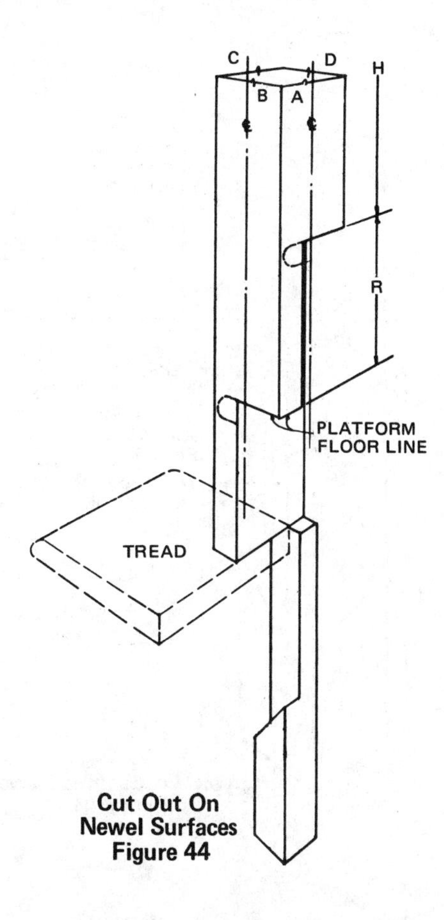

Cut Out On
Newel Surfaces
Figure 44

1. When the layout is complete, chisel or cut out the sections in the same general way as in cutting the sections of the starting newel.

2. Cut out the sections so that the newel, when fitted to the stairs, will center with both the stringers and risers.

NOTE: It may be necessary to make a trial assembly of the newel to the stairs several times before it fits perfectly.

3. Cut the bottom end square or trim it as desired.

4. When the newel is properly fitted, temporarily nail it to the stair members and wall and brace it in a plumb position.

NOTE: This newel generally has to be extensively cut. In some instances where the newel may be weakened by cutting, the return nosings of the treads might better be notched rather than to weaken the post.

How To Lay Out And Fit The Landing Newel

Figure 45 shows the second floor landing newel layout in relation to the stringer and the riser of the upper flight at the second floor line. To lay out this newel, the bottom of the newel is dropped 4 inches below the bottom edge of the upper flight stringer the same as the bottom end of the newel in figure 41. This location is marked on the wall and from this point the vertical distance to the top of the top tread in the upper flight is measured. See figure 46. This locates the top edge for the return nosing for this tread.

The other lines on this surface are laid out for the thickness of the stringer and the fit of the newel against the wall in a similar manner to side C of the angle newel in figure 41.

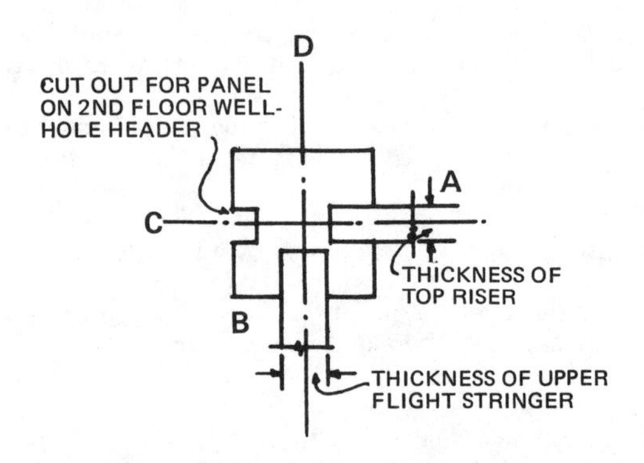

Landing Newel Diagram
Figure 45

The nosing lines on side A of the newel in figure 46 are laid out for the top nosing and riser of the upper flight at the second floor line exactly the same as those on the side A, figure 44. The opposite sides C and D are laid out as shown in figure 47. On the side C a panel similar in width to the stringer is applied to the wall of the header of the well hole. The slanting cut shows where the newel fits to the bottom of the stringer of this flight. The lines showing how the panel of the well hole fits into the newel are shown as they would be laid out on surface C.

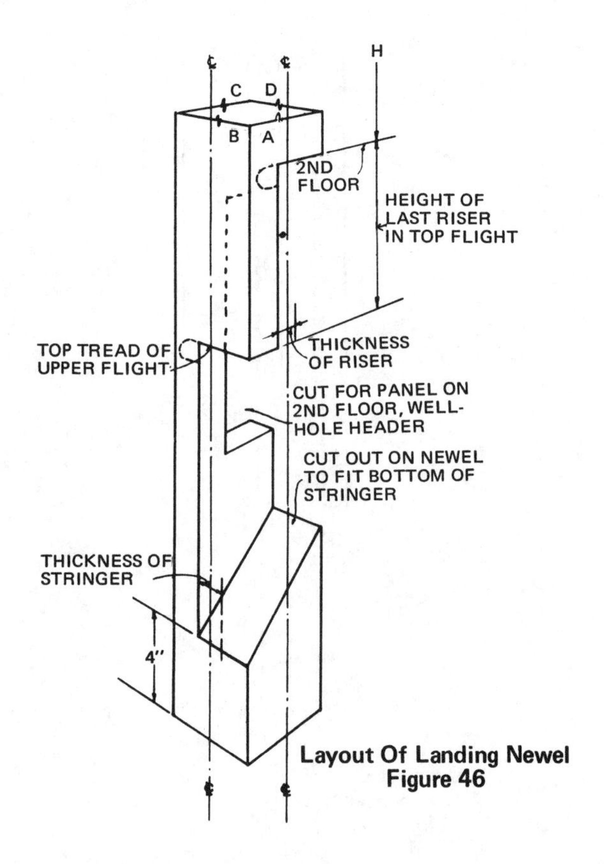

Layout Of Landing Newel
Figure 46

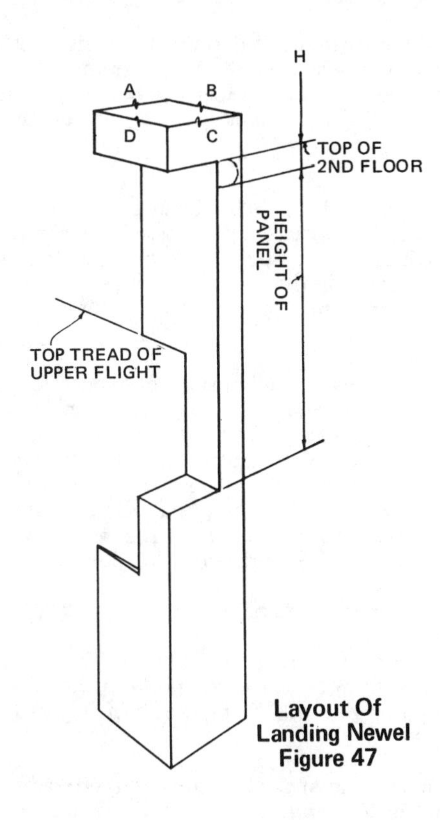

Layout Of Landing Newel Figure 47

The panel cut in the face C is similar to the cut for the stringers in the other newels. The second floor line as located on the side A, figure 46 is transferred to side C of figure 47. The drop of 4 inches below the stringer is the same as the drop below the bottom edge of the stringer in side B, figure 46. The other lines on this side are also similar except that the nosing return on side C, figure 47 may be omitted where the edge of the second floor projects over the edge of the panel. See figure 38. The height of the handrails from the top of the top tread at the second floor is shown at H, figure 47.

The fitting of this newel to the stair top nosing and floor is similar to that of the landing newel except that one side is fitted around a panel board that is nailed to the face of the stairwell header at the second floor line.

1. Select a panel board 1 1/16 inch thick and approximately as wide as the stair stringer and as long as the header of the second floor between the newel and end wall of the well hole.

2. Fit this board to the wall of the header. Keep the top edge in line with the bottom of the finished floor. Nail it in this position and erect the landing newel on the stairs in the same manner as in erecting the angle newel.

3. Cut and fit the finish floor nosing, allowing it to project over the panel about 1½ inch. Fit moulding below it and to the panel.

NOTE: If the panel is to be butted against the newel, the

newel is erected first and the panel and second floor nosing is fitted to the post.

If there are more newel posts in the well hole, they may be fitted and erected in the same general manner as described above for posts.

How To Fit Handrails

A handrail is the top member of the balustrade. Rails may be obtained in many shapes and sizes from millwork companies. The two member type is shown in B, figure 48. The larger of these two members is ploughed out to receive the thickness of the balusters. The other member is a form of fillet which is cut and fitted into the ploughed section between the balusters. Rails of the type shown in B, figure 48 may be used at both the top and the bottom of the balusters on open stringers without return treads. The open stringer with return treads uses only the top rail because the bottom of the balusters are set into the tread surfaces.

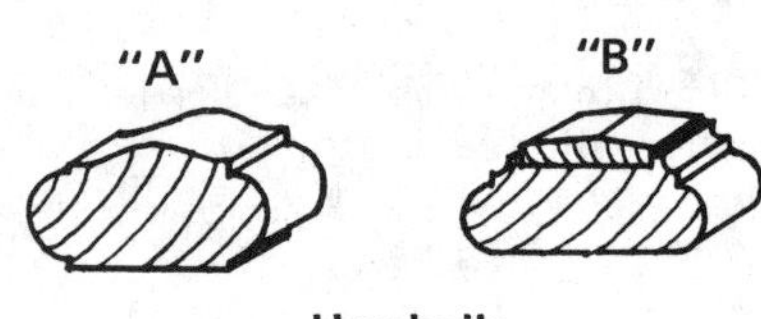

Handrails
Figure 48

In laying out the handrail for this type of stairs, the balusters must be properly spaced. Since there are several arrangements for spacing balusters and placing the newels on the treads it is difficult to describe a definite method of spacing. However, the spacing of the balusters on the treads in relation to the faces of the risers as shown in figure 38 will give a general idea of baluster placement on the tread and handrail.

Balusters are made in many shapes, sizes and lengths and may be round or square. For open stairwork they are generally provided with a dowel on the bottom end for fitting into a hole bored in the tread. The same arrangement is provided at the top end for fitting the balusters into the handrail.

After the centers of the balusters have been located on the treads and the underside of the handrail, small holes are bored into the handrail and the treads to receive the dowels on the top and bottom of the balusters. The handrail is then fitted between the newel post and is fastened with special lag screws or nailed in place with finishing nails. Before fitting the handrail, be sure the newel posts are plumbed both ways and are braced in position.

1. Select the handrail and balusters.

2. Lay the top rail along the tops of the treads and against the sides of the newel posts as shown by dotted lines in figure 49. Temporarily nail it in this position.

3. Space the centers of the balusters on the tread surfaces and transfer these points to the rail. Use a steel square placed on the treads to transfer these points.

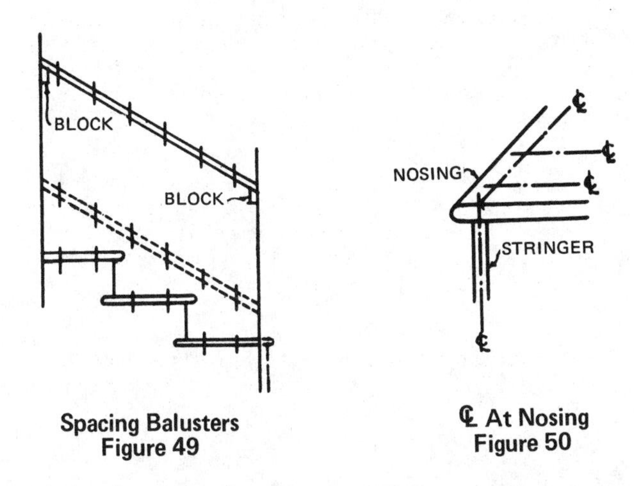

Spacing Balusters
Figure 49

℄ At Nosing
Figure 50

4. Center the points on the width of the handrails and also directly over the center line of the stringer on the treads. See figure 50.

5. Bore holes to fit the dowels of the balusters at the center points on the treads and on the underside of the handrail.

6. Tack blocks of wood to the newels to support the handrail at the correct height from the top and bottom treads. See figure 49.

7. Lay the handrail on these blocks and against the surfaces of the newel to which it is to be fitted. Mark the length of the handrail and cut it off.

8. Fit the balusters into the treads and mark their lengths where they intersect the handrail. Cut the balusters to length and fit them into the holes in the handrail.

9. Make a trial assembly of the handrail and balusters and mark the handrail centers on the newels. Check the whole assembly to be sure it fits and it is plumb and straight.

10. Remove the top rail and glue the bottom dowels in the holes in the treads. Apply glue to the top dowels of the balusters and replace the top rail, fitting the balusters into their proper holes in the handrail.

11. Nail or bolt the handrail to the newels and toenail the balusters in the underside of the handrail so that they cannot turn.

12. Nail the newel posts permanently to the stairs and fit mouldings along the underside of the open stringers.

13. Sand all rough surfaces and corners and set the nails. Cover the stairs with building paper and cleats to protect the surfaces from injury.

NOTE: In erecting the handrail and balusters on a box stringer, the procedure is similar with the addition of a bottom member of the handrail as shown in B, figure 48. This member is nailed to the top edge of the stringer and square balusters are cut at an angle to fit into the top and bottom rail members. Fillet blocks are then fitted between the balusters at the bottom rail. In some cases the member in B, figure 48 is also used as a top rail and the balusters are fitted in the same manner as those of the bottom rail.

Chapter 5

USE OF THE STAIR TABLES

•

USEFUL INFORMATION

•

STAIR LAYOUT TABLES

Traditional stair building practice requires the use of dividers and exhaustive trial and error experimentation to equalize riser heights. The tables in this volume permit the stair designer to select any one of several riser and tread combinations for any total rise. The designer can use his judgment to select the rise and run combination that is best for the use intended.

Assume that you are planning to build a stairway where the ceiling height above the finished floor is 8 feet and the floor joists, subfloor and finish floor of the upper story add 9 1/8 inches. The total floor to floor rise should be 8 feet 9 1/8 inches. Careful practice requires that the total rise be measured where the stairs are to be built since adjustments during construction often cause the actual floor to floor height to vary from the design height. Assume that the floor to floor rise is actually 8 feet 9 1/8 inches. If you turn to page 235 you will see that there are five different tread and riser combinations listed for a floor to floor rise of 8 feet 9 1/8 inches. Either 13, 14, 15, 16, or 17 risers may be used depending on the requirements of the job. Which of the five combinations should you select? If the stairway is being built in a basement or if the total run of the stairway cannot exceed 10 feet, then the stairway with 13 risers 8 1/16 inches high may be the best for the job. Be aware of code limitations, however. Many building codes limit riser heights to 8 inches even in basement stairs. If the stairway will receive heavy use and the space is adequate for a 15 foot 1 inch total stairway length, the stairway with 17 risers 6 3/16 inches high may be best. When the stairway is at right angles to the ceiling joists, a stairway requiring a smaller well opening is called for. Notice that the stairway with 13 risers requires the smallest well opening. Where the space for the stairway is not limited and the stairs will be in general residential use, an angle of incline between 33 and 37 degrees is best. Referring again to the example on page 235, both the 14 and 15 riser stairway fall between 33 degrees and 37 degrees. The 14 riser stairway has an angle of incline of 36 degrees 52 minutes; the 15 riser stairway has an angle of incline of 33 degrees 41 minutes. Either of these two stairways could be selected for general residential use where no headroom or space limitations exist. Notice, however, that the 14 riser stairway uses less materials than the 15 riser stairway. The carriage length of the 14 riser stairway is 13 feet 6½ inches compared to 14 feet 8 21/32 inches for the 15 riser stairway.

Another factor in selecting the right stair dimensions for your job is the availability of materials. Most tread material is 11½ inches wide. If the only tread material available is 11½ inches wide and 1¼ inches is required for nosing on each tread, the tread width should not exceed 10¼ inches. In the 8 foot 9 1/8 inch floor to floor rise example, only the 13 and 14 riser stairways have less than a 10¼ inch tread width.

Finally, when convenient, it is better to select a tread and riser combination that has the smallest "overage" or "underage". Many tread and riser combinations do not reach the exact floor to floor rise desired. The overage or underage in the following tables is never more than ½ inch and can be made unnoticeable in the finished stairway by fitting the last several risers 1/8 inch over or under the listed riser height. However, it is much easier to select a tread and riser combination with the smallest "overage" or "underage" when several choices are available. The height of every riser and the width of every tread should be the same within a tolerance of ¼ inch when the stairway is finished.

Anyone familiar with the design and construction of stairs should be able to use the tables in this volume without detailed explanation. However, even experienced stair builders may want to refer to the definitions given below. Figure 51 illustrates the dimensions listed for a stairway with a floor to floor rise of 8 feet 9 1/8 inches and 14 risers.

The *Floor to Floor Rise* is the actual rise required where the stairs are to be built after the finish flooring has been installed on both floors. Generally the finished floor will be ¼ inch to ½ inch above the level of the subfloor. Resilient flooring will add about 1/8 inch and underlayment will add between 1/8 inch and ¼ inch to the subfloor level. Hardwood flooring is usually 5/16 inch or 25/32 inch thick. Most carpeting is about 3/8 inch thick including the pad. Of course, where both the lower and upper floors are to be finished in the same or similar material, the floor to floor rise may be taken as the subfloor to subfloor rise.

The *Number of Risers* includes both the first and last rise of the stairway even though the last rise may be against the headers as illustrated in figure 51. The upper end of the stairway may be anchored in any of three ways as illustrated in figure 16. Most stairs meet the header as in figure 16C because any other method lengthens the run of the stairway and increases the headroom and carriage material required.

The *Height of Each Riser* is measured from the top of one tread to the top of the next tread. This is the same measurement as the vertical distance between one horizontal cutout and the next in a cutout stringer stairway. It is often called the "unit rise" and should be between 6 inches and 8¼ inches. This dimension must be within ¼ inch the same for each step of the stairway.

The *Total Height of Risers* is equivalent to the height of each riser multiplied by the total number of risers.

Risers Overage or Underage is the difference between the *Floor to Floor Rise* and the *Total Height of Risers.* Many times the riser height does not divide evenly into the floor to floor rise and a small fraction of an inch is left over. In the tables in this book, the difference between the *Floor to Floor Rise* and the *Total Height of Risers* never exceeds ½ inch. When the stairs are built, any overage less than ¼ inch can usually be left in the top or bottom step without producing any noticeable irregularity in the stairway. Where the underage or overage approaches ½ inch, several risers may be increased or reduced 1/8 inch so that each stair will have very nearly the same rise. *Overage* (+) is the height of all risers over the *Floor to Floor Rise. Underage* (-) is the total height of risers less than the *Floor to Floor Rise.*

The *Number of Treads* is one less than the number of risers since the top "tread" is usually the landing or the upper floor. See figure 16C. Where a full tread or a partial tread is used at the top of the stairway as in figures 16A and 16B, the number of treads and risers will be equal. The addition of a full or partial tread at the top of the stairway will add an equivalent horizontal distance to the total run of

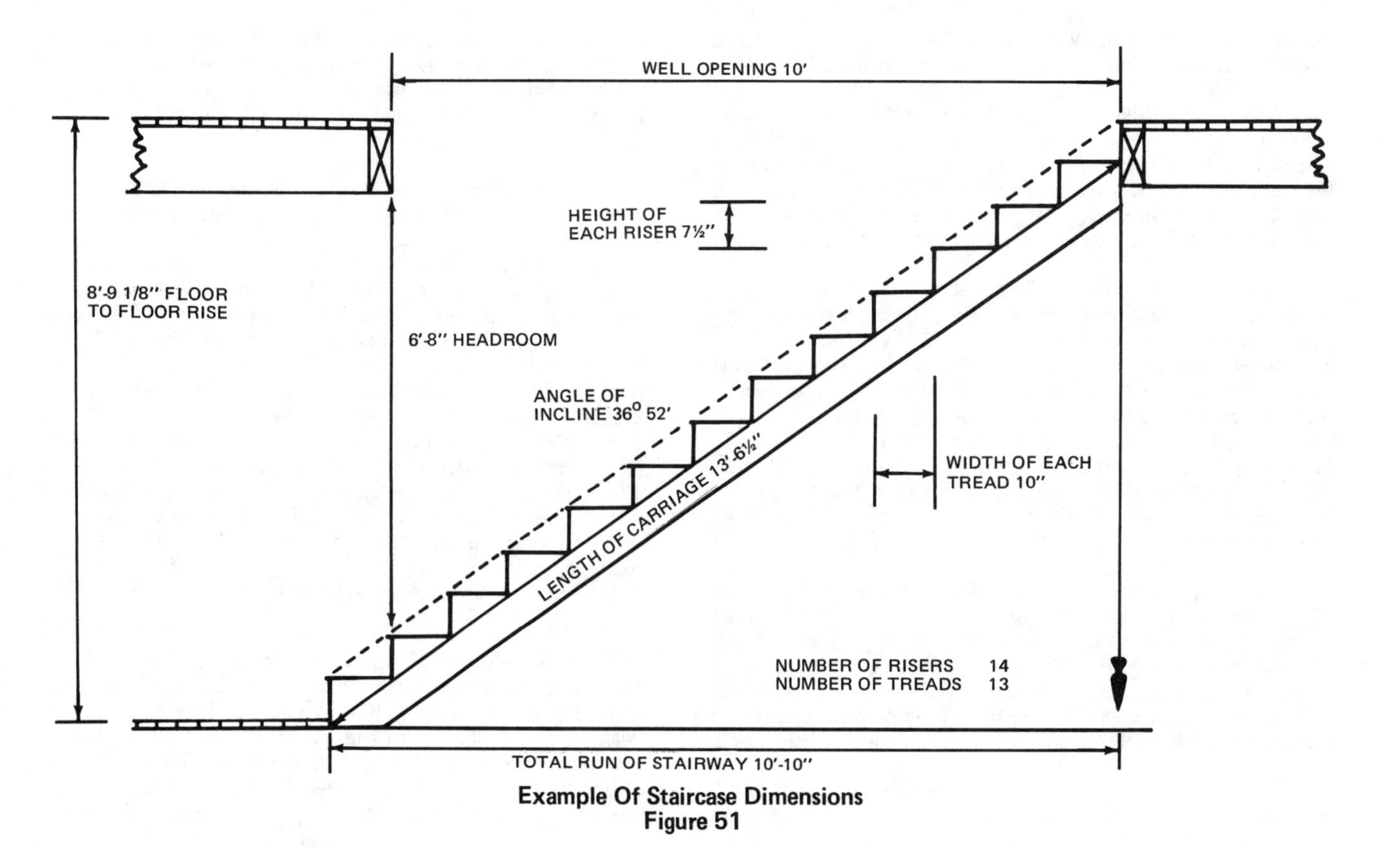

Example Of Staircase Dimensions
Figure 51

the stairway and the well opening required to maintain 6 foot 8 inch headroom. It does not change the angle.

The *Width of Each Tread* is always the distance which, when added to the riser height, equals 17½ inches. It is measured from the front of one riser to the front of the next riser and is the same as the horizontal measurement of each cutout stringer of an enclosed stringer stairway. The width of each tread does not include the nosing on the tread when nosing is used.

The *Total Run of Stairway* is equal to the *Width of Each Tread* times the *Number of Treads.* It is the horizontal distance from directly below the front edge of the header to which the carriage is nailed to the front edge of the bottom rise of the carriage. When riser material is attached to the front edge of each rise, the total run is not changed if all riser material is the same thickness. When a full tread or a partial tread is used at the top of the stairway as illustrated in figure 16A and 16B, the total run of the stairway is extended by the width of the tread added.

Maximum Run With 6 foot 8 inch Headroom is the maximum horizontal distance permitted for the bottom end of the stairway before headroom is reduced below 6 feet 8 inches in a room with an 8 foot ceiling. It is measured from the front edge of the bottom riser. For most stairways, only one or two treads can rise toward an 8 foot ceiling before additional headroom is required.

The *Well Opening to Maintain 6'-8" Headroom* is the length of the well opening required to permit a 6 foot 8 inch vertical clearance from any tread. It assumes a thickness of 10" from top of the floor to the bottom of the ceiling below. The well opening is measured from the front edge of the header at the top of the stairway. When a full or partial tread is used at the top of the stairway, the well opening must be increased by the width of the full or partial tread.

The *Angle of Incline* of the stairway is determined by the rise and run of each stair of the stairway. For most purposes an angle of incline between 33 and 37 degrees is best.

The *Length of Carriage* is measured along a line parallel to the angle of incline from the point where the bottom riser meets the lower floor to the point where the line reaches the well header. The length listed will be correct for all carriages framed with top rise flush against the well header as illustrated in figure 16C. Where a full or partial tread extends the top floor level beyond the well header as in figures 16A and 16B use table 52 to compute the additional material needed. Opposite the rise and run combination appropriate for your stairway is a factor. Multiply this by the distance the upper floor is extended beyond the well header. The answer is the additional carriage length needed.

For example, in figure 53 one 4 inch tread at the upper floor level extends beyond the well header. How much longer should the carriage be? From table 52 we find that for a rise and tread combination of 7½ and 10 inches the factor is 1.25. Multiplying 1.25 times 4 inches, we get 5 inches. The carriage length for this particular stairway is 5 inches longer than the carriage length listed in the tables. Dimension "X" in figure 53 is 5 inches, the added carriage

Carriage Extension Factors

RISER	TREAD	FACTOR	RISER	TREAD	FACTOR
6"	11 1/2"	1.13	7 3/16"	10 5/16"	1.22
6 1/16"	11 7/16"	1.13	7 1/4"	10 1/4"	1.22
6 1/8"	11 3/8"	1.14	7 5/16"	10 3/16"	1.23
6 3/16"	11 5/16"	1.14	7 3/8"	10 1/8"	1.24
6 1/4"	11 1/4"	1.14	7 7/16"	10 1/16"	1.24
6 5/16"	11 3/16"	1.15	7 1/2"	10"	1.25
6 3/8"	11 1/8"	1.15	7 9/16"	9 15/16"	1.26
6 7/16"	11 1/16"	1.16	7 5/8"	9 7/8"	1.26
6 1/2"	11"	1.16	7 11/16"	9 13/16"	1.27
6 9/16"	10 15/16"	1.17	7 3/4"	9 3/4"	1.28
6 5/8"	10 7/8"	1.17	7 13/16"	9 11/16"	1.28
6 11/16"	10 13/16"	1.18	7 7/8"	9 5/8"	1.29
6 3/4"	10 3/4"	1.18	7 15/16"	9 9/16"	1.30
6 13/16"	10 11/16"	1.19	8"	9 1/2"	1.31
6 7/8"	10 5/8"	1.19	8 1/16"	9 7/16"	1.31
6 15/16"	10 9/16"	1.20	8 1/8"	9 3/8"	1.32
7"	10 1/2"	1.20	8 3/16"	9 5/16"	1.33
7 1/16"	10 7/16"	1.21	8 1/4"	9 1/4"	1.34
7 1/8"	10 3/8"	1.21			

Factors For Carriage Extensions
Table 52

length. Remember that there will usually be 18 to 20 inches of "waste" cut from the end of the carriage when the carriage is cut to size and order materials accordingly.

L. F. of Riser per Inch of Stairway Width

L. F. of Tread per Inch of Stairway Width

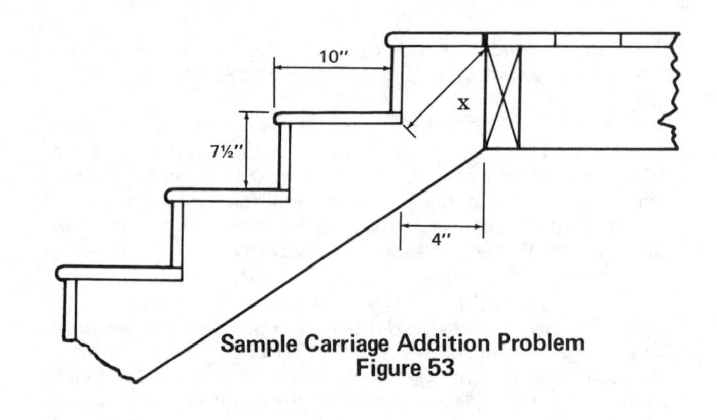

Sample Carriage Addition Problem
Figure 53

To find the amount of tread or riser material needed, multiply the factor listed by the stairway width in inches. The answer is the linear feet of tread and riser material needed to cover the stairway. No waste is included in these figures and between 5% and 25% should be added depending on the length of treads and risers available and the width of the stairway.

Framing Square Settings for Carriage Cuts. The stair rise is usually set on the tongue of the square and the stair run is then set on the body. See figure 14.

USEFUL INFORMATION

Estimating Stair Construction Costs

The labor required to frame a stairway and install treads and risers will vary widely from one job to the next. However, for estimating purposes the figures below will be good indicators of the time required for the types of stairways listed. These figures include only productive time. No allowance has been made for supervision, poor access or unusual conditions. Highly repetitive jobs or work done by specialized crews will require less time than indicated below. These figures are based on work by skilled craftsmen working with modern power tools. Adjust these figures where other situations exist. All figures are for straight runs with only the landings indicated.

Type Of Work	Man Hours Required
Basement stairs with open risers, no landing, cleat or dado joint treads. Includes one handrail. From job cut materials. Based on 8 foot rise.	8.0
Basement stairs with one intermediate landing, from job cut materials including treads and risers. With simple handrail. Per floor of rise.	11.0
Main stairs, notched or full stringer, job cut materials, with simple handrail, 8 to 10 foot rise.	20.0
Add for landing and 2 short flights.	3.0
Add for newels, rails and balusters, per linear foot of horizontal run.	.4
Deduct for precut materials.	3.0
Add for landing and two short flights.	2.0
Ornate main stairs, housed, to 48" wide, 12 foot rise, job cut materials, with newels, handrail and balusters.	60.0

STAIR LAYOUT TABLES

FLOOR TO FLOOR RISE	3'0"	3'0"	3' 1/8"	3' 1/8"	3' 1/4"	3' 1/4"	3' 3/8"	3' 3/8"	3' 1/2"	3' 1/2"
NUMBER OF RISERS	5	6	5	6	5	6	5	6	5	6
HEIGHT OF EACH RISER	7 3/16"	6"	7 1/4"	6"	7 1/4"	6 1/16"	7 1/4"	6 1/16"	7 5/16"	6 1/16"
TOTAL HEIGHT OF RISERS	2'11 15/16"	3'	3' 1/4"	3'	3' 1/4"	3' 3/8"	3' 1/4"	3' 3/8"	3' 9/16"	3' 3/8"
RISERS OVERAGE (+) OR UNDERAGE (—)	−1/16"	0	+1/8"	−1/8"	0	+1/8"	−1/8"	0	+1/16"	−1/8"
NUMBER OF TREADS	4	5	4	5	4	5	4	5	4	5
WIDTH OF EACH TREAD	10 5/16"	11 1/2"	10 1/4"	11 1/2"	10 1/4"	11 7/16"	10 1/4"	11 7/16"	10 3/16"	11 7/16"
TOTAL RUN OF STAIRWAY	3'5 1/4"	4'9 1/2"	3'5"	4'9 1/2"	3'5"	4'9 3/16"	3'5"	4'9 3/16"	3'4 3/4"	4'9 3/16"
MAXIMUM RUN WITH 6'-8" HEADROOM	1'9 7/8"	1'11"	1'8 1/2"	1'11"	1'8 1/2"	1'10 7/8"	1'8 1/2"	1'10 7/8"	1'8 3/8"	1'10 7/8"
ANGLE OF INCLINE	34°53'	27°33'	35°16'	27°33'	35°16'	27°56'	35°16'	27°56'	35°40'	27°56'
LENGTH OF CARRIAGE	4'2 9/32"	5'4 27/32"	4'2 7/32"	5'4 27/32"	4'2 7/32"	5'4 23/32"	4'2 7/32"	5'4 23/32"	4'2 3/32"	5'4 23/32"
L. F. OF RISER PER INCH OF STAIRWAY WIDTH	.416	.5	.416	.5	.416	.5	.416	.5	.416	.5
L. F. OF TREAD PER INCH OF STAIRWAY WIDTH	.333	.416	.333	.416	.333	.416	.333	.416	.333	.416
FRAMING SQUARE SETTINGS FOR CARRIAGE CUTS — TONGUE	7 3/16"	6"	7 1/4"	6"	7 1/4"	6 1/16"	7 1/4"	6 1/16"	7 5/16"	6 1/16"
FRAMING SQUARE SETTINGS FOR CARRIAGE CUTS — BODY	10 5/16"	11 1/2"	10 1/4"	11 1/2"	10 1/4"	11 7/16"	10 1/4"	11 7/16"	10 3/16"	11 7/16"

FLOOR TO FLOOR RISE	3'5/8"	3'5/8"	3'3/4"	3'3/4"	3'7/8"	3'7/8"	3'1"	3'1"	3'1 1/8"	3'1 1/8"
NUMBER OF RISERS	5	6	5	6	5	6	5	6	5	6
HEIGHT OF EACH RISER	7 5/16"	6 1/8"	7 3/8"	6 1/8"	7 3/8"	6 1/8"	7 3/8"	6 3/16"	7 7/16"	6 3/16"
TOTAL HEIGHT OF RISERS	3'9/16"	3'3/4"	3'7/8"	3'3/4"	3'7/8"	3'3/4"	3'7/8"	3'1 1/8"	3'1 3/16"	3'1 1/8"
RISERS OVERAGE (+) OR UNDERAGE (—)	−1/16"	+1/8"	+1/8"	0	0	−1/8"	−1/8"	+1/8"	+1/16"	0
NUMBER OF TREADS	4	5	4	5	4	5	4	5	4	5
WIDTH OF EACH TREAD	10 3/16"	11 3/8"	10 1/8"	11 3/8"	10 1/8"	11 3/8"	10 1/8"	11 5/16"	10 1/16"	11 5/16"
TOTAL RUN OF STAIRWAY	3'4 3/4"	4'8 7/8"	3'4 1/2"	4'8 7/8"	3'4 1/2"	4'8 7/8"	3'4 1/2"	4'8 9/16"	3'4 1/4"	4'8 9/16"
MAXIMUM RUN WITH 6'-8" HEADROOM	1'8 3/8"	1'10 3/4"	1'8 1/4"	1'10 3/4"	1'8 1/4"	1'10 3/4"	1'8 1/4"	1'10 5/8"	1'8 1/8"	1'10 5/8"
ANGLE OF INCLINE	35°40'	28°18'	36°4'	28°18'	36°4'	28°18'	36°4'	28°41'	36°28'	28°41'
LENGTH OF CARRIAGE	4'2 3/32"	5'4 19/32"	4'2 3/32"	5'4 19/32"	4'2 3/32"	5'4 19/32"	4'2 3/32"	5'4 15/32"	4'2 1/16"	5'4 15/32"
L. F. OF RISER PER INCH OF STAIRWAY WIDTH	.416	.5	.416	.5	.416	.5	.416	.5	.416	.5
L. F. OF TREAD PER INCH OF STAIRWAY WIDTH	.333	.416	.333	.416	.333	.416	.333	.416	.333	.416
FRAMING SQUARE SETTINGS FOR CARRIAGE CUTS — TONGUE	7 5/16"	6 1/8"	7 3/8"	6 1/8"	7 3/8"	6 1/8"	7 3/8"	6 3/16"	7 7/16"	6 3/16"
FRAMING SQUARE SETTINGS FOR CARRIAGE CUTS — BODY	10 3/16"	11 3/8"	10 1/8"	11 3/8"	10 1/8"	11 3/8"	10 1/8"	11 5/16"	10 1/16"	11 5/16"

FLOOR TO FLOOR RISE		3'1 1/4"	3'1 1/4"	3'1 3/8"	3'1 3/8"	3'1 1/2"	3'1 1/2"	3'1 5/8"	3'1 5/8"	3'1 3/4"	3'1 3/4"
NUMBER OF RISERS		5	6	5	6	5	6	5	6	5	6
HEIGHT OF EACH RISER		7 7/16"	6 3/16"	7 1/2"	6 1/4"	7 1/2"	6 1/4"	7 1/2"	6 1/4"	7 9/16"	6 5/16"
TOTAL HEIGHT OF RISERS		3'1 3/16"	3'1 1/8"	3'1 1/2"	3'1 1/2"	3'1 1/2"	3'1 1/2"	3'1 1/2"	3'1 1/2"	3'1 13/16"	3'1 7/8"
RISERS OVERAGE (+) OR UNDERAGE (—)		−1/16"	−1/8"	+1/8"	+1/8"	0	0	−1/8"	−1/8"	+1/16"	+1/8"
NUMBER OF TREADS		4	5	4	5	4	5	4	5	4	5
WIDTH OF EACH TREAD		10 1/16"	11 5/16"	10"	11 1/4"	10"	11 1/4"	10"	11 1/4"	9 15/16"	11 3/16"
TOTAL RUN OF STAIRWAY		3'4 1/4"	4'8 9/16"	3'4"	4'8 1/4"	3'4"	4'8 1/4"	3'4"	4'8 1/4"	3'3 3/4"	4'7 15/16"
MAXIMUM RUN WITH 6'-8" HEADROOM		1'8 1/8"	1'10 5/8"	1'8"	1'10 1/2"	1'8"	1'10 1/2"	1'8"	1'10 1/2"	1'7 7/8"	1'10 3/8"
ANGLE OF INCLINE		36°28'	28°41'	36°52'	29°3'	36°52'	29°3'	36°52'	29°3'	37°16'	29°26'
LENGTH OF CARRIAGE		4'2 1/16"	5'4 15/32"	4'2"	5'4 11/32"	4'2"	5'4 11/32"	4'2"	5'4 11/32"	4'1 15/16"	5'4 7/32"
L. F. OF RISER PER INCH OF STAIRWAY WIDTH		.416	.5	.416	.5	.416	.5	.416	.5	.416	.5
L. F. OF TREAD PER INCH OF STAIRWAY WIDTH		.333	.416	.333	.416	.333	.416	.333	.416	.333	.416
FRAMING SQUARE SETTINGS FOR CARRIAGE CUTS	TONGUE	7 7/16"	6 3/16"	7 1/2"	6 1/4"	7 1/2"	6 1/4"	7 1/2"	6 1/4"	7 9/16"	6 5/16"
	BODY	10 1/16"	11 5/16"	10"	11 1/4"	10"	11 1/4"	10"	11 1/4"	9 15/16"	11 3/16"

FLOOR TO FLOOR RISE	3'1 7/8"	3'1 7/8"	3'2"	3'2"	3'2 1/8"	3'2 1/8"	3'2 1/4"	3'2 1/4"	3'2 3/8"	3'2 3/8"
NUMBER OF RISERS	5	6	5	6	5	6	5	6	5	6
HEIGHT OF EACH RISER	7 9/16"	6 5/16"	7 5/8"	6 5/16"	7 5/8"	6 3/8"	7 5/8"	6 3/8"	7 11/16"	6 3/8"
TOTAL HEIGHT OF RISERS	3'1 13/16"	3'1 7/8"	3'2 1/8"	3'1 7/8"	3'2 1/8"	3'2 1/4"	3'2 1/8"	3'2 1/4"	3'2 7/16"	3'2 1/4"
RISERS OVERAGE (+) OR UNDERAGE (—)	−1/16"	0	+1/8"	−1/8"	0	+1/8"	−1/8"	0	+1/16"	−1/8"
NUMBER OF TREADS	4	5	4	5	4	5	4	5	4	5
WIDTH OF EACH TREAD	9 15/16"	11 3/16"	9 7/8"	11 3/16"	9 7/8"	11 1/8"	9 7/8"	11 1/8"	9 13/16"	11 1/8"
TOTAL RUN OF STAIRWAY	3'3 3/4"	4'7 15/16"	3'3 1/2"	4'7 15/16"	3'3 1/2"	4'7 5/8"	3'3 1/2"	4'7 5/8"	3'3 1/4"	4'7 5/8"
MAXIMUM RUN WITH 6'-8" HEADROOM	1'7 7/8"	1'10 3/8"	1'7 3/4"	1'10 3/8"	1'7 3/4"	1'10 1/4"	1'7 3/4"	1'10 1/4"	1'7 5/8"	1'10 1/4"
ANGLE OF INCLINE	37°16'	29°26'	37°40'	29°26'	37°40'	29°49'	37°40'	29°49'	38°5'	29°49'
LENGTH OF CARRIAGE	4'1 15/16"	5'4 7/32"	4'1 29/32"	5'4 7/32"	4'1 29/32"	5'4 1/8"	4'1 29/32"	5'4 1/8"	4'1 27/32"	5'4 1/8"
L. F. OF RISER PER INCH OF STAIRWAY WIDTH	.416	.5	.416	.5	.416	.5	.416	.5	.416	.5
L. F. OF TREAD PER INCH OF STAIRWAY WIDTH	.333	.416	.333	.416	.333	.416	.333	.416	.333	.416
FRAMING SQUARE SETTINGS FOR CARRIAGE CUTS — TONGUE	7 9/16"	6 5/16"	7 5/8"	6 5/16"	7 5/8"	6 3/8"	7 5/8"	6 3/8"	7 11/16"	6 3/8"
FRAMING SQUARE SETTINGS FOR CARRIAGE CUTS — BODY	9 15/16"	11 3/16"	9 7/8"	11 3/16"	9 7/8"	11 1/8"	9 7/8"	11 1/8"	9 13/16"	11 1/8"

FLOOR TO FLOOR RISE	3'2 1/2"	3'2 1/2"	3'2 5/8"	3'2 5/8"	3'2 3/4"	3'2 3/4"	3'2 7/8"	3'2 7/8"	3'3"	3'3"
NUMBER OF RISERS	5	6	5	6	5	6	5	6	5	6
HEIGHT OF EACH RISER	7 11/16"	6 7/16"	7 3/4"	6 7/16"	7 3/4"	6 7/16"	7 3/4"	6 1/2"	7 13/16"	6 1/2"
TOTAL HEIGHT OF RISERS	3'2 7/16"	3'2 5/8"	3'2 3/4"	3'2 5/8"	3'2 3/4"	3'2 5/8"	3'2 3/4"	3'3"	3'3 1/16"	3'3"
RISERS OVERAGE (+) OR UNDERAGE (—)	−1/16"	+1/8"	+1/8"	0	0	−1/8"	−1/8"	+1/8"	+1/16"	0
NUMBER OF TREADS	4	5	4	5	4	5	4	5	4	5
WIDTH OF EACH TREAD	9 13/16"	11 1/16"	9 3/4"	11 1/16"	9 3/4"	11 1/16"	9 3/4"	11"	9 11/16"	11"
TOTAL RUN OF STAIRWAY	3'3 1/4"	4'7 5/16"	3'3"	4'7 5/16"	3'3"	4'7 5/16"	3'3"	4'7"	3'2 3/4"	4'7"
MAXIMUM RUN WITH 6'-8" HEADROOM	1'7 5/8"	1'10 1/8"	1'7 1/2"	1'10 1/8"	1'7 1/2"	1'10 1/8"	1'7 1/2"	1'10"	1'7 3/8"	1'10"
ANGLE OF INCLINE	38°5'	30°12'	28°29'	30°12'	28°29'	30°12'	28°29'	30°35'	28°53'	30°35'
LENGTH OF CARRIAGE	4'1 27/32"	5'4"	4'1 13/16"	5'4"	4'1 13/16"	5'4"	4'1 13/16"	5'3 7/8"	4'1 25/32"	5'3 7/8"
L. F. OF RISER PER INCH OF STAIRWAY WIDTH	.416	.5	.416	.5	.416	.5	.416	.5	.416	.5
L. F. OF TREAD PER INCH OF STAIRWAY WIDTH	.333	.416	.333	.416	.333	.416	.333	.416	.333	.416
FRAMING SQUARE SETTINGS FOR CARRIAGE CUTS — TONGUE	7 11/16"	6 7/16"	7 3/4"	6 7/16"	7 3/4"	6 7/16"	7 3/4"	6 1/2"	7 13/16"	6 1/2"
FRAMING SQUARE SETTINGS FOR CARRIAGE CUTS — BODY	9 13/16"	11 1/16"	9 3/4"	11 1/16"	9 3/4"	11 1/16"	9 3/4"	11"	9 11/16"	11"

FLOOR TO FLOOR RISE	3'3 1/8"	3'3 1/8"	3'3 1/4"	3'3 1/4"	3'3 3/8"	3'3 3/8"	3'3 1/2"	3'3 1/2"	3'3 5/8"	3'3 5/8"
NUMBER OF RISERS	5	6	5	6	5	6	5	6	5	6
HEIGHT OF EACH RISER	7 13/16"	6 1/2"	7 7/8"	6 9/16"	7 7/8"	6 9/16"	7 7/8"	6 9/16"	7 15/16"	6 5/8"
TOTAL HEIGHT OF RISERS	3'3 1/16"	3'3"	3'3 3/8"	3'3 3/8"	3'3 3/8"	3'3 3/8"	3'3 3/8"	3'3 3/8"	3'3 11/16"	3'3 3/4"
RISERS OVERAGE (+) OR UNDERAGE (—)	−1/16"	−1/8"	+1/8"	+1/8"	0	0	−1/8"	−1/8"	+1/16"	+1/8"
NUMBER OF TREADS	4	5	4	5	4	5	4	5	4	5
WIDTH OF EACH TREAD	9 11/16"	11"	9 5/8"	10 15/16"	9 5/8"	10 15/16"	9 5/8"	10 15/16"	9 9/16"	10 7/8"
TOTAL RUN OF STAIRWAY	3'2 3/4"	4'7"	3'2 1/2"	4'6 11/16"	3'2 1/2"	4'6 11/16"	3'2 1/2"	4'6 11/16"	3'2 1/4"	4'6 3/8"
MAXIMUM RUN WITH 6'-8" HEADROOM	1'7 3/8"	1'10"	1'7 1/4"	1'9 7/8"	1'7 1/4"	1'9 7/8"	1'7 1/4"	1'9 7/8"	1'7 1/8"	1'9 3/4"
ANGLE OF INCLINE	38°53'	30°35'	39°17'	30°58'	39°17'	30°58'	39°17'	30°58'	39°42'	31°21'
LENGTH OF CARRIAGE	4'1 25/32"	5'3 7/8"	4'1 3/4"	5'3 25/32"	4'1 3/4"	5'3 25/32"	4'1 3/4"	5'3 25/32"	4'1 23/32"	5'3 21/32"
L. F. OF RISER PER INCH OF STAIRWAY WIDTH	.416	.5	.416	.5	.416	.5	.416	.5	.416	.5
L. F. OF TREAD PER INCH OF STAIRWAY WIDTH	.333	.416	.333	.416	.333	.416	.333	.416	.333	.416
FRAMING SQUARE SETTINGS FOR CARRIAGE CUTS — TONGUE	7 13/16"	6 1/2"	7 7/8"	6 9/16"	7 7/8"	6 9/16"	7 7/8"	6 9/16"	7 15/16"	6 5/8"
FRAMING SQUARE SETTINGS FOR CARRIAGE CUTS — BODY	9 11/16"	11"	9 5/8"	10 15/16"	9 5/8"	10 15/16"	9 5/8"	10 15/16"	9 9/16"	10 7/8"

FLOOR TO FLOOR RISE		3'3 3/4"	3'3 3/4"	3'3 7/8"	3'3 7/8"	3'4"	3'4"	3'4 1/8"	3'4 1/8"	3'4 1/4"	3'4 1/4"
NUMBER OF RISERS		5	6	5	6	5	6	5	6	5	6
HEIGHT OF EACH RISER		7 15/16"	6 5/8"	8"	6 5/8"	8"	6 11/16"	8"	6 11/16"	8 1/16"	6 11/16"
TOTAL HEIGHT OF RISERS		3'3 11/16"	3'3 3/4"	3'4"	3'3 3/4"	3'4"	3'4 1/8"	3'4"	3'4 1/8"	3'4 5/16"	3'4 1/8"
RISERS OVERAGE (+) OR UNDERAGE (—)		−1/16"	0	+1/8"	−1/8"	0	+1/8"	−1/8"	0	+1/16"	−1/8"
NUMBER OF TREADS		4	5	4	5	4	5	4	5	4	5
WIDTH OF EACH TREAD		9 9/16"	10 7/8"	9 1/2"	10 7/8"	9 1/2"	10 13/16"	9 1/2"	10 13/16"	9 7/16"	10 13/16"
TOTAL RUN OF STAIRWAY		3'2 1/4"	4'6 3/8"	3'2"	4'6 3/8"	3'2"	4'6 1/16"	3'2"	4'6 1/16"	3'1 3/4"	4'6 1/16"
MAXIMUM RUN WITH 6'-8" HEADROOM		1'7 1/8"	1'9 3/4"	9 1/2"	1'9 3/4"	9 1/2"	1'9 5/8"	9 1/2"	1'9 5/8"	9 7/16"	1'9 5/8"
ANGLE OF INCLINE		39°42'	31°21'	40°6'	31°21'	40°6'	31°44'	40°6'	31°44'	40°30'	31°44'
LENGTH OF CARRIAGE		4'1 23/32"	5'3 21/32"	4'1 11/16"	5'3 21/32"	4'1 11/16"	5'3 9/16"	4'1 11/16"	5'3 9/16"	4'1 21/32"	5'3 9/16"
L. F. OF RISER PER INCH OF STAIRWAY WIDTH		.416	.5	.416	.5	.416	.5	.416	.5	.416	.5
L. F. OF TREAD PER INCH OF STAIRWAY WIDTH		.333	.416	.333	.416	.333	.416	.333	.416	.333	.416
FRAMING SQUARE SETTINGS FOR CARRIAGE CUTS	TONGUE	7 15/16"	6 5/8"	8"	6 5/8"	8"	6 11/16"	8"	6 11/16"	8 1/16"	6 11/16"
	BODY	9 9/16"	10 7/8"	9 1/2"	10 7/8"	9 1/2"	10 13/16"	9 1/2"	10 13/16"	9 7/16"	10 13/16"

FLOOR TO FLOOR RISE		3'4 3/8"	3'4 3/8"	3'4 1/2"	3'4 1/2"	3'4 5/8"	3'4 5/8"	3'4 3/4"	3'4 3/4"	3'4 7/8"	3'4 7/8"
NUMBER OF RISERS		5	6	5	6	5	6	5	6	5	6
HEIGHT OF EACH RISER		8 1/16"	6 3/4"	8 1/8"	6 3/4"	8 1/8"	6 3/4"	8 1/8"	6 13/16"	8 3/16"	6 13/16"
TOTAL HEIGHT OF RISERS		3'4 5/16"	3'4 1/2"	3'4 5/8"	3'4 1/2"	3'4 5/8"	3'4 1/2"	3'4 5/8"	3'4 7/8"	3'4 15/16"	3'4 7/8"
RISERS OVERAGE (+) OR UNDERAGE (—)		−1/16"	+1/8"	+1/8"	0	0	−1/8"	−1/8"	+1/8"	+1/16"	0
NUMBER OF TREADS		4	5	4	5	4	5	4	5	4	5
WIDTH OF EACH TREAD		9 7/16"	10 3/4"	9 3/8"	10 3/4"	9 3/8"	10 3/4"	9 3/8"	10 11/16"	9 5/16"	10 11/16"
TOTAL RUN OF STAIRWAY		3'1 3/4"	4'5 3/4"	3'1 1/2"	4'5 3/4"	3'1 1/2"	4'5 3/4"	3'1 1/2"	4'5 7/16"	3'1 1/4"	4'5 7/16"
MAXIMUM RUN WITH 6'-8" HEADROOM		9 7/16"	1'9 1/2"	9 3/8"	1'9 1/2"	9 3/8"	1'9 1/2"	9 3/8"	1'9 3/8"	9 5/16"	1'9 3/8"
ANGLE OF INCLINE		40°30'	32°7'	40°55'	32°7'	40°55'	32°7'	40°55'	32°31'	41°19'	32°31'
LENGTH OF CARRIAGE		4'1 21/32"	5'3 15/16"	4'1 5/8"	5'3 15/16"	4'1 5/8"	5'3 15/32"	4'1 5/8"	5'3 3/8"	4'1 19/32"	5'3 3/8"
L. F. OF RISER PER INCH OF STAIRWAY WIDTH		.416	.5	.416	.5	.416	.5	.416	.5	.416	.5
L. F. OF TREAD PER INCH OF STAIRWAY WIDTH		.333	.416	.333	.416	.333	.416	.333	.416	.333	.416
FRAMING SQUARE SETTINGS FOR CARRIAGE CUTS	TONGUE	8 1/16"	6 3/4"	8 1/8"	6 3/4"	8 1/8"	6 3/4"	8 1/8"	6 13/16"	8 3/16"	6 13/16"
	BODY	9 7/16"	10 3/4"	9 3/8"	10 3/4"	9 3/8"	10 3/4"	9 3/8"	10 11/16"	9 5/16"	10 11/16"

FLOOR TO FLOOR RISE		3'5"	3'5"	3'5 1/8"	3'5 1/8"	3'5 1/4"	3'5 1/4"	3'5 3/8"	3'5 3/8"	3'5 1/2"	3'5 5/8"
NUMBER OF RISERS		5	6	5	6	5	6	5	6	6	6
HEIGHT OF EACH RISER		8 3/16"	6 13/16"	8 1/4"	6 7/8"	8 1/4"	6 7/8"	8 1/4"	6 7/8"	6 15/16"	6 15/16"
TOTAL HEIGHT OF RISERS		3'4 15/16"	3'4 7/8"	3'5 1/4"	3'5 1/4"	3'5 1/4"	3'5 1/4"	3'5 1/4"	3'5 1/4"	3'5 5/8"	3'5 5/8"
RISERS OVERAGE (+) OR UNDERAGE (—)		−1/16"	−1/8"	+1/8"	+1/8"	0	0	−1/8"	−1/8"	+1/8"	0
NUMBER OF TREADS		4	5	4	5	4	5	4	5	5	5
WIDTH OF EACH TREAD		9 5/16"	10 11/16"	9 1/4"	10 5/8"	9 1/4"	10 5/8"	9 1/4"	10 5/8"	10 9/16"	10 9/16"
TOTAL RUN OF STAIRWAY		3'1 1/4"	4'5 7/16"	3'1"	4'5 1/8"	3'1"	4'5 1/8"	3'1"	4'5 1/8"	4'4 13/16"	4'4 13/16"
MAXIMUM RUN WITH 6'-8" HEADROOM		9 5/16"	1'9 3/8"	9 1/4"	1'9 1/4"	9 1/4"	1'9 1/4"	9 1/4"	1'9 1/4"	1'9 1/8"	1'9 1/8"
ANGLE OF INCLINE		41°19'	32°31'	41°44'	32°54'	41°44'	32°54'	41°44'	32°54'	33°18'	33°18'
LENGTH OF CARRIAGE		4'1 19/32"	5'3 3/8"	4'1 9/16"	5'3 9/32"	4'1 9/16"	5'3 9/32"	4'1 9/16"	5'3 9/32"	5'3 3/16"	5'3 3/16"
L. F. OF RISER PER INCH OF STAIRWAY WIDTH		.416	.5	.416	.5	.416	.5	.416	.5	.5	.5
L. F. OF TREAD PER INCH OF STAIRWAY WIDTH		.333	.416	.333	.416	.333	.416	.333	.416	.416	.416
FRAMING SQUARE SETTINGS FOR CARRIAGE CUTS	TONGUE	8 3/16"	6 13/16"	8 1/4"	6 7/8"	8 1/4"	6 7/8"	8 1/4"	6 7/8"	6 15/16"	6 15/16"
	BODY	9 5/16"	10 11/16"	9 1/4"	10 5/8"	9 1/4"	10 5/8"	9 1/4"	10 5/8"	10 9/16"	10 9/16"

FLOOR TO FLOOR RISE	3'5$\frac{3}{4}$"	3'5$\frac{7}{8}$"	3'5$\frac{7}{8}$"	3'6"	3'6"	3'6$\frac{1}{8}$"	3'6$\frac{1}{8}$"	3'6$\frac{1}{4}$"	3'6$\frac{1}{4}$"	3'6$\frac{3}{8}$"
NUMBER OF RISERS	6	6	7	6	7	6	7	6	7	6
HEIGHT OF EACH RISER	6$\frac{15}{16}$"	7"	6"	7"	6"	7"	6"	7$\frac{1}{16}$"	6$\frac{1}{16}$"	7$\frac{1}{16}$"
TOTAL HEIGHT OF RISERS	3'5$\frac{5}{8}$"	3'6"	3'6"	3'6"	3'6"	3'6"	3'6"	3'6$\frac{3}{8}$"	3'6$\frac{7}{16}$"	3'6$\frac{3}{8}$"
RISERS OVERAGE (+) OR UNDERAGE (—)	$-\frac{1}{8}$"	$+\frac{1}{8}$"	$+\frac{1}{8}$"	0	0	$-\frac{1}{8}$"	$-\frac{1}{8}$"	$+\frac{1}{8}$"	$+\frac{3}{16}$"	0
NUMBER OF TREADS	5	5	6	5	6	5	6	5	6	5
WIDTH OF EACH TREAD	10$\frac{9}{16}$"	10$\frac{1}{2}$"	11$\frac{1}{2}$"	10$\frac{1}{2}$"	11$\frac{1}{2}$"	10$\frac{1}{2}$"	11$\frac{1}{2}$"	10$\frac{7}{16}$"	11$\frac{7}{16}$"	10$\frac{7}{16}$"
TOTAL RUN OF STAIRWAY	4'4$\frac{13}{16}$"	4'4$\frac{1}{2}$"	5'9"	4'4$\frac{1}{2}$"	5'9"	4'4$\frac{1}{2}$"	5'9"	4'4$\frac{3}{16}$"	5'8$\frac{5}{8}$"	4'4$\frac{3}{16}$"
MAXIMUM RUN WITH 6'-8" HEADROOM	1'9$\frac{1}{8}$"	1'9"	1'11"	1'9"	1'11"	1'9"	1'11"	1'8$\frac{7}{8}$"	1'10$\frac{7}{8}$"	1'8$\frac{7}{8}$"
ANGLE OF INCLINE	33°18'	33°41'	27°33'	33°41'	27°33'	33°41'	27°33'	34°5'	27°56'	34°5'
LENGTH OF CARRIAGE	5'3$\frac{3}{16}$"	5'3$\frac{3}{32}$"	6'5$\frac{13}{16}$"	5'3$\frac{3}{32}$"	6'5$\frac{13}{16}$"	5'3$\frac{3}{32}$"	6'5$\frac{13}{16}$"	5'3"	6'5$\frac{21}{32}$"	5'3"
L. F. OF RISER PER INCH OF STAIRWAY WIDTH	.5	.5	.583	.5	.583	.5	.583	.5	.583	.5
L. F. OF TREAD PER INCH OF STAIRWAY WIDTH	.416	.416	.5	.416	.5	.416	.5	.416	.5	.416
FRAMING SQUARE SETTINGS FOR CARRIAGE CUTS — TONGUE	6$\frac{15}{16}$"	7"	6"	7"	6"	7"	6"	7$\frac{1}{16}$"	6$\frac{1}{16}$"	7$\frac{1}{16}$"
FRAMING SQUARE SETTINGS FOR CARRIAGE CUTS — BODY	10$\frac{9}{16}$"	10$\frac{1}{2}$"	11$\frac{1}{2}$"	10$\frac{1}{2}$"	11$\frac{1}{2}$"	10$\frac{1}{2}$"	11$\frac{1}{2}$"	10$\frac{7}{16}$"	11$\frac{7}{16}$"	10$\frac{7}{16}$"

FLOOR TO FLOOR RISE	3'6 3/8"	3'6 1/2"	3'6 1/2"	3'6 5/8"	3'6 5/8"	3'6 3/4"	3'6 3/4"	3'6 7/8"	3'6 7/8"	3'7"
NUMBER OF RISERS	7	6	7	6	7	6	7	6	7	6
HEIGHT OF EACH RISER	6 1/16"	7 1/16"	6 1/16"	7 1/8"	6 1/16"	7 1/8"	6 1/8"	7 1/8"	6 1/8"	7 3/16"
TOTAL HEIGHT OF RISERS	3'6 7/16"	3'6 3/8"	3'6 7/16"	3'6 3/4"	3'6 7/16"	3'6 3/4"	3'6 7/8"	3'6 3/4"	3'6 7/8"	3'7 1/8"
RISERS OVERAGE (+) OR UNDERAGE (—)	+1/16"	−1/8"	−1/16"	+1/8"	−3/16"	0	+1/8"	−1/8"	0	+1/8"
NUMBER OF TREADS	6	5	6	5	6	5	6	5	6	5
WIDTH OF EACH TREAD	11 7/16"	10 7/16"	11 7/16"	10 3/8"	11 7/16"	10 3/8"	11 3/8"	10 3/8"	11 3/8"	10 5/16"
TOTAL RUN OF STAIRWAY	5'8 5/8"	4'4 3/16"	5'8 5/8"	4'3 7/8"	5'8 5/8"	4'3 7/8"	5'8 1/4"	4'3 7/8"	5'8 1/4"	4'3 9/16"
MAXIMUM RUN WITH 6'-8" HEADROOM	1'10 7/8"	1'8 7/8"	1'10 7/8"	1'8 3/4"	1'10 7/8"	1'8 3/4"	1'10 3/4"	1'8 3/4"	1'10 3/4"	1'8 5/8"
ANGLE OF INCLINE	27°56'	34°5'	27°56'	34°29'	27°56'	34°29'	28°18'	34°29'	28°18'	34°53'
LENGTH OF CARRIAGE	6'5 21/32"	5'3"	6'5 21/32"	5'2 15/16"	6'5 21/32"	5'2 15/16"	6'5 1/2"	5'2 15/16"	6'5 1/2"	5'2 27/32"
L. F. OF RISER PER INCH OF STAIRWAY WIDTH	.583	.5	.583	.5	.583	.5	.583	.5	.583	.5
L. F. OF TREAD PER INCH OF STAIRWAY WIDTH	.5	.416	.5	.416	.5	.416	.5	.416	.5	.416
FRAMING SQUARE SETTINGS FOR CARRIAGE CUTS — TONGUE	6 1/16"	7 1/16"	6 1/16"	7 1/8"	6 1/16"	7 1/8"	6 1/8"	7 1/8"	6 1/8"	7 3/16"
FRAMING SQUARE SETTINGS FOR CARRIAGE CUTS — BODY	11 7/16"	10 7/16"	11 7/16"	10 3/8"	11 7/16"	10 3/8"	11 3/8"	10 3/8"	11 3/8"	10 5/16"

FLOOR TO FLOOR RISE		3'7"	3'7 1/8"	3'7 1/8"	3'7 1/4"	3'7 1/4"	3'7 3/8"	3'7 3/8"	3'7 1/2"	3'7 1/2"	3'7 5/8"
NUMBER OF RISERS		7	6	7	6	7	6	7	6	7	6
HEIGHT OF EACH RISER		6 1/8"	7 3/16"	6 3/16"	7 3/16"	6 3/16"	7 1/4"	6 3/16"	7 1/4"	6 3/16"	7 1/4"
TOTAL HEIGHT OF RISERS		3'6 7/8"	3'7 1/8"	3'7 5/16"	3'7 1/8"	3'7 5/16"	3'7 1/2"	3'7 5/16"	3'7 1/2"	3'7 5/16"	3'7 1/2"
RISERS OVERAGE (+) OR UNDERAGE (—)		−1/8"	0	+3/16"	−1/8"	+1/16"	+1/8"	−1/16"	0	−3/16"	−1/8"
NUMBER OF TREADS		6	5	6	5	6	5	6	5	6	5
WIDTH OF EACH TREAD		11 3/8"	10 5/16"	11 5/16"	10 5/16"	11 5/16"	10 1/4"	11 5/16"	10 1/4"	11 5/16"	10 1/4"
TOTAL RUN OF STAIRWAY		5'8 1/4"	4'3 9/16"	5'7 7/8"	4'3 9/16"	5'7 7/8"	4'3 1/4"	5'7 7/8"	4'3 1/4"	5'7 7/8"	4'3 1/4"
MAXIMUM RUN WITH 6'-8" HEADROOM		1'10 3/4"	1'8 5/8"	1'10 5/8"	1'8 5/8"	1'10 5/8"	1'8 1/2"	1'10 5/8"	1'8 1/2"	1'10 5/8"	1'8 1/2"
ANGLE OF INCLINE		28°18'	34°53'	28°41'	34°53'	28°41'	35°16'	28°41'	35°16'	28°41'	35°16'
LENGTH OF CARRIAGE		6'5 1/2"	5'2 27/32"	6'5 3/8"	5'2 27/32"	6'5 3/8"	5'2 25/32"	6'5 3/8"	5'2 25/32"	6'5 3/8"	5'2 25/32"
L. F. OF RISER PER INCH OF STAIRWAY WIDTH		.583	.5	.583	.5	.583	.5	.583	.5	.583	.5
L. F. OF TREAD PER INCH OF STAIRWAY WIDTH		.5	.416	.5	.416	.5	.416	.5	.416	.5	.416
FRAMING SQUARE SETTINGS FOR CARRIAGE CUTS	TONGUE	6 1/8"	7 3/16"	6 3/16"	7 3/16"	6 3/16"	7 1/4"	6 3/16"	7 1/4"	6 3/16"	7 1/4"
	BODY	11 3/8"	10 5/16"	11 5/16"	10 5/16"	11 5/16"	10 1/4"	11 5/16"	10 1/4"	11 5/16"	10 1/4"

FLOOR TO FLOOR RISE	3'7$\frac{5}{8}$"	3'7$\frac{3}{4}$"	3'7$\frac{3}{4}$"	3'7$\frac{7}{8}$"	3'7$\frac{7}{8}$"	3'8"	3'8"	3'8$\frac{1}{8}$"	3'8$\frac{1}{8}$"	3'8$\frac{1}{4}$"
NUMBER OF RISERS	7	6	7	6	7	6	7	6	7	6
HEIGHT OF EACH RISER	6$\frac{1}{4}$"	7$\frac{5}{16}$"	6$\frac{1}{4}$"	7$\frac{5}{16}$"	6$\frac{1}{4}$"	7$\frac{5}{16}$"	6$\frac{5}{16}$"	7$\frac{3}{8}$"	6$\frac{5}{16}$"	7$\frac{3}{8}$"
TOTAL HEIGHT OF RISERS	3'7$\frac{3}{4}$"	3'7$\frac{7}{8}$"	3'7$\frac{3}{4}$"	3'7$\frac{7}{8}$"	3'7$\frac{3}{4}$"	3'7$\frac{7}{8}$"	3'8$\frac{3}{16}$"	3'8$\frac{1}{4}$"	3'8$\frac{3}{16}$"	3'8$\frac{1}{4}$"
RISERS OVERAGE (+) OR UNDERAGE (—)	+$\frac{1}{8}$"	+$\frac{1}{8}$"	0	0	−$\frac{1}{8}$"	−$\frac{1}{8}$"	+$\frac{3}{16}$"	+$\frac{1}{8}$"	+$\frac{1}{16}$"	0
NUMBER OF TREADS	6	5	6	5	6	5	6	5	6	5
WIDTH OF EACH TREAD	11$\frac{1}{4}$"	10$\frac{3}{16}$"	11$\frac{1}{4}$"	10$\frac{3}{16}$"	11$\frac{1}{4}$"	10$\frac{3}{16}$"	11$\frac{3}{16}$"	10$\frac{1}{8}$"	11$\frac{3}{16}$"	10$\frac{1}{8}$"
TOTAL RUN OF STAIRWAY	5'7$\frac{1}{2}$"	4'2$\frac{15}{16}$"	5'7$\frac{1}{2}$"	4'2$\frac{15}{16}$"	5'7$\frac{1}{2}$"	4'2$\frac{15}{16}$"	5'7$\frac{1}{8}$"	4'2$\frac{5}{8}$"	5'7$\frac{1}{8}$"	4'2$\frac{5}{8}$"
MAXIMUM RUN WITH 6'-8" HEADROOM	1'10$\frac{1}{2}$"	1'8$\frac{3}{8}$"	1'10$\frac{1}{2}$"	1'8$\frac{3}{8}$"	1'10$\frac{1}{2}$"	1'8$\frac{3}{8}$"	1'10$\frac{3}{8}$"	1'8$\frac{1}{4}$"	1'10$\frac{3}{8}$"	1'8$\frac{1}{4}$"
ANGLE OF INCLINE	29°3'	35°40'	29°3'	35°40'	29°3'	35°40'	29°26'	36°4'	29°26'	36°4'
LENGTH OF CARRIAGE	6'5$\frac{7}{32}$"	5'2$\frac{11}{16}$"	6'5$\frac{7}{32}$"	5'2$\frac{11}{16}$"	6'5$\frac{7}{32}$"	5'2$\frac{11}{16}$"	6'5$\frac{1}{16}$"	5'2$\frac{5}{8}$"	6'5$\frac{1}{16}$"	5'2$\frac{5}{8}$"
L. F. OF RISER PER INCH OF STAIRWAY WIDTH	.583	.5	.583	.5	.583	.5	.583	.5	.583	.5
L. F. OF TREAD PER INCH OF STAIRWAY WIDTH	.5	.416	.5	.416	.5	.416	.5	.416	.5	.416
FRAMING SQUARE SETTINGS FOR CARRIAGE CUTS TONGUE	6$\frac{1}{4}$"	7$\frac{5}{16}$"	6$\frac{1}{4}$"	7$\frac{5}{16}$"	6$\frac{1}{4}$"	7$\frac{5}{16}$"	6$\frac{5}{16}$"	7$\frac{3}{8}$"	6$\frac{5}{16}$"	7$\frac{3}{8}$"
BODY	11$\frac{1}{4}$"	10$\frac{3}{16}$"	11$\frac{1}{4}$"	10$\frac{3}{16}$"	11$\frac{1}{4}$"	10$\frac{3}{16}$"	11$\frac{3}{16}$"	10$\frac{1}{8}$"	11$\frac{3}{16}$"	10$\frac{1}{8}$"

FLOOR TO FLOOR RISE		3'8 1/4"	3'8 3/8"	3'8 3/8"	3'8 1/2"	3'8 1/2"	3'8 5/8"	3'8 5/8"	3'8 3/4"	3'8 3/4"	3'8 7/8"
NUMBER OF RISERS		7	6	7	6	7	6	7	6	7	6
HEIGHT OF EACH RISER		6 5/16"	7 3/8"	6 5/16"	7 7/16"	6 3/8"	7 7/16"	6 3/8"	7 7/16"	6 3/8"	7 1/2"
TOTAL HEIGHT OF RISERS		3'8 3/16"	3'8 1/4"	3'8 3/16"	3'8 5/8"	3'8 5/8"	3'8 5/8"	3'8 5/8"	3'8 5/8"	3'8 5/8"	3'9"
RISERS OVERAGE (+) OR UNDERAGE (—)		−1/16"	−1/8"	−3/16"	+1/8"	+1/8"	0	0	−1/8"	−1/8"	+1/8"
NUMBER OF TREADS		6	5	6	5	6	5	6	5	6	5
WIDTH OF EACH TREAD		11 3/16"	10 1/8"	11 3/16"	10 1/16"	11 1/8"	10 1/16"	11 1/8"	10 1/16"	11 1/8"	10"
TOTAL RUN OF STAIRWAY		5'7 1/8"	4'2 5/8"	5'7 1/8"	4'2 5/16"	5'6 3/4"	4'2 5/16"	5'6 3/4"	4'2 5/16"	5'6 3/4"	4'2"
MAXIMUM RUN WITH 6'-8" HEADROOM		1'10 3/8"	1'8 1/4"	1'10 3/8"	1'8 1/8"	1'10 1/4"	1'8 1/8"	1'10 1/4"	1'8 1/8"	1'10 1/4"	1'8"
ANGLE OF INCLINE		29°26'	36°4'	29°26'	36°28'	29°49'	36°28'	29°49'	36°28'	29°49'	36°52'
LENGTH OF CARRIAGE		6'5 1/16"	5'2 5/8"	6'5 1/16"	5'2 9/16"	6'4 15/16"	5'2 9/16"	6'4 15/16"	5'2 9/16"	6'4 15/16"	5'2 1/2"
L. F. OF RISER PER INCH OF STAIRWAY WIDTH		.583	.5	.583	.5	.583	.5	.583	.5	.583	.5
L. F. OF TREAD PER INCH OF STAIRWAY WIDTH		.5	.416	.5	.416	.5	.416	.5	.416	.5	.416
FRAMING SQUARE SETTINGS FOR CARRIAGE CUTS	TONGUE	6 5/16"	7 3/8"	6 5/16"	7 7/16"	6 3/8"	7 7/16"	6 3/8"	7 7/16"	6 3/8"	7 1/2"
	BODY	11 3/16"	10 1/8"	11 3/16"	10 1/16"	11 1/8"	10 1/16"	11 1/8"	10 1/16"	11 1/8"	10"

FLOOR TO FLOOR RISE	3'8 7/8"	3'9"	3'9"	3'9 1/8"	3'9 1/8"	3'9 1/4"	3'9 1/4"	3'9 3/8"	3'9 3/8"	3'9 1/2"
NUMBER OF RISERS	7	6	7	6	7	6	7	6	7	6
HEIGHT OF EACH RISER	6 7/16"	7 1/2"	6 7/16"	7 1/2"	6 7/16"	7 9/16"	6 7/16"	7 9/16"	6 1/2"	7 9/16"
TOTAL HEIGHT OF RISERS	3'9 1/16"	3'9"	3'9 1/16"	3'9"	3'9 1/16"	3'9 3/8"	3'9 1/16"	3'9 3/8"	3'9 1/2"	3'9 3/8"
RISERS OVERAGE (+) OR UNDERAGE (—)	+3/16"	0	+1/16"	−1/8"	−1/16"	+1/8"	−3/16"	0	+1/8"	−1/8"
NUMBER OF TREADS	6	5	6	5	6	5	6	5	6	5
WIDTH OF EACH TREAD	11 1/16"	10"	11 1/16"	10"	11 1/16"	9 15/16"	11 1/16"	9 15/16"	11"	9 15/16"
TOTAL RUN OF STAIRWAY	5'6 3/8"	4'2"	5'6 3/8"	4'2"	5'6 3/8"	4'1 11/16"	5'6 3/8"	4'1 11/16"	5'6"	4'1 11/16"
MAXIMUM RUN WITH 6'-8" HEADROOM	1'10 1/8"	1'8"	1'10 1/8"	1'8"	1'10 1/8"	1'7 7/8"	1'10 1/8"	1'7 7/8"	1'10"	1'7 7/8"
ANGLE OF INCLINE	30°12'	36°52'	30°12'	36°52'	30°12'	37°16'	30°12'	37°16'	30°35'	37°16'
LENGTH OF CARRIAGE	6'4 25/32"	5'2 1/2"	6'4 25/32"	5'2 1/2"	6'4 25/32"	5'2 7/16"	6'4 25/32"	5'2 7/16"	6'4 21/32"	5'2 7/16"
L. F. OF RISER PER INCH OF STAIRWAY WIDTH	.583	.5	.583	.5	.583	.5	.583	.5	.583	.5
L. F. OF TREAD PER INCH OF STAIRWAY WIDTH	.5	.416	.5	.416	.5	.416	.5	.416	.5	.416
FRAMING SQUARE SETTINGS FOR CARRIAGE CUTS — TONGUE	6 7/16"	7 1/2"	6 7/16"	7 1/2"	6 7/16"	7 9/16"	6 7/16"	7 9/16"	6 1/2"	7 9/16"
FRAMING SQUARE SETTINGS FOR CARRIAGE CUTS — BODY	11 1/16"	10"	11 1/16"	10"	11 1/16"	9 15/16"	11 1/16"	9 15/16"	11"	9 15/16"

FLOOR TO FLOOR RISE	3'9 1/2"	3'9 5/8"	3'9 5/8"	3'9 3/4"	3'9 3/4"	3'9 7/8"	3'9 7/8"	3'10"	3'10"	3'10 1/8"
NUMBER OF RISERS	7	6	7	6	7	6	7	6	7	6
HEIGHT OF EACH RISER	6 1/2"	7 5/8"	6 1/2"	7 5/8"	6 9/16"	7 5/8"	6 9/16"	7 11/16"	6 9/16"	7 11/16"
TOTAL HEIGHT OF RISERS	3'9 1/2"	3'9 3/4"	3'9 1/2"	3'9 3/4"	3'9 15/16"	3'9 3/4"	3'9 15/16"	3'10 1/8"	3'9 15/16"	3'10 1/8"
RISERS OVERAGE (+) OR UNDERAGE (—)	0	+1/8"	−1/8"	0	+3/16"	−1/8"	+1/16"	+1/8"	−1/16"	0
NUMBER OF TREADS	6	5	6	5	6	5	6	5	6	5
WIDTH OF EACH TREAD	11"	9 7/8"	11"	9 7/8"	10 15/16"	9 7/8"	10 15/16"	9 13/16"	10 15/16"	9 13/16"
TOTAL RUN OF STAIRWAY	5'6"	4'1 3/8"	5'6"	4'1 3/8"	5'5 5/8"	4'1 3/8"	5'5 5/8"	4'1 1/16"	5'5 5/8"	4'1 1/16"
MAXIMUM RUN WITH 6'-8" HEADROOM	1'10"	1'7 3/4"	1'10"	1'7 3/4"	1'9 7/8"	1'7 3/4"	1'9 7/8"	1'7 5/8"	1'9 7/8"	1'7 5/8"
ANGLE OF INCLINE	30°35'	37°40'	30°35'	37°40'	30°58'	37°40'	30°58'	38°5'	30°58'	38°5'
LENGTH OF CARRIAGE	6'4 21/32"	5'2 3/8"	6'4 21/32"	5'2 3/8"	6'4 17/32"	5'2 3/8"	6'4 17/32"	5'2 5/16"	6'4 17/32"	5'2 5/16"
L. F. OF RISER PER INCH OF STAIRWAY WIDTH	.583	.5	.583	.5	.583	.5	.583	.5	.583	.5
L. F. OF TREAD PER INCH OF STAIRWAY WIDTH	.5	.416	.5	.416	.5	.416	.5	.416	.5	.416
FRAMING SQUARE SETTINGS FOR CARRIAGE CUTS — TONGUE	6 1/2"	7 5/8"	6 1/2"	7 5/8"	6 9/16"	7 5/8"	6 9/16"	7 11/16"	6 9/16"	7 11/16"
BODY	11"	9 7/8"	11"	9 7/8"	10 15/16"	9 7/8"	10 15/16"	9 13/16"	10 15/16"	9 13/16"

FLOOR TO FLOOR RISE	3'10 1/8"	3'10 1/4"	3'10 1/4"	3'10 3/8"	3'10 3/8"	3'10 1/2"	3'10 1/2"	3'10 5/8"	3'10 5/8"	3'10 3/4"
NUMBER OF RISERS	7	6	7	6	7	6	7	6	7	6
HEIGHT OF EACH RISER	6 9/16"	7 11/16"	6 5/8"	7 3/4"	6 5/8"	7 3/4"	6 5/8"	7 3/4"	6 11/16"	7 13/16"
TOTAL HEIGHT OF RISERS	3'9 15/16"	3'10 1/8"	3'10 3/8"	3'10 1/2"	3'10 3/8"	3'10 1/2"	3'10 3/8"	3'10 1/2"	3'10 13/16"	3'10 7/8"
RISERS OVERAGE (+) OR UNDERAGE (—)	−3/16"	−1/8"	+1/8"	+1/8"	0	0	−1/8"	−1/8"	+3/16"	+1/8"
NUMBER OF TREADS	6	5	6	5	6	5	6	5	6	5
WIDTH OF EACH TREAD	10 15/16"	9 13/16"	10 7/8"	9 3/4"	10 7/8"	9 3/4"	10 7/8"	9 3/4"	10 13/16"	9 11/16"
TOTAL RUN OF STAIRWAY	5'5 3/8"	4'1 1/16"	5'5 1/4"	4' 3/4"	5'5 1/4"	4' 3/4"	5'5 1/4"	4' 3/4"	5'4 7/8"	4' 7/16"
MAXIMUM RUN WITH 6'-8" HEADROOM	1'9 7/8"	1'7 5/8"	1'9 3/4"	1'7 1/2"	1'9 3/4"	1'7 1/2"	1'9 3/4"	1'7 1/2"	1'9 5/8"	1'7 3/8"
ANGLE OF INCLINE	30°58'	38°5'	31°21'	38°29'	31°21'	38°29'	31°21'	38°29'	31°44'	38°53'
LENGTH OF CARRIAGE	6'4 17/32"	5'2 5/16"	6'4 13/32"	5'2 9/32"	6'4 13/32"	5'2 9/32"	6'4 13/16"	5'2 9/32"	6'4 9/32"	5'2 7/32"
L. F. OF RISER PER INCH OF STAIRWAY WIDTH	.583	.5	.583	.5	.583	.5	.583	.5	.583	.5
L. F. OF TREAD PER INCH OF STAIRWAY WIDTH	.5	.416	.5	.416	.5	.416	.5	.416	.5	.416
FRAMING SQUARE SETTINGS FOR CARRIAGE CUTS — TONGUE	6 9/16"	7 11/16"	6 5/8"	7 3/4"	6 5/8"	7 3/4"	6 5/8"	7 3/4"	6 11/16"	7 13/16"
FRAMING SQUARE SETTINGS FOR CARRIAGE CUTS — BODY	10 15/16"	9 13/16"	10 7/8"	9 3/4"	10 7/8"	9 3/4"	10 7/8"	9 3/4"	10 13/16"	9 11/16"

FLOOR TO FLOOR RISE		3'10 3/4"	3'10 7/8"	3'10 7/8"	3'11"	3'11"	3'11 1/8"	3'11 1/8"	3'11 1/4"	3'11 1/4"	3'11 3/8"
NUMBER OF RISERS		7	6	7	6	7	6	7	6	7	6
HEIGHT OF EACH RISER		6 11/16"	7 13/16"	6 11/16"	7 13/16"	6 11/16"	7 7/8"	6 3/4"	7 7/8"	6 3/4"	7 7/8"
TOTAL HEIGHT OF RISERS		3'10 13/16"	3'10 7/8"	3'10 13/16"	3'10 7/8"	3'10 13/16"	3'11 1/4"	3'11 1/4"	3'11 1/4"	3'11 1/4"	3'11 1/4"
RISERS OVERAGE (+) OR UNDERAGE (—)		+1/16"	0	−1/16"	−1/8"	−3/16"	+1/8"	+1/8"	0	0	−1/8"
NUMBER OF TREADS		6	5	6	5	6	5	6	5	6	5
WIDTH OF EACH TREAD		10 13/16"	9 11/16"	10 13/16"	9 11/16"	10 13/16"	9 5/8"	10 3/4"	9 5/8"	10 3/4"	9 5/8"
TOTAL RUN OF STAIRWAY		5'4 7/8"	4' 7/16"	5'4 7/8"	4' 7/16"	5'4 7/8"	4' 1/8"	5'4 1/2"	4' 1/8"	5'4 1/2"	4' 1/8"
MAXIMUM RUN WITH 6'-8" HEADROOM		1'9 5/8"	1'7 3/8"	1'9 5/8"	1'7 3/8"	1'9 5/8"	1'7 1/4"	1'9 1/2"	1'7 1/4"	1'9 1/2"	1'7 1/4"
ANGLE OF INCLINE		31°44'	38°53'	31°44'	38°53'	31°44'	39°17'	32°7'	39°17'	32°7'	39°17'
LENGTH OF CARRIAGE		6'4 9/32"	5'2 7/32"	6'4 9/32"	5'2 7/32"	6'4 9/32"	5'2 3/16"	6'4 5/32"	5'2 3/16"	6'4 5/32"	5'2 3/16"
L. F. OF RISER PER INCH OF STAIRWAY WIDTH		.583	.5	.583	.5	.583	.5	.583	.5	.583	.5
L. F. OF TREAD PER INCH OF STAIRWAY WIDTH		.5	.416	.5	.416	.5	.416	.5	.416	.5	.416
FRAMING SQUARE SETTINGS FOR CARRIAGE CUTS	TONGUE	6 11/16"	7 13/16"	6 11/16"	7 13/16"	6 11/16"	7 7/8"	6 3/4"	7 7/8"	6 3/4"	7 7/8"
	BODY	10 13/16"	9 11/16"	10 13/16"	9 11/16"	10 13/16"	9 5/8"	10 3/4"	9 5/8"	10 3/4"	9 5/8"

FLOOR TO FLOOR RISE	3'11 3/8"	3'11 1/2"	3'11 1/2"	3'11 5/8"	3'11 5/8"	3'11 3/4"	3'11 3/4"	3'11 7/8"	3'11 7/8"	4'
NUMBER OF RISERS	7	6	7	6	7	6	7	6	7	6
HEIGHT OF EACH RISER	6 3/4"	7 15/16"	6 13/16"	7 15/16"	6 13/16"	7 15/16"	6 13/16"	8"	6 13/16"	8"
TOTAL HEIGHT OF RISERS	3'11 1/4"	3'11 5/8"	3'11 11/16"	3'11 5/8"	3'11 11/16"	3'11 5/8"	3'11 11/16"	4'	3'11 11/16"	4'
RISERS OVERAGE (+) OR UNDERAGE (—)	−1/8"	+1/8"	+3/16"	0	+1/16"	−1/8"	−1/16"	+1/8"	−3/16"	0
NUMBER OF TREADS	6	5	6	5	6	5	6	5	6	5
WIDTH OF EACH TREAD	10 3/4"	9 9/16"	10 11/16"	9 9/16"	10 11/16"	9 9/16"	10 11/16"	9 1/2"	10 11/16"	9 1/2"
TOTAL RUN OF STAIRWAY	5'4 1/2"	3'11 13/16"	5'4 1/8"	3'11 13/16"	5'4 1/8"	3'11 13/16"	5'4 1/8"	3'11 1/2"	5'4 1/8"	3'11 1/2"
MAXIMUM RUN WITH 6'-8" HEADROOM	1'9 1/2"	1'7 1/8"	1'9 3/8"	1'7 1/8"	1'9 3/8"	1'7 1/8"	1'9 3/8"	9 1/2"	1'9 3/8"	9 1/2"
ANGLE OF INCLINE	32°7'	39°42'	32°31'	39°42'	32°31'	39°42'	32°31'	40°6'	32°31'	40°6'
LENGTH OF CARRIAGE	6'4 5/32"	5'2 1/8"	6'4 1/32"	5'2 1/8"	6'4 1/32"	5'2 1/8"	6'4 1/32"	5'2 3/32"	6'4 1/32"	5'2 3/32"
L. F. OF RISER PER INCH OF STAIRWAY WIDTH	.583	.5	.583	.5	.583	.5	.583	.5	.583	.5
L. F. OF TREAD PER INCH OF STAIRWAY WIDTH	.5	.416	.5	.416	.5	.416	.5	.416	.5	.416
FRAMING SQUARE SETTINGS FOR CARRIAGE CUTS — TONGUE	6 3/4"	7 15/16"	6 13/16"	7 15/16"	6 13/16"	7 15/16"	6 13/16"	8"	6 13/16"	8"
FRAMING SQUARE SETTINGS FOR CARRIAGE CUTS — BODY	10 3/4"	9 9/16"	10 11/16"	9 9/16"	10 11/16"	9 9/16"	10 11/16"	9 1/2"	10 11/16"	9 1/2"

FLOOR TO FLOOR RISE	4'	4'	4'$\frac{1}{8}$"	4'$\frac{1}{8}$"	4'$\frac{1}{8}$"	4'$\frac{1}{4}$"	4'$\frac{1}{4}$"	4'$\frac{1}{4}$"	4'$\frac{3}{8}$"	4'$\frac{3}{8}$"
NUMBER OF RISERS	7	8	6	7	8	6	7	8	6	7
HEIGHT OF EACH RISER	6$\frac{7}{8}$"	6"	8"	6$\frac{7}{8}$"	6"	8$\frac{1}{16}$"	6$\frac{7}{8}$"	6$\frac{1}{16}$"	8$\frac{1}{16}$"	6$\frac{15}{16}$"
TOTAL HEIGHT OF RISERS	4'$\frac{1}{8}$"	4'	4'	4'$\frac{1}{8}$"	4'	4'$\frac{3}{8}$"	4'$\frac{1}{8}$"	4'$\frac{1}{2}$"	4'$\frac{3}{8}$"	4'$\frac{9}{16}$"
RISERS OVERAGE (+) OR UNDERAGE (—)	+$\frac{1}{8}$"	0	−$\frac{1}{8}$"	0	−$\frac{1}{8}$"	+$\frac{1}{8}$"	−$\frac{1}{8}$"	+$\frac{1}{4}$"	0	+$\frac{3}{16}$"
NUMBER OF TREADS	6	7	5	6	7	5	6	7	5	6
WIDTH OF EACH TREAD	10$\frac{5}{8}$"	11$\frac{1}{2}$"	9$\frac{1}{2}$"	10$\frac{5}{8}$"	11$\frac{1}{2}$"	9$\frac{7}{16}$"	10$\frac{5}{8}$"	11$\frac{7}{16}$"	9$\frac{7}{16}$"	10$\frac{9}{16}$"
TOTAL RUN OF STAIRWAY	5'3$\frac{3}{4}$"	6'8$\frac{1}{2}$"	3'11$\frac{1}{2}$"	5'3$\frac{3}{4}$"	6'8$\frac{1}{2}$"	3'11$\frac{3}{16}$"	5'3$\frac{3}{4}$"	6'8$\frac{1}{16}$"	3'11$\frac{3}{16}$"	5'3$\frac{3}{8}$"
MAXIMUM RUN WITH 6'-8" HEADROOM	1'9$\frac{1}{4}$"	1'11"	9$\frac{1}{2}$"	1'9$\frac{1}{4}$"	1'11"	9$\frac{7}{16}$"	1'9$\frac{1}{4}$"	1'10$\frac{7}{8}$"	9$\frac{7}{16}$"	1'9$\frac{1}{8}$"
ANGLE OF INCLINE	32°54'	27°33'	40°6'	32°54'	27°33'	40°30'	32°54'	27°56'	40°30'	33°18'
LENGTH OF CARRIAGE	6'3$\frac{15}{16}$"	7'6$\frac{3}{16}$"	5'2$\frac{3}{32}$"	6'3$\frac{15}{16}$"	7'6$\frac{3}{16}$"	5'2$\frac{1}{16}$"	6'3$\frac{15}{16}$"	7'6$\frac{5}{8}$"	5'2$\frac{1}{16}$"	6'3$\frac{13}{16}$"
L. F. OF RISER PER INCH OF STAIRWAY WIDTH	.583	.666	.5	.583	.666	.5	.583	.666	.5	.583
L. F. OF TREAD PER INCH OF STAIRWAY WIDTH	.5	.583	.416	.5	.583	.416	.5	.583	.416	.5
FRAMING SQUARE SETTINGS FOR CARRIAGE CUTS — TONGUE	6$\frac{7}{8}$"	6"	8"	6$\frac{7}{8}$"	6"	8$\frac{1}{16}$"	6$\frac{7}{8}$"	6$\frac{1}{16}$"	8$\frac{1}{16}$"	6$\frac{15}{16}$"
FRAMING SQUARE SETTINGS FOR CARRIAGE CUTS — BODY	10$\frac{5}{8}$"	11$\frac{1}{2}$"	9$\frac{1}{2}$"	10$\frac{5}{8}$"	11$\frac{1}{2}$"	9$\frac{7}{16}$"	10$\frac{5}{8}$"	11$\frac{7}{16}$"	9$\frac{7}{16}$"	10$\frac{9}{16}$"

FLOOR TO FLOOR RISE	4' 3/8"	4' 1/2"	4' 1/2"	4' 1/2"	4' 5/8"	4' 5/8"	4' 5/8"	4' 3/4"	4' 3/4"	4' 3/4"
NUMBER OF RISERS	8	6	7	8	6	7	8	6	7	8
HEIGHT OF EACH RISER	6 1/16"	8 1/16"	6 15/16"	6 1/16"	8 1/8"	6 15/16"	6 1/16"	8 1/8"	6 15/16"	6 1/8"
TOTAL HEIGHT OF RISERS	4' 1/2"	4' 3/8"	4' 9/16"	4' 1/2"	4' 3/4"	4' 9/16"	4' 1/2"	4' 3/4"	4' 9/16"	4' 1"
RISERS OVERAGE (+) OR UNDERAGE (—)	+1/8"	−1/8"	+1/16"	0	+1/8"	−1/16"	−1/8"	0	−3/16"	+1/4"
NUMBER OF TREADS	7	5	6	7	5	6	7	5	6	7
WIDTH OF EACH TREAD	11 7/16"	9 7/16"	10 9/16"	11 7/16"	9 3/8"	10 9/16"	11 7/16"	9 3/8"	10 9/16"	11 3/8"
TOTAL RUN OF STAIRWAY	6' 8 1/16"	3' 11 3/16"	5' 3 3/8"	6' 8 1/16"	3' 10 7/8"	5' 3 3/8"	6' 8 1/16"	3' 10 7/8"	5' 3 3/8"	6' 7 5/8"
MAXIMUM RUN WITH 6'-8" HEADROOM	1' 10 7/8"	9 7/16"	1' 9 1/8"	1' 10 7/8"	9 3/8"	1' 9 1/8"	1' 10 7/8"	9 3/8"	1' 9 1/8"	1' 10 3/4"
ANGLE OF INCLINE	27°56'	40°30'	33°18'	27°56'	40°55'	33°18'	27°56'	40°55'	33°18'	28°18'
LENGTH OF CARRIAGE	7' 6 5/8"	5' 2 1/16"	6' 3 13/16"	7' 6 5/8"	5' 2 1/32"	6' 3 13/16"	7' 6 5/8"	5' 2 1/32"	6' 3 13/16"	7' 6 7/16"
L. F. OF RISER PER INCH OF STAIRWAY WIDTH	.666	.5	.583	.666	.5	.583	.666	.5	.583	.666
L. F. OF TREAD PER INCH OF STAIRWAY WIDTH	.583	.416	.5	.583	.416	.5	.583	.416	.5	.583
FRAMING SQUARE SETTINGS FOR CARRIAGE CUTS — TONGUE	6 1/16"	8 1/16"	6 15/16"	6 1/16"	8 1/8"	6 15/16"	6 1/16"	8 1/8"	6 15/16"	6 1/8"
FRAMING SQUARE SETTINGS FOR CARRIAGE CUTS — BODY	11 7/16"	9 7/16"	10 9/16"	11 7/16"	9 3/8"	10 9/16"	11 7/16"	9 3/8"	10 9/16"	11 3/8"

FLOOR TO FLOOR RISE	4'7/8"	4'7/8"	4'7/8"	4'1"	4'1"	4'1"	4'1 1/8"	4'1 1/8"	4'1 1/8"	4'1 1/4"
NUMBER OF RISERS	6	7	8	6	7	8	6	7	8	6
HEIGHT OF EACH RISER	8 1/8"	7"	6 1/8"	8 3/16"	7"	6 1/8"	8 3/16"	7"	6 1/8"	8 3/16"
TOTAL HEIGHT OF RISERS	4'3/4"	4'1"	4'1"	4'1 1/8"	4'1"	4'1"	4'1 1/8"	4'1"	4'1"	4'1 1/8"
RISERS OVERAGE (+) OR UNDERAGE (—)	−1/8"	+1/8"	+1/8"	+1/8"	0	0	0	−1/8"	−1/8"	−1/8"
NUMBER OF TREADS	5	6	7	5	6	7	5	6	7	5
WIDTH OF EACH TREAD	9 3/8"	10 1/2"	11 3/8"	9 5/16"	10 1/2"	11 3/8"	9 5/16"	10 1/2"	11 3/8"	9 5/16"
TOTAL RUN OF STAIRWAY	3'10 7/8"	5'3"	6'7 5/8"	3'10 9/16"	5'3"	6'7 5/8"	3'10 9/16"	5'3"	6'7 5/8"	3'10 9/16"
MAXIMUM RUN WITH 6'-8" HEADROOM	9 3/8"	1'9"	1'10 3/4"	9 5/16"	1'9"	1'10 3/4"	9 5/16"	1'9"	1'10 3/4"	9 5/16"
ANGLE OF INCLINE	40°55'	33°41'	28°18'	41°19'	33°41'	28°18'	41°19'	33°41'	28°18'	41°19'
LENGTH OF CARRIAGE	5'2 1/32"	6'3 23/32"	7'6 7/16"	5'2"	6'3 23/32"	7'6 7/16"	5'2"	6'3 23/32"	7'6 7/16"	5'2"
L. F. OF RISER PER INCH OF STAIRWAY WIDTH	.5	.583	.666	.5	.583	.666	.5	.583	.666	.5
L. F. OF TREAD PER INCH OF STAIRWAY WIDTH	.416	.5	.583	.416	.5	.583	.416	.5	.583	.416
FRAMING SQUARE SETTINGS FOR CARRIAGE CUTS — TONGOE	8 1/8"	7"	6 1/8"	8 3/16"	7"	6 1/8"	8 3/16"	7"	6 1/8"	8 3/16"
FRAMING SQUARE SETTINGS FOR CARRIAGE CUTS — BODY	9 3/8"	10 1/2"	11 3/8"	9 5/16"	10 1/2"	11 3/8"	9 5/16"	10 1/2"	11 3/8"	9 5/16"

FLOOR TO FLOOR RISE		4'1 1/4"	4'1 1/4"	4'1 3/8"	4'1 3/8"	4'1 3/8"	4'1 1/2"	4'1 1/2"	4'1 1/2"	4'1 5/8"	4'1 5/8"
NUMBER OF RISERS		7	8	6	7	8	6	7	8	6	7
HEIGHT OF EACH RISER		7 1/16"	6 3/16"	8 1/4"	7 1/16"	6 3/16"	8 1/4"	7 1/16"	6 3/16"	8 1/4"	7 1/16"
TOTAL HEIGHT OF RISERS		4'1 7/16"	4'1 1/2"	4'1 1/2"	4'1 7/16"	4'1 1/2"	4'1 1/2"	4'1 7/16"	4'1 1/2"	4'1 1/2"	4'1 7/16"
RISERS OVERAGE (+) OR UNDERAGE (—)		+3/16"	+1/4"	+1/8"	+1/16"	+1/8"	0	−1/16"	0	−1/8"	−3/16"
NUMBER OF TREADS		6	7	5	6	7	5	6	7	5	6
WIDTH OF EACH TREAD		10 7/16"	11 5/16"	9 1/4"	10 7/16"	11 5/16"	9 1/4"	10 7/16"	11 5/16"	9 1/4"	10 7/16"
TOTAL RUN OF STAIRWAY		5'2 5/8"	6'7 3/16"	3'10 1/4"	5'2 5/8"	6'7 3/16"	3'10 1/4"	5'2 5/8"	6'7 3/16"	3'10 1/4"	5'2 5/8"
MAXIMUM RUN WITH 6'-8" HEADROOM		1'8 7/8"	1'10 5/8"	9 1/4"	1'8 7/8"	1'10 5/8"	9 1/4"	1'8 7/8"	1'10 5/8"	9 1/4"	1'8 7/8"
ANGLE OF INCLINE		34°5'	28°41'	41°44'	34°5'	28°41'	41°44'	34°5'	28°41'	41°44'	34°5'
LENGTH OF CARRIAGE		6'3 5/8"	7'6 1/4"	5'1 31/32"	6'3 5/8"	7'6 1/4"	5'1 31/32"	6'3 5/8"	7'6 1/4"	5'1 31/32"	6'3 5/8"
L. F. OF RISER PER INCH OF STAIRWAY WIDTH		.583	.666	.5	.583	.666	.5	.583	.666	.5	.583
L. F. OF TREAD PER INCH OF STAIRWAY WIDTH		.5	.583	.416	.5	.583	.416	.5	.583	.416	.5
FRAMING SQUARE SETTINGS FOR CARRIAGE CUTS	TONGUE	7 1/16"	6 3/16"	8 1/4"	7 1/16"	6 3/16"	8 1/4"	7 1/16"	6 3/16"	8 1/4"	7 1/16"
	BODY	10 7/16"	11 5/16"	9 1/4"	10 7/16"	11 5/16"	9 1/4"	10 7/16"	11 5/16"	9 1/4"	10 7/16"

FLOOR TO FLOOR RISE		4'1 5/8"	4'1 3/4"	4'1 3/4"	4'1 7/8"	4'1 7/8"	4'2"	4'2"	4'2 1/8"	4'2 1/8"	4'2 1/4"
NUMBER OF RISERS		8	7	8	7	8	7	8	7	8	7
HEIGHT OF EACH RISER		6 3/16"	7 1/8"	6 1/4"	7 1/8"	6 1/4"	7 1/8"	6 1/4"	7 3/16"	6 1/4"	7 3/16"
TOTAL HEIGHT OF RISERS		4'1 1/2"	4'1 7/8"	4'2"	4'1 7/8"	4'2"	4'1 7/8"	4'2"	4'2 5/16"	4'2"	4'2 5/16"
RISERS OVERAGE (+) OR UNDERAGE (—)		−1/8"	+1/8"	+1/4"	0	+1/8"	−1/8"	0	+3/16"	−1/8"	+1/16"
NUMBER OF TREADS		7	6	7	6	7	6	7	6	7	6
WIDTH OF EACH TREAD		11 5/16"	10 3/8"	11 1/4"	10 3/8"	11 1/4"	10 3/8"	11 1/4"	10 5/16"	11 1/4"	10 5/16"
TOTAL RUN OF STAIRWAY		6'7 3/16"	5'2 1/4"	6'6 3/4"	5'2 1/4"	6'6 3/4"	5'2 1/4"	6'6 3/4"	5'1 7/8"	6'6 3/4"	5'1 7/8"
MAXIMUM RUN WITH 6'-8" HEADROOM		1'10 5/8"	1'8 3/4"	1'10 1/2"	1'8 3/4"	1'10 1/2"	1'8 3/4"	1'10 1/2"	1'8 5/8"	1'10 1/2"	1'8 5/8"
ANGLE OF INCLINE		28°41'	34°29'	29°3'	34°29'	29°3'	34°29'	29°3'	34°53'	29°3'	34°53'
LENGTH OF CARRIAGE		7'6 1/4"	6'3 1/2"	7'6 3/32"	6'3 1/2"	7'6 3/32"	6'3 1/2"	7'6 3/32"	6'3 13/16"	7'6 3/32"	6'3 13/16"
L. F. OF RISER PER INCH OF STAIRWAY WIDTH		.666	.583	.666	.583	.666	.583	.666	.583	.666	.583
L. F. OF TREAD PER INCH OF STAIRWAY WIDTH		.583	.5	.583	.5	.583	.5	.583	.5	.583	.5
FRAMING SQUARE SETTINGS FOR CARRIAGE CUTS	TONGUE	6 3/16"	7 1/8"	6 1/4"	7 1/8"	6 1/4"	7 1/8"	6 1/4"	7 3/16"	6 1/4"	7 3/16"
	BODY	11 5/16"	10 3/8"	11 1/4"	10 3/8"	11 1/4"	10 3/8"	11 1/4"	10 5/16"	11 1/4"	10 5/16"

FLOOR TO FLOOR RISE	4'2 1/4"	4'2 3/8"	4'2 3/8"	4'2 1/2"	4'2 1/2"	4'2 5/8"	4'2 5/8"	4'2 3/4"	4'2 3/4"	4'2 7/8"
NUMBER OF RISERS	8	7	8	7	8	7	8	7	8	7
HEIGHT OF EACH RISER	6 5/16"	7 3/16"	6 5/16"	7 3/16"	6 5/16"	7 1/4"	6 5/16"	7 1/4"	6 3/8"	7 1/4"
TOTAL HEIGHT OF RISERS	4'2 1/2"	4'2 5/16"	4'2 1/2"	4'2 5/16"	4'2 1/2"	4'2 3/4"	4'2 1/2"	4'2 3/4"	4'3"	4'2 3/4"
RISERS OVERAGE (+) OR UNDERAGE (—)	+1/4"	−1/16"	+1/8"	−3/16"	0	+1/8"	−1/8"	0	+1/4"	−1/8"
NUMBER OF TREADS	7	6	7	6	7	6	7	6	7	6
WIDTH OF EACH TREAD	11 3/16"	10 5/16"	11 3/16"	10 5/16"	11 3/16"	10 1/4"	11 3/16"	10 1/4"	11 1/8"	10 1/4"
TOTAL RUN OF STAIRWAY	6'6 5/16"	5'1 7/8"	6'6 5/16"	5'1 7/8"	6'6 5/16"	5'1 1/2"	6'6 5/16"	5'1 1/2"	6'5 7/8"	5'1 1/2"
MAXIMUM RUN WITH 6'-8" HEADROOM	1'10 3/8"	1'8 5/8"	1'10 3/8"	1'8 5/8"	1'10 3/8"	1'8 1/2"	1'10 3/8"	1'8 1/2"	1'10 1/4"	1'8 1/2"
ANGLE OF INCLINE	29°26'	34°53'	29°26'	34°53'	29°26'	35°16'	29°26'	35°16'	29°49'	35°16'
LENGTH OF CARRIAGE	7'5 29/32"	6'3 13/16"	7'5 29/32"	6'3 13/16"	7'5 29/32"	6'3 11/32"	7'5 29/32"	6'3 11/32"	7'5 3/4"	6'3 11/32"
L. F. OF RISER PER INCH OF STAIRWAY WIDTH	.666	.583	.666	.583	.666	.583	.666	.583	.666	.583
L. F. OF TREAD PER INCH OF STAIRWAY WIDTH	.583	.5	.583	.5	.583	.5	.583	.5	.583	.5
FRAMING SQUARE SETTINGS FOR CARRIAGE CUTS — TONGUE	6 5/16"	7 3/16"	6 5/16"	7 3/16"	6 5/16"	7 1/4"	6 5/16"	7 1/4"	6 3/8"	7 1/4"
FRAMING SQUARE SETTINGS FOR CARRIAGE CUTS — BODY	11 3/16"	10 5/16"	11 3/16"	10 5/16"	11 3/16"	10 1/4"	11 3/16"	10 1/4"	11 1/8"	10 1/4"

FLOOR TO FLOOR RISE	4'2 7/8"	4'3"	4'3"	4'3 1/8"	4'3 1/8"	4'3 1/4"	4'3 1/4"	4'3 3/8"	4'3 3/8"	4'3 1/2"
NUMBER OF RISERS	8	7	8	7	8	7	8	7	8	7
HEIGHT OF EACH RISER	6 3/8"	7 5/16"	6 3/8"	7 5/16"	6 3/8"	7 5/16"	6 7/16"	7 5/16"	6 7/16"	7 3/8"
TOTAL HEIGHT OF RISERS	4'3"	4'3 3/16"	4'3"	4'3 3/16"	4'3"	4'3 3/16"	4'3 1/2"	4'3 3/16"	4'3 1/2"	4'3 5/8"
RISERS OVERAGE (+) OR UNDERAGE (—)	+1/8"	+3/16"	0	+1/16"	−1/8"	−1/16"	+1/4"	−3/16"	+1/8"	+1/8"
NUMBER OF TREADS	7	6	7	6	7	6	7	6	7	6
WIDTH OF EACH TREAD	11 1/8"	10 3/16"	11 1/8"	10 3/16"	11 1/8"	10 3/16"	11 1/16"	10 3/16"	11 1/16"	10 1/8"
TOTAL RUN OF STAIRWAY	6'5 7/8"	5'1 1/8"	6'5 7/8"	5'1 1/8"	6'5 7/8"	5'1 1/8"	6'5 7/16"	5'1 1/8"	6'5 7/16"	5' 3/4"
MAXIMUM RUN WITH 6'-8" HEADROOM	1'10 1/4"	1'8 3/8"	1'10 1/4"	1'8 3/8"	1'10 1/4"	1'8 3/8"	1'10 1/8"	1'8 3/8"	1'10 1/8"	1'8 1/4"
ANGLE OF INCLINE	29°49'	35°40'	29°49'	35°40'	29°49'	35°40'	30°12'	35°40'	30°12'	36°4'
LENGTH OF CARRIAGE	7'5 3/4"	6'3 7/32"	7'5 3/4"	6'3 7/32"	7'5 3/4"	6'3 7/32"	7'5 19/32"	6'3 7/32"	7'5 19/32"	6'3 5/32"
L. F. OF RISER PER INCH OF STAIRWAY WIDTH	.666	.583	.666	.583	.666	.583	.666	.583	.666	.583
L. F. OF TREAD PER INCH OF STAIRWAY WIDTH	.583	.5	.583	.5	.583	.5	.583	.5	.583	.5
FRAMING SQUARE SETTINGS FOR CARRIAGE CUTS — TONGUE	6 3/8"	7 5/16"	6 3/8"	7 5/16"	6 3/8"	7 5/16"	6 7/16"	7 5/16"	6 7/16"	7 3/8"
FRAMING SQUARE SETTINGS FOR CARRIAGE CUTS — BODY	11 1/8"	10 3/16"	11 1/8"	10 3/16"	11 1/8"	10 3/16"	11 1/16"	10 3/16"	11 1/16"	10 1/8"

FLOOR TO FLOOR RISE	4'3 1/2"	4'3 5/8"	4'3 5/8"	4'3 3/4"	4'3 3/4"	4'3 7/8"	4'3 7/8"	4'4"	4'4"	4'4 1/8"
NUMBER OF RISERS	8	7	8	7	8	7	8	7	8	7
HEIGHT OF EACH RISER	6 7/16"	7 3/8"	6 7/16"	7 3/8"	6 1/2"	7 7/16"	6 1/2"	7 7/16"	6 1/2"	7 7/16"
TOTAL HEIGHT OF RISERS	4'3 1/2"	4'3 5/8"	4'3 1/2"	4'3 5/8"	4'4"	4'4 1/16"	4'4"	4'4 1/16"	4'4"	4'4 1/16"
RISERS OVERAGE (+) OR UNDERAGE (—)	0	0	−1/8"	−1/8"	+1/4"	+3/16"	+1/8"	+1/16"	0	−1/16"
NUMBER OF TREADS	7	6	7	6	7	6	7	6	7	6
WIDTH OF EACH TREAD	11 1/16"	10 1/8"	11 1/16"	10 1/8"	11"	10 1/16"	11"	10 1/16"	11"	10 1/16"
TOTAL RUN OF STAIRWAY	6'5 7/16"	5'3/4"	6'5 7/16"	5'3/4"	6'5"	5'3/8"	6'5"	5'3/8"	6'5"	5'3/8"
MAXIMUM RUN WITH 6'-8" HEADROOM	1'10 1/8"	1'8 1/4"	1'10 1/8"	1'8 1/4"	1'10"	1'8 1/8"	1'10"	1'8 1/8"	1'10"	1'8 1/8"
ANGLE OF INCLINE	30°12'	36°4'	30°12'	36°4'	30°35'	36°28'	30°35'	36°28'	30°35'	36°28'
LENGTH OF CARRIAGE	7'5 19/32"	6'3 5/32"	7'5 19/32"	6'3 5/32"	7'5 7/16"	6'3 1/16"	7'5 7/16"	6'3 1/16"	7'5 7/16"	6'3 1/16"
L. F. OF RISER PER INCH OF STAIRWAY WIDTH	.666	.583	.666	.583	.666	.583	.666	.583	.666	.583
L. F. OF TREAD PER INCH OF STAIRWAY WIDTH	.583	.5	.583	.5	.583	.5	.583	.5	.583	.5
FRAMING SQUARE SETTINGS FOR CARRIAGE CUTS — TONGUE	6 7/16"	7 3/8"	6 7/16"	7 3/8"	6 1/2"	7 7/16"	6 1/2"	7 7/16"	6 1/2"	7 7/16"
FRAMING SQUARE SETTINGS FOR CARRIAGE CUTS — BODY	11 1/16"	10 1/8"	11 1/16"	10 1/8"	11"	10 1/16"	11"	10 1/16"	11"	10 1/16"

FLOOR TO FLOOR RISE	4'4 1/8"	4'4 1/4"	4'4 1/4"	4'4 3/8"	4'4 3/8"	4'4 1/2"	4'4 1/2"	4'4 5/8"	4'4 5/8"	4'4 3/4"
NUMBER OF RISERS	8	7	8	7	8	7	8	7	8	7
HEIGHT OF EACH RISER	6 1/2"	7 7/16"	6 9/16"	7 1/2"	6 9/16"	7 1/2"	6 9/16"	7 1/2"	6 9/16"	7 9/16"
TOTAL HEIGHT OF RISERS	4'4"	4'4 1/16"	4'4 1/2"	4'4 1/2"	4'4 1/2"	4'4 1/2"	4'4 1/2"	4'4 1/2"	4'4 1/2"	4'4 15/16"
RISERS OVERAGE (+) OR UNDERAGE (—)	−1/8"	−3/16"	+1/4"	+1/8"	+1/8"	0	0	−1/8"	−1/8"	+3/16"
NUMBER OF TREADS	7	6	7	6	7	6	7	6	7	6
WIDTH OF EACH TREAD	11"	10 1/16"	10 15/16"	10"	10 15/16"	10"	10 15/16"	10"	10 15/16"	9 15/16"
TOTAL RUN OF STAIRWAY	6'5"	5' 3/8"	6'4 9/16"	5'	6'4 9/16"	5'	6'4 9/16"	5'	6'4 9/16"	4'11 5/8"
MAXIMUM RUN WITH 6'-8" HEADROOM	1'10"	1'8 1/8"	1'9 7/8"	1'8"	1'9 7/8"	1'8"	1'9 7/8"	1'8"	1'9 7/8"	1'7 7/8"
ANGLE OF INCLINE	30°35'	36°28'	30°58'	36°52'	30°58'	36°52'	30°58'	36°52'	30°58'	37°16'
LENGTH OF CARRIAGE	7'5 7/16"	6'3 1/16"	7'5 9/32"	6'3"	7'5 9/32"	6'3"	7'5 9/32"	6'3"	7'5 9/32"	6'2 15/16"
L. F. OF RISER PER INCH OF STAIRWAY WIDTH	.666	.583	.666	.583	.666	.583	.666	.583	.666	.583
L. F. OF TREAD PER INCH OF STAIRWAY WIDTH	.583	.5	.583	.5	.583	.5	.583	.5	.583	.5
FRAMING SQUARE SETTINGS FOR CARRIAGE CUTS — TONGUE	6 1/2"	7 7/16"	6 9/16"	7 1/2"	6 9/16"	7 1/2"	6 9/16"	7 1/2"	6 9/16"	7 9/16"
FRAMING SQUARE SETTINGS FOR CARRIAGE CUTS — BODY	11"	10 1/16"	10 15/16"	10"	10 15/16"	10"	10 15/16"	10"	10 15/16"	9 15/16"

FLOOR TO FLOOR RISE	4'4 3/4"	4'4 7/8"	4'4 7/8"	4'5"	4'5"	4'5 1/8"	4'5 1/8"	4'5 1/4"	4'5 1/4"	4'5 3/8"
NUMBER OF RISERS	8	7	8	7	8	7	8	7	8	7
HEIGHT OF EACH RISER	6 5/8"	7 9/16"	6 5/8"	7 9/16"	6 5/8"	7 9/16"	6 5/8"	7 5/8"	6 11/16"	7 5/8"
TOTAL HEIGHT OF RISERS	4'5"	4'4 15/16"	4'5"	4'4 15/16"	4'5"	4'4 15/16"	4'5"	4'5 3/8"	4'5 1/2"	4'5 3/8"
RISERS OVERAGE (+) OR UNDERAGE (—)	+1/4"	+1/16"	+1/8"	−1/16"	0	−3/16"	−1/8"	+1/8"	+1/4"	0
NUMBER OF TREADS	7	6	7	6	7	6	7	6	7	6
WIDTH OF EACH TREAD	10 7/8"	9 15/16"	10 7/8"	9 15/16"	10 7/8"	9 15/16"	10 7/8"	9 7/8"	10 13/16"	9 7/8"
TOTAL RUN OF STAIRWAY	6'4 1/8"	4'11 5/8"	6'4 1/8"	4'11 5/8"	6'4 1/8"	4'11 5/8"	6'4 1/8"	4'11 1/4"	6'3 11/16"	4'11 1/4"
MAXIMUM RUN WITH 6'-8" HEADROOM	1'9 3/4"	1'7 7/8"	1'9 3/4"	1'7 7/8"	1'9 3/4"	1'7 7/8"	1'9 3/4"	1'7 3/4"	1'9 5/8"	1'7 3/4"
ANGLE OF INCLINE	31°21'	37°16'	31°21'	37°16'	31°21'	37°16'	31°21'	37°40'	31°44'	37°40'
LENGTH OF CARRIAGE	7'5 1/8"	6'2 15/16"	7'5 1/8"	6'2 15/16"	7'5 1/8"	6'2 15/16"	7'5 1/8"	6'2 27/32"	7'5"	6'2 27/32"
L. F. OF RISER PER INCH OF STAIRWAY WIDTH	.666	.583	.666	.583	.666	.583	.666	.583	.666	.583
L. F. OF TREAD PER INCH OF STAIRWAY WIDTH	.583	.5	.583	.5	.583	.6	.583	.5	.583	.5
FRAMING SQUARE SETTINGS FOR CARRIAGE CUTS — TONGUE	6 5/8"	7 9/16"	6 5/8"	7 9/16"	6 5/8"	7 9/16"	6 5/8"	7 5/8"	6 11/16"	7 5/8"
BODY	10 7/8"	9 15/16"	10 7/8"	9 15/16"	10 7/8"	9 15/16"	10 7/8"	9 7/8"	10 13/16"	9 7/8"

FLOOR TO FLOOR RISE	4'5 3/8"	4'5 1/2"	4'5 1/2"	4'5 5/8"	4'5 5/8"	4'5 3/4"	4'5 3/4"	4'5 7/8"	4'5 7/8"	4'6"
NUMBER OF RISERS	8	7	8	7	8	7	8	7	8	7
HEIGHT OF EACH RISER	6 11/16"	7 5/8"	6 11/16"	7 11/16"	6 11/16"	7 11/16"	6 3/4"	7 11/16"	6 3/4"	7 11/16"
TOTAL HEIGHT OF RISERS	4'5 1/2"	4'5 3/8"	4'5 1/2"	4'5 13/16"	4'5 1/2"	4'5 13/16"	4'6"	4'5 13/16"	4'6"	4'5 13/16"
RISERS OVERAGE (+) OR UNDERAGE (—)	+1/8"	−1/8"	0	+3/16"	−1/8"	+1/16"	+1/4"	−1/16"	+1/8"	−3/16"
NUMBER OF TREADS	7	6	7	6	7	6	7	6	7	6
WIDTH OF EACH TREAD	10 13/16"	9 7/8"	10 13/16"	9 13/16"	10 13/16"	9 13/16"	10 3/4"	9 13/16"	10 3/4"	9 13/16"
TOTAL RUN OF STAIRWAY	6'3 11/16"	4'11 1/4"	6'3 11/16"	4'10 7/8"	6'3 11/16"	4'10 7/8"	6'3 1/4"	4'10 7/8"	6'3 1/4"	4'10 7/8"
MAXIMUM RUN WITH 6'-8" HEADROOM	1'9 5/8"	1'7 3/4"	1'9 5/8"	1'7 5/8"	1'9 5/8"	1'7 5/8"	1'9 1/2"	1'7 5/8"	1'9 1/2"	1'7 5/8"
ANGLE OF INCLINE	31°44'	37°40'	31°44'	38°5'	31°44'	38°5'	32°7'	38°5'	32°7'	38°5'
LENGTH OF CARRIAGE	7'5"	6'2 27/32"	7'5"	6'2 25/32"	7'5"	6'2 25/32"	7'4 27/32"	6'2 25/32"	7'4 27/32"	6'2 25/32"
L. F. OF RISER PER INCH OF STAIRWAY WIDTH	.666	.583	.666	.583	.666	.583	.666	.583	.666	.583
L. F. OF TREAD PER INCH OF STAIRWAY WIDTH	.583	.5	.583	.5	.583	.5	.583	.5	.583	.5
FRAMING SQUARE SETTINGS FOR CARRIAGE CUTS — TONGUE	6 11/16"	7 5/8"	6 11/16"	7 11/16"	6 11/16"	7 11/16"	6 3/4"	7 11/16"	6 3/4"	7 11/16"
FRAMING SQUARE SETTINGS FOR CARRIAGE CUTS — BODY	10 13/16"	9 7/8"	10 13/16"	9 13/16"	10 13/16"	9 13/16"	10 3/4"	9 13/16"	10 3/4"	9 13/16"

FLOOR TO FLOOR RISE	4'6"	4'6"	4'6$\frac{1}{8}$"	4'6$\frac{1}{8}$"	4'6$\frac{1}{8}$"	4'6$\frac{1}{4}$"	4'6$\frac{1}{4}$"	4'6$\frac{1}{4}$"	4'6$\frac{3}{8}$"	4'6$\frac{3}{8}$"
NUMBER OF RISERS	8	9	7	8	9	7	8	9	7	8
HEIGHT OF EACH RISER	6$\frac{3}{4}$"	6"	7$\frac{3}{4}$"	6$\frac{3}{4}$"	6"	7$\frac{3}{4}$"	6$\frac{13}{16}$"	6"	7$\frac{3}{4}$"	6$\frac{13}{16}$"
TOTAL HEIGHT OF RISERS	4'6"	4'6"	4'6$\frac{1}{4}$"	4'6"	4'6"	4'6$\frac{1}{4}$"	4'6$\frac{1}{2}$"	4'6"	4'6$\frac{1}{4}$"	4'6$\frac{1}{2}$"
RISERS OVERAGE (+) OR UNDERAGE (—)	0	0	+$\frac{1}{8}$"	−$\frac{1}{8}$"	−$\frac{1}{8}$"	0	+$\frac{1}{4}$"	−$\frac{1}{4}$"	−$\frac{1}{8}$"	+$\frac{1}{8}$"
NUMBER OF TREADS	7	8	6	7	8	6	7	8	6	7
WIDTH OF EACH TREAD	10$\frac{3}{4}$"	11$\frac{1}{2}$"	9$\frac{3}{4}$"	10$\frac{3}{4}$"	11$\frac{1}{2}$"	9$\frac{3}{4}$"	10$\frac{11}{16}$"	11$\frac{1}{2}$"	9$\frac{3}{4}$"	10$\frac{11}{16}$"
TOTAL RUN OF STAIRWAY	6'3$\frac{1}{4}$"	7'8"	4'10$\frac{1}{2}$"	6'3$\frac{1}{4}$"	7'8"	4'10$\frac{1}{2}$"	6'2$\frac{13}{16}$"	7'8"	4'10$\frac{1}{2}$"	6'2$\frac{13}{16}$"
MAXIMUM RUN WITH 6'-8" HEADROOM	1'9$\frac{1}{2}$"	1'11"	1'7$\frac{1}{2}$"	1'9$\frac{1}{2}$"	1'11"	1'7$\frac{1}{2}$"	1'9$\frac{3}{8}$"	1'11"	1'7$\frac{1}{2}$"	1'9$\frac{3}{8}$"
ANGLE OF INCLINE	32°7'	27°33'	38°29'	32°7'	27°33'	38°29'	32°31'	27°33'	38°29'	32°31'
LENGTH OF CARRIAGE	7'4$\frac{27}{32}$"	8'7$\frac{25}{32}$"	6'2$\frac{23}{32}$"	7'4$\frac{27}{32}$"	8'7$\frac{25}{32}$"	6'2$\frac{23}{32}$"	7'4$\frac{23}{32}$"	8'7$\frac{25}{32}$"	6'2$\frac{23}{32}$"	7'4$\frac{23}{32}$"
L. F. OF RISER PER INCH OF STAIRWAY WIDTH	.666	.75	.583	.666	.75	.583	.666	.75	.583	.666
L. F. OF TREAD PER INCH OF STAIRWAY WIDTH	.583	.666	.5	.583	.666	.5	.583	.666	.5	.583
FRAMING SQUARE SETTINGS FOR CARRIAGE CUTS — TONGUE	6$\frac{3}{4}$"	6"	7$\frac{3}{4}$"	6$\frac{3}{4}$"	6"	7$\frac{3}{4}$"	6$\frac{13}{16}$"	6"	7$\frac{3}{4}$"	6$\frac{13}{16}$"
BODY	10$\frac{3}{4}$"	11$\frac{1}{2}$"	9$\frac{3}{4}$"	10$\frac{3}{4}$"	11$\frac{1}{2}$"	9$\frac{3}{4}$"	10$\frac{11}{16}$"	11$\frac{1}{2}$"	9$\frac{3}{4}$"	10$\frac{11}{16}$"

FLOOR TO FLOOR RISE	4'6 3/8"	4'6 1/2"	4'6 1/2"	4'6 1/2"	4'6 5/8"	4'6 5/8"	4'6 5/8"	4'6 3/4"	4'6 3/4"	4'6 3/4"
NUMBER OF RISERS	9	7	8	9	7	8	9	7	8	9
HEIGHT OF EACH RISER	6 1/16"	7 13/16"	6 13/16"	6 1/16"	7 13/16"	6 13/16"	6 1/16"	7 13/16"	6 13/16"	6 1/16"
TOTAL HEIGHT OF RISERS	4'6 9/16"	4'6 11/16"	4'6 1/2"	4'6 9/16"	4'6 11/16"	4'6 1/2"	4'6 9/16"	4'6 11/16"	4'6 1/2"	4'6 9/16"
RISERS OVERAGE (+) OR UNDERAGE (—)	+3/16"	+3/16"	0	+1/16"	+1/16"	−1/8"	−1/16"	−1/16"	−1/4"	−3/16"
NUMBER OF TREADS	8	6	7	8	6	7	8	6	7	8
WIDTH OF EACH TREAD	11 7/16"	9 11/16"	10 11/16"	11 7/16"	9 11/16"	10 11/16"	11 7/16"	9 11/16"	10 11/16"	11 7/16"
TOTAL RUN OF STAIRWAY	7'7 1/2"	4'10 1/8"	6'2 13/16"	7'7 1/2"	4'10 1/8"	6'2 13/16"	7'7 1/2"	4'10 1/8"	6'2 13/16"	7'7 1/2"
MAXIMUM RUN WITH 6'-8" HEADROOM	1'10 7/8"	1'7 3/8"	1'9 3/8"	1'10 7/8"	1'7 3/8"	1'9 3/8"	1'10 7/8"	1'7 3/8"	1'9 3/8"	1'10 7/8"
ANGLE OF INCLINE	27°56'	38°53'	32°31'	27°56'	38°53'	32°31'	27°56'	38°53'	32°31'	27°56'
LENGTH OF CARRIAGE	8'7 9/16"	6'2 21/32"	7'4 23/32"	8'7 9/16"	6'2 21/32"	7'4 23/32"	8'7 9/16"	6'2 21/32"	7'4 23/32"	8'7 9/16"
L. F. OF RISER PER INCH OF STAIRWAY WIDTH	.75	.583	.666	.75	.583	.666	.75	.583	.666	.75
L. F. OF TREAD PER INCH OF STAIRWAY WIDTH	.666	.5	.583	.666	.5	.583	.666	.5	.583	.666
FRAMING SQUARE SETTINGS FOR CARRIAGE CUTS — TONGUE	6 1/16"	7 13/16"	6 13/16"	6 1/16"	7 13/16"	6 13/16"	6 1/16"	7 13/16"	6 13/16"	6 1/16"
FRAMING SQUARE SETTINGS FOR CARRIAGE CUTS — BODY	11 7/16"	9 11/16"	10 11/16"	11 7/16"	9 11/16"	10 11/16"	11 7/16"	9 11/16"	10 11/16"	11 7/16"

FLOOR TO FLOOR RISE	4'6 7/8"	4'6 7/8"	4'6 7/8"	4'7"	4'7"	4'7"	4'7 1/8"	4'7 1/8"	4'7 1/8"	4'7 1/4"
NUMBER OF RISERS	7	8	9	7	8	9	7	8	9	7
HEIGHT OF EACH RISER	7 13/16"	6 7/8"	6 1/8"	7 7/8"	6 7/8"	6 1/8"	7 7/8"	6 7/8"	6 1/8"	7 7/8"
TOTAL HEIGHT OF RISERS	4'6 11/16"	4'7"	4'7 1/8"	4'7 1/8"	4'7"	4'7 1/8"	4'7 1/8"	4'7"	4'7 1/8"	4'7 1/8"
RISERS OVERAGE (+) OR UNDERAGE (—)	−3/16"	+1/8"	+1/4"	+1/8"	0	+1/8"	0	−1/8"	0	−1/8"
NUMBER OF TREADS	6	7	8	6	7	8	6	7	8	6
WIDTH OF EACH TREAD	9 11/16"	10 5/8"	11 3/8"	9 5/8"	10 5/8"	11 3/8"	9 5/8"	10 5/8"	11 3/8"	9 5/8"
TOTAL RUN OF STAIRWAY	4'10 1/8"	6'2 3/8"	7'7"	4'9 3/4"	6'2 3/8"	7'7"	4'9 3/4"	6'2 3/8"	7'7"	4'9 3/4"
MAXIMUM RUN WITH 6'-8" HEADROOM	1'7 3/8"	1'9 1/4"	1'8 3/4"	1'7 1/4"	1'9 1/4"	1'8 3/4"	1'7 1/4"	1'9 1/4"	1'8 3/4"	1'7 1/4"
ANGLE OF INCLINE	38°53'	32°54'	28°18'	39°17'	32°54'	28°18'	39°17'	32°54'	28°18'	39°17'
LENGTH OF CARRIAGE	6'2 21/32"	7'4 19/32"	8'7 11/32"	6'2 5/8"	7'4 19/32"	8'7 11/32"	6'2 5/8"	7'4 19/32"	8'7 11/32"	6'2 5/8"
L. F. OF RISER PER INCH OF STAIRWAY WIDTH	.583	.666	.75	.583	.666	.75	.583	.666	.75	.583
L. F. OF TREAD PER INCH OF STAIRWAY WIDTH	.5	.583	.666	.5	.583	.666	.5	.583	.666	.5
FRAMING SQUARE SETTINGS FOR CARRIAGE CUTS — TONGUE	7 13/16"	6 7/8"	6 1/8"	7 7/8"	6 7/8"	6 1/8"	7 7/8"	6 7/8"	6 1/8"	7 7/8"
BODY	9 11/16"	10 5/8"	11 3/8"	9 5/8"	10 5/8"	11 3/8"	9 5/8"	10 5/8"	11 3/8"	9 5/8"

FLOOR TO FLOOR RISE	4'7 1/4"	4'7 1/4"	4'7 3/8"	4'7 3/8"	4'7 3/8"	4'7 1/2"	4'7 1/2"	4'7 1/2"	4'7 5/8"	4'7 5/8"
NUMBER OF RISERS	8	9	7	8	9	7	8	9	7	8
HEIGHT OF EACH RISER	6 15/16"	6 1/8"	7 15/16"	6 15/16"	6 1/8"	7 15/16"	6 15/16"	6 3/16"	7 15/16"	6 15/16"
TOTAL HEIGHT OF RISERS	4'7 1/2"	4'7 1/8"	4'7 9/16"	4'7 1/2"	4'7 1/8"	4'7 9/16"	4'7 1/2"	4'7 11/16"	4'7 9/16"	4'7 1/2"
RISERS OVERAGE (+) OR UNDERAGE (—)	+1/4"	−1/8"	+3/16"	+1/8"	−1/4"	+1/16"	0	+3/16"	−1/16"	−1/8"
NUMBER OF TREADS	7	8	6	7	8	6	7	8	6	7
WIDTH OF EACH TREAD	10 9/16"	11 3/8"	9 9/16"	10 9/16"	11 3/8"	9 9/16"	10 9/16"	11 5/16"	9 9/16"	10 9/16"
TOTAL RUN OF STAIRWAY	6'1 15/16"	7'7"	4'9 3/8"	6'1 15/16"	7'7"	4'9 3/8"	6'1 15/16"	7'6 1/2"	4'9 3/8"	6'1 15/16"
MAXIMUM RUN WITH 6'-8" HEADROOM	1'9 1/8"	1'8 3/4"	1'7 1/8"	1'9 1/8"	1'8 3/4"	1'7 1/8"	1'9 1/8"	1'10 5/8"	1'7 1/8"	1'9 1/8"
ANGLE OF INCLINE	33°18'	28°18'	39°42'	33°18'	28°18'	39°42'	33°18'	28°41'	39°42'	33°18'
LENGTH OF CARRIAGE	7'4 15/32"	8'7 11/32"	6'2 9/16"	7'4 15/32"	8'7 11/32"	6'2 9/16"	7'4 15/32"	8'7 5/32"	6'2 9/16"	7'4 15/32"
L. F. OF RISER PER INCH OF STAIRWAY WIDTH	.666	.75	.583	.666	.75	.583	.666	.75	.583	.666
L. F. OF TREAD PER INCH OF STAIRWAY WIDTH	.583	.666	.5	.583	.666	.5	.583	.666	.5	.583
FRAMING SQUARE SETTINGS FOR CARRIAGE CUTS — TONGUE	6 15/16"	6 1/8"	7 15/16"	6 15/16"	6 1/8"	7 15/16"	6 15/16"	6 3/16"	7 15/16"	6 15/16"
FRAMING SQUARE SETTINGS FOR CARRIAGE CUTS — BODY	10 9/16"	11 3/8"	9 9/16"	10 9/16"	11 3/8"	9 9/16"	10 9/16"	11 5/16"	9 9/16"	10 9/16"

FLOOR TO FLOOR RISE		4'7 5/8"	4'7 3/4"	4'7 3/4"	4'7 3/4"	4'7 7/8"	4'7 7/8"	4'7 7/8"	4'8"	4'8"	4'8"
NUMBER OF RISERS		9	7	8	9	7	8	9	7	8	9
HEIGHT OF EACH RISER		6 3/16"	7 15/16"	7"	6 3/16"	8"	7"	6 3/16"	8"	7"	6 1/4"
TOTAL HEIGHT OF RISERS		4'7 11/16"	4'7 9/16"	4'8"	4'7 11/16"	4'8"	4'8"	4'7 11/16"	4'8"	4'8"	4'8 1/4"
RISERS OVERAGE (+) OR UNDERAGE (—)		+1/16"	−3/16"	+1/4"	−1/16"	+1/8"	+1/8"	−3/16"	0	0	+1/4"
NUMBER OF TREADS		8	6	7	8	6	7	8	6	7	8
WIDTH OF EACH TREAD		11 5/16"	9 9/16"	10 1/2"	11 5/16"	9 1/2"	10 1/2"	11 5/16"	9 1/2"	10 1/2"	11 1/4"
TOTAL RUN OF STAIRWAY		7'6 1/2"	4'9 3/8"	6'1 1/2"	7'6 1/2"	4'9"	6'1 1/2"	7'6 1/2"	4'9"	6'1 1/2"	7'6"
MAXIMUM RUN WITH 6'-8" HEADROOM		1'10 5/8"	1'7 1/8"	1'9"	1'10 5/8"	9 1/2"	1'9"	1'10 5/8"	9 1/2"	1'9"	1'10 1/2"
ANGLE OF INCLINE		28°41'	39°42'	33°41'	28°41'	40°6'	33°41'	28°41'	40°6'	33°41'	29°3'
LENGTH OF CARRIAGE		8'7 5/32"	6'2 9/16"	7'4 11/32"	8'7 5/32"	6'2 17/32"	7'4 11/32"	8'7 5/32"	6'2 17/32"	7'4 11/32"	8'6 31/32"
L. F. OF RISER PER INCH OF STAIRWAY WIDTH		.75	.583	.666	.75	.583	.666	.75	.583	.666	.75
L. F. OF TREAD PER INCH OF STAIRWAY WIDTH		.666	.5	.583	.666	.5	.583	.666	.5	.583	.666
FRAMING SQUARE SETTINGS FOR CARRIAGE CUTS	TONGUE	6 3/16"	7 15/16"	7"	6 3/16"	8"	7"	6 3/16"	8"	7"	6 1/4"
	BODY	11 5/16"	9 9/16"	10 1/2"	11 5/16"	9 1/2"	10 1/2"	11 5/16"	9 1/2"	10 1/2"	11 1/4"

FLOOR TO FLOOR RISE	4'8 1/8"	4'8 1/8"	4'8 1/8"	4'8 1/4"	4'8 1/4"	4'8 1/4"	4'8 3/8"	4'8 3/8"	4'8 3/8"	4'8 1/2"
NUMBER OF RISERS	7	8	9	7	8	9	7	8	9	7
HEIGHT OF EACH RISER	8"	7"	6 1/4"	8 1/16"	7 1/16"	6 1/4"	8 1/16"	7 1/16"	6 1/4"	8 1/16"
TOTAL HEIGHT OF RISERS	4'8"	4'8"	4'8 1/4"	4'8 7/16"	4'8 1/2"	4'8 1/4"	4'8 7/16"	4'8 1/2"	4'8 1/4"	4'8 7/16"
RISERS OVERAGE (+) OR UNDERAGE (—)	−1/8"	−1/8"	+1/8"	+3/16"	+1/4"	0	+1/16"	+1/8"	−1/8"	−1/16"
NUMBER OF TREADS	6	7	8	6	7	8	6	7	8	6
WIDTH OF EACH TREAD	9 1/2"	10 1/2"	11 1/4"	9 7/16"	10 7/16"	11 1/4"	9 7/16"	10 7/16"	11 1/4"	9 7/16"
TOTAL RUN OF STAIRWAY	4'9"	6'1 1/2"	7'6"	4'8 5/8"	6'1 1/16"	7'6"	4'8 5/8"	6'1 1/16"	7'6"	4'8 5/8"
MAXIMUM RUN WITH 6'-8" HEADROOM	9 1/2"	1'9"	1'10 1/2"	9 7/16"	1'8 7/8"	1'10 1/2"	9 7/16"	1'8 7/8"	1'10 1/2"	9 7/16"
ANGLE OF INCLINE	40°6'	33°41'	29°3'	40°30'	34°5'	29°3'	40°30'	34°5'	29°3'	40°30'
LENGTH OF CARRIAGE	6'2 17/32"	7'4 11/32"	8'6 31/32"	6'2 15/16"	7'4 7/32"	8'6 31/32"	6'2 15/16"	7'4 7/32"	8'6 31/32"	6'2 15/32"
L. F. OF RISER PER INCH OF STAIRWAY WIDTH	.583	.666	.75	.583	.666	.75	.583	.666	.75	.583
L. F. OF TREAD PER INCH OF STAIRWAY WIDTH	.5	.583	.666	.5	.583	.666	.5	.583	.666	.5
FRAMING SQUARE SETTINGS FOR CARRIAGE CUTS — TONGUE	8"	7"	6 1/4"	8 1/16"	7 1/16"	6 1/4"	8 1/16"	7 1/16"	6 1/4"	8 1/16"
FRAMING SQUARE SETTINGS FOR CARRIAGE CUTS — BODY	9 1/2"	10 1/2"	11 1/4"	9 7/16"	10 7/16"	11 1/4"	9 7/16"	10 7/16"	11 1/4"	9 7/16"

FLOOR TO FLOOR RISE		4'8 1/2"	4'8 1/2"	4'8 5/8"	4'8 5/8"	4'8 5/8"	4'8 3/4"	4'8 3/4"	4'8 3/4"	4'8 7/8"	4'8 7/8"
NUMBER OF RISERS		8	9	7	8	9	7	8	9	7	8
HEIGHT OF EACH RISER		7 1/16"	6 1/4"	8 1/16"	7 1/16"	6 5/16"	8 1/8"	7 1/8"	6 5/16"	8 1/8"	7 1/8"
TOTAL HEIGHT OF RISERS		4'8 1/2"	4'8 1/4"	4'8 7/16"	4'8 1/2"	4'8 13/16"	4'8 7/8"	4'9"	4'8 13/16"	4'8 7/8"	4'9"
RISERS OVERAGE (+) OR UNDERAGE (—)		0	-1/4"	-3/16"	-1/8"	+3/16"	+1/8"	+1/4"	+1/16"	0	+1/8"
NUMBER OF TREADS		7	8	6	7	8	6	7	8	6	7
WIDTH OF EACH TREAD		10 7/16"	11 1/4"	9 7/16"	10 7/16"	11 3/16"	9 3/8"	10 3/8"	11 3/16"	9 3/8"	10 3/8"
TOTAL RUN OF STAIRWAY		6'1 1/16"	7'6"	4'8 5/8"	6'1 1/16"	7'5 1/2"	4'8 1/4"	6' 5/8"	7'5 1/2"	4'8 1/4"	6' 5/8"
MAXIMUM RUN WITH 6'-8" HEADROOM		1'8 7/8"	1'10 1/2"	9 7/16"	1'8 7/8"	1'10 3/8"	9 3/8"	1'8 3/4"	1'10 3/8"	9 3/8"	1'8 3/4"
ANGLE OF INCLINE		34°5'	29°3'	40°30'	34°5'	29°26'	40°55'	34°29'	29°26'	40°55'	34°29'
LENGTH OF CARRIAGE		7'4 7/32"	8'6 31/32"	6'2 15/32"	7'4 7/32"	8'6 3/4"	6'2 7/16"	7'4 3/32"	8'6 3/4"	6'2 7/16"	7'4 3/32"
L. F. OF RISER PER INCH OF STAIRWAY WIDTH		.666	.75	.583	.666	.75	.583	.666	.75	.583	.666
L. F. OF TREAD PER INCH OF STAIRWAY WIDTH		.583	.666	.5	.583	.666	.5	.583	.666	.5	.583
FRAMING SQUARE SETTINGS FOR CARRIAGE CUTS	TONGUE	7 1/16"	6 1/4"	8 1/16"	7 1/16"	6 5/16"	8 1/8"	7 1/8"	6 5/16"	8 1/8"	7 1/8"
	BODY	10 7/16"	11 1/4"	9 7/16"	10 7/16"	11 3/16"	9 3/8"	10 3/8"	11 3/16"	9 3/8"	10 3/8"

FLOOR TO FLOOR RISE	4'8 7/8"	4'9"	4'9"	4'9"	4'9 1/8"	4'9 1/8"	4'9 1/8"	4'9 1/4"	4'9 1/4"	4'9 1/4"
NUMBER OF RISERS	9	7	8	9	7	8	9	7	8	9
HEIGHT OF EACH RISER	6 5/16"	8 1/8"	7 1/8"	6 5/16"	8 3/16"	7 1/8"	6 3/8"	8 3/16"	7 3/16"	6 3/8"
TOTAL HEIGHT OF RISERS	4'8 13/16"	4'8 7/8"	4'9"	4'8 13/16"	4'9 5/16"	4'9"	4'9 3/8"	4'9 5/16"	4'9 1/2"	4'9 3/8"
RISERS OVERAGE (+) OR UNDERAGE (—)	−1/16"	−1/8"	0	−3/16"	+3/16"	−1/8"	+1/4"	+1/16"	+1/4"	+1/8"
NUMBER OF TREADS	8	6	7	8	6	7	8	6	7	8
WIDTH OF EACH TREAD	11 3/16"	9 3/8"	10 3/8"	11 3/16"	9 5/16"	10 3/8"	11 1/8"	9 5/16"	10 5/16"	11 1/8"
TOTAL RUN OF STAIRWAY	7'5 1/2"	4'8 1/4"	6' 5/8"	7'5 1/2"	4'7 7/8"	6' 5/8"	7'5"	4'7 7/8"	6' 3/16"	7'5"
MAXIMUM RUN WITH 6'-8" HEADROOM	1'10 3/8"	9 3/8"	1'8 3/4"	1'10 3/8"	9 5/16"	1'8 3/4"	1'10 1/4"	9 5/16"	1'8 5/8"	1'10 1/4"
ANGLE OF INCLINE	29°26'	40°55'	34°29'	29°26'	41°29'	34°29'	29°49'	41°19'	34°53'	29°49'
LENGTH OF CARRIAGE	8'6 3/4"	6'2 7/16"	7'4 3/32"	8'6 3/4"	6'2 13/32"	7'4 3/32"	8'6 9/16"	6'2 13/32"	7'4"	8'6 9/16"
L. F. OF RISER PER INCH OF STAIRWAY WIDTH	.75	.583	.666	.75	.583	.666	.75	.583	.666	.75
L. F. OF TREAD PER INCH OF STAIRWAY WIDTH	.666	.5	.583	.666	.5	.583	.666	.5	.583	.666
FRAMING SQUARE SETTINGS FOR CARRIAGE CUTS — TONGUE	6 5/16"	8 1/8"	7 1/8"	6 5/16"	8 3/16"	7 1/8"	6 3/8"	8 3/16"	7 3/16"	6 3/8"
FRAMING SQUARE SETTINGS FOR CARRIAGE CUTS — BODY	11 3/16"	9 3/8"	10 3/8"	11 3/16"	9 5/16"	10 3/8"	11 1/8"	9 5/16"	10 5/16"	11 1/8"

FLOOR TO FLOOR RISE		4'9 3/8"	4'9 3/8"	4'9 3/8"	4'9 1/2"	4'9 1/2"	4'9 1/2"	4'9 5/8"	4'9 5/8"	4'9 5/8"	4'9 3/4"
NUMBER OF RISERS		7	8	9	7	8	9	7	8	9	7
HEIGHT OF EACH RISER		8 3/16"	7 3/16"	6 3/8"	8 3/16"	7 3/16"	6 3/8"	8 1/4"	7 3/16"	6 3/8"	8 1/4"
TOTAL HEIGHT OF RISERS		4'9 5/16"	4'9 1/2"	4'9 3/8"	4'9 5/16"	4'9 1/2"	4'9 3/8"	4'9 3/4"	4'9 1/2"	4'9 3/8"	4'9 3/4"
RISERS OVERAGE (+) OR UNDERAGE (—)		−1/16"	+1/8"	0	−3/16"	0	−1/8"	+1/8"	−1/8"	−1/4"	0
NUMBER OF TREADS		6	7	8	6	7	8	6	7	8	6
WIDTH OF EACH TREAD		9 5/16"	10 5/16"	11 1/8"	9 5/16"	10 5/16"	11 1/8"	9 1/4"	10 5/16"	11 1/8"	9 1/4"
TOTAL RUN OF STAIRWAY		4'7 7/8"	6' 3/16"	7'5"	4'7 7/8"	6' 3/16"	7'5"	4'7 1/2"	6' 3/16"	7'5"	4'7 1/2"
MAXIMUM RUN WITH 6'-8" HEADROOM		9 5/16"	1'8 5/8"	1'10 1/4"	9 5/16"	1'8 5/8"	1'10 1/4"	9 1/4"	1'8 5/8"	1'10 1/4"	9 1/4"
ANGLE OF INCLINE		41°19'	34°53'	29°49'	41°19'	34°53'	29°49'	41°44'	34°53'	29°49'	41°44'
LENGTH OF CARRIAGE		6'2 13/32"	7'4"	8'6 9/16"	6'2 13/32"	7'4"	8'6 9/16"	6'2 3/8"	7'4"	8'6 9/16"	6'2 3/8"
L. F. OF RISER PER INCH OF STAIRWAY WIDTH		.583	.666	.75	.583	.666	.75	.583	.666	.75	.583
L. F. OF TREAD PER INCH OF STAIRWAY WIDTH		.5	.583	.666	.5	.583	.666	.5	.583	.666	.5
FRAMING SQUARE SETTINGS FOR CARRIAGE CUTS	TONGUE	8 3/16"	7 3/16"	6 3/8"	8 3/16"	7 3/16"	6 3/8"	8 1/4"	7 3/16"	6 3/8"	8 1/4"
	BODY	9 5/16"	10 5/16"	11 1/8"	9 5/16"	10 5/16"	11 1/8"	9 1/4"	10 5/16"	11 1/8"	9 1/4"

FLOOR TO FLOOR RISE	4'9 3/4"	4'9 3/4"	4'9 7/8"	4'9 7/8"	4'9 7/8"	4'10"	4'10"	4'10 1/8"	4'10 1/8"	4'10 1/4"
NUMBER OF RISERS	8	9	7	8	9	8	9	8	9	8
HEIGHT OF EACH RISER	7 1/4"	6 7/16"	8 1/4"	7 1/4"	6 7/16"	7 1/4"	6 7/16"	7 1/4"	6 7/16"	7 5/16"
TOTAL HEIGHT OF RISERS	4'10"	4'9 15/16"	4'9 3/4"	4'10"	4'9 15/16"	4'10"	4'9 15/16"	4'10"	4'9 15/16"	4'10 1/2"
RISERS OVERAGE (+) OR UNDERAGE (—)	+1/4"	+3/16"	−1/8"	+1/8"	+1/16"	0	−1/16"	−1/8"	−3/16"	+1/4"
NUMBER OF TREADS	7	8	6	7	8	7	8	7	8	7
WIDTH OF EACH TREAD	10 1/4"	11 1/16"	9 1/4"	10 1/4"	11 1/16"	10 1/4"	11 1/16"	10 1/4"	11 1/16"	10 3/16"
TOTAL RUN OF STAIRWAY	5'11 3/4"	7'4 1/2"	4'7 1/2"	5'11 3/4"	7'4 1/2"	5'11 3/4"	7'4 1/2"	5'11 3/4"	7'4 1/2"	5'11 5/16"
MAXIMUM RUN WITH 6'-8" HEADROOM	1'8 1/2"	1'10 1/8"	9 1/4"	1'8 1/2"	1'10 1/8"	1'8 1/2"	1'10 1/8"	1'8 1/2"	1'10 1/8"	1'8 3/8"
ANGLE OF INCLINE	35°16'	30°12'	41°44'	35°16'	30°12'	35°16'	30°12'	35°16'	30°12'	35°40'
LENGTH OF CARRIAGE	7'3 7/8"	8'6 13/32"	6'2 3/8"	7'3 7/8"	8'6 13/32"	7'3 7/8"	8'6 13/32"	7'3 7/8"	8'6 13/32"	7'3 25/32"
L. F. OF RISER PER INCH OF STAIRWAY WIDTH	.666	.75	.583	.666	.75	.666	.75	.666	.75	.666
L. F. OF TREAD PER INCH OF STAIRWAY WIDTH	.583	.666	.5	.583	.666	.583	.666	.583	.666	.583
FRAMING SQUARE SETTINGS FOR CARRIAGE CUTS — TONGUE	7 1/4"	6 7/16"	8 1/4"	7 1/4"	6 7/16"	7 1/4"	6 7/16"	7 1/4"	6 7/16"	7 5/16"
FRAMING SQUARE SETTINGS FOR CARRIAGE CUTS — BODY	10 1/4"	11 1/16"	9 1/4"	10 1/4"	11 1/16"	10 1/4"	11 1/16"	10 1/4"	11 1/16"	10 3/16"

FLOOR TO FLOOR RISE		4'10 1/4"	4'10 3/8"	4'10 3/8"	4'10 1/2"	4'10 1/2"	4'10 5/8"	4'10 5/8"	4'10 3/4"	4'10 3/4"	4'10 7/8"
NUMBER OF RISERS		9	8	9	8	9	8	9	8	9	8
HEIGHT OF EACH RISER		6 1/2"	7 5/16"	6 1/2"	7 5/16"	6 1/2"	7 5/16"	6 1/2"	7 3/8"	6 1/2"	7 3/8"
TOTAL HEIGHT OF RISERS		4'10 1/2"	4'10 1/2"	4'10 1/2"	4'10 1/2"	4'10 1/2"	4'10 1/2"	4'10 1/2"	4'11"	4'10 1/2"	4'11"
RISERS OVERAGE (+) OR UNDERAGE (—)		+1/4"	+1/8"	+1/8"	0	0	−1/8"	−1/8"	+1/4"	−1/4"	+1/8"
NUMBER OF TREADS		8	7	8	7	8	7	8	7	8	7
WIDTH OF EACH TREAD		11"	10 3/16"	11"	10 3/16"	11"	10 3/16"	11"	10 1/8"	11"	10 1/8"
TOTAL RUN OF STAIRWAY		7'4"	5'11 5/16"	7'4"	5'11 5/16"	7'4"	5'11 5/16"	7'4"	5'10 7/8"	7'4"	5'10 7/8"
MAXIMUM RUN WITH 6'-8" HEADROOM		1'10"	1'8 3/8"	1'10"	1'8 3/8"	1'10"	1'8 3/8"	1'10"	1'8 1/4"	1'10"	1'8 1/4"
ANGLE OF INCLINE		30°35'	35°40'	30°35'	35°40'	30°35'	35°40'	30°35'	36°4'	30°35'	36°4'
LENGTH OF CARRIAGE		8'6 7/32"	7'3 25/32"	8'6 7/32"	7'3 25/32"	8'6 7/32"	7'3 25/32"	8'6 7/32"	7'3 11/16"	8'6 7/32"	7'3 11/16"
L. F. OF RISER PER INCH OF STAIRWAY WIDTH		.75	.666	.75	.666	.75	.666	.75	.666	.75	.666
L. F. OF TREAD PER INCH OF STAIRWAY WIDTH		.666	.583	.666	.583	.666	.583	.666	.583	.666	.583
FRAMING SQUARE SETTINGS FOR CARRIAGE CUTS	TONGUE	6 1/2"	7 5/16"	6 1/2"	7 5/16"	6 1/2"	7 5/16"	6 1/2"	7 3/8"	6 1/2"	7 3/8"
	BODY	11"	10 3/16"	11"	10 3/16"	11"	10 3/16"	11"	10 1/8"	11"	10 1/8"

FLOOR TO FLOOR RISE		4'10 7/8"	4'11"	4'11"	4'11 1/8"	4'11 1/8"	4'11 1/4"	4'11 1/4"	4'11 3/8"	4'11 3/8"	4'11 1/2"
NUMBER OF RISERS		9	8	9	8	9	8	9	8	9	8
HEIGHT OF EACH RISER		6 9/16"	7 3/8"	6 9/16"	7 3/8"	6 9/16"	7 7/16"	6 9/16"	7 7/16"	6 5/8"	7 7/16"
TOTAL HEIGHT OF RISERS		4'11 1/16"	4'11"	4'11 1/16"	4'11"	4'11 1/16"	4'11 1/2"	4'11 1/16"	4'11 1/2"	4'11 5/8"	4'11 1/2"
RISERS OVERAGE (+) OR UNDERAGE (—)		+3/16"	0	+1/16"	−1/8"	−1/16"	+1/4"	−3/16"	+1/8"	+1/4"	0
NUMBER OF TREADS		8	7	8	7	8	7	8	7	8	7
WIDTH OF EACH TREAD		10 15/16"	10 1/8"	10 15/16"	10 1/8"	10 15/16"	10 1/16"	10 15/16"	10 1/16"	10 7/8"	10 1/16"
TOTAL RUN OF STAIRWAY		7'3 1/2"	5'10 7/8"	7'3 1/2"	5'10 7/8"	7'3 1/2"	5'10 7/16"	7'3 1/2"	5'10 7/16"	7'3"	5'10 7/16"
MAXIMUM RUN WITH 6'-8" HEADROOM		1'9 7/8"	1'8 1/4"	1'9 7/8"	1'8 1/4"	1'9 7/8"	1'8 1/8"	1'9 7/8"	1'8 1/8"	1'9 3/4"	1'8 1/8"
ANGLE OF INCLINE		30°58'	36°4'	30°58'	36°4'	30°58'	36°28'	30°58'	36°28'	31°21'	36°28'
LENGTH OF CARRIAGE		8'6 1/32"	7'3 11/16"	8'6 1/32"	7'3 11/16"	8'6 1/32"	7'3 19/32"	8'6 1/32"	7'3 19/32"	8'5 7/8"	7'3 19/32"
L. F. OF RISER PER INCH OF STAIRWAY WIDTH		.75	.666	.75	.666	.75	.666	.75	.666	.75	.666
L. F. OF TREAD PER INCH OF STAIRWAY WIDTH		.666	.583	.666	.583	.666	.583	.666	.583	.666	.583
FRAMING SQUARE SETTINGS FOR CARRIAGE CUTS	TONGUE	6 9/16"	7 3/8"	6 9/16"	7 3/8"	6 9/16"	7 7/16"	6 9/16"	7 7/16"	6 5/8"	7 7/16"
	BODY	10 15/16"	10 1/8"	10 15/16"	10 1/8"	10 15/16"	10 1/16"	10 15/16"	10 1/16"	10 7/8"	10 1/16"

FLOOR TO FLOOR RISE	4'11 1/2"	4'11 5/8"	4'11 5/8"	4'11 3/4"	4'11 3/4"	4'11 7/8"	4'11 7/8"	5'	5'	5'
NUMBER OF RISERS	9	8	9	8	9	8	9	8	9	10
HEIGHT OF EACH RISER	6 5/8"	7 7/16"	6 5/8"	7 1/2"	6 5/8"	7 1/2"	6 5/8"	7 1/2"	6 11/16"	6"
TOTAL HEIGHT OF RISERS	4'11 5/8"	4'11 1/2"	4'11 5/8"	5'	4'11 5/8"	5'	4'11 5/8"	5'	5' 3/16"	5'
RISERS OVERAGE (+) OR UNDERAGE (—)	+1/8"	−1/8"	0	+1/4"	−1/8"	+1/8"	−1/4"	0	+3/16"	0
NUMBER OF TREADS	8	7	8	7	8	7	8	7	8	9
WIDTH OF EACH TREAD	10 7/8"	10 1/16"	10 7/8"	10"	10 7/8"	10"	10 7/8"	10"	10 13/16"	11 1/2"
TOTAL RUN OF STAIRWAY	7'3"	5'10 7/16"	7'3"	5'10"	7'3"	5'10"	7'3"	5'10"	7'2 1/2"	8'7 1/2"
MAXIMUM RUN WITH 6'-8" HEADROOM	1'9 3/4"	1'8 1/8"	1'9 3/4"	1'8"	1'9 3/4"	1'8"	1'9 3/4"	1'8"	1'9 5/8"	1'11"
ANGLE OF INCLINE	31°21'	36°28'	31°21'	36°52'	31°21'	36°52'	31°21'	36°52'	31°44'	27°33'
LENGTH OF CARRIAGE	8'5 7/8"	7'3 19/32"	8'5 7/8"	7'3 1/2"	8'5 7/8"	7'3 1/2"	8'5 7/8"	7'3 1/2"	8'5 23/32"	9'8 3/4"
L. F. OF RISER PER INCH OF STAIRWAY WIDTH	.75	.666	.75	.666	.75	.666	.75	.666	.75	.833
L. F. OF TREAD PER INCH OF STAIRWAY WIDTH	.666	.583	.666	.583	.666	.583	.666	.583	.666	.75
FRAMING SQUARE SETTINGS FOR CARRIAGE CUTS — TONGUE	6 5/8"	7 7/16"	6 5/8"	7 1/2"	6 5/8"	7 1/2"	6 5/8"	7 1/2"	6 11/16"	6"
FRAMING SQUARE SETTINGS FOR CARRIAGE CUTS — BODY	10 7/8"	10 1/16"	10 7/8"	10"	10 7/8"	10"	10 7/8"	10"	10 13/16"	11 1/2"

FLOOR TO FLOOR RISE	5'$\frac{1}{8}$"	5'$\frac{1}{8}$"	5'$\frac{1}{8}$"	5'$\frac{1}{4}$"	5'$\frac{1}{4}$"	5'$\frac{1}{4}$"	5'$\frac{3}{8}$"	5'$\frac{3}{8}$"	5'$\frac{3}{8}$"	5'$\frac{1}{2}$"
NUMBER OF RISERS	8	9	10	8	9	10	8	9	10	8
HEIGHT OF EACH RISER	7$\frac{1}{2}$"	6$\frac{11}{16}$"	6"	7$\frac{9}{16}$"	6$\frac{11}{16}$"	6"	7$\frac{9}{16}$"	6$\frac{11}{16}$"	6$\frac{1}{16}$"	7$\frac{9}{16}$"
TOTAL HEIGHT OF RISERS	5'	5'$\frac{3}{16}$"	5'	5'$\frac{1}{2}$"	5'$\frac{3}{16}$"	5'	5'$\frac{1}{2}$"	5'$\frac{3}{16}$"	5'$\frac{5}{8}$"	5'$\frac{1}{2}$"
RISERS OVERAGE (+) OR UNDERAGE (—)	−$\frac{1}{8}$"	+$\frac{1}{16}$"	−$\frac{1}{8}$"	+$\frac{1}{4}$"	−$\frac{1}{16}$"	−$\frac{1}{4}$"	+$\frac{1}{8}$"	−$\frac{3}{16}$"	+$\frac{1}{4}$"	0
NUMBER OF TREADS	7	8	9	7	8	9	7	8	9	7
WIDTH OF EACH TREAD	10"	10$\frac{13}{16}$"	11$\frac{1}{2}$"	9$\frac{15}{16}$"	10$\frac{13}{16}$"	11$\frac{1}{2}$"	9$\frac{15}{16}$"	10$\frac{13}{16}$"	11$\frac{7}{16}$"	9$\frac{15}{16}$"
TOTAL RUN OF STAIRWAY	5'10"	7'2$\frac{1}{2}$"	8'7$\frac{1}{2}$"	5'9$\frac{9}{16}$"	7'2$\frac{1}{2}$"	8'7$\frac{1}{2}$"	5'9$\frac{9}{16}$"	7'2$\frac{1}{2}$"	8'6$\frac{15}{16}$"	5'9$\frac{9}{16}$"
MAXIMUM RUN WITH 6'-8" HEADROOM	1'8"	1'9$\frac{5}{8}$"	1'11"	1'7$\frac{7}{8}$"	1'9$\frac{5}{8}$"	1'11"	1'7$\frac{7}{8}$"	1'9$\frac{5}{8}$"	1'10$\frac{7}{8}$"	1'7$\frac{7}{8}$"
ANGLE OF INCLINE	36°52'	31°44'	27°33'	37°16'	31°44'	27°33'	37°16'	31°44'	27°56'	37°16'
LENGTH OF CARRIAGE	7'3$\frac{1}{2}$"	8'5$\frac{23}{32}$"	9'8$\frac{3}{4}$"	7'3$\frac{13}{32}$"	8'5$\frac{23}{32}$"	9'8$\frac{3}{4}$"	7'3$\frac{13}{16}$"	8'5$\frac{23}{32}$"	9'8$\frac{1}{2}$"	7'3$\frac{13}{32}$"
L. F. OF RISER PER INCH OF STAIRWAY WIDTH	.666	.75	.833	.666	.75	.833	.666	.75	.833	.666
L. F. OF TREAD PER INCH OF STAIRWAY WIDTH	.583	.666	.75	.583	.666	.75	.583	.666	.75	.583
FRAMING SQUARE SETTINGS FOR CARRIAGE CUTS — TONGUE	7$\frac{1}{2}$"	6$\frac{11}{16}$"	6"	7$\frac{9}{16}$"	6$\frac{11}{16}$"	6"	7$\frac{9}{16}$"	6$\frac{11}{16}$"	6$\frac{1}{16}$"	7$\frac{9}{16}$"
FRAMING SQUARE SETTINGS FOR CARRIAGE CUTS — BODY	10"	10$\frac{13}{16}$"	11$\frac{1}{2}$"	9$\frac{15}{16}$"	10$\frac{13}{16}$"	11$\frac{1}{2}$"	9$\frac{15}{16}$"	10$\frac{13}{16}$"	11$\frac{7}{16}$"	9$\frac{15}{16}$"

FLOOR TO FLOOR RISE	$5'\frac{1}{2}''$	$5'\frac{1}{2}''$	$5'\frac{5}{8}''$	$5'\frac{5}{8}''$	$5'\frac{5}{8}''$	$5'\frac{3}{4}''$	$5'\frac{3}{4}''$	$5'\frac{3}{4}''$	$5'\frac{7}{8}''$	$5'\frac{7}{8}''$
NUMBER OF RISERS	9	10	8	9	10	8	9	10	8	9
HEIGHT OF EACH RISER	$6\frac{3}{4}''$	$6\frac{1}{16}''$	$7\frac{9}{16}''$	$6\frac{3}{4}''$	$6\frac{1}{16}''$	$7\frac{5}{8}''$	$6\frac{3}{4}''$	$6\frac{1}{16}''$	$7\frac{5}{8}''$	$6\frac{3}{4}''$
TOTAL HEIGHT OF RISERS	$5'\frac{3}{4}''$	$5'\frac{5}{8}''$	$5'\frac{1}{2}''$	$5'\frac{3}{4}''$	$5'\frac{5}{8}''$	$5'1''$	$5'\frac{3}{4}''$	$5'\frac{5}{8}''$	$5'1''$	$5'\frac{3}{4}''$
RISERS OVERAGE (+) OR UNDERAGE (—)	$+\frac{1}{4}''$	$+\frac{1}{8}''$	$-\frac{1}{8}''$	$+\frac{1}{8}''$	0	$+\frac{1}{4}''$	0	$-\frac{1}{8}''$	$+\frac{1}{8}''$	$-\frac{1}{8}''$
NUMBER OF TREADS	8	9	7	8	9	7	8	9	7	8
WIDTH OF EACH TREAD	$10\frac{3}{4}''$	$11\frac{7}{16}''$	$9\frac{15}{16}''$	$10\frac{3}{4}''$	$11\frac{7}{16}''$	$9\frac{7}{8}''$	$10\frac{3}{4}''$	$11\frac{7}{16}''$	$9\frac{7}{8}''$	$10\frac{3}{4}''$
TOTAL RUN OF STAIRWAY	$7'2''$	$8'6\frac{15}{16}''$	$5'9\frac{9}{16}''$	$7'2''$	$8'6\frac{15}{16}''$	$5'9\frac{1}{8}''$	$7'2''$	$8'6\frac{15}{16}''$	$5'9\frac{1}{8}''$	$7'2''$
MAXIMUM RUN WITH 6'-8" HEADROOM	$1'9\frac{1}{2}''$	$1'10\frac{7}{8}''$	$1'7\frac{7}{8}''$	$1'9\frac{1}{2}''$	$1'10\frac{7}{8}''$	$1'7\frac{3}{4}''$	$1'9\frac{1}{2}''$	$1'10\frac{7}{8}''$	$1'7\frac{3}{4}''$	$1'9\frac{1}{2}''$
ANGLE OF INCLINE	32°7'	27°56'	37°16'	32°7'	27°56'	37°40'	32°7'	27°56'	37°40'	32°7'
LENGTH OF CARRIAGE	$8'5\frac{9}{16}''$	$9'8\frac{1}{2}''$	$7'3\frac{13}{32}''$	$8'5\frac{9}{16}''$	$9'8\frac{1}{2}''$	$7'3\frac{11}{32}''$	$8'5\frac{9}{16}''$	$9'8\frac{1}{2}''$	$7'3\frac{11}{32}''$	$8'5\frac{9}{16}''$
L. F. OF RISER PER INCH OF STAIRWAY WIDTH	.75	.833	.666	.75	.833	.666	.75	.833	.666	.75
L. F. OF TREAD PER INCH OF STAIRWAY WIDTH	.666	.75	.583	.666	.75	.583	.666	.75	.583	.666
FRAMING SQUARE SETTINGS FOR CARRIAGE CUTS — TONGUE	$6\frac{3}{4}''$	$6\frac{1}{16}''$	$7\frac{9}{16}''$	$6\frac{3}{4}''$	$6\frac{1}{16}''$	$7\frac{5}{8}''$	$6\frac{3}{4}''$	$6\frac{1}{16}''$	$7\frac{5}{8}''$	$6\frac{3}{4}''$
FRAMING SQUARE SETTINGS FOR CARRIAGE CUTS — BODY	$10\frac{3}{4}''$	$11\frac{7}{16}''$	$9\frac{15}{16}''$	$10\frac{3}{4}''$	$11\frac{7}{16}''$	$9\frac{7}{8}''$	$10\frac{3}{4}''$	$11\frac{7}{16}''$	$9\frac{7}{8}''$	$10\frac{3}{4}''$

FLOOR TO FLOOR RISE	5' 7/8"	5'1"	5'1"	5'1"	5'1 1/8"	5'1 1/8"	5'1 1/8"	5'1 1/4"	5'1 1/4"	5'1 1/4"
NUMBER OF RISERS	10	8	9	10	8	9	10	8	9	10
HEIGHT OF EACH RISER	6 1/16"	7 5/8"	6 3/4"	6 1/8"	7 5/8"	6 13/16"	6 1/8"	7 11/16"	6 13/16"	6 1/8"
TOTAL HEIGHT OF RISERS	5' 5/8"	5'1"	5' 3/4"	5'1 1/4"	5'1"	5'1 5/16"	5'1 1/4"	5'1 1/2"	5'1 5/16"	5'1 1/4"
RISERS OVERAGE (+) OR UNDERAGE (—)	−1/4"	0	−1/4"	+1/4"	−1/8"	+3/16"	+1/8"	+1/4"	+1/16"	0
NUMBER OF TREADS	9	7	8	9	7	8	9	7	8	9
WIDTH OF EACH TREAD	11 7/16"	9 7/8"	10 3/4"	11 3/8"	9 7/8"	10 11/16"	11 3/8"	9 13/16"	10 11/16"	11 3/8"
TOTAL RUN OF STAIRWAY	8'6 15/16"	5'9 1/8"	7'2"	8'6 3/8"	5'9 1/8"	7'1 1/2"	8'6 3/8"	5'8 11/16"	7'1 1/2"	8'6 3/8"
MAXIMUM RUN WITH 6'-8" HEADROOM	1'10 7/8"	1'7 3/4"	1'9 1/2"	1'10 3/4"	1'7 3/4"	1'9 3/8"	1'10 3/4"	1'7 5/8"	1'9 3/8"	1'10 3/4"
ANGLE OF INCLINE	27°56'	37°40'	32°7'	28°18'	37°40'	32°31'	28°18'	38°5'	32°31'	28°18'
LENGTH OF CARRIAGE	9'8 1/2"	7'3 11/32"	8'5 9/16"	9'8 9/32"	7'3 11/32"	8'5 13/32"	9'8 9/32"	7'3 1/4"	8'5 13/32"	9'8 9/32"
L. F. OF RISER PER INCH OF STAIRWAY WIDTH	.833	.666	.75	.833	.666	.75	.833	.666	.75	.833
L. F. OF TREAD PER INCH OF STAIRWAY WIDTH	.75	.583	.666	.75	.583	.666	.75	.583	.666	.75
FRAMING SQUARE SETTINGS FOR CARRIAGE CUTS — TONGUE	6 1/16"	7 5/8"	6 3/4"	6 1/8"	7 5/8"	6 13/16"	6 1/8"	7 11/16"	6 13/16"	6 1/8"
FRAMING SQUARE SETTINGS FOR CARRIAGE CUTS — BODY	11 7/16"	9 7/8"	10 3/4"	11 3/8"	9 7/8"	10 11/16"	11 3/8"	9 13/16"	10 11/16"	11 3/8"

FLOOR TO FLOOR RISE		5'1 3/8"	5'1 3/8"	5'1 3/8"	5'1 1/2"	5'1 1/2"	5'1 1/2"	5'1 5/8"	5'1 5/8"	5'1 5/8"	5'1 3/4"
NUMBER OF RISERS		8	9	10	8	9	10	8	9	10	8
HEIGHT OF EACH RISER		7 11/16"	6 13/16"	6 1/8"	7 11/16"	6 13/16"	6 1/8"	7 11/16"	6 7/8"	6 3/16"	7 3/4"
TOTAL HEIGHT OF RISERS		5'1 1/2"	5'1 5/16"	5'1 1/4"	5'1 1/2"	5'1 5/16"	5'1 1/4"	5'1 1/2"	5'1 7/8"	5'1 7/8"	5'2"
RISERS OVERAGE (+) OR UNDERAGE (—)		+1/8"	−1/16"	−1/8"	0	−3/16"	−1/4"	−1/8"	+1/4"	+1/4"	+1/4"
NUMBER OF TREADS		7	8	9	7	8	9	7	8	9	7
WIDTH OF EACH TREAD		9 13/16"	10 11/16"	11 3/8"	9 13/16"	10 11/16"	11 3/8"	9 13/16"	10 5/8"	11 5/16"	9 3/4"
TOTAL RUN OF STAIRWAY		5'8 11/16"	7'1 1/2"	8'6 3/8"	5'8 11/16"	7'1 1/2"	8'6 3/8"	5'8 11/16"	7'1"	8'5 13/16"	5'8 1/4"
MAXIMUM RUN WITH 6'-8" HEADROOM		1'7 5/8"	1'9 3/8"	1'10 3/4"	1'7 5/8"	1'9 3/8"	1'10 3/4"	1'7 5/8"	1'9 1/4"	1'10 5/8"	1'7 1/2"
ANGLE OF INCLINE		38°5'	32°31'	28°18'	38°5'	32°31'	28°18'	38°5'	32°54'	28°41'	38°29'
LENGTH OF CARRIAGE		7'3 1/4"	8'5 13/32"	9'8 9/32"	7'3 1/4"	8'5 13/32"	9'8 9/32"	7'3 1/4"	8'5 1/4"	9'8 1/16"	7'3 3/16"
L. F. OF RISER PER INCH OF STAIRWAY WIDTH		.666	.75	.833	.666	.75	.833	.666	.75	.833	.666
L. F. OF TREAD PER INCH OF STAIRWAY WIDTH		.583	.666	.75	.583	.666	.75	.583	.666	.75	.583
FRAMING SQUARE SETTINGS FOR CARRIAGE CUTS	TONGUE	7 11/16"	6 13/16"	6 1/8"	7 11/16"	6 13/16"	6 1/8"	7 11/16"	6 7/8"	6 3/16"	7 3/4"
	BODY	9 13/16"	10 11/16"	11 3/8"	9 13/16"	10 11/16"	11 3/8"	9 13/16"	10 5/8"	11 5/16"	9 3/4"

FLOOR TO FLOOR RISE		5'1 3/4"	5'1 3/4"	5'1 7/8"	5'1 7/8"	5'1 7/8"	5'2"	5'2"	5'2"	5'2 1/8"	5'2 1/8"
NUMBER OF RISERS		9	10	8	9	10	8	9	10	8	9
HEIGHT OF EACH RISER		6 7/8"	6 3/16"	7 3/4"	6 7/8"	6 3/16"	7 3/4"	6 7/8"	6 3/16"	7 3/4"	6 7/8"
TOTAL HEIGHT OF RISERS		5'1 7/8"	5'1 7/8"	5'2"	5'1 7/8"	5'1 7/8"	5'2"	5'1 7/8"	5'1 7/8"	5'2"	5'1 7/8"
RISERS OVERAGE (+) OR UNDERAGE (—)		+1/8"	+1/8"	+1/8"	0	0	0	−1/8"	−1/8"	−1/8"	−1/4"
NUMBER OF TREADS		8	9	7	8	9	7	8	9	7	8
WIDTH OF EACH TREAD		10 5/8"	11 5/16"	9 3/4"	10 5/8"	11 5/16"	9 3/4"	10 5/8"	11 5/16"	9 3/4"	10 5/8"
TOTAL RUN OF STAIRWAY		7'1"	8'5 13/16"	5'8 1/4"	7'1"	8'5 13/16"	5'8 1/4"	7'1"	8'5 13/16"	5'8 1/4"	7'1"
MAXIMUM RUN WITH 6'-8" HEADROOM		1'9 1/4"	1'10 5/8"	1'7 1/2"	1'9 1/4"	1'10 5/8"	1'7 1/2"	1'9 1/4"	1'10 5/8"	1'7 1/2"	1'9 1/4"
ANGLE OF INCLINE		32°54'	28°41'	38°29'	32°54'	28°41'	38°29'	32°54'	28°41'	38°29'	32°54'
LENGTH OF CARRIAGE		8'5 1/4"	9'8 1/16"	7'3 3/16"	8'5 1/4"	9'8 1/16"	7'3 3/16"	8'5 1/4"	9'8 1/16"	7'3 3/16"	8'5 1/4"
L. F. OF RISER PER INCH OF STAIRWAY WIDTH		.75	.833	.666	.75	.833	.666	.75	.833	.666	.75
L. F. OF TREAD PER INCH OF STAIRWAY WIDTH		.666	.75	.583	.666	.75	.583	.666	.75	.583	.666
FRAMING SQUARE SETTINGS FOR CARRIAGE CUTS	TONGUE	6 7/8"	6 3/16"	7 3/4"	6 7/8"	6 3/16"	7 3/4"	6 7/8"	6 3/16"	7 3/4"	6 7/8"
	BODY	10 5/8"	11 5/16"	9 3/4"	10 5/8"	11 5/16"	9 3/4"	10 5/8"	11 5/16"	9 3/4"	10 5/8"

FLOOR TO FLOOR RISE	5'2 1/8"	5'2 1/4"	5'2 1/4"	5'2 1/4"	5'2 3/8"	5'2 3/8"	5'2 3/8"	5'2 1/2"	5'2 1/2"	5'2 1/2"
NUMBER OF RISERS	10	8	9	10	8	9	10	8	9	10
HEIGHT OF EACH RISER	6 3/16"	7 13/16"	6 15/16"	6 1/4"	7 13/16"	6 15/16"	6 1/4"	7 13/16"	6 15/16"	6 1/4"
TOTAL HEIGHT OF RISERS	5'1 7/8"	5'2 1/2"	5'2 7/16"	5'2 1/2"	5'2 1/2"	5'2 7/16"	5'2 1/2"	5'2 1/2"	5'2 7/16"	5'2 1/2"
RISERS OVERAGE (+) OR UNDERAGE (—)	−1/4"	+1/4"	+3/16"	+1/4"	+1/8"	+1/16"	+1/8"	0	−1/16"	0
NUMBER OF TREADS	9	7	8	9	7	8	9	7	8	9
WIDTH OF EACH TREAD	11 5/16"	9 11/16"	10 9/16"	11 1/4"	9 11/16"	10 9/16"	11 1/4"	9 11/16"	10 9/16"	11 1/4"
TOTAL RUN OF STAIRWAY	8'5 13/16"	5'7 13/16"	7'1/2"	8'5 1/4"	5'7 13/16"	7'1/2"	8'5 1/4"	5'7 13/16"	7'1/2"	8'5 1/4"
MAXIMUM RUN WITH 6'-8" HEADROOM	1'10 5/8"	1'7 3/8"	1'9 1/8"	1'10 1/2"	1'7 3/8"	1'9 1/8"	1'10 1/2"	1'7 3/8"	1'9 1/8"	1'10 1/2"
ANGLE OF INCLINE	28°41'	38°53'	33°18'	29°3'	38°53'	33°18'	29°3'	38°53'	33°18'	29°3'
LENGTH OF CARRIAGE	9'8 1/16"	7'3 1/8"	8'5 3/32"	9'7 13/16"	7'3 1/8"	8'5 3/32"	9'7 13/16"	7'3 1/8"	8'5 3/32"	9'7 13/16"
L. F. OF RISER PER INCH OF STAIRWAY WIDTH	.833	.666	.75	.833	.666	.75	.833	.666	.75	.833
L. F. OF TREAD PER INCH OF STAIRWAY WIDTH	.75	.583	.666	.75	.583	.666	.75	.583	.666	.75
FRAMING SQUARE SETTINGS FOR CARRIAGE CUTS — TONGUE	6 3/16"	7 13/16"	6 15/16"	6 1/4"	7 13/16"	6 15/16"	6 1/4"	7 13/16"	6 15/16"	6 1/4"
FRAMING SQUARE SETTINGS FOR CARRIAGE CUTS — BODY	11 5/16"	9 11/16"	10 9/16"	11 1/4"	9 11/16"	10 9/16"	11 1/4"	9 11/16"	10 9/16"	11 1/4"

FLOOR TO FLOOR RISE		5'2 5/8"	5'2 5/8"	5'2 5/8"	5'2 3/4"	5'2 3/4"	5'2 3/4"	5'2 7/8"	5'2 7/8"	5'2 7/8"	5'3"
NUMBER OF RISERS		8	9	10	8	9	10	8	9	10	8
HEIGHT OF EACH RISER		7 13/16"	6 15/16"	6 1/4"	7 7/8"	7"	6 1/4"	7 7/8"	7"	6 5/16"	7 7/8"
TOTAL HEIGHT OF RISERS		5'2 1/2"	5'2 7/16"	5'2 1/2"	5'3"	5'3"	5'2 1/2"	5'3"	5'3"	5'3 1/8"	5'3"
RISERS OVERAGE (+) OR UNDERAGE (—)		−1/8"	−3/16"	−1/8"	+1/4"	+1/4"	−1/4"	+1/8"	+1/8"	+1/4"	0
NUMBER OF TREADS		7	8	9	7	8	9	7	8	9	7
WIDTH OF EACH TREAD		9 11/16"	10 9/16"	11 1/4"	9 5/8"	10 1/2"	11 1/4"	9 5/8"	10 1/2"	11 3/16"	9 5/8"
TOTAL RUN OF STAIRWAY		5'7 13/16"	7 1/2"	8'5 1/4"	5'7 3/8"	7"	8'5 1/4"	5'7 3/8"	7"	8'4 11/16"	5'7 3/8"
MAXIMUM RUN WITH 6'-8" HEADROOM		1'7 3/8"	1'9 1/8"	1'10 1/2"	1'7 1/4"	1'9"	1'10 1/2"	1'7 1/4"	1'9"	1'10 3/8"	1'7 1/4"
ANGLE OF INCLINE		38°53'	33°18'	29°3'	39°17'	33°41'	29°3'	39°17'	33°41'	29°26'	39°17'
LENGTH OF CARRIAGE		7'3 1/8"	8'5 3/32"	9'7 13/16"	7'3 1/16"	8'4 31/32"	9'7 13/16"	7'3 1/16"	8'4 31/32"	9'7 19/32"	7'3 1/16"
L. F. OF RISER PER INCH OF STAIRWAY WIDTH		.666	.75	.833	.666	.75	.833	.666	.75	.833	.666
L. F. OF TREAD PER INCH OF STAIRWAY WIDTH		.583	.666	.75	.583	.555	.75	.583	.666	.75	.583
FRAMING SQUARE SETTINGS FOR CARRIAGE CUTS	TONGUE	7 13/16"	6 15/16"	6 1/4"	7 7/8"	7"	6 1/4"	7 7/8"	7"	6 5/16"	7 7/8"
	BODY	9 11/16"	10 9/16"	11 1/4"	9 5/8"	10 1/2"	11 1/4"	9 5/8"	10 1/2"	11 3/16"	9 5/8"

FLOOR TO FLOOR RISE		5'3"	5'3"	5'3 1/8"	5'3 1/8"	5'3 1/8"	5'3 1/4"	5'3 1/4"	5'3 1/4"	5'3 3/8"	5'3 3/8"
NUMBER OF RISERS		9	10	8	9	10	8	9	10	8	9
HEIGHT OF EACH RISER		7"	6 5/16"	7 7/8"	7"	6 5/16"	7 15/16"	7"	6 5/16"	7 15/16"	7 1/16"
TOTAL HEIGHT OF RISERS		5'3"	5'3 1/8"	5'3"	5'3"	5'3 1/8"	5'3 1/2"	5'3"	5'3 1/8"	5'3 1/2"	5'3 9/16"
RISERS OVERAGE (+) OR UNDERAGE (—)		0	+1/8"	−1/8"	−1/8"	0	+1/4"	−1/4"	−1/8"	+1/8"	+3/16"
NUMBER OF TREADS		8	9	7	8	9	7	8	9	7	8
WIDTH OF EACH TREAD		10 1/2"	11 3/16"	9 5/8"	10 1/2"	11 3/16"	9 9/16"	10 1/2"	11 3/16"	9 9/16"	10 7/16"
TOTAL RUN OF STAIRWAY		7'	8'4 11/16"	5'7 3/8"	7'	8'4 11/16"	5'6 15/16"	7'	8'4 11/16"	5'6 15/16"	6'11 1/2"
MAXIMUM RUN WITH 6'-8" HEADROOM		1'9"	1'10 3/8"	1'7 1/4"	1'9"	1'10 3/8"	1'7 1/8"	1'9"	1'10 3/8"	1'7 1/8"	1'8 7/8"
ANGLE OF INCLINE		33°41'	29°26'	39°17'	33°41'	29°26'	39°42'	33°41'	29°26'	39°42'	34°5'
LENGTH OF CARRIAGE		8'4 31/32"	9'7 19/32"	7'3 1/16"	8'4 31/32"	9'7 19/32"	7'3"	8'4 31/32"	9'7 19/32"	7'3"	8'4 13/16"
L. F. OF RISER PER INCH OF STAIRWAY WIDTH		.75	.833	.666	.75	.833	.666	.75	.833	.666	.75
L. F. OF TREAD PER INCH OF STAIRWAY WIDTH		.666	.75	.583	.666	.75	.583	.666	.75	.583	.666
FRAMING SQUARE SETTINGS FOR CARRIAGE CUTS	TONGUE	7"	6 5/16"	7 7/8"	7"	6 5/16"	7 15/16"	7"	6 5/16"	7 15/16"	7 1/16"
	BODY	10 1/2"	11 3/16"	9 5/8"	10 1/2"	11 3/16"	9 9/16"	10 1/2"	11 3/16"	9 9/16"	10 7/16"

FLOOR TO FLOOR RISE	5'3 3/8"	5'3 1/2"	5'3 1/2"	5'3 1/2"	5'3 5/8"	5'3 5/8"	5'3 5/8"	5'3 3/4"	5'3 3/4"	5'3 3/4"
NUMBER OF RISERS	10	8	9	10	8	9	10	8	9	10
HEIGHT OF EACH RISER	6 5/16"	7 15/16"	7 1/16"	6 3/8"	7 15/16"	7 1/16"	6 3/8"	8"	7 1/16"	6 3/8"
TOTAL HEIGHT OF RISERS	5'3 1/8"	5'3 1/2"	5'3 9/16"	5'3 3/4"	5'3 1/2"	5'3 9/16"	5'3 3/4"	5'4"	5'3 9/16"	5'3 3/4"
RISERS OVERAGE (+) OR UNDERAGE (—)	−1/4"	0	+1/16"	+1/4"	−1/8"	−1/16"	+1/8"	+1/4"	−3/16"	0
NUMBER OF TREADS	9	7	8	9	7	8	9	7	8	9
WIDTH OF EACH TREAD	11 3/16"	9 9/16"	10 7/16"	11 1/8"	9 9/16"	10 7/16"	11 1/8"	9 1/2"	10 7/16"	11 1/8"
TOTAL RUN OF STAIRWAY	8'4 11/16"	5'6 15/16"	6'11 1/2"	8'4 1/8"	5'6 15/16"	6'11 1/2"	8'4 1/8"	5'6 1/2"	6'11 1/2"	8'4 1/8"
MAXIMUM RUN WITH 6'-8" HEADROOM	1'10 3/8"	1'7 1/8"	1'8 7/8"	1'10 1/4"	1'7 1/8"	1'8 7/8"	1'10 1/4"	9 1/2"	1'8 7/8"	1'10 1/4"
ANGLE OF INCLINE	29°26'	39°42'	34°5'	29°49'	39°42'	34°5'	29°49'	40°6'	34°5'	29°49'
LENGTH OF CARRIAGE	9'7 19/32"	7'3"	8'4 13/16"	9'7 13/32"	7'3"	8'4 13/16"	9'7 13/32"	7'2 15/16"	8'4 13/16"	9'7 13/32"
L. F. OF RISER PER INCH OF STAIRWAY WIDTH	.833	.666	.75	.833	.666	.75	.833	.666	.75	.833
L. F. OF TREAD PER INCH OF STAIRWAY WIDTH	.75	.583	.666	.75	.583	.666	.75	.583	.666	.75
FRAMING SQUARE SETTINGS FOR CARRIAGE CUTS — TONGUE	6 5/16"	7 15/16"	7 1/16"	6 3/8"	7 15/16"	7 1/16"	6 3/8"	8"	7 1/16"	6 3/8"
FRAMING SQUARE SETTINGS FOR CARRIAGE CUTS — BODY	11 3/16"	9 9/16"	10 7/16"	11 1/8"	9 9/16"	10 7/16"	11 1/8"	9 1/2"	10 7/16"	11 1/8"

FLOOR TO FLOOR RISE	5'3 7/8"	5'3 7/8"	5'3 7/8"	5'4"	5'4"	5'4"	5'4 1/8"	5'4 1/8"	5'4 1/8"	5'4 1/4"
NUMBER OF RISERS	8	9	10	8	9	10	8	9	10	8
HEIGHT OF EACH RISER	8"	7 1/8"	6 3/8"	8"	7 1/8"	6 3/8"	8"	7 1/8"	6 7/16"	8 1/16"
TOTAL HEIGHT OF RISERS	5'4"	5'4 1/8"	5'3 3/4"	5'4"	5'4 1/8"	5'3 3/4"	5'4"	5'4 1/8"	5'4 3/8"	5'4 1/2"
RISERS OVERAGE (+) OR UNDERAGE (—)	+1/8"	+1/4"	−1/8"	0	+1/8"	−1/4"	−1/8"	0	+1/4"	+1/4"
NUMBER OF TREADS	7	8	9	7	8	9	7	8	9	7
WIDTH OF EACH TREAD	9 1/2"	10 3/8"	11 1/8"	9 1/2"	10 3/8"	11 1/8"	9 1/2"	10 3/8"	11 1/16"	9 7/16"
TOTAL RUN OF STAIRWAY	5'6 1/2"	6'11"	8'4 1/8"	5'6 1/2"	6'11"	8'4 1/8"	5'6 1/2"	6'11"	8'3 9/16"	5'6 1/16"
MAXIMUM RUN WITH 6'-8" HEADROOM	9 1/2"	1'8 3/4"	1'10 1/4"	9 1/2"	1'8 3/4"	1'10 1/4"	9 1/2"	1'8 3/4"	1'10 1/8"	9 7/16"
ANGLE OF INCLINE	40°6'	34°29'	29°49'	40°6'	34°29'	29°49'	40°6'	34°29'	30°12'	40°30'
LENGTH OF CARRIAGE	7'2 15/16"	8'4 11/16"	9'7 13/32"	7'2 15/16"	8'4 11/16"	9'7 13/32"	7'2 15/16"	8'4 11/16"	9'7 3/16"	7'2 7/8"
L. F. OF RISER PER INCH OF STAIRWAY WIDTH	.666	.75	.833	.666	.75	.833	.666	.75	.833	.666
L. F. OF TREAD PER INCH OF STAIRWAY WIDTH	.583	.666	.75	.583	.666	.75	.583	.666	.75	.583
FRAMING SQUARE SETTINGS FOR CARRIAGE CUTS — TONGUE	8"	7 1/8"	6 3/8"	8"	7 1/8"	6 3/8"	8"	7 1/8"	6 7/16"	8 1/16"
FRAMING SQUARE SETTINGS FOR CARRIAGE CUTS — BODY	9 1/2"	10 3/8"	11 1/8"	9 1/2"	10 3/8"	11 1/8"	9 1/2"	10 3/8"	11 1/16"	9 7/16"

FLOOR TO FLOOR RISE		5'4 1/4"	5'4 1/4"	5'4 3/8"	5'4 3/8"	5'4 3/8"	5'4 1/2"	5'4 1/2"	5'4 1/2"	5'4 5/8"	5'4 5/8"
NUMBER OF RISERS		9	10	8	9	10	8	9	10	8	9
HEIGHT OF EACH RISER		7 1/8"	6 7/16"	8 1/16"	7 1/8"	6 7/16"	8 1/16"	7 3/16"	6 7/16"	8 1/16"	7 3/16"
TOTAL HEIGHT OF RISERS		5'4 1/8"	5'4 3/8"	5'4 1/2"	5'4 1/8"	5'4 3/8"	5'4 1/2"	5'4 11/16"	5'4 3/8"	5'4 1/2"	5'4 11/16"
RISERS OVERAGE (+) OR UNDERAGE (—)		−1/8"	+1/8"	+1/8"	−1/4"	0	0	+3/16"	−1/8"	−1/8"	+1/16"
NUMBER OF TREADS		8	9	7	8	9	7	8	9	7	8
WIDTH OF EACH TREAD		10 3/8"	11 1/16"	9 7/16"	10 3/8"	11 1/16"	9 7/16"	10 5/16"	11 1/16"	9 7/16"	10 5/16"
TOTAL RUN OF STAIRWAY		6'11"	8'3 9/16"	5'6 1/16"	6'11"	8'3 9/16"	5'6 1/16"	6'10 1/2"	8'3 9/16"	5'6 1/16"	6'10 1/2"
MAXIMUM RUN WITH 6'-8" HEADROOM		1'8 3/4"	1'10 1/8"	9 7/16"	1'8 3/4"	1'10 1/8"	9 7/16"	1'8 5/8"	1'10 1/8"	9 7/16"	1'8 5/8"
ANGLE OF INCLINE		34°29'	30°12'	40°30'	34°29'	30°12'	40°30'	34°53'	30°12'	40°30'	34°53'
LENGTH OF CARRIAGE		8'4 11/16"	9'7 3/16"	7'2 7/8"	8'4 11/16"	9'7 3/16"	7'2 7/8"	8'4 9/16"	9'7 3/16"	7'2 7/8"	8'4 9/16"
L. F. OF RISER PER INCH OF STAIRWAY WIDTH		.75	.833	.666	.75	.833	.666	.75	.833	.666	.75
L. F. OF TREAD PER INCH OF STAIRWAY WIDTH		.666	.75	.583	.666	.75	.583	.666	.75	.583	.666
FRAMING SQUARE SETTINGS FOR CARRIAGE CUTS	TONGUE	7 1/8"	6 7/16"	8 1/16"	7 1/8"	6 7/16"	8 1/16"	7 3/16"	6 7/16"	8 1/16"	7 3/16"
	BODY	10 3/8"	11 1/16"	9 7/16"	10 3/8"	11 1/16"	9 7/16"	10 5/16"	11 1/16"	9 7/16"	10 5/16"

FLOOR TO FLOOR RISE		5'4$\frac{5}{8}$"	5'4$\frac{3}{4}$"	5'4$\frac{3}{4}$"	5'4$\frac{3}{4}$"	5'4$\frac{7}{8}$"	5'4$\frac{7}{8}$"	5'4$\frac{7}{8}$"	5'5"	5'5"	5'5"
NUMBER OF RISERS		10	8	9	10	8	9	10	8	9	10
HEIGHT OF EACH RISER		6$\frac{7}{16}$"	8$\frac{1}{8}$"	7$\frac{3}{16}$"	6$\frac{1}{2}$"	8$\frac{1}{8}$"	7$\frac{3}{16}$"	6$\frac{1}{2}$"	8$\frac{1}{8}$"	7$\frac{1}{4}$"	6$\frac{1}{2}$"
TOTAL HEIGHT OF RISERS		5'4$\frac{3}{8}$"	5'5"	5'4$\frac{11}{16}$"	5'5"	5'5"	5'4$\frac{11}{16}$"	5'5"	5'5"	5'5$\frac{1}{4}$"	5'5"
RISERS OVERAGE (+) OR UNDERAGE (—)		−$\frac{1}{4}$"	+$\frac{1}{4}$"	−$\frac{1}{16}$"	+$\frac{1}{4}$"	+$\frac{1}{8}$"	−$\frac{3}{16}$"	+$\frac{1}{8}$"	0	+$\frac{1}{4}$"	0
NUMBER OF TREADS		9	7	8	9	7	8	9	7	8	9
WIDTH OF EACH TREAD		11$\frac{1}{16}$"	9$\frac{3}{8}$"	10$\frac{5}{16}$"	11"	9$\frac{3}{8}$"	10$\frac{5}{16}$"	11"	9$\frac{3}{8}$"	10$\frac{1}{4}$"	11"
TOTAL RUN OF STAIRWAY		8'3$\frac{9}{16}$"	5'5$\frac{5}{8}$"	6'10$\frac{1}{2}$"	8'3"	5'5$\frac{5}{8}$"	6'10$\frac{1}{2}$"	8'3"	5'5$\frac{5}{8}$"	6'10"	8'3"
MAXIMUM RUN WITH 6'-8" HEADROOM		1'10$\frac{1}{8}$"	9$\frac{3}{8}$"	1'8$\frac{5}{8}$"	1'10"	9$\frac{3}{8}$"	1'8$\frac{5}{8}$"	1'10"	9$\frac{3}{8}$"	1'8$\frac{1}{2}$"	1'10"
ANGLE OF INCLINE		30°12'	40°55'	34°53'	30°35'	40°55'	34°53'	30°35'	40°55'	35°16'	30°35'
LENGTH OF CARRIAGE		9'7$\frac{3}{16}$"	7'2$\frac{27}{32}$"	8'4$\frac{9}{16}$"	9'7"	7'2$\frac{27}{32}$"	8'4$\frac{9}{16}$"	9'7"	7'2$\frac{27}{32}$"	8'4$\frac{7}{16}$"	9'7"
L. F. OF RISER PER INCH OF STAIRWAY WIDTH		.833	.666	.75	.833	.666	.75	.833	.666	.75	.833
L. F. OF TREAD PER INCH OF STAIRWAY WIDTH		.75	.583	.666	.75	.583	.666	.75	.583	.666	.75
FRAMING SQUARE SETTINGS FOR CARRIAGE CUTS	TONGUE	6$\frac{7}{16}$"	8$\frac{1}{8}$"	7$\frac{3}{16}$"	6$\frac{1}{2}$"	8$\frac{1}{8}$"	7$\frac{3}{16}$"	6$\frac{1}{2}$"	8$\frac{1}{8}$"	7$\frac{1}{4}$"	6$\frac{1}{2}$"
	BODY	11$\frac{1}{16}$"	9$\frac{3}{8}$"	10$\frac{5}{16}$"	11"	9$\frac{3}{8}$"	10$\frac{5}{16}$"	11"	9$\frac{3}{8}$"	10$\frac{1}{4}$"	11"

FLOOR TO FLOOR RISE		5'5 1/8"	5'5 1/8"	5'5 1/8"	5'5 1/4"	5'5 1/4"	5'5 1/4"	5'5 3/8"	5'5 3/8"	5'5 3/8"	5'5 1/2"
NUMBER OF RISERS		8	9	10	8	9	10	8	9	10	8
HEIGHT OF EACH RISER		8 1/8"	7 1/4"	6 1/2"	8 3/16"	7 1/4"	6 1/2"	8 3/16"	7 1/4"	6 9/16"	8 3/16"
TOTAL HEIGHT OF RISERS		5'5"	5'5 1/4"	5'5"	5'5 1/2"	5'5 1/4"	5'5"	5'5 1/2"	5'5 1/4"	5'5 5/8"	5'5 1/2"
RISERS OVERAGE (+) OR UNDERAGE (—)		−1/8"	+1/8"	−1/8"	+1/4"	0	−1/4"	+1/8"	−1/8"	+1/4"	0
NUMBER OF TREADS		7	8	9	7	8	9	7	8	9	7
WIDTH OF EACH TREAD		9 3/8"	10 1/4"	11"	9 5/16"	10 1/4"	11"	9 5/16"	10 1/4"	10 15/16"	9 5/16"
TOTAL RUN OF STAIRWAY		5'5 5/8"	6'10"	8'3"	5'5 3/16"	6'10"	8'3"	5'5 3/16"	6'10"	8'2 7/16"	5'5 3/16"
MAXIMUM RUN WITH 6'-8" HEADROOM		9 3/8"	1'8 1/2"	1'10"	9 5/16"	1'8 1/2"	1'10"	9 5/16"	1'8 1/2"	1'9 7/8"	9 5/16"
ANGLE OF INCLINE		40°55'	35°16'	30°35'	41°19'	35°16'	30°35'	41°19'	35°16'	30°58'	41°19'
LENGTH OF CARRIAGE		7'2 27/32"	8'4 7/16"	9'7"	7'2 13/16"	8'4 7/16"	9'7"	7'2 13/16"	8'4 7/16"	9'6 13/16"	7'2 13/16"
L. F. OF RISER PER INCH OF STAIRWAY WIDTH		.666	.75	.833	.666	.75	.833	.666	.75	.833	.666
L. F. OF TREAD PER INCH OF STAIRWAY WIDTH		.583	.666	.75	.583	.666	.75	.583	.666	.75	.583
FRAMING SQUARE SETTINGS FOR CARRIAGE CUTS	TONGUE	8 1/8"	7 1/4"	6 1/2"	8 3/16"	7 1/4"	6 1/2"	8 3/16"	7 1/4"	6 9/16"	8 3/16"
	BODY	9 3/8"	10 1/4"	11"	9 5/16"	10 1/4"	11"	9 5/16"	10 1/4"	10 15/16"	9 5/16"

FLOOR TO FLOOR RISE	5'5 1/2"	5'5 1/2"	5'5 5/8"	5'5 5/8"	5'5 5/8"	5'5 3/4"	5'5 3/4"	5'5 3/4"	5'5 7/8"	5'5 7/8"
NUMBER OF RISERS	9	10	8	9	10	8	9	10	8	9
HEIGHT OF EACH RISER	7 1/4"	6 9/16"	8 3/16"	7 5/16"	6 9/16"	8 1/4"	7 5/16"	6 9/16"	8 1/4"	7 5/16"
TOTAL HEIGHT OF RISERS	5'5 1/4"	5'5 5/8"	5'5 1/2"	5'5 13/16"	5'5 5/8"	5'6"	5'5 13/16"	5'5 5/8"	5'6"	5'5 13/16"
RISERS OVERAGE (+) OR UNDERAGE (—)	−1/4"	+1/8"	−1/8"	+3/16"	0	+1/4"	+1/16"	−1/8"	+1/8"	−1/16"
NUMBER OF TREADS	8	9	7	8	9	7	8	9	7	8
WIDTH OF EACH TREAD	10 1/4"	10 15/16"	9 5/16"	10 3/16"	10 15/16"	9 1/4"	10 3/16"	10 15/16"	9 1/4"	10 3/16"
TOTAL RUN OF STAIRWAY	6'10"	8'2 7/16"	5'5 3/16"	6'9 1/2"	8'2 7/16"	5'4 3/4"	6'9 1/2"	8'2 7/16"	5'4 3/4"	6'9 1/2"
MAXIMUM RUN WITH 6'-8" HEADROOM	1'8 1/2"	1'9 7/8"	9 5/16"	1'8 3/8"	1'9 7/8"	9 1/4"	1'8 3/8"	1'9 7/8"	9 1/4"	1'8 3/8"
ANGLE OF INCLINE	35°16'	30°58'	41°19'	35°40'	30°58'	41°44'	35°40'	30°58'	41°44'	35°40'
LENGTH OF CARRIAGE	8'4 7/16"	9'6 13/16"	7'2 13/16"	8'4 5/16"	9'6 13/16"	7'2 3/4"	8'4 5/16"	9'6 13/16"	7'2 3/4"	8'4 5/16"
L. F. OF RISER PER INCH OF STAIRWAY WIDTH	.75	.833	.666	.75	.833	.666	.75	.833	.666	.75
L. F. OF TREAD PER INCH OF STAIRWAY WIDTH	.666	.75	.583	.666	.75	.583	.666	.75	.583	.666
FRAMING SQUARE SETTINGS FOR CARRIAGE CUTS — TONGUE	7 1/4"	6 9/16"	8 3/16"	7 5/16"	6 9/16"	8 1/4"	7 5/16"	6 9/16"	8 1/4"	7 5/16"
FRAMING SQUARE SETTINGS FOR CARRIAGE CUTS — BODY	10 1/4"	10 15/16"	9 5/16"	10 3/16"	10 15/16"	9 1/4"	10 3/16"	10 15/16"	9 1/4"	10 3/16"

FLOOR TO FLOOR RISE	5'5 7/8"	5'6"	5'6"	5'6"	5'6"	5'6 1/8"	5'6 1/8"	5'6 1/8"	5'6 1/8"	5'6 1/4"
NUMBER OF RISERS	10	8	9	10	11	8	9	10	11	9
HEIGHT OF EACH RISER	6 9/16"	8 1/4"	7 5/16"	6 5/8"	6"	8 1/4"	7 3/8"	6 5/8"	6"	7 3/8"
TOTAL HEIGHT OF RISERS	5'5 5/8"	5'6"	5'5 13/16"	5'6 1/4"	5'6"	5'6"	5'6 3/8"	5'6 1/4"	5'6"	5'6 3/8"
RISERS OVERAGE (+) OR UNDERAGE (—)	−1/4"	0	−3/16"	+1/4"	0	−1/8"	+1/4"	+1/8"	−1/8"	+1/8"
NUMBER OF TREADS	9	7	8	9	10	7	8	9	10	8
WIDTH OF EACH TREAD	10 15/16"	9 1/4"	10 3/16"	10 7/8"	11 1/2"	9 1/4"	10 1/8"	10 7/8"	11 1/2"	10 1/8"
TOTAL RUN OF STAIRWAY	8'2 7/16"	5'4 3/4"	6'9 1/2"	8'1 7/8"	9'7"	5'4 3/4"	6'9"	8'1 7/8"	9'7"	6'9"
MAXIMUM RUN WITH 6'-8" HEADROOM	1'9 7/8"	9 1/4"	1'8 3/8"	1'9 3/4"	1'11"	9 1/4"	1'8 1/4"	1'9 3/4"	1'11"	1'8 1/4"
ANGLE OF INCLINE	30°58'	41°44'	35°40'	31°21'	27°33'	41°44'	36°4'	31°21'	27°33'	36°4'
LENGTH OF CARRIAGE	9'6 13/16"	7'2 3/4"	8'4 5/16"	9'6 19/32"	10'9 23/32"	7'2 3/4"	8'4 7/32"	9'6 19/32"	10'9 23/32"	8'4 7/32"
L. F. OF RISER PER INCH OF STAIRWAY WIDTH	.833	.666	.75	.833	.916	.666	.833	.833	.916	.833
L. F. OF TREAD PER INCH OF STAIRWAY WIDTH	.75	.583	.666	.75	.833	.583	.666	.75	.833	.666
FRAMING SQUARE SETTINGS FOR CARRIAGE CUTS — TONGUE	6 9/16"	8 1/4"	7 5/16"	6 5/8"	6"	8 1/4"	7 3/8"	6 5/8"	6"	7 3/8"
FRAMING SQUARE SETTINGS FOR CARRIAGE CUTS — BODY	10 15/16"	9 1/4"	10 3/16"	10 7/8"	11 1/2"	9 1/4"	10 1/8"	10 7/8"	11 1/2"	10 1/8"

FLOOR TO FLOOR RISE		5'6 1/4"	5'6 1/4"	5'6 3/8"	5'6 3/8"	5'6 3/8"	5'6 1/2"	5'6 1/2"	5'6 1/2"	5'6 5/8"	5'6 5/8"
NUMBER OF RISERS		10	11	9	10	11	9	10	11	9	10
HEIGHT OF EACH RISER		6 5/8"	6"	7 3/8"	6 5/8"	6 1/16"	7 3/8"	6 5/8"	6 1/16"	7 3/8"	6 11/16"
TOTAL HEIGHT OF RISERS		5'6 1/4"	5'6"	5'6 3/8"	5'6 1/4"	5'6 11/16"	5'6 3/8"	5'6 1/4"	5'6 11/16"	5'6 3/8"	5'6 7/8"
RISERS OVERAGE (+) OR UNDERAGE (—)		0	−1/4"	0	−1/8"	+5/16"	−1/8"	−1/4"	+3/16"	−1/4"	+1/4"
NUMBER OF TREADS		9	10	8	9	10	8	9	10	8	9
WIDTH OF EACH TREAD		10 7/8"	11 1/2"	10 1/8"	10 7/8"	11 7/16"	10 1/8"	10 7/8"	11 7/16"	10 1/8"	10 13/16"
TOTAL RUN OF STAIRWAY		8'1 7/8"	9'7"	6'9"	8'1 7/8"	9'6 3/8"	6'9"	8'1 7/8"	9'6 3/8"	6'9"	8'1 5/16"
MAXIMUM RUN WITH 6'-8" HEADROOM		1'9 3/4"	1'11"	1'8 1/4"	1'9 3/4"	1'10 7/8"	1'8 1/4"	1'9 3/4"	1'10 7/8"	1'8 1/4"	1'9 5/8"
ANGLE OF INCLINE		31°21'	27°33'	36°4'	31°21'	27°56'	36°4'	31°21'	27°56'	36°4'	31°44'
LENGTH OF CARRIAGE		9'6 19/32"	10'9 23/32"	8'4 7/32"	9'6 19/32"	10'9 7/16"	8'4 7/32"	9'6 19/32"	10'9 7/16"	8'4 7/32"	9'6 13/32"
L. F. OF RISER PER INCH OF STAIRWAY WIDTH		.833	.916	.75	.833	.916	.75	.833	.916	.75	.833
L. F. OF TREAD PER INCH OF STAIRWAY WIDTH		.75	.833	.666	.75	.833	.666	.75	.833	.666	.75
FRAMING SQUARE SETTINGS FOR CARRIAGE CUTS	TONGUE	6 5/8"	6"	7 3/8"	6 5/8"	6 1/16"	7 3/8"	6 5/8"	6 1/16"	7 3/8"	6 11/16"
	BODY	10 7/8"	11 1/2"	10 1/8"	10 7/8"	11 7/16"	10 1/8"	10 7/8"	11 7/16"	10 1/8"	10 13/16"

FLOOR TO FLOOR RISE		5'$6\frac{5}{8}$"	5'$6\frac{3}{4}$"	5'$6\frac{3}{4}$"	5'$6\frac{3}{4}$"	5'$6\frac{7}{8}$"	5'$6\frac{7}{8}$"	5'$6\frac{7}{8}$"	5'7"	5'7"	5'7"
NUMBER OF RISERS		11	9	10	11	9	10	11	9	10	11
HEIGHT OF EACH RISER		$6\frac{1}{16}$"	$7\frac{7}{16}$"	$6\frac{11}{16}$"	$6\frac{1}{16}$"	$7\frac{7}{16}$"	$6\frac{11}{16}$"	$6\frac{1}{16}$"	$7\frac{7}{16}$"	$6\frac{11}{16}$"	$6\frac{1}{16}$"
TOTAL HEIGHT OF RISERS		5'$6\frac{11}{16}$"	5'$6\frac{15}{16}$"	5'$6\frac{7}{8}$"	5'$6\frac{11}{16}$"	5'$6\frac{15}{16}$"	5'$6\frac{7}{8}$"	5'$6\frac{11}{16}$"	5'$6\frac{15}{16}$"	5'$6\frac{7}{8}$"	5'$6\frac{11}{16}$"
RISERS OVERAGE (+) OR UNDERAGE (—)		$+\frac{1}{16}$"	$+\frac{3}{16}$"	$+\frac{1}{8}$"	$-\frac{1}{16}$"	$+\frac{1}{16}$"	0	$-\frac{3}{16}$"	$-\frac{1}{16}$"	$-\frac{1}{8}$"	$-\frac{5}{16}$"
NUMBER OF TREADS		10	8	9	10	8	9	10	8	9	10
WIDTH OF EACH TREAD		$11\frac{7}{16}$"	$10\frac{1}{16}$"	$10\frac{13}{16}$"	$11\frac{7}{16}$"	$10\frac{1}{16}$"	$10\frac{13}{16}$"	$11\frac{7}{16}$"	$10\frac{1}{16}$"	$10\frac{13}{16}$"	$11\frac{7}{16}$"
TOTAL RUN OF STAIRWAY		9'$6\frac{3}{8}$"	6'$8\frac{1}{2}$"	8'$1\frac{5}{16}$"	9'$6\frac{3}{8}$"	6'$8\frac{1}{2}$"	8'$1\frac{5}{16}$"	9'$6\frac{3}{8}$"	6'$8\frac{1}{2}$"	8'$1\frac{5}{16}$"	9'$6\frac{3}{8}$"
MAXIMUM RUN WITH 6'-8" HEADROOM		1'$10\frac{7}{8}$"	1'$8\frac{1}{8}$"	1'$9\frac{5}{8}$"	1'$10\frac{7}{8}$"	1'$8\frac{1}{8}$"	1'$9\frac{5}{8}$"	1'$10\frac{7}{8}$"	1'$8\frac{1}{8}$"	1'$9\frac{5}{8}$"	1'$10\frac{7}{8}$"
ANGLE OF INCLINE		27°56'	36°28'	31°44'	27°56'	36°28'	31°44'	27°56'	36°28'	31°44'	27°56'
LENGTH OF CARRIAGE		10'$9\frac{7}{16}$"	8'$4\frac{3}{32}$"	9'$6\frac{13}{32}$"	10'$9\frac{7}{16}$"	8'$4\frac{3}{32}$"	9'$6\frac{13}{32}$"	10'$9\frac{7}{16}$"	8'$4\frac{3}{32}$"	9'$6\frac{13}{32}$"	10'$9\frac{7}{16}$"
L. F. OF RISER PER INCH OF STAIRWAY WIDTH		.916	.75	.833	.916	.75	.833	.916	.75	.833	.916
L. F. OF TREAD PER INCH OF STAIRWAY WIDTH		.833	.666	.75	.833	.666	.75	.833	.666	.75	.833
FRAMING SQUARE SETTINGS FOR CARRIAGE CUTS	TONGUE	$6\frac{1}{16}$"	$7\frac{7}{16}$"	$6\frac{11}{16}$"	$6\frac{1}{16}$"	$7\frac{7}{16}$"	$6\frac{11}{16}$"	$6\frac{1}{16}$"	$7\frac{7}{16}$"	$6\frac{11}{16}$"	$6\frac{1}{16}$"
	BODY	$11\frac{7}{16}$"	$10\frac{1}{16}$"	$10\frac{13}{16}$"	$11\frac{7}{16}$"	$10\frac{1}{16}$"	$10\frac{13}{16}$"	$11\frac{7}{16}$"	$10\frac{1}{16}$"	$10\frac{13}{16}$"	$11\frac{7}{16}$"

FLOOR TO FLOOR RISE	5'7 1/8"	5'7 1/8"	5'7 1/8"	5'7 1/4"	5'7 1/4"	5'7 1/4"	5'7 3/8"	5'7 3/8"	5'7 3/8"	5'7 1/2"
NUMBER OF RISERS	9	10	11	9	10	11	9	10	11	9
HEIGHT OF EACH RISER	7 7/16"	6 11/16"	6 1/8"	7 1/2"	6 3/4"	6 1/8"	7 1/2"	6 3/4"	6 1/8"	7 1/2"
TOTAL HEIGHT OF RISERS	5'6 15/16"	5'6 7/8"	5'7 3/8"	5'7 1/2"	5'7 1/2"	5'7 3/8"	5'7 1/2"	5'7 1/2"	5'7 3/8"	5'7 1/2"
RISERS OVERAGE (+) OR UNDERAGE (—)	−3/16"	−1/4"	+1/4"	+1/4"	+1/4"	+1/8"	+1/8"	+1/8"	0	0
NUMBER OF TREADS	8	9	10	8	9	10	8	9	10	8
WIDTH OF EACH TREAD	10 1/16"	10 13/16"	11 3/8"	10"	10 3/4"	11 3/8"	10"	10 3/4"	11 3/8"	10"
TOTAL RUN OF STAIRWAY	6'8 1/2"	8'1 5/16"	9'5 3/4"	6'8"	8' 3/4"	9'5 3/4"	6'8"	8' 3/4"	9'5 3/4"	6'8"
MAXIMUM RUN WITH 6'-8" HEADROOM	1'8 5/8"	1'9 5/8"	1'10 3/4"	1'8"	1'9 1/2"	1'10 3/4"	1'8"	1'9 1/2"	1'10 3/4"	1'8"
ANGLE OF INCLINE	36°28'	31°44'	28°18'	36°52'	32°7'	28°18'	36°52'	32°7'	28°18'	36°52'
LENGTH OF CARRIAGE	8'4 3/32"	9'6 13/32"	10'9 3/16"	8'4"	9'6 1/4"	10'9 3/16"	8'4"	9'6 1/4"	10'9 3/16"	8'4"
L. F. OF RISER PER INCH OF STAIRWAY WIDTH	.75	.833	.916	.75	.833	.916	.75	.833	.916	.75
L. F. OF TREAD PER INCH OF STAIRWAY WIDTH	.666	.75	.833	.666	.75	.833	.666	.75	.833	.666
FRAMING SQUARE SETTINGS FOR CARRIAGE CUTS — TONGUE	7 7/16"	6 11/16"	6 1/8"	7 1/2"	6 3/4"	6 1/8"	7 1/2"	6 3/4"	6 1/8"	7 1/2"
FRAMING SQUARE SETTINGS FOR CARRIAGE CUTS — BODY	10 1/16"	10 13/16"	11 3/8"	10"	10 3/4"	11 3/8"	10"	10 3/4"	11 3/8"	10"

FLOOR TO FLOOR RISE	5'7 1/2"	5'7 1/2"	5'7 5/8"	5'7 5/8"	5'7 5/8"	5'7 3/4"	5'7 3/4"	5'7 3/4"	5'7 7/8"	5'7 7/8"
NUMBER OF RISERS	10	11	9	10	11	9	10	11	9	10
HEIGHT OF EACH RISER	6 3/4"	6 1/8"	7 1/2"	6 3/4"	6 1/8"	7 1/2"	6 3/4"	6 3/16"	7 9/16"	6 13/16"
TOTAL HEIGHT OF RISERS	5'7 1/2"	5'7 3/8"	5'7 1/2"	5'7 1/2"	5'7 3/8"	5'7 1/2"	5'7 1/2"	5'8 1/16"	5'8 1/16"	5'8 1/8"
RISERS OVERAGE (+) OR UNDERAGE (—)	0	−1/8"	−1/8"	−1/8"	−1/4"	−1/4"	−1/4"	+5/16"	+3/16"	+1/4"
NUMBER OF TREADS	9	10	8	9	10	8	9	10	8	9
WIDTH OF EACH TREAD	10 3/4"	11 3/8"	10"	10 3/4"	11 3/8"	10"	10 3/4"	11 5/16"	9 15/16"	10 11/16"
TOTAL RUN OF STAIRWAY	8' 3/4"	9'5 3/4"	6'8"	8' 3/4"	9'5 3/4"	6'8"	8' 3/4"	9'5 1/8"	6'7 1/2"	8' 3/16"
MAXIMUM RUN WITH 6'-8" HEADROOM	1'9 1/2"	1'10 3/4"	1'8"	1'9 1/2"	1'10 3/4"	1'8"	1'9 1/2"	1'10 5/8"	1'7 7/8"	1'9 3/8"
ANGLE OF INCLINE	32°7'	28°18'	36°52'	32°7'	28°18'	36°52'	32°7'	28°41'	37°16'	32°31'
LENGTH OF CARRIAGE	9'6 1/4"	10'9 3/16"	8'4"	9'6 1/4"	10'9 3/16"	8'4"	9'6 1/4"	10'8 15/16"	8'3 29/32"	9'6 1/16"
L. F. OF RISER PER INCH OF STAIRWAY WIDTH	.833	.916	.75	.833	.916	.75	.833	.916	.75	.833
L. F. OF TREAD PER INCH OF STAIRWAY WIDTH	.75	.833	.666	.75	.833	.666	.75	.833	.666	.75
FRAMING SQUARE SETTINGS FOR CARRIAGE CUTS — TONGUE	6 3/4"	6 1/8"	7 1/2"	6 3/4"	6 1/8"	7 1/2"	6 3/4"	6 3/16"	7 9/16"	6 13/16"
FRAMING SQUARE SETTINGS FOR CARRIAGE CUTS — BODY	10 3/4"	11 3/8"	10"	10 3/4"	11 3/8"	10"	10 3/4"	11 5/16"	9 15/16"	10 11/16"

FLOOR TO FLOOR RISE		5'7 7/8"	5'8"	5'8"	5'8"	5'8 1/8"	5'8 1/8"	5'8 1/8"	5'8 1/4"	5'8 1/4"	5'8 1/4"
NUMBER OF RISERS		11	9	10	11	9	10	11	9	10	11
HEIGHT OF EACH RISER		6 3/16"	7 9/16"	6 13/16"	6 3/16"	7 9/16"	6 13/16"	6 3/16"	7 9/16"	6 13/16"	6 3/16"
TOTAL HEIGHT OF RISERS		5'8 1/16"	5'8 1/16"	5'8 1/8"	5'8 1/16"	5'8 1/16"	5'8 1/8"	5'8 1/16"	5'8 1/16"	5'8 1/8"	5'8 1/16"
RISERS OVERAGE (+) OR UNDERAGE (—)		+3/16"	+1/16"	+1/8"	+1/16"	−1/16"	0	−1/16"	−3/16"	−1/8"	−3/16"
NUMBER OF TREADS		10	8	9	10	8	9	10	8	9	10
WIDTH OF EACH TREAD		11 5/16"	9 15/16"	10 11/16"	11 5/16"	9 15/16"	10 11/16"	11 5/16"	9 15/16"	10 11/16"	11 5/16"
TOTAL RUN OF STAIRWAY		9'5 1/8"	6'7 1/2"	8'3/16"	9'5 1/8"	6'7 1/2"	8'3/16"	9'5 1/8"	6'7 1/2"	8'3/16"	9'5 1/8"
MAXIMUM RUN WITH 6'-8" HEADROOM		1'10 5/8"	1'7 7/8"	1'9 3/8"	1'10 5/8"	1'7 7/8"	1'9 3/8"	1'10 5/8"	1'7 7/8"	1'9 3/8"	1'10 5/8"
ANGLE OF INCLINE		28°41'	37°16'	32°31'	28°41'	37°16'	32°31'	28°41'	37°16'	32°31'	28°41'
LENGTH OF CARRIAGE		10'8 15/16"	8'3 29/32"	9'6 1/16"	10'8 15/16"	8'3 29/32"	9'6 1/16"	10'8 15/16"	8'3 29/32"	9'6 1/16"	10'8 15/16"
L. F. OF RISER PER INCH OF STAIRWAY WIDTH		.916	.75	.833	.916	.75	.833	.916	.75	.833	.916
L. F. OF TREAD PER INCH OF STAIRWAY WIDTH		.833	.666	.75	.833	.666	.75	.833	.666	.75	.833
FRAMING SQUARE SETTINGS FOR CARRIAGE CUTS	TONGUE	6 3/16"	7 9/16"	6 13/16"	6 3/16"	7 9/16"	6 13/16"	6 3/16"	7 9/16"	6 13/16"	6 3/16"
	BODY	11 5/16"	9 15/16"	10 11/16"	11 5/16"	9 15/16"	10 11/16"	11 5/16"	9 15/16"	10 11/16"	11 5/16"

FLOOR TO FLOOR RISE	5'8 3/8"	5'8 3/8"	5'8 3/8"	5'8 1/2"	5'8 1/2"	5'8 1/2"	5'8 5/8"	5'8 5/8"	5'8 5/8"	5'8 3/4"
NUMBER OF RISERS	9	10	11	9	10	11	9	10	11	9
HEIGHT OF EACH RISER	7 5/8"	6 13/16"	6 3/16"	7 5/8"	6 7/8"	6 1/4"	7 5/8"	6 7/8"	6 1/4"	7 5/8"
TOTAL HEIGHT OF RISERS	5'8 5/8"	5'8 1/8"	5'8 1/16"	5'8 5/8"	5'8 3/4"	5'8 3/4"	5'8 5/8"	5'8 3/4"	5'8 3/4"	5'8 5/8"
RISERS OVERAGE (+) OR UNDERAGE (—)	+1/4"	−1/4"	−5/16"	+1/8"	+1/4"	+1/4"	0	+1/8"	+1/8"	−1/8"
NUMBER OF TREADS	8	9	10	8	9	10	8	9	10	8
WIDTH OF EACH TREAD	9 7/8"	10 11/16"	11 5/16"	9 7/8"	10 5/8"	11 1/4"	9 7/8"	10 5/8"	11 1/4"	9 7/8"
TOTAL RUN OF STAIRWAY	6'7"	8' 3/16"	9'5 1/8"	6'7"	7'11 5/8"	9'4 1/2"	6'7"	7'11 5/8"	9'4 1/2"	6'7"
MAXIMUM RUN WITH 6'-8" HEADROOM	1'7 3/4"	1'9 3/8"	1'10 5/8"	1'7 3/4"	1'9 1/4"	1'10 1/2"	1'7 3/4"	1'9 1/4"	1'10 1/2"	1'7 3/4"
ANGLE OF INCLINE	37°40'	32°31'	28°41'	37°40'	32°54'	29°3'	37°40'	32°54'	29°3'	37°40'
LENGTH OF CARRIAGE	8'3 13/16"	9'6 1/16"	10'8 15/16"	8'3 13/16"	9'5 29/32"	10'8 11/16"	8'3 13/16"	9'5 29/32"	10'8 11/16"	8'3 13/16"
L. F. OF RISER PER INCH OF STAIRWAY WIDTH	.75	.833	.916	.75	.833	.916	.75	.833	.916	.75
L. F. OF TREAD PER INCH OF STAIRWAY WIDTH	.666	.75	.833	.666	.75	.833	.666	.75	.833	.666
FRAMING SQUARE SETTINGS FOR CARRIAGE CUTS — TONGUE	7 5/8"	6 13/16"	6 3/16"	7 5/8"	6 7/8"	6 1/4"	7 5/8"	6 7/8"	6 1/4"	7 5/8"
FRAMING SQUARE SETTINGS FOR CARRIAGE CUTS — BODY	9 7/8"	10 11/16"	11 5/16"	9 7/8"	10 5/8"	11 1/4"	9 7/8"	10 5/8"	11 1/4"	9 7/8"

FLOOR TO FLOOR RISE	5'8 3/4"	5'8 3/4"	5'8 7/8"	5'8 7/8"	5'8 7/8"	5'9"	5'9"	5'9"	5'9 1/8"	5'9 1/8"
NUMBER OF RISERS	10	11	9	10	11	9	10	11	9	10
HEIGHT OF EACH RISER	6 7/8"	6 1/4"	7 5/8"	6 7/8"	6 1/4"	7 11/16"	6 7/8"	6 1/4"	7 11/16"	6 15/16"
TOTAL HEIGHT OF RISERS	5'8 3/4"	5'8 3/4"	5'8 5/8"	5'8 3/4"	5'8 3/4"	5'9 3/16"	5'8 3/4"	5'8 3/4"	5'9 3/16"	5'9 3/8"
RISERS OVERAGE (+) OR UNDERAGE (—)	0	0	−1/4"	−1/8"	−1/8"	+3/16"	−1/4"	−1/4"	+1/16"	+1/4"
NUMBER OF TREADS	9	10	8	9	10	8	9	10	8	9
WIDTH OF EACH TREAD	10 5/8"	11 1/4"	9 7/8"	10 5/8"	11 1/4"	9 13/16"	10 5/8"	11 1/4"	9 13/16"	10 9/16"
TOTAL RUN OF STAIRWAY	7'11 5/8"	9'4 1/2"	6'7"	7'11 5/8"	9'4 1/2"	6'6 1/2"	7'11 5/8"	9'4 1/2"	6'6 1/2"	7'11 1/16"
MAXIMUM RUN WITH 6'-8" HEADROOM	1'9 1/4"	1'10 1/2"	1'7 3/4"	1'9 1/4"	1'10 1/2"	1'7 5/8"	1'9 1/4"	1'10 1/2"	1'7 5/8"	1'9 1/8"
ANGLE OF INCLINE	32°54'	29°3'	37°40'	32°54'	29°3'	38°5'	32°54'	29°3'	38°5'	33°18'
LENGTH OF CARRIAGE	9'5 29/32"	10'8 11/16"	8'3 13/16"	9'5 29/32"	10'8 11/16"	8'3 23/32"	9'5 29/32"	10'8 11/16"	8'3 23/32"	9'5 23/32"
L. F. OF RISER PER INCH OF STAIRWAY WIDTH	.833	.916	.75	.833	.916	.75	.833	.916	.75	.833
L. F. OF TREAD PER INCH OF STAIRWAY WIDTH	.75	.833	.666	.75	.833	.666	.75	.833	.666	.75
FRAMING SQUARE SETTINGS FOR CARRIAGE CUTS TONGUE	6 7/8"	6 1/4"	7 5/8"	6 7/8"	6 1/4"	7 11/16"	6 7/8"	6 1/4"	7 11/16"	6 15/16"
BODY	10 5/8"	11 1/4"	9 7/8"	10 5/8"	11 1/4"	9 13/16"	10 5/8"	11 1/4"	9 13/16"	10 9/16"

FLOOR TO FLOOR RISE		5'9 1/8"	5'9 1/4"	5'9 1/4"	5'9 1/4"	5'9 3/8"	5'9 3/8"	5'9 3/8"	5'9 1/2"	5'9 1/2"	5'9 1/2"
NUMBER OF RISERS		11	9	10	11	9	10	11	9	10	11
HEIGHT OF EACH RISER		6 5/16"	7 11/16"	6 15/16"	6 5/16"	7 11/16"	6 15/16"	6 5/16"	7 3/4"	6 15/16"	6 5/16"
TOTAL HEIGHT OF RISERS		5'9 7/16"	5'9 3/16"	5'9 3/8"	5'9 7/16"	5'9 3/16"	5'9 3/8"	5'9 7/16"	5'9 3/4"	5'9 3/8"	5'9 7/16"
RISERS OVERAGE (+) OR UNDERAGE (—)		+5/16"	−1/16"	+1/8"	+3/16"	−3/16"	0	+1/16"	+1/4"	−1/8"	−1/16"
NUMBER OF TREADS		10	8	9	10	8	9	10	8	9	10
WIDTH OF EACH TREAD		11 3/16"	9 13/16"	10 9/16"	11 3/16"	9 13/16"	10 9/16"	11 3/16"	9 3/4"	10 9/16"	11 3/16"
TOTAL RUN OF STAIRWAY		9'3 7/8"	6'6 1/2"	7'11 1/16"	9'3 7/8"	6'6 1/2"	7'11 1/16"	9'3 7/8"	6'6"	7'11 1/16"	9'3 7/8"
MAXIMUM RUN WITH 6'-8" HEADROOM		1'10 3/8"	1'7 5/8"	1'9 1/8"	1'10 3/8"	1'7 5/8"	1'9 1/8"	1'10 3/8"	1'7 1/2"	1'9 1/8"	1'10 3/8"
ANGLE OF INCLINE		29°26'	38°5'	33°18'	29°26'	38°5'	33°18'	29°26'	38°29'	33°18'	29°26'
LENGTH OF CARRIAGE		10'8 15/32"	8'3 23/32"	9'5 23/32"	10'8 15/32"	8'3 23/32"	9'5 23/32"	10'8 15/32"	8'3 5/8"	9'5 23/32"	10'8 15/32"
L. F. OF RISER PER INCH OF STAIRWAY WIDTH		.916	.75	.833	.916	.75	.833	.916	.75	.833	.916
L. F. OF TREAD PER INCH OF STAIRWAY WIDTH		.833	.666	.75	.833	.666	.75	.833	.666	.75	.833
FRAMING SQUARE SETTINGS FOR CARRIAGE CUTS	TONGUE	6 5/16"	7 11/16"	6 15/16"	6 5/16"	7 11/16"	6 15/16"	6 5/16"	7 3/4"	6 15/16"	6 5/16"
	BODY	11 3/16"	9 13/16"	10 9/16"	11 3/16"	9 13/16"	10 9/16"	11 3/16"	9 3/4"	10 9/16"	11 3/16"

FLOOR TO FLOOR RISE		5'9 5/8"	5'9 5/8"	5'9 5/8"	5'9 3/4"	5'9 3/4"	5'9 3/4"	5'9 7/8"	5'9 7/8"	5'9 7/8"	5'10"
NUMBER OF RISERS		9	10	11	9	10	11	9	10	11	9
HEIGHT OF EACH RISER		7 3/4"	6 15/16"	6 5/16"	7 3/4"	7"	6 5/16"	7 3/4"	7"	6 3/8"	7 3/4"
TOTAL HEIGHT OF RISERS		5'9 3/4"	5'9 3/8"	5'9 7/16"	5'9 3/4"	5'10"	5'9 7/16"	5'9 3/4"	5'10"	5'10 1/8"	5'9 3/4"
RISERS OVERAGE (+) OR UNDERAGE (—)		+1/8"	−1/4"	−3/16"	0	+1/4"	−5/16"	−1/8"	+1/8"	+1/4"	−1/4"
NUMBER OF TREADS		8	9	10	8	9	10	8	9	10	8
WIDTH OF EACH TREAD		9 3/4"	10 9/16"	11 3/16"	9 3/4"	10 1/2"	11 3/16"	9 3/4"	10 1/2"	11 1/8"	9 3/4"
TOTAL RUN OF STAIRWAY		6'6"	7'11 1/16"	9'3 7/8"	6'6"	7'10 1/2"	9'3 7/8"	6'6"	7'10 1/2"	9'3 1/4"	6'6"
MAXIMUM RUN WITH 6'-8" HEADROOM		1'7 1/2"	1'9 1/8"	1'10 3/8"	1'7 1/2"	1'9"	1'10 3/8"	1'7 1/2"	1'9"	1'10 1/4"	1'7 1/2"
ANGLE OF INCLINE		38°29'	33°18'	29°26'	38°29'	33°41'	29°26'	38°29'	33°41'	29°49'	38°29'
LENGTH OF CARRIAGE		8'3 5/8"	9'5 23/32"	10'8 15/32"	8'3 5/8"	9'5 9/16"	10'8 15/32"	8'3 5/8"	9'5 9/16"	10'8 7/32"	8'3 5/8"
L. F. OF RISER PER INCH OF STAIRWAY WIDTH		.75	.833	.916	.75	.833	.916	.75	.833	.916	.75
L. F. OF TREAD PER INCH OF STAIRWAY WIDTH		.666	.75	.833	.666	.75	.833	.666	.75	.833	.666
FRAMING SQUARE SETTINGS FOR CARRIAGE CUTS	TONGUE	7 3/4"	6 15/16"	6 5/16"	7 3/4"	7"	6 5/16"	7 3/4"	7"	6 3/8"	7 3/4"
	BODY	9 3/4"	10 9/16"	11 3/16"	9 3/4"	10 1/2"	11 3/16"	9 3/4"	10 1/2"	11 1/8"	9 3/4"

FLOOR TO FLOOR RISE	5'10"	5'10"	5'10 1/8"	5'10 1/8"	5'10 1/8"	5'10 1/4"	5'10 1/4"	5'10 1/4"	5'10 3/8"	5'10 3/8"
NUMBER OF RISERS	10	11	9	10	11	9	10	11	9	10
HEIGHT OF EACH RISER	7"	6 3/8"	7 13/16"	7"	6 3/8"	7 13/16"	7"	6 3/8"	7 13/16"	7 1/16"
TOTAL HEIGHT OF RISERS	5'10"	5'10 1/8"	5'10 5/16"	5'10"	5'10 1/8"	5'10 5/16"	5'10"	5'10 1/8"	5'10 5/16"	5'10 5/8"
RISERS OVERAGE (+) OR UNDERAGE (—)	0	+1/8"	+3/16"	−1/8"	0	+1/16"	−1/4"	−1/8"	−1/16"	+1/4"
NUMBER OF TREADS	9	10	8	9	10	8	9	10	8	9
WIDTH OF EACH TREAD	10 1/2"	11 1/8"	9 11/16"	10 1/2"	11 1/8"	9 11/16"	10 1/2"	11 1/8"	9 11/16"	10 7/16"
TOTAL RUN OF STAIRWAY	7'10 1/2"	9'3 1/4"	6'5 1/2"	7'10 1/2"	9'3 1/4"	6'5 1/2"	7'10 1/2"	9'3 1/4"	6'5 1/2"	7'9 15/16"
MAXIMUM RUN WITH 6'-8" HEADROOM	1'9"	1'10 1/4"	1'7 3/8"	1'9"	1'10 1/4"	1'7 3/8"	1'9"	1'10 1/4"	1'7 3/8"	1'8 7/8"
ANGLE OF INCLINE	33°41'	29°49'	38°53'	33°41'	29°49'	38°53'	33°41'	29°49'	38°53'	34°5'
LENGTH OF CARRIAGE	9'5 9/16"	10'8 7/32"	8'3 9/16"	9'5 9/16"	10'8 7/32"	8'3 9/16"	9'5 9/16"	10'8 7/32"	8'3 9/16"	9'5 13/32"
L. F. OF RISER PER INCH OF STAIRWAY WIDTH	.833	.916	.75	.833	.916	.75	.833	.916	.75	.833
L. F. OF TREAD PER INCH OF STAIRWAY WIDTH	.75	.833	.666	.75	.833	.666	.75	.833	.666	.75
FRAMING SQUARE SETTINGS FOR CARRIAGE CUTS — TONGUE	7"	6 3/8"	7 13/16"	7"	6 3/8"	7 13/16"	7"	6 3/8"	7 13/16"	7 1/16"
FRAMING SQUARE SETTINGS FOR CARRIAGE CUTS — BODY	10 1/2"	11 1/8"	9 11/16"	10 1/2"	11 1/8"	9 11/16"	10 1/2"	11 1/8"	9 11/16"	10 7/16"

FLOOR TO FLOOR RISE	5'10 3/8"	5'10 1/2"	5'10 1/2"	5'10 1/2"	5'10 5/8"	5'10 5/8"	5'10 5/8"	5'10 3/4"	5'10 3/4"	5'10 3/4"
NUMBER OF RISERS	11	9	10	11	9	10	11	9	10	11
HEIGHT OF EACH RISER	6 3/8"	7 13/16"	7 1/16"	6 7/16"	7 7/8"	7 1/16"	6 7/16"	7 7/8"	7 1/16"	6 7/16"
TOTAL HEIGHT OF RISERS	5'10 1/8"	5'10 5/16"	5'10 5/8"	5'10 13/16"	5'10 7/8"	5'10 5/8"	5'10 13/16"	5'10 7/8"	5'10 5/8"	5'10 13/16"
RISERS OVERAGE (+) OR UNDERAGE (—)	−1/4"	−3/16"	+1/8"	+5/16"	+1/4"	0	+3/16"	+1/8"	−1/8"	+1/16"
NUMBER OF TREADS	10	8	9	10	8	9	10	8	9	10
WIDTH OF EACH TREAD	11 1/8"	9 11/16"	10 7/16"	11 1/16"	9 5/8"	10 7/16"	11 1/16"	9 5/8"	10 7/16"	11 1/16"
TOTAL RUN OF STAIRWAY	9'3 1/4"	6'5 1/2"	7'9 15/16"	9'2 5/8"	6'5"	7'9 15/16"	9'2 5/8"	6'5"	7'9 15/16"	9'2 5/8"
MAXIMUM RUN WITH 6'-8" HEADROOM	1'10 1/4"	1'7 3/8"	1'8 7/8"	1'10 1/8"	1'7 1/4"	1'8 7/8"	1'10 1/8"	1'7 1/4"	1'8 7/8"	1'10 1/8"
ANGLE OF INCLINE	29°49'	38°53'	34°5'	30°12'	39°17'	34°5'	30°12'	39°17'	34°5'	30°12'
LENGTH OF CARRIAGE	10'8 7/32"	8'3 9/16"	9'5 13/32"	10'8"	8'3 1/2"	9'5 13/32"	10'8"	8'3 1/2"	9'5 13/32"	10'8"
L. F. OF RISER PER INCH OF STAIRWAY WIDTH	.916	.75	.833	.916	.75	.833	.916	.75	.833	.916
L. F. OF TREAD PER INCH OF STAIRWAY WIDTH	.833	.666	.75	.833	.666	.75	.833	.666	.75	.833
FRAMING SQUARE SETTINGS FOR CARRIAGE CUTS — TONGUE	6 3/8"	7 13/16"	7 1/16"	6 7/16"	7 7/8"	7 1/16"	6 7/16"	7 7/8"	7 1/16"	6 7/16"
FRAMING SQUARE SETTINGS FOR CARRIAGE CUTS — BODY	11 1/8"	9 11/16"	10 7/16"	11 1/16"	9 5/8"	10 7/16"	11 1/16"	9 5/8"	10 7/16"	11 1/16"

FLOOR TO FLOOR RISE		5'10 7/8"	5'10 7/8"	5'10 7/8"	5'11"	5'11"	5'11"	5'11 1/8"	5'11 1/8"	5'11 1/8"	5'11 1/4"
NUMBER OF RISERS		9	10	11	9	10	11	9	10	11	9
HEIGHT OF EACH RISER		7 7/8"	7 1/16"	6 7/16"	7 7/8"	7 1/8"	6 7/16"	7 7/8"	7 1/8"	6 7/16"	7 15/16"
TOTAL HEIGHT OF RISERS		5'10 7/8"	5'10 5/8"	5'10 13/16"	5'10 7/8"	5'11 1/4"	5'10 13/16"	5'10 7/8"	5'11 1/4"	5'10 13/16"	5'11 7/16"
RISERS OVERAGE (+) OR UNDERAGE (—)		0	−1/4"	−1/16"	−1/8"	+1/4"	−3/16"	−1/4"	+1/8"	−5/16"	+3/16"
NUMBER OF TREADS		8	9	10	8	9	10	8	9	10	8
WIDTH OF EACH TREAD		9 5/8"	10 7/16"	11 1/16"	9 5/8"	10 3/8"	11 1/16"	9 5/8"	10 3/8"	11 1/16"	9 9/16"
TOTAL RUN OF STAIRWAY		6'5"	7'9 15/16"	9'2 5/8"	6'5"	7'9 3/8"	9'2 5/8"	6'5"	7'9 3/8"	9'2 5/8"	6'4 1/2"
MAXIMUM RUN WITH 6'-8" HEADROOM		1'7 1/4"	1'8 7/8"	1'10 1/8"	1'7 1/4"	1'8 3/4"	1'10 1/8"	1'7 1/4"	1'8 3/4"	1'10 1/8"	1'7 1/8"
ANGLE OF INCLINE		39°17'	34°5'	30°12'	39°17'	34°29'	30°12'	39°17'	34°29'	30°12'	39°42'
LENGTH OF CARRIAGE		8'3 1/2"	9'5 13/32"	10'8"	8'3 1/2"	9'5 9/32"	10'8"	8'3 1/2"	9'5 9/32"	10'8"	8'3 13/32"
L. F. OF RISER PER INCH OF STAIRWAY WIDTH		.75	.833	.916	.75	.833	.916	.75	.833	.916	.75
L. F. OF TREAD PER INCH OF STAIRWAY WIDTH		.666	.75	.833	.666	.75	.833	.666	.75	.833	.666
FRAMING SQUARE SETTINGS FOR CARRIAGE CUTS	TONGUE	7 7/8"	7 1/16"	6 7/16"	7 7/8"	7 1/8"	6 7/16"	7 7/8"	7 1/8"	6 7/16"	7 15/16"
	BODY	9 5/8"	10 7/16"	11 1/16"	9 5/8"	10 3/8"	11 1/16"	9 5/8"	10 3/8"	11 1/16"	9 9/16"

FLOOR TO FLOOR RISE		5'11 1/4"	5'11 1/4"	5'11 3/8"	5'11 3/8"	5'11 3/8"	5'11 1/2"	5'11 1/2"	5'11 1/2"	5'11 5/8"	5'11 5/8"
NUMBER OF RISERS		10	11	9	10	11	9	10	11	9	10
HEIGHT OF EACH RISER		7 1/8"	6 1/2"	7 15/16"	7 1/8"	6 1/2"	7 15/16"	7 1/8"	6 1/2"	7 15/16"	7 3/16"
TOTAL HEIGHT OF RISERS		5'11 1/4"	5'11 1/2"	5'11 7/16"	5'11 1/4"	5'11 1/2"	5'11 7/16"	5'11 1/4"	5'11 1/2"	5'11 7/16"	5'11 7/8"
RISERS OVERAGE (+) OR UNDERAGE (—)		0	+1/4"	+1/16"	−1/8"	+1/8"	−1/16"	−1/4"	0	−3/16"	+1/4"
NUMBER OF TREADS		9	10	8	9	10	8	9	10	8	9
WIDTH OF EACH TREAD		10 3/8"	11"	9 9/16"	10 3/8"	11"	9 9/16"	10 3/8"	11"	9 9/16"	10 5/16"
TOTAL RUN OF STAIRWAY		7'9 3/8"	9'2"	6'4 1/2"	7'9 3/8"	9'2"	6'4 1/2"	7'9 3/8"	9'2"	6'4 1/2"	7'8 13/16"
MAXIMUM RUN WITH 6'-8" HEADROOM		1'8 3/4"	1'10"	1'7 1/8"	1'8 3/4"	1'10"	1'7 1/8"	1'8 3/4"	1'10"	1'7 1/8"	1'8 5/8"
ANGLE OF INCLINE		34°29'	30°35'	39°42'	34°29'	30°35'	39°42'	34°29'	30°35'	39°42'	34°53'
LENGTH OF CARRIAGE		9'5 9/32"	10'7 25/32"	8'3 13/32"	9'5 9/32"	10'7 25/32"	8'3 13/32"	9'5 9/32"	10'7 25/32"	8'3 13/32"	9'5 1/8"
L. F. OF RISER PER INCH OF STAIRWAY WIDTH		.833	.916	.75	.833	.916	.75	.833	.916	.75	.833
L. F. OF TREAD PER INCH OF STAIRWAY WIDTH		.75	.833	.666	.75	.833	.666	.75	.833	.666	.75
FRAMING SQUARE SETTINGS FOR CARRIAGE CUTS	TONGUE	7 1/8"	6 1/2"	7 15/16"	7 1/8"	6 1/2"	7 15/16"	7 1/8"	6 1/2"	7 15/16"	7 3/16"
	BODY	10 3/8"	11"	9 9/16"	10 3/8"	11"	9 9/16"	10 3/8"	11"	9 9/16"	10 5/16"

FLOOR TO FLOOR RISE		5'11 5/8"	5'11 3/4"	5'11 3/4"	5'11 3/4"	5'11 7/8"	5'11 7/8"	5'11 7/8"	6'	6'	6'
NUMBER OF RISERS		11	9	10	11	9	10	11	9	10	11
HEIGHT OF EACH RISER		6 1/2"	8"	7 3/16"	6 1/2"	8"	7 3/16"	6 9/16"	8"	7 3/16"	6 9/16"
TOTAL HEIGHT OF RISERS		5'11 1/2"	6'	5'11 7/8"	5'11 1/2"	6'	5'11 7/8"	6' 3/16"	6'	5'11 7/8"	6' 3/16"
RISERS OVERAGE (+) OR UNDERAGE (—)		−1/8"	+1/4"	+1/8"	−1/4"	+1/8"	0	+5/16"	0	−1/8"	+3/16"
NUMBER OF TREADS		10	8	9	10	8	9	10	8	9	10
WIDTH OF EACH TREAD		11"	9 1/2"	10 5/16"	11"	9 1/2"	10 5/16"	10 15/16"	9 1/2"	10 5/16"	10 15/16"
TOTAL RUN OF STAIRWAY		9'2"	6'4"	7'8 13/16"	9'2"	6'4"	7'8 13/16"	9'1 3/8"	6'4"	7'8 13/16"	9'1 3/8"
MAXIMUM RUN WITH 6'-8" HEADROOM		1'10"	9 1/2"	1'8 5/8"	1'10"	9 1/2"	1'8 5/8"	1'9 7/8"	9 1/2"	1'8 5/8"	1'9 7/8"
ANGLE OF INCLINE		30°35'	40°6'	34°53'	30°35'	40°6'	34°53'	30°58'	40°6'	34°53'	30°58'
LENGTH OF CARRIAGE		10'7 25/32"	8'3 11/32"	9'5 1/8"	10'7 25/32"	8'3 11/32"	9'5 1/8"	10'7 9/16"	8'3 11/32"	9'5 1/8"	10'7 9/16"
L. F. OF RISER PER INCH OF STAIRWAY WIDTH		.916	.75	.833	.916	.75	.833	.916	.75	.833	.916
L. F. OF TREAD PER INCH OF STAIRWAY WIDTH		.833	.666	.75	.833	.666	.75	.833	.666	.75	.833
FRAMING SQUARE SETTINGS FOR CARRIAGE CUTS	TONGUE	6 1/2"	8"	7 3/16"	6 1/2"	8"	7 3/16"	6 9/16"	8"	7 3/16"	6 9/16"
	BODY	11"	9 1/2"	10 5/16"	11"	9 1/2"	10 5/16"	10 15/16"	9 1/2"	10 5/16"	10 15/16"

FLOOR TO FLOOR RISE	6'	6' 1/8"	6' 1/8"	6' 1/8"	6' 1/8"	6' 1/4"	6' 1/4"	6' 1/4"	6' 1/4"	6' 3/8"
NUMBER OF RISERS	12	9	10	11	12	9	10	11	12	9
HEIGHT OF EACH RISER	6"	8"	7 3/16"	6 9/16"	6"	8"	7 1/4"	6 9/16"	6"	8 1/16"
TOTAL HEIGHT OF RISERS	6'	6'	5' 11 7/8"	6' 3/16"	6'	6'	6' 1/2"	6' 3/16"	6'	6' 9/16"
RISERS OVERAGE (+) OR UNDERAGE (—)	0	−1/8"	−1/4"	+1/16"	−1/8"	−1/4"	+1/4"	−1/16"	−1/4"	+3/16"
NUMBER OF TREADS	11	8	9	10	11	8	9	10	11	8
WIDTH OF EACH TREAD	11 1/2"	9 1/2"	10 5/16"	10 15/16"	11 1/2"	9 1/2"	10 1/4"	10 15/16"	11 1/2"	9 7/16"
TOTAL RUN OF STAIRWAY	10' 6 1/2"	6' 4"	7' 8 13/16"	9' 1 3/8"	10' 6 1/2"	6' 4"	7' 8 1/4"	9' 1 3/8"	10' 6 1/2"	6' 3 1/2"
MAXIMUM RUN WITH 6'-8" HEADROOM	1' 11"	9 1/2"	1' 8 5/8"	1' 9 7/8"	1' 11"	9 1/2"	1' 8 1/2"	1' 9 7/8"	1' 11"	9 7/16"
ANGLE OF INCLINE	27°33'	40°6'	34°53'	30°58'	27°33'	40°6'	35°16'	30°58'	27°33'	40°30'
LENGTH OF CARRIAGE	11' 10 11/16"	8' 3 11/32"	9' 5 1/8"	10' 7 9/16"	11' 10 11/16"	8' 3 11/32"	9' 5"	10' 7 9/16"	11' 10 11/16"	8' 3 5/16"
L. F. OF RISER PER INCH OF STAIRWAY WIDTH	1.0	.75	.833	.916	1.0	75	.833	.916	1.0	.75
L. F. OF TREAD PER INCH OF STAIRWAY WIDTH	.916	.666	.75	.833	.916	.666	.75	.833	.916	.666
FRAMING SQUARE SETTINGS FOR CARRIAGE CUTS — TONGUE	6"	8"	7 3/16"	6 9/16"	6"	8"	7 1/4"	6 9/16"	6"	8 1/16"
FRAMING SQUARE SETTINGS FOR CARRIAGE CUTS — BODY	11 1/2"	9 1/2"	10 5/16"	10 15/16"	11 1/2"	9 1/2"	10 1/4"	10 15/16"	11 1/2"	9 7/16"

FLOOR TO FLOOR RISE	6'3/8"	6'3/8"	6'3/8"	6'1/2"	6'1/2"	6'1/2"	6'1/2"	6'5/8"	6'5/8"	6'5/8"
NUMBER OF RISERS	10	11	12	9	10	11	12	9	10	11
HEIGHT OF EACH RISER	7 1/4"	6 9/16"	6 1/16"	8 1/16"	7 1/4"	6 9/16"	6 1/16"	8 1/16"	7 1/4"	6 5/8"
TOTAL HEIGHT OF RISERS	6'1/2"	6'3/16"	6'3/4"	6'9/16"	6'1/2"	6'3/16"	6'3/4"	6'9/16"	6'1/2"	6'7/8"
RISERS OVERAGE (+) OR UNDERAGE (—)	+1/8"	−3/16"	+3/8"	+1/16"	0	−5/16"	+1/4"	−1/16"	−1/8"	+1/4"
NUMBER OF TREADS	9	10	11	8	9	10	11	8	9	10
WIDTH OF EACH TREAD	10 1/4"	10 15/16"	11 7/16"	9 7/16"	10 1/4"	10 15/16"	11 7/16"	9 7/16"	10 1/4"	10 7/8"
TOTAL RUN OF STAIRWAY	7'8 1/4"	9'1 3/8"	10'5 13/16"	6'3 1/2"	7'8 1/4"	9'1 3/8"	10'5 13/16"	6'3 1/2"	7'8 1/4"	9'3/4"
MAXIMUM RUN WITH 6'-8" HEADROOM	1'8 1/2"	1'9 7/8"	1'10 7/8"	9 7/16"	1'8 1/2"	1'9 7/8"	1'10 7/8"	9 7/16"	1'8 1/2"	1'9 3/4"
ANGLE OF INCLINE	35°16'	30°58'	27°56'	40°30'	35°16'	30°58'	27°56'	40°30'	35°16'	31°21'
LENGTH OF CARRIAGE	9'5"	10'7 9/16"	11'10 13/32"	8'3 5/16"	9'5"	10'7 9/16"	11'10 13/32"	8'3 5/16"	9'5"	10'7 11/32"
L. F. OF RISER PER INCH OF STAIRWAY WIDTH	.833	.916	1.0	.75	.833	.916	1.0	.75	.833	.916
L. F. OF TREAD PER INCH OF STAIRWAY WIDTH	.75	.833	.916	.666	.75	.833	.916	.666	.75	.833
FRAMING SQUARE SETTINGS FOR CARRIAGE CUTS — TONGUE	7 1/4"	6 9/16"	6 1/16"	8 1/16"	7 1/4"	6 9/16"	6 1/16"	8 1/16"	7 1/4"	6 5/8"
FRAMING SQUARE SETTINGS FOR CARRIAGE CUTS — BODY	10 1/4"	10 15/16"	11 7/16"	9 7/16"	10 1/4"	10 15/16"	11 7/16"	9 7/16"	10 1/4"	10 7/8"

FLOOR TO FLOOR RISE	6' 5/8"	6' 3/4"	6' 3/4"	6' 3/4"	6' 3/4"	6' 7/8"	6' 7/8"	6' 7/8"	6' 7/8"	6' 1"
NUMBER OF RISERS	12	9	10	11	12	9	10	11	12	9
HEIGHT OF EACH RISER	6 1/16"	8 1/16"	7 1/4"	6 5/8"	6 1/16"	8 1/8"	7 5/16"	6 5/8"	6 1/16"	8 1/8"
TOTAL HEIGHT OF RISERS	6' 3/4"	6' 9/16"	6' 1/2"	6' 7/8"	6' 3/4"	6' 1 1/8"	6' 1 1/8"	6' 7/8"	6' 3/4"	6' 1 1/8"
RISERS OVERAGE (+) OR UNDERAGE (—)	+1/8"	−3/16"	−1/4"	+1/8"	0	+1/4"	+1/4"	0	−1/8"	+1/8"
NUMBER OF TREADS	11	8	9	10	11	8	9	10	11	8
WIDTH OF EACH TREAD	11 7/16"	9 7/16"	10 1/4"	10 7/8"	11 7/16"	9 3/8"	10 3/16"	10 7/8"	11 7/16"	9 3/8"
TOTAL RUN OF STAIRWAY	10' 5 13/16"	6' 3 1/2"	7' 8 1/4"	9' 3/4"	10' 5 13/16"	6' 3"	7' 7 11/16"	9' 3/4"	10' 5 13/16"	6' 3"
MAXIMUM RUN WITH 6'-8" HEADROOM	1' 10 7/8"	9 7/16"	1' 8 1/2"	1' 9 3/4"	1' 10 7/8"	9 3/8"	1' 8 3/8"	1' 9 3/4"	1' 10 7/8"	9 3/8"
ANGLE OF INCLINE	27°56'	40°30'	35°16'	31°21'	27°56'	40°55'	35°40'	31°21'	27°56'	40°55'
LENGTH OF CARRIAGE	11' 10 13/32"	8' 3 5/16"	9' 5"	10' 7 11/32"	11' 10 13/16"	8' 3 1/4"	9' 4 7/8"	10' 7 11/32"	11' 10 13/32"	8' 3 1/4"
L. F. OF RISER PER INCH OF STAIRWAY WIDTH	1.0	.75	.833	.916	1.0	.75	.833	.916	1.0	.75
L. F. OF TREAD PER INCH OF STAIRWAY WIDTH	.916	.666	.75	.833	.916	.666	.75	.833	.916	.666
FRAMING SQUARE SETTINGS FOR CARRIAGE CUTS — TONGUE	6 1/16"	8 1/16"	7 1/4"	6 5/8"	6 1/16"	8 1/8"	7 5/16"	6 5/8"	6 1/16"	8 1/8"
FRAMING SQUARE SETTINGS FOR CARRIAGE CUTS — BODY	11 7/16"	9 7/16"	10 1/4"	10 7/8"	11 7/16"	9 3/8"	10 3/16"	10 7/8"	11 7/16"	9 3/8"

FLOOR TO FLOOR RISE	6'1"	6'1"	6'1"	6'1 1/8"	6'1 1/8"	6'1 1/8"	6'1 1/8"	6'1 1/4"	6'1 1/4"	6'1 1/4"
NUMBER OF RISERS	10	11	12	9	10	11	12	9	10	11
HEIGHT OF EACH RISER	7 5/16"	6 5/8"	6 1/16"	8 1/8"	7 5/16"	6 5/8"	6 1/8"	8 1/8"	7 5/16"	6 11/16"
TOTAL HEIGHT OF RISERS	6'1 1/8"	6'7/8"	6'3/4"	6'1 1/8"	6'1 1/8"	6'7/8"	6'1 1/2"	6'1 1/8"	6'1 1/8"	6'1 9/16"
RISERS OVERAGE (+) OR UNDERAGE (—)	+1/8"	−1/8"	−1/4"	0	0	−1/4"	+3/8"	−1/8"	−1/8"	+5/16"
NUMBER OF TREADS	9	10	11	8	9	10	11	8	9	10
WIDTH OF EACH TREAD	10 3/16"	10 7/8"	11 7/16"	9 3/8"	10 3/16"	10 7/8"	11 3/8"	9 3/8"	10 3/16"	10 13/16"
TOTAL RUN OF STAIRWAY	7'7 11/16"	9'3/4"	10'5 13/16"	6'3"	7'7 11/16"	9'3/4"	10'5 1/8"	6'3"	7'7 11/16"	9'1/8"
MAXIMUM RUN WITH 6'-8" HEADROOM	1'8 3/8"	1'9 3/4"	1'10 7/8"	9 3/8"	1'8 3/8"	1'9 3/4"	1'10 3/4"	9 3/8"	1'8 3/8"	1'9 5/8"
ANGLE OF INCLINE	35°40'	31°21'	27°56'	40°55'	35°40'	31°21'	28°18'	40°55'	35°40'	31°44'
LENGTH OF CARRIAGE	9'4 7/8"	10'7 11/32"	11'10 13/32"	8'3 1/4"	9'4 7/8"	10'7 11/32"	11'10 1/8"	8'3 1/4"	9'4 7/8"	10'7 1/8"
L. F. OF RISER PER INCH OF STAIRWAY WIDTH	.833	.916	1.0	.75	.833	.916	1.0	.75	.833	.916
L. F. OF TREAD PER INCH OF STAIRWAY WIDTH	.75	.833	.916	.666	.75	.833	.916	.666	.75	.833
FRAMING SQUARE SETTINGS FOR CARRIAGE CUTS — TONGUE	7 5/16"	6 5/8"	6 1/16"	8 1/8"	7 5/16"	6 5/8"	6 1/8"	8 1/8"	7 5/16"	6 11/16"
FRAMING SQUARE SETTINGS FOR CARRIAGE CUTS — BODY	10 3/16"	10 7/8"	11 7/16"	9 3/8"	10 3/16"	10 7/8"	11 3/8"	9 3/8"	10 3/16"	10 13/16"

FLOOR TO FLOOR RISE		6'1 1/4"	6'1 3/8"	6'1 3/8"	6'1 3/8"	6'1 3/8"	6'1 1/2"	6'1 1/2"	6'1 1/2"	6'1 1/2"	6'1 5/8"
NUMBER OF RISERS		12	9	10	11	12	9	10	11	12	9
HEIGHT OF EACH RISER		6 1/8"	8 1/8"	7 5/16"	6 11/16"	6 1/8"	8 3/16"	7 3/8"	6 11/16"	6 1/8"	8 3/16"
TOTAL HEIGHT OF RISERS		6'1 1/2"	6'1 1/8"	6'1 1/8"	6'1 9/16"	6'1 1/2"	6'1 11/16"	6'1 3/4"	6'1 9/16"	6'1 1/2"	6'1 11/16"
RISERS OVERAGE (+) OR UNDERAGE (—)		+1/4"	−1/4"	−1/4"	+3/16"	+1/8"	+3/16"	+1/4"	+1/16"	0	+1/16"
NUMBER OF TREADS		11	8	9	10	11	8	9	10	11	8
WIDTH OF EACH TREAD		11 3/8"	9 3/8"	10 3/16"	10 13/16"	11 3/8"	9 5/16"	10 1/8"	10 13/16"	11 3/8"	9 5/16"
TOTAL RUN OF STAIRWAY		10'5 1/8"	6'3"	7'7 11/16"	9' 1/8"	10'5 1/8"	6'2 1/2"	7'7 1/8"	9' 1/8"	10'5 1/8"	6'2 1/2"
MAXIMUM RUN WITH 6'-8" HEADROOM		1'10 3/4"	9 3/8"	1'8 3/8"	1'9 5/8"	1'10 3/4"	9 5/16"	1'8 1/4"	1'9 5/8"	1'10 3/4"	9 5/16"
ANGLE OF INCLINE		28°18'	40°55'	35°40'	31°44'	28°18'	41°19'	36°4'	31°44'	28°18'	41°19'
LENGTH OF CARRIAGE		11'10 1/8"	8'3 1/4"	9'4 7/8"	10'7 1/8"	11'10 1/8"	8'3 3/16"	9'4 3/4"	10'7 1/8"	11'10 1/8"	8'3 3/16"
L. F. OF RISER PER INCH OF STAIRWAY WIDTH		1.0	.75	.833	.916	1.0	.75	.833	.916	1.0	.75
L. F. OF TREAD PER INCH OF STAIRWAY WIDTH		.916	.666	.75	.833	.916	.666	.75	.833	.916	.666
FRAMING SQUARE SETTINGS FOR CARRIAGE CUTS	TONGUE	6 1/8"	8 1/8"	7 5/16"	6 11/16"	6 1/8"	8 3/16"	7 3/8"	6 11/16"	6 1/8"	8 3/16"
	BODY	11 3/8"	9 3/8"	10 3/16"	10 13/16"	11 3/8"	9 5/16"	10 1/8"	10 13/16"	11 3/8"	9 5/16"

FLOOR TO FLOOR RISE	6'1 5/8"	6'1 5/8"	6'1 5/8"	6'1 3/4"	6'1 3/4"	6'1 3/4"	6'1 3/4"	6'1 7/8"	6'1 7/8"	6'1 7/8"
NUMBER OF RISERS	10	11	12	9	10	11	12	9	10	11
HEIGHT OF EACH RISER	7 3/8"	6 11/16"	6 1/8"	8 3/16"	7 3/8"	6 11/16"	6 1/8"	8 3/16"	7 3/8"	6 11/16"
TOTAL HEIGHT OF RISERS	6'1 3/4"	6'1 9/16"	6'1 1/2"	6'1 11/16"	6'1 3/4"	6'1 9/16"	6'1 1/2"	6'1 11/16"	6'1 3/4"	6'1 9/16"
RISERS OVERAGE (+) OR UNDERAGE (—)	+1/8"	−1/16"	−1/8"	−1/16"	0	−3/16"	−1/4"	−3/16"	−1/8"	−5/16"
NUMBER OF TREADS	9	10	11	8	9	10	11	8	9	10
WIDTH OF EACH TREAD	10 1/8"	10 13/16"	11 3/8"	9 5/16"	10 1/8"	10 13/16"	11 3/8"	9 5/16"	10 1/8"	10 13/16"
TOTAL RUN OF STAIRWAY	7'7 1/8"	9' 1/8"	10'5 1/8"	6'2 1/2"	7'7 1/8"	9' 1/8"	10'5 1/8"	6'2 1/2"	7'7 1/8"	9' 1/8"
MAXIMUM RUN WITH 6'-8" HEADROOM	1'8 1/4"	1'9 5/8"	1'10 3/4"	9 5/16"	1'8 1/4"	1'9 5/8"	1'10 3/4"	9 5/16"	1'8 1/4"	1'9 5/8"
ANGLE OF INCLINE	36°4'	31°44'	28°18'	41°19'	36°4'	31°44'	28°18'	41°19'	36°4'	31°44'
LENGTH OF CARRIAGE	9'4 3/4"	10'7 1/8"	11'10 1/8"	8'3 3/16"	9'4 3/4"	10'7 1/8"	11'10 1/8"	8'3 3/16"	9'4 3/4"	10'7 1/8"
L. F. OF RISER PER INCH OF STAIRWAY WIDTH	.833	.916	1.0	.75	.833	.916	1.0	.75	.833	.916
L. F. OF TREAD PER INCH OF STAIRWAY WIDTH	.75	.833	.916	.666	.75	.833	.916	.666	.75	.833
FRAMING SQUARE SETTINGS FOR CARRIAGE CUTS — TONGUE	7 3/8"	6 11/16"	6 1/8"	8 3/16"	7 3/8"	6 11/16"	6 1/8"	8 3/16"	7 3/8"	6 11/16"
FRAMING SQUARE SETTINGS FOR CARRIAGE CUTS — BODY	10 1/8"	10 13/16"	11 3/8"	9 5/16"	10 1/8"	10 13/16"	11 3/8"	9 5/16"	10 1/8"	10 13/16"

FLOOR TO FLOOR RISE		6'1$\frac{7}{8}$"	6'2"	6'2"	6'2"	6'2"	6'2$\frac{1}{8}$"	6'2$\frac{1}{8}$"	6'2$\frac{1}{8}$"	6'2$\frac{1}{8}$"	6'2$\frac{1}{4}$"
NUMBER OF RISERS		12	9	10	11	12	9	10	11	12	9
HEIGHT OF EACH RISER		6$\frac{3}{16}$"	8$\frac{1}{4}$"	7$\frac{3}{8}$"	6$\frac{3}{4}$"	6$\frac{3}{16}$"	8$\frac{1}{4}$"	7$\frac{7}{16}$"	6$\frac{3}{4}$"	6$\frac{3}{16}$"	8$\frac{1}{4}$"
TOTAL HEIGHT OF RISERS		6'2$\frac{1}{4}$"	6'2$\frac{1}{4}$"	6'1$\frac{3}{4}$"	6'2$\frac{1}{4}$"	6'2$\frac{1}{4}$"	6'2$\frac{1}{4}$"	6'2$\frac{3}{8}$"	6'2$\frac{1}{4}$"	6'2$\frac{1}{4}$"	6'2$\frac{1}{4}$"
RISERS OVERAGE (+) OR UNDERAGE (—)		+$\frac{3}{8}$"	+$\frac{1}{4}$"	−$\frac{1}{4}$"	+$\frac{1}{4}$"	+$\frac{1}{4}$"	+$\frac{1}{8}$"	+$\frac{1}{4}$"	+$\frac{1}{8}$"	+$\frac{1}{8}$"	0
NUMBER OF TREADS		11	8	9	10	11	8	9	10	11	8
WIDTH OF EACH TREAD		11$\frac{5}{16}$"	9$\frac{1}{4}$"	10$\frac{1}{8}$"	10$\frac{3}{4}$"	11$\frac{5}{16}$"	9$\frac{1}{4}$"	10$\frac{1}{16}$"	10$\frac{3}{4}$"	11$\frac{5}{16}$"	9$\frac{1}{4}$"
TOTAL RUN OF STAIRWAY		10'4$\frac{7}{16}$"	6'2"	7'7$\frac{1}{8}$"	8'11$\frac{1}{2}$"	10'4$\frac{7}{16}$"	6'2"	7'6$\frac{9}{16}$"	8'11$\frac{1}{2}$"	10'4$\frac{7}{16}$"	6'2"
MAXIMUM RUN WITH 6'-8" HEADROOM		1'10$\frac{5}{8}$"	9$\frac{1}{4}$"	1'8$\frac{1}{4}$"	1'9$\frac{1}{2}$"	1'10$\frac{5}{8}$"	9$\frac{1}{4}$"	1'8$\frac{1}{8}$"	1'9$\frac{1}{2}$"	1'10$\frac{5}{8}$"	9$\frac{1}{4}$"
ANGLE OF INCLINE		28°41'	41°44'	36°4'	32°7'	28°41'	41°44'	36°28'	32°7'	28°41'	41°44'
LENGTH OF CARRIAGE		11'9$\frac{27}{32}$"	8'3$\frac{5}{32}$"	9'4$\frac{3}{4}$"	10'6$\frac{15}{16}$"	11'9$\frac{27}{32}$"	8'3$\frac{5}{32}$"	9'4$\frac{5}{8}$"	10'6$\frac{15}{16}$"	11'9$\frac{27}{32}$"	8'3$\frac{5}{32}$"
L. F. OF RISER PER INCH OF STAIRWAY WIDTH		1.0	.75	.833	.916	1.0	.75	.833	.916	1.0	.75
L. F. OF TREAD PER INCH OF STAIRWAY WIDTH		.916	.666	.75	.833	.916	.666	.75	.833	.916	.666
FRAMING SQUARE SETTINGS FOR CARRIAGE CUTS	TONGUE	6$\frac{3}{16}$"	8$\frac{1}{4}$"	7$\frac{3}{8}$"	6$\frac{3}{4}$"	6$\frac{3}{16}$"	8$\frac{1}{4}$"	7$\frac{7}{16}$"	6$\frac{3}{4}$"	6$\frac{3}{16}$"	8$\frac{1}{4}$"
	BODY	11$\frac{5}{16}$"	9$\frac{1}{4}$"	10$\frac{1}{8}$"	10$\frac{3}{4}$"	11$\frac{5}{16}$"	9$\frac{1}{4}$"	10$\frac{1}{16}$"	10$\frac{3}{4}$"	11$\frac{5}{16}$"	9$\frac{1}{4}$"

FLOOR TO FLOOR RISE	6'2 1/4"	6'2 1/4"	6'2 1/4"	6'2 3/8"	6'2 3/8"	6'2 3/8"	6'2 3/8"	6'2 1/2"	6'2 1/2"	6'2 1/2"
NUMBER OF RISERS	10	11	12	9	10	11	12	9	10	11
HEIGHT OF EACH RISER	7 7/16"	6 3/4"	6 3/16"	8 1/4"	7 7/16"	6 3/4"	6 3/16"	8 1/4"	7 7/16"	6 3/4"
TOTAL HEIGHT OF RISERS	6'2 3/8"	6'2 1/4"	6'2 1/4"	6'2 1/4"	6'2 3/8"	6'2 1/4"	6'2 1/4"	6'2 1/4"	6'2 3/8"	6'2 1/4"
RISERS OVERAGE (+) OR UNDERAGE (—)	+1/8"	0	0	−1/8"	0	−1/8"	−1/8"	−1/4"	−1/8"	−1/4"
NUMBER OF TREADS	9	10	11	8	9	10	11	8	9	10
WIDTH OF EACH TREAD	10 1/16"	10 3/4"	11 5/16"	9 1/4"	10 1/16"	10 3/4"	11 5/16"	9 1/4"	10 1/16"	10 3/4"
TOTAL RUN OF STAIRWAY	7'6 9/16"	8'11 1/2"	10'4 7/16"	6'2"	7'6 9/16"	8'11 1/2"	10'4 7/16"	6'2"	7'6 9/16"	8'11 1/2"
MAXIMUM RUN WITH 6'-8" HEADROOM	1'8 1/8"	1'9 1/2"	1'10 5/8"	9 1/4"	1'8 1/8"	1'9 1/2"	1'10 5/8"	9 1/4"	1'8 1/8"	1'9 1/2"
ANGLE OF INCLINE	36°28'	32°7'	28°41'	41°44'	36°28'	32°7'	28°41'	41°44'	36°28'	32°7'
LENGTH OF CARRIAGE	9'4 5/8"	10'6 15/16"	11'9 27/32"	8'3 5/32"	9'4 5/8"	10'6 15/16"	11'9 27/32"	8'3 5/32"	9'4 5/8"	10'6 15/16"
L. F. OF RISER PER INCH OF STAIRWAY WIDTH	.833	.916	1.0	.75	.833	.916	1.0	.75	.833	.916
L. F. OF TREAD PER INCH OF STAIRWAY WIDTH	.75	.833	.916	.666	.75	.833	.916	.666	.75	.833
FRAMING SQUARE SETTINGS FOR CARRIAGE CUTS — TONGUE	7 7/16"	6 3/4"	6 3/16"	8 1/4"	7 7/16"	6 3/4"	6 3/16"	8 1/4"	7 7/16"	6 3/4"
FRAMING SQUARE SETTINGS FOR CARRIAGE CUTS — BODY	10 1/16"	10 3/4"	11 5/16"	9 1/4"	10 1/16"	10 3/4"	11 5/16"	9 1/4"	10 1/16"	10 3/4"

FLOOR TO FLOOR RISE	6'2 1/2"	6'2 5/8"	6'2 5/8"	6'2 5/8"	6'2 3/4"	6'2 3/4"	6'2 3/4"	6'2 7/8"	6'2 7/8"	6'2 7/8"
NUMBER OF RISERS	12	10	11	12	10	11	12	10	11	12
HEIGHT OF EACH RISER	6 3/16"	7 7/16"	6 13/16"	6 1/4"	7 1/2"	6 13/16"	6 1/4"	7 1/2"	6 13/16"	6 1/4"
TOTAL HEIGHT OF RISERS	6'2 1/4"	6'2 3/8"	6'2 15/16"	6'3"	6'3"	6'2 15/16"	6'3"	6'3"	6'2 15/16"	6'3"
RISERS OVERAGE (+) OR UNDERAGE (—)	−1/4"	−1/4"	+5/16"	+3/8"	+1/4"	+3/16"	+1/4"	+1/8"	+1/16"	+1/8"
NUMBER OF TREADS	11	9	10	11	9	10	11	9	10	11
WIDTH OF EACH TREAD	11 5/16"	10 1/16"	10 11/16"	11 1/4"	10"	10 11/16"	11 1/4"	10"	10 11/16"	11 1/4"
TOTAL RUN OF STAIRWAY	10'4 7/16"	7'6 9/16"	8'10 7/8"	10'3 3/4"	7'6"	8'10 7/8"	10'3 3/4"	7'6"	8'10 7/8"	10'3 3/4"
MAXIMUM RUN WITH 6'-8" HEADROOM	1'10 5/8"	1'8 1/8"	1'9 3/8"	1'10 1/2"	1'8"	1'9 3/8"	1'10 1/2"	1'8"	1'9 3/8"	1'10 1/2"
ANGLE OF INCLINE	28°41'	36°28'	32°31'	29°3'	36°52'	32°31'	29°3'	36°52'	32°31'	29°3'
LENGTH OF CARRIAGE	11'9 27/32"	9'4 5/8"	10'6 3/4"	11'9 9/16"	9'4 1/2"	10'6 3/4"	11'9 9/16"	9'4 1/2"	10'6 3/4"	11'9 9/16"
L. F. OF RISER PER INCH OF STAIRWAY WIDTH	1.0	.833	.916	1.0	.833	.916	1.0	.833	.916	1.0
L. F. OF TREAD PER INCH OF STAIRWAY WIDTH	.916	.75	.833	.916	.75	.833	.916	.75	.833	.916
FRAMING SQUARE SETTINGS FOR CARRIAGE CUTS — TONGUE	6 3/16"	7 7/16"	6 13/16"	6 1/4"	7 1/2"	6 13/16"	6 1/4"	7 1/2"	6 13/16"	6 1/4"
FRAMING SQUARE SETTINGS FOR CARRIAGE CUTS — BODY	11 5/16"	10 1/16"	10 11/16"	11 1/4"	10"	10 11/16"	11 1/4"	10"	10 11/16"	11 1/4"

FLOOR TO FLOOR RISE	6'3"	6'3"	6'3"	6'3 1/8"	6'3 1/8"	6'3 1/8"	6'3 1/4"	6'3 1/4"	6'3 1/4"	6'3 3/8"
NUMBER OF RISERS	10	11	12	10	11	12	10	11	12	10
HEIGHT OF EACH RISER	7 1/2"	6 13/16"	6 1/4"	7 1/2"	6 13/16"	6 1/4"	7 1/2"	6 13/16"	6 1/4"	7 9/16"
TOTAL HEIGHT OF RISERS	6'3"	6'2 15/16"	6'3"	6'3"	6'2 15/16"	6'3"	6'3"	6'2 15/16"	6'3"	6'3 5/8"
RISERS OVERAGE (+) OR UNDERAGE (—)	0	−1/16"	0	−1/8"	−3/16"	−1/8"	−1/4"	−5/16"	−1/4"	+1/4"
NUMBER OF TREADS	9	10	11	9	10	11	9	10	11	9
WIDTH OF EACH TREAD	10"	10 11/16"	11 1/4"	10"	10 11/16"	11 1/4"	10"	10 11/16"	11 1/4"	9 15/16"
TOTAL RUN OF STAIRWAY	7'6"	8'10 7/8"	10'3 3/4"	7'6"	8'10 7/8"	10'3 3/4"	7'6"	8'10 7/8"	10'3 3/4"	7'5 7/16"
MAXIMUM RUN WITH 6'-8" HEADROOM	1'8"	1'9 3/8"	1'10 1/2"	1'8"	1'9 3/8"	1'10 1/2"	1'8"	1'9 3/8"	1'10 1/2"	1'7 7/8"
ANGLE OF INCLINE	36°52'	32°31'	29°3'	36°52'	32°31'	29°3'	36°52'	32°31'	29°3'	37°16'
LENGTH OF CARRIAGE	9'4 1/2"	10'6 3/4"	11'9 9/16"	9'4 1/2"	10'6 3/4"	11'9 9/16"	9'4 1/2"	10'6 3/4"	11'9 9/16"	9'4 3/8"
L. F. OF RISER PER INCH OF STAIRWAY WIDTH	.833	.916	1.0	.833	.916	1.0	.833	.916	1.0	.833
L. F. OF TREAD PER INCH OF STAIRWAY WIDTH	.75	.833	.916	.75	.833	.916	.75	.833	.916	.75
FRAMING SQUARE SETTINGS FOR CARRIAGE CUTS — TONGUE	7 1/2"	6 13/16"	6 1/4"	7 1/2"	6 13/16"	6 1/4"	7 1/2"	6 13/16"	6 1/4"	7 9/16"
BODY	10"	10 11/16"	11 1/4"	10"	10 11/16"	11 1/4"	10"	10 11/16"	11 1/4"	9 15/16"

FLOOR TO FLOOR RISE	6'3$\frac{3}{8}$"	6'3$\frac{3}{8}$"	6'3$\frac{1}{2}$"	6'3$\frac{1}{2}$"	6'3$\frac{1}{2}$"	6'3$\frac{5}{8}$"	6'3$\frac{5}{8}$"	6'3$\frac{5}{8}$"	6'3$\frac{3}{4}$"	6'3$\frac{3}{4}$"
NUMBER OF RISERS	11	12	10	11	12	10	11	12	10	11
HEIGHT OF EACH RISER	6$\frac{7}{8}$"	6$\frac{5}{16}$"	7$\frac{9}{16}$"	6$\frac{7}{8}$"	6$\frac{5}{16}$"	7$\frac{9}{16}$"	6$\frac{7}{8}$"	6$\frac{5}{16}$"	7$\frac{9}{16}$"	6$\frac{7}{8}$"
TOTAL HEIGHT OF RISERS	6'3$\frac{5}{8}$"	6'3$\frac{3}{4}$"	6'3$\frac{5}{8}$"	6'3$\frac{5}{8}$"	6'3$\frac{3}{4}$"	6'3$\frac{5}{8}$"	6'3$\frac{5}{8}$"	6'3$\frac{3}{4}$"	6'3$\frac{5}{8}$"	6'3$\frac{5}{8}$"
RISERS OVERAGE (+) OR UNDERAGE (—)	+$\frac{1}{4}$"	+$\frac{3}{8}$"	+$\frac{1}{8}$"	+$\frac{1}{8}$"	+$\frac{1}{4}$"	0	0	+$\frac{1}{8}$"	−$\frac{1}{8}$"	−$\frac{1}{8}$"
NUMBER OF TREADS	10	11	9	10	11	9	10	11	9	10
WIDTH OF EACH TREAD	10$\frac{5}{8}$"	11$\frac{3}{16}$"	9$\frac{15}{16}$"	10$\frac{5}{8}$"	11$\frac{3}{16}$"	9$\frac{15}{16}$"	10$\frac{5}{8}$"	11$\frac{3}{16}$"	9$\frac{15}{16}$"	10$\frac{5}{8}$"
TOTAL RUN OF STAIRWAY	8'10$\frac{1}{4}$"	10'3$\frac{1}{16}$"	7'5$\frac{7}{16}$"	8'10$\frac{1}{4}$"	10'3$\frac{1}{16}$"	7'5$\frac{7}{16}$"	8'10$\frac{1}{4}$"	10'3$\frac{1}{16}$"	7'5$\frac{7}{16}$"	8'10$\frac{1}{4}$"
MAXIMUM RUN WITH 6'-8" HEADROOM	1'9$\frac{1}{4}$"	1'10$\frac{3}{8}$"	1'7$\frac{7}{8}$"	1'9$\frac{1}{4}$"	1'10$\frac{3}{8}$"	1'7$\frac{7}{8}$"	1'9$\frac{1}{4}$"	1'10$\frac{3}{8}$"	1'7$\frac{7}{8}$"	1'9$\frac{1}{4}$"
ANGLE OF INCLINE	32°54'	29°26'	37°16'	32°54'	29°26'	37°16'	32°54'	29°26'	37°16'	32°54'
LENGTH OF CARRIAGE	10'6$\frac{9}{16}$"	11'9$\frac{5}{16}$"	9'4$\frac{3}{8}$"	10'6$\frac{9}{16}$"	11'9$\frac{5}{16}$"	9'4$\frac{3}{8}$"	10'6$\frac{9}{16}$"	11'9$\frac{5}{16}$"	9'4$\frac{3}{8}$"	10'6$\frac{9}{16}$"
L. F. OF RISER PER INCH OF STAIRWAY WIDTH	.916	1.0	.833	.916	1.0	.833	.916	1.0	.833	.916
L. F. OF TREAD PER INCH OF STAIRWAY WIDTH	.833	.916	.75	.833	.916	.75	.833	.916	.75	.833
FRAMING SQUARE SETTINGS FOR CARRIAGE CUTS — TONGUE	6$\frac{7}{8}$"	6$\frac{5}{16}$"	7$\frac{9}{16}$"	6$\frac{7}{8}$"	6$\frac{5}{16}$"	7$\frac{9}{16}$"	6$\frac{7}{8}$"	6$\frac{5}{16}$"	7$\frac{9}{16}$"	6$\frac{7}{8}$"
BODY	10$\frac{5}{8}$"	11$\frac{3}{16}$"	9$\frac{15}{16}$"	10$\frac{5}{8}$"	11$\frac{3}{16}$"	9$\frac{15}{16}$"	10$\frac{5}{8}$"	11$\frac{3}{16}$"	9$\frac{15}{16}$"	10$\frac{5}{8}$"

FLOOR TO FLOOR RISE		6'3 3/4"	6'3 7/8"	6'3 7/8"	6'3 7/8"	6'4"	6'4"	6'4"	6'4 1/8"	6'4 1/8"	6'4 1/8"
NUMBER OF RISERS		12	10	11	12	10	11	12	10	11	12
HEIGHT OF EACH RISER		6 5/16"	7 9/16"	6 7/8"	6 5/16"	7 5/8"	6 15/16"	6 5/16"	7 5/8"	6 15/16"	6 3/8"
TOTAL HEIGHT OF RISERS		6'3 3/4"	6'3 5/8"	6'3 5/8"	6'3 3/4"	6'4 1/4"	6'4 5/16"	6'3 3/4"	6'4 1/4"	6'4 5/16"	6'4 1/2"
RISERS OVERAGE (+) OR UNDERAGE (—)		0	−1/4"	−1/4"	−1/8"	+1/4"	+5/16"	−1/4"	+1/8"	+3/16"	+3/8"
NUMBER OF TREADS		11	9	10	11	9	10	11	9	10	11
WIDTH OF EACH TREAD		11 3/16"	9 15/16"	10 5/8"	11 3/16"	9 7/8"	10 9/16"	11 3/16"	9 7/8"	10 9/16"	11 1/8"
TOTAL RUN OF STAIRWAY		10'3 1/16"	7'5 7/16"	8'10 1/4"	10'3 1/16"	7'4 7/8"	8'9 5/8"	10'3 1/16"	7'4 7/8"	8'9 5/8"	10'2 3/8"
MAXIMUM RUN WITH 6'-8" HEADROOM		1'10 3/8"	1'7 7/8"	1'9 1/4"	1'10 3/8"	1'7 3/4"	1'9 1/8"	1'10 3/8"	1'7 3/4"	1'9 1/8"	1'10 1/4"
ANGLE OF INCLINE		29°26'	37°16'	32°54'	29°26'	37°40'	33°18'	29°26'	37°40'	33°18'	29°49'
LENGTH OF CARRIAGE		11'9 5/16"	9'4 3/8"	10'6 9/16"	11'9 5/16"	9'4 9/32"	10'6 3/8"	11'9 5/16"	9'4 9/32"	10'6 3/8"	11'9 1/32"
L. F. OF RISER PER INCH OF STAIRWAY WIDTH		1.0	.833	.916	1.0	.833	.916	1.0	.833	.916	1.0
L. F. OF TREAD PER INCH OF STAIRWAY WIDTH		.916	.75	.833	.916	.75	.833	.916	.75	.833	.916
FRAMING SQUARE SETTINGS FOR CARRIAGE CUTS	TONGUE	6 5/16"	7 9/16"	6 7/8"	6 5/16"	7 5/8"	6 15/16"	6 5/16"	7 5/8"	6 15/16"	6 3/8"
	BODY	11 3/16"	9 15/16"	10 5/8"	11 3/16"	9 7/8"	10 9/16"	11 3/16"	9 7/8"	10 9/16"	11 1/8"

FLOOR TO FLOOR RISE	6'4 1/4"	6'4 1/4"	6'4 1/4"	6'4 3/8"	6'4 3/8"	6'4 3/8"	6'4 1/2"	6'4 1/2"	6'4 1/2"	6'4 5/8"
NUMBER OF RISERS	10	11	12	10	11	12	10	11	12	10
HEIGHT OF EACH RISER	7 5/8"	6 15/16"	6 3/8"	7 5/8"	6 15/16"	6 3/8"	7 5/8"	6 15/16"	6 3/8"	7 11/16"
TOTAL HEIGHT OF RISERS	6'4 1/4"	6'4 5/16"	6'4 1/2"	6'4 1/4"	6'4 5/16"	6'4 1/2"	6'4 1/4"	6'4 5/16"	6'4 1/2"	6'4 7/8"
RISERS OVERAGE (+) OR UNDERAGE (—)	0	+1/16"	+1/4"	−1/8"	−1/16"	+1/8"	−1/4"	−3/16"	0	+1/4"
NUMBER OF TREADS	9	10	11	9	10	11	9	10	11	9
WIDTH OF EACH TREAD	9 7/8"	10 9/16"	11 1/8"	9 7/8"	10 9/16"	11 1/8"	9 7/8"	10 9/16"	11 1/8"	9 13/16"
TOTAL RUN OF STAIRWAY	7'4 7/8"	8'9 5/8"	10'2 3/8"	7'4 7/8"	8'9 5/8"	10'2 3/8"	7'4 7/8"	8'9 5/8"	10'2 3/8"	7'4 5/16"
MAXIMUM RUN WITH 6'-8" HEADROOM	1'7 3/4"	1'9 1/8"	1'10 1/4"	1'7 3/4"	1'9 1/8"	1'10 1/4"	1'7 3/4"	1'9 1/8"	1'10 1/4"	1'7 5/8"
ANGLE OF INCLINE	37°40'	33°18'	29°49'	37°40'	33°18'	29°49'	37°40'	33°18'	29°49'	38°5'
LENGTH OF CARRIAGE	9'4 9/32"	10'6 3/8"	11'9 1/32"	9'4 9/32"	10'6 3/8"	11'9 1/32"	9'4 9/32"	10'6 3/8"	11'9 1/32"	9'4 3/16"
L. F. OF RISER PER INCH OF STAIRWAY WIDTH	.833	.916	1.0	.833	.916	1.0	.833	.916	1.0	.833
L. F. OF TREAD PER INCH OF STAIRWAY WIDTH	.75	.833	.916	.75	.833	.916	.75	.833	.916	.75
FRAMING SQUARE SETTINGS FOR CARRIAGE CUTS — TONGUE	7 5/8"	6 15/16"	6 3/8"	7 5/8"	6 15/16"	6 3/8"	7 5/8"	6 15/16"	6 3/8"	7 11/16"
FRAMING SQUARE SETTINGS FOR CARRIAGE CUTS — BODY	9 7/8"	10 9/16"	11 1/8"	9 7/8"	10 9/16"	11 1/8"	9 7/8"	10 9/16"	11 1/8"	9 13/16"

FLOOR TO FLOOR RISE	6'4 5/8"	6'4 5/8"	6'4 3/4"	6'4 3/4"	6'4 3/4"	6'4 7/8"	6'4 7/8"	6'4 7/8"	6'5"	6'5"
NUMBER OF RISERS	11	12	10	11	12	10	11	12	10	11
HEIGHT OF EACH RISER	6 15/16"	6 3/8"	7 11/16"	7"	6 3/8"	7 11/16"	7"	6 7/16"	7 11/16"	7"
TOTAL HEIGHT OF RISERS	6'4 5/16"	6'4 1/2"	6'4 7/8"	6'5"	6'4 1/2"	6'4 7/8"	6'5"	6'5 1/4"	6'4 7/8"	6'5"
RISERS OVERAGE (+) OR UNDERAGE (—)	−5/16"	−1/8"	+1/8"	+1/4"	−1/4"	0	+1/8"	+3/8"	−1/8"	0
NUMBER OF TREADS	10	11	9	10	11	9	10	11	9	10
WIDTH OF EACH TREAD	10 9/16"	11 1/8"	9 13/16"	10 1/2"	11 1/8"	9 13/16"	10 1/2"	11 1/16"	9 13/16"	10 1/2"
TOTAL RUN OF STAIRWAY	8'9 5/8"	10'2 3/8"	7'4 5/16"	8'9"	10'2 3/8"	7'4 5/16"	8'9"	10'1 11/16"	7'4 5/16"	8'9"
MAXIMUM RUN WITH 6'-8" HEADROOM	1'9 1/8"	1'10 1/4"	1'7 5/8"	1'9"	1'10 1/4"	1'7 5/8"	1'9"	1'10 1/8"	1'7 5/8"	1'9"
ANGLE OF INCLINE	33°18'	29°49'	38°5'	33°41'	29°49'	38°5'	33°41'	30°12'	38°5'	33°41'
LENGTH OF CARRIAGE	10'6 3/8"	11'9 1/32"	9'4 3/16"	10'6 3/16"	11'9 1/32"	9'4 3/16"	10'6 3/16"	11'8 25/32"	9'4 3/16"	10'6 3/16"
L. F. OF RISER PER INCH OF STAIRWAY WIDTH	.916	1.0	.833	.916	1.0	.833	.916	1.0	.833	.916
L. F. OF TREAD PER INCH OF STAIRWAY WIDTH	.833	.916	.75	.833	.916	.75	.833	.916	.75	.833
FRAMING SQUARE SETTINGS FOR CARRIAGE CUTS — TONGUE	6 15/16"	6 3/8"	7 11/16"	7"	6 3/8"	7 11/16"	7"	6 7/16"	7 11/16"	7"
FRAMING SQUARE SETTINGS FOR CARRIAGE CUTS — BODY	10 9/16"	11 1/8"	9 13/16"	10 1/2"	11 1/8"	9 13/16"	10 1/2"	11 1/16"	9 13/16"	10 1/2"

FLOOR TO FLOOR RISE	6'5"	6'5 1/8"	6'5 1/8"	6'5 1/8"	6'5 1/4"	6'5 1/4"	6'5 1/4"	6'5 3/8"	6'5 3/8"	6'5 3/8"
NUMBER OF RISERS	12	10	11	12	10	11	12	10	11	12
HEIGHT OF EACH RISER	6 7/16"	7 11/16"	7"	6 7/16"	7 3/4"	7"	6 7/16"	7 3/4"	7 1/16"	6 7/16"
TOTAL HEIGHT OF RISERS	6'5 1/4"	6'4 7/8"	6'5"	6'5 1/4"	6'5 1/2"	6'5"	6'5 1/4"	6'5 1/2"	6'5 11/16"	6'5 1/4"
RISERS OVERAGE (+) OR UNDERAGE (—)	+1/4"	−1/4"	−1/8"	+1/8"	+1/4"	−1/4"	0	+1/8"	+5/16"	−1/8"
NUMBER OF TREADS	11	9	10	11	9	10	11	9	10	11
WIDTH OF EACH TREAD	11 1/16"	9 13/16"	10 1/2"	11 1/16"	9 3/4"	10 1/2"	11 1/16"	9 3/4"	10 7/16"	11 1/16"
TOTAL RUN OF STAIRWAY	10'1 11/16"	7'4 5/16"	8'9"	10'1 11/16"	7'3 3/4"	8'9"	10'1 11/16"	7'3 3/4"	8'8 3/8"	10'1 11/16"
MAXIMUM RUN WITH 6'-8" HEADROOM	1'10 1/8"	1'7 5/8"	1'9"	1'10 1/8"	1'7 1/2"	1'9"	1'10 1/8"	1'7 1/2"	1'8 7/8"	1'10 1/8"
ANGLE OF INCLINE	30°12'	38°5'	33°41'	30°12'	38°29'	33°41'	30°12'	38°29'	34°5'	30°12'
LENGTH OF CARRIAGE	11'8 25/32"	9'4 3/16"	10'6 3/16"	11'8 25/32"	9'4 3/32"	10'6 3/16"	11'8 25/32"	9'4 3/32"	10'6 1/32"	11'8 25/32"
L. F. OF RISER PER INCH OF STAIRWAY WIDTH	1.0	.833	.916	1.0	.833	.916	1.0	.833	.916	1.0
L. F. OF TREAD PER INCH OF STAIRWAY WIDTH	.916	.75	.833	.916	.75	.833	.916	.75	.833	.916
FRAMING SQUARE SETTINGS FOR CARRIAGE CUTS — TONGUE	6 7/16"	7 11/16"	7"	6 7/16"	7 3/4"	7"	6 7/16"	7 3/4"	7 1/16"	6 7/16"
FRAMING SQUARE SETTINGS FOR CARRIAGE CUTS — BODY	11 1/16"	9 13/16"	10 1/2"	11 1/16"	9 3/4"	10 1/2"	11 1/16"	9 3/4"	10 7/16"	11 1/16"

FLOOR TO FLOOR RISE	6'5 1/2"	6'5 1/2"	6'5 1/2"	6'5 5/8"	6'5 5/8"	6'5 5/8"	6'5 3/4"	6'5 3/4"	6'5 3/4"	6'5 7/8"
NUMBER OF RISERS	10	11	12	10	11	12	10	11	12	10
HEIGHT OF EACH RISER	7 3/4"	7 1/16"	6 7/16"	7 3/4"	7 1/16"	6 1/2"	7 3/4"	7 1/16"	6 1/2"	7 13/16"
TOTAL HEIGHT OF RISERS	6'5 1/2"	6'5 11/16"	6'5 1/4"	6'5 1/2"	6'5 11/16"	6'6"	6'5 1/2"	6'5 11/16"	6'6"	6'6 1/8"
RISERS OVERAGE (+) OR UNDERAGE (—)	0	+3/16"	−1/4"	−1/8"	+1/16"	+3/8"	−1/4"	−1/16"	+1/4"	+1/4"
NUMBER OF TREADS	9	10	11	9	10	11	9	10	11	9
WIDTH OF EACH TREAD	9 3/4"	10 7/16"	11 1/16"	9 3/4"	10 7/16"	11"	9 3/4"	10 7/16"	11"	9 11/16"
TOTAL RUN OF STAIRWAY	7'3 3/4"	8'8 3/8"	10'1 11/16"	7'3 3/4"	8'8 3/8"	10'1"	7'3 3/4"	8'8 3/8"	10'1"	7'3 3/16"
MAXIMUM RUN WITH 6'-8" HEADROOM	1'7 1/2"	1'8 7/8"	1'10 1/8"	1'7 1/2"	1'8 7/8"	1'10"	1'7 1/2"	1'8 7/8"	1'10"	1'7 3/8"
ANGLE OF INCLINE	38°29'	34°5'	30°12'	38°29'	34°5'	30°35'	38°29'	34°5'	30°35'	38°53'
LENGTH OF CARRIAGE	9'4 3/32"	10'6 1/32"	11'8 25/32"	9'4 3/32"	10'6 1/32"	11'8 17/32"	9'4 3/32"	10'6 1/32"	11'8 17/32"	9'4"
L. F. OF RISER PER INCH OF STAIRWAY WIDTH	.833	.916	1.0	.833	.916	1.0	.833	.916	1.0	.833
L. F. OF TREAD PER INCH OF STAIRWAY WIDTH	.75	.833	.916	.75	.833	.916	.75	.833	.916	.75
FRAMING SQUARE SETTINGS FOR CARRIAGE CUTS — TONGUE	7 3/4"	7 1/16"	6 7/16"	7 3/4"	7 1/16"	6 1/2"	7 3/4"	7 1/16"	6 1/2"	7 13/16"
FRAMING SQUARE SETTINGS FOR CARRIAGE CUTS — BODY	9 3/4"	10 7/16"	11 1/16"	9 3/4"	10 7/16"	11"	9 3/4"	10 7/16"	11"	9 11/16"

FLOOR TO FLOOR RISE	6'5 7/8"	6'5 7/8"	6'6"	6'6"	6'6"	6'6"	6'6 1/8"	6'6 1/8"	6'6 1/8"	6'6 1/8"
NUMBER OF RISERS	11	12	10	11	12	13	10	11	12	13
HEIGHT OF EACH RISER	7 1/16"	6 1/2"	7 13/16"	7 1/16"	6 1/2"	6"	7 13/16"	7 1/8"	6 1/2"	6"
TOTAL HEIGHT OF RISERS	6'5 11/16"	6'6"	6'6 1/8"	6'5 11/16"	6'6"	6'6"	6'6 1/8"	6'6 3/8"	6'6"	6'6"
RISERS OVERAGE (+) OR UNDERAGE (—)	−3/16"	+1/8"	+1/8"	−5/16"	0	0	0	+1/4"	−1/8"	−1/8"
NUMBER OF TREADS	10	11	9	10	11	12	9	10	11	12
WIDTH OF EACH TREAD	10 7/16"	11"	9 11/16"	10 7/16"	11"	11 1/2"	9 11/16"	10 3/8"	11"	11 1/2"
TOTAL RUN OF STAIRWAY	8'8 3/8"	10'1"	7'3 3/16"	8'8 3/8"	10'1"	11'6"	7'3 3/16"	8'7 3/4"	10'1"	11'6"
MAXIMUM RUN WITH 6'-8" HEADROOM	1'8 7/8"	1'10"	1'7 3/8"	1'8 7/8"	1'10"	1'11"	1'7 3/8"	1'8 3/4"	1'10"	1'11"
ANGLE OF INCLINE	34°5'	30°35'	38°53'	34°5'	30°35'	27°33'	38°53'	34°29'	30°35'	27°33'
LENGTH OF CARRIAGE	10'6 1/32"	11'8 17/32"	9'4"	10'6 1/32"	11'8 17/32"	12'11 21/32"	9'4"	10'5 27/32"	11'8 17/32"	12'11 21/32"
L. F. OF RISER PER INCH OF STAIRWAY WIDTH	.916	1.0	.833	.916	1.0	1.083	.833	.916	1.0	1.083
L. F. OF TREAD PER INCH OF STAIRWAY WIDTH	.833	.916	.75	.833	.916	1.0	.75	.833	.916	1.0
FRAMING SQUARE SETTINGS FOR CARRIAGE CUTS — TONGUE	7 1/16"	6 1/2"	7 13/16"	7 1/16"	6 1/2"	6"	7 13/16"	7 1/8"	6 1/2"	6"
FRAMING SQUARE SETTINGS FOR CARRIAGE CUTS — BODY	10 7/16"	11"	9 11/16"	10 7/16"	11"	11 1/2"	9 11/16"	10 3/8"	11"	11 1/2"

FLOOR TO FLOOR RISE		6'6 1/4"	6'6 1/4"	6'6 1/4"	6'6 1/4"	6'6 3/8"	6'6 3/8"	6'6 3/8"	6'6 3/8"	6'6 1/2"	6'6 1/2"
NUMBER OF RISERS		10	11	12	13	10	11	12	13	10	11
HEIGHT OF EACH RISER		7 13/16"	7 1/8"	6 1/2"	6"	7 13/16"	7 1/8"	6 9/16"	6"	7 7/8"	7 1/8"
TOTAL HEIGHT OF RISERS		6'6 1/8"	6'6 3/8"	6'6"	6'6"	6'6 1/8"	6'6 3/8"	6'6 3/4"	6'6"	6'6 3/4"	6'6 3/8"
RISERS OVERAGE (+) OR UNDERAGE (—)		−1/8"	+1/8"	−1/4"	−1/4"	−1/4"	0	+3/8"	−3/8"	+1/4"	−1/8"
NUMBER OF TREADS		9	10	11	12	9	10	11	12	9	10
WIDTH OF EACH TREAD		9 11/16"	10 3/8"	11"	11 1/2"	9 11/16"	10 3/8"	10 15/16"	11 1/2"	9 5/8"	10 3/8"
TOTAL RUN OF STAIRWAY		7'3 3/16"	8'7 3/4"	10'1"	11'6"	7'3 3/16"	8'7 3/4"	10' 5/16"	11'6"	7'2 5/8"	8'7 3/4"
MAXIMUM RUN WITH 6'-8" HEADROOM		1'7 3/8"	1'8 3/4"	1'10"	1'11"	1'7 3/8"	1'8 3/4"	1'9 7/8"	1'11"	1'7 1/4"	1'8 3/4"
ANGLE OF INCLINE		38°53'	34°29'	30°35'	27°33'	38°53'	34°29'	30°58'	27°33'	39°17'	34°29'
LENGTH OF CARRIAGE		9'4"	10'5 27/32"	11'8 17/32"	12'11 21/32"	9'4"	10'5 27/32"	11'8 5/16"	12'11 21/32"	9'3 15/16"	10'5 27/32"
L. F. OF RISER PER INCH OF STAIRWAY WIDTH		.833	.916	1.0	1.083	.833	.916	1.0	1.083	.833	.916
L. F. OF TREAD PER INCH OF STAIRWAY WIDTH		.75	.833	.916	1.0	.75	.833	.916	1.0	.75	.833
FRAMING SQUARE SETTINGS FOR CARRIAGE CUTS	TONGUE	7 13/16"	7 1/8"	6 1/2"	6"	7 13/16"	7 1/8"	6 9/16"	6"	7 7/8"	7 1/8"
	BODY	9 11/16"	10 3/8"	11"	11 1/2"	9 11/16"	10 3/8"	10 15/16"	11 1/2"	9 5/8"	10 3/8"

FLOOR TO FLOOR RISE		6'6 1/2"	6'6 1/2"	6'6 5/8"	6'6 5/8"	6'6 5/8"	6'6 5/8"	6'6 3/4"	6'6 3/4"	6'6 3/4"	6'6 3/4"
NUMBER OF RISERS		12	13	10	11	12	13	10	11	12	13
HEIGHT OF EACH RISER		6 9/16"	6 1/16"	7 7/8"	7 1/8"	6 9/16"	6 1/16"	7 7/8"	7 3/16"	6 9/16"	6 1/16"
TOTAL HEIGHT OF RISERS		6'6 3/4"	6'6 13/16"	6'6 3/4"	6'6 3/8"	6'6 3/4"	6'6 13/16"	6'6 3/4"	6'7 1/16"	6'6 3/4"	6'6 13/16"
RISERS OVERAGE (+) OR UNDERAGE (—)		+1/4"	+5/16"	+1/8"	−1/4"	+1/8"	+3/16"	0	+5/16"	0	+1/16"
NUMBER OF TREADS		11	12	9	10	11	12	9	10	11	12
WIDTH OF EACH TREAD		10 15/16"	11 7/16"	9 5/8"	10 3/8"	10 15/16"	11 7/16"	9 5/8"	10 5/16"	10 15/16"	11 7/16"
TOTAL RUN OF STAIRWAY		10' 5/16"	11'5 1/4"	7'2 5/8"	8'7 3/4"	10' 5/16"	11'5 1/4"	7'2 5/8"	8'7 1/8"	10' 5/16"	11'5 1/4"
MAXIMUM RUN WITH 6'-8" HEADROOM		1'9 7/8"	1'10 7/8"	1'7 1/4"	1'8 3/4"	1'9 7/8"	1'10 7/8"	1'7 1/4"	1'8 5/8"	1'9 7/8"	1'10 7/8"
ANGLE OF INCLINE		30°58'	27°56'	39°17'	34°29'	30°58'	27°56'	39°17'	34°53'	30°58'	27°56'
LENGTH OF CARRIAGE		11'8 5/16"	12'11 11/32"	9'3 15/16"	10'5 27/32"	11'8 5/16"	12'11 11/32"	9'3 15/16"	10'5 11/16"	11'8 5/16"	12'11 11/32"
L. F. OF RISER PER INCH OF STAIRWAY WIDTH		1.0	1.083	.833	.916	1.0	1.083	.833	.916	1.0	1.083
L. F. OF TREAD PER INCH OF STAIRWAY WIDTH		.916	1.0	.75	.833	.916	1.0	.75	.833	.916	1.0
FRAMING SQUARE SETTINGS FOR CARRIAGE CUTS	TONGUE	6 9/16"	6 1/16"	7 7/8"	7 1/8"	6 9/16"	6 1/16"	7 7/8"	7 3/16"	6 9/16"	6 1/16"
	BODY	10 15/16"	11 7/16"	9 5/8"	10 3/8"	10 15/16"	11 7/16"	9 5/8"	10 5/16"	10 15/16"	11 7/16"

FLOOR TO FLOOR RISE	6'6 7/8"	6'6 7/8"	6'6 7/8"	6'6 7/8"	6'7"	6'7"	6'7"	6'7"	6'7 1/8"	6'7 1/8"
NUMBER OF RISERS	10	11	12	13	10	11	12	13	10	11
HEIGHT OF EACH RISER	7 7/8"	7 3/16"	6 9/16"	6 1/16"	7 7/8"	7 3/16"	6 9/16"	6 1/16"	7 15/16"	7 3/16"
TOTAL HEIGHT OF RISERS	6'6 3/4"	6'7 1/16"	6'6 3/4"	6'6 13/16"	6'6 3/4"	6'7 1/16"	6'6 3/4"	6'6 13/16"	6'7 3/8"	6'7 1/16"
RISERS OVERAGE (+) OR UNDERAGE (—)	−1/8"	+3/16"	−1/8"	−1/16"	−1/4"	+1/16"	−1/4"	−3/16"	+1/4"	−1/16"
NUMBER OF TREADS	9	10	11	12	9	10	11	12	9	10
WIDTH OF EACH TREAD	9 5/8"	10 5/16"	10 15/16"	11 7/16"	9 5/8"	10 5/16"	10 15/16"	11 7/16"	9 9/16"	10 5/16"
TOTAL RUN OF STAIRWAY	7'2 5/8"	8'7 1/8"	10' 5/16"	11'5 1/4"	7'2 5/8"	8'7 1/8"	10' 5/16"	11'5 1/4"	7'2 1/16"	8'7 1/8"
MAXIMUM RUN WITH 6'-8" HEADROOM	1'7 1/4"	1'8 5/8"	1'9 7/8"	1'10 7/8"	1'7 1/4"	1'8 5/8"	1'9 7/8"	1'10 7/8"	1'7 1/8"	1'8 5/8"
ANGLE OF INCLINE	39°17'	34°53'	30°58'	27°56'	39°17'	34°53'	30°58'	27°56'	39°42'	34°53'
LENGTH OF CARRIAGE	9'3 15/16"	10'5 11/16"	11'8 5/16"	12'11 11/32"	9'3 15/16"	10'5 11/16"	11'8 5/16"	12'11 11/32"	9'3 27/32"	10'5 11/16"
L. F. OF RISER PER INCH OF STAIRWAY WIDTH	.833	.916	1.0	1.083	.833	.916	1.0	1.083	.833	.916
L. F. OF TREAD PER INCH OF STAIRWAY WIDTH	.75	.833	.916	1.0	.75	.833	.916	1.0	.75	.833
FRAMING SQUARE SETTINGS FOR CARRIAGE CUTS — TONGUE	7 7/8"	7 3/16"	6 9/16"	6 1/16"	7 7/8"	7 3/16"	6 9/16"	6 1/16"	7 15/16"	7 3/16"
FRAMING SQUARE SETTINGS FOR CARRIAGE CUTS — BODY	9 5/8"	10 5/16"	10 15/16"	11 7/16"	9 5/8"	10 5/16"	10 15/16"	11 7/16"	9 9/16"	10 5/16"

FLOOR TO FLOOR RISE		6'7 1/8"	6'7 1/8"	6'7 1/4"	6'7 1/4"	6'7 1/4"	6'7 1/4"	6'7 3/8"	6'7 3/8"	6'7 3/8"	6'7 3/8"
NUMBER OF RISERS		12	13	10	11	12	13	10	11	12	13
HEIGHT OF EACH RISER		6 5/8"	6 1/16"	7 15/16"	7 3/16"	6 5/8"	6 1/8"	7 15/16"	7 3/16"	6 5/8"	6 1/8"
TOTAL HEIGHT OF RISERS		6'7 1/2"	6'6 13/16"	6'7 3/8"	6'7 1/16"	6'7 1/2"	6'7 5/8"	6'7 3/8"	6'7 1/16"	6'7 1/2"	6'7 5/8"
RISERS OVERAGE (+) OR UNDERAGE (—)		+3/8"	−5/16"	+1/8"	−3/16"	+1/4"	+3/8"	0	−5/16"	+1/8"	+1/4"
NUMBER OF TREADS		11	12	9	10	11	12	9	10	11	12
WIDTH OF EACH TREAD		10 7/8"	11 7/16"	9 9/16"	10 5/16"	10 7/8"	11 3/8"	9 9/16"	10 5/16"	10 7/8"	11 3/8"
TOTAL RUN OF STAIRWAY		9'11 5/8"	11'5 1/4"	7'2 1/16"	8'7 1/8"	9'11 5/8"	11'4 1/2"	7'2 1/16"	8'7 1/8"	9'11 5/8"	11'4 1/2"
MAXIMUM RUN WITH 6'-8" HEADROOM		1'9 3/4"	1'10 7/8"	1'7 1/8"	1'8 5/8"	1'9 3/4"	1'10 3/4"	1'7 1/8"	1'8 5/8"	1'9 3/4"	1'10 3/4"
ANGLE OF INCLINE		31°21'	27°56'	39°42'	34°53'	31°21'	28°18'	39°42'	34°53'	31°21'	28°18'
LENGTH OF CARRIAGE		11'8 1/16"	12'11 11/32"	9'3 27/32"	10'5 11/16"	11'8 1/16"	12'11 1/32"	9'3 27/32"	10'5 11/16"	11'8 1/16"	12'11 1/32"
L. F. OF RISER PER INCH OF STAIRWAY WIDTH		1.0	1.083	.833	.916	1.0	1.083	.833	.916	1.0	1.083
L. F. OF TREAD PER INCH OF STAIRWAY WIDTH		.916	1.0	.75	.833	.916	1.0	.75	.833	.916	1.0
FRAMING SQUARE SETTINGS FOR CARRIAGE CUTS	TONGUE	6 5/8"	6 1/16"	7 15/16"	7 3/16"	6 5/8"	6 1/8"	7 15/16"	7 3/16"	6 5/8"	6 1/8"
	BODY	10 7/8"	11 7/16"	9 9/16"	10 5/16"	10 7/8"	11 3/8"	9 9/16"	10 5/16"	10 7/8"	11 3/8"

FLOOR TO FLOOR RISE	6'7$\frac{1}{2}$"	6'7$\frac{1}{2}$"	6'7$\frac{1}{2}$"	6'7$\frac{1}{2}$"	6'7$\frac{5}{8}$"	6'7$\frac{5}{8}$"	6'7$\frac{5}{8}$"	6'7$\frac{5}{8}$"	6'7$\frac{3}{4}$"	6'7$\frac{3}{4}$"
NUMBER OF RISERS	10	11	12	13	10	11	12	13	10	11
HEIGHT OF EACH RISER	7$\frac{15}{16}$"	7$\frac{1}{4}$"	6$\frac{5}{8}$"	6$\frac{1}{8}$"	7$\frac{15}{16}$"	7$\frac{1}{4}$"	6$\frac{5}{8}$"	6$\frac{1}{8}$"	8"	7$\frac{1}{4}$"
TOTAL HEIGHT OF RISERS	6'7$\frac{3}{8}$"	6'7$\frac{3}{4}$"	6'7$\frac{1}{2}$"	6'7$\frac{5}{8}$"	6'7$\frac{3}{8}$"	6'7$\frac{3}{4}$"	6'7$\frac{1}{2}$"	6'7$\frac{5}{8}$"	6'8"	6'7$\frac{3}{4}$"
RISERS OVERAGE (+) OR UNDERAGE (—)	$-\frac{1}{8}$"	$+\frac{1}{4}$"	0	$+\frac{1}{8}$"	$-\frac{1}{4}$"	$+\frac{1}{8}$"	$-\frac{1}{8}$"	0	$+\frac{1}{4}$"	0
NUMBER OF TREADS	9	10	11	12	9	10	11	12	9	10
WIDTH OF EACH TREAD	9$\frac{9}{16}$"	10$\frac{1}{4}$"	10$\frac{7}{8}$"	11$\frac{3}{8}$"	9$\frac{9}{16}$"	10$\frac{1}{4}$"	10$\frac{7}{8}$"	11$\frac{3}{8}$"	9$\frac{1}{2}$"	10$\frac{1}{4}$"
TOTAL RUN OF STAIRWAY	7'2$\frac{1}{16}$"	8'6$\frac{1}{2}$"	9'11$\frac{5}{8}$"	11'4$\frac{1}{2}$"	7'2$\frac{1}{16}$"	8'6$\frac{1}{2}$"	9'11$\frac{5}{8}$"	11'4$\frac{1}{2}$"	7'1$\frac{1}{2}$"	8'6$\frac{1}{2}$"
MAXIMUM RUN WITH 6'-8" HEADROOM	1'7$\frac{1}{8}$"	1'8$\frac{1}{2}$"	1'9$\frac{3}{4}$"	1'10$\frac{3}{4}$"	1'7$\frac{1}{8}$"	1'8$\frac{1}{2}$"	1'9$\frac{3}{4}$"	1'10$\frac{3}{4}$"	9$\frac{1}{2}$"	1'8$\frac{1}{2}$"
ANGLE OF INCLINE	39°42'	35°16'	31°21'	28°18'	39°42'	35°16'	31°21'	28°18'	40°6'	35°16'
LENGTH OF CARRIAGE	9'3$\frac{27}{32}$"	10'5$\frac{9}{16}$"	11'8$\frac{1}{16}$"	12'11$\frac{1}{32}$"	9'3$\frac{27}{32}$"	10'5$\frac{9}{16}$"	11'8$\frac{1}{16}$"	12'11$\frac{1}{32}$"	9'3$\frac{25}{32}$"	10'5$\frac{9}{16}$"
L. F. OF RISER PER INCH OF STAIRWAY WIDTH	.833	.916	1.0	1.083	.833	.916	1.0	1.083	.833	.916
L. F. OF TREAD PER INCH OF STAIRWAY WIDTH	.75	.833	.916	1.0	.75	.833	.916	1.0	.75	.833
FRAMING SQUARE SETTINGS FOR CARRIAGE CUTS — TONGUE	7$\frac{15}{16}$"	7$\frac{1}{4}$"	6$\frac{5}{8}$"	6$\frac{1}{8}$"	7$\frac{15}{16}$"	7$\frac{1}{4}$"	6$\frac{5}{8}$"	6$\frac{1}{8}$"	8"	7$\frac{1}{4}$"
FRAMING SQUARE SETTINGS FOR CARRIAGE CUTS — BODY	9$\frac{9}{16}$"	10$\frac{1}{4}$"	10$\frac{7}{8}$"	11$\frac{3}{8}$"	9$\frac{9}{16}$"	10$\frac{1}{4}$"	10$\frac{7}{8}$"	11$\frac{3}{8}$"	9$\frac{1}{2}$"	10$\frac{1}{4}$"

FLOOR TO FLOOR RISE	6'7 3/4"	6'7 3/4"	6'7 7/8"	6'7 7/8"	6'7 7/8"	6'7 7/8"	6'8"	6'8"	6'8"	6'8"
NUMBER OF RISERS	12	13	10	11	12	13	10	11	12	13
HEIGHT OF EACH RISER	6 5/8"	6 1/8"	8"	7 1/4"	6 11/16"	6 1/8"	8"	7 1/4"	6 11/16"	6 1/8"
TOTAL HEIGHT OF RISERS	6'7 1/2"	6'7 5/8"	6'8"	6'7 3/4"	6'8 1/4"	6'7 5/8"	6'8"	6'7 3/4"	6'8 1/4"	6'7 5/8"
RISERS OVERAGE (+) OR UNDERAGE (—)	−1/4"	−1/8"	+1/8"	−1/8"	+3/8"	−1/4"	0	−1/4"	+1/4"	−3/8"
NUMBER OF TREADS	11	12	9	10	11	12	9	10	11	12
WIDTH OF EACH TREAD	10 7/8"	11 3/8"	9 1/2"	10 1/4"	10 13/16"	11 3/8"	9 1/2"	10 1/4"	10 13/16"	11 3/8"
TOTAL RUN OF STAIRWAY	9'11 5/8"	11'4 1/2"	7'1 1/2"	8'6 1/2"	9'10 15/16"	11'4 1/2"	7'1 1/2"	8'6 1/2"	9'10 15/16"	11'4 1/2"
MAXIMUM RUN WITH 6'-8" HEADROOM	1'9 3/4"	1'10 3/4"	9 1/2"	1'8 1/2"	1'9 5/8"	1'10 3/4"	9 1/2"	1'8 1/2"	1'9 5/8"	1'10 3/4"
ANGLE OF INCLINE	31°21'	28°18'	40°6'	35°16'	31°44'	28°18'	40°6'	35°16'	31°44'	28°18'
LENGTH OF CARRIAGE	11'8 1/16"	12'11 1/32"	9'3 25/32"	10'5 9/16"	11'7 27/32"	12'11 1/32"	9'3 25/32"	10'5 9/16"	11'7 27/32"	12'11 1/32"
L. F. OF RISER PER INCH OF STAIRWAY WIDTH	1.0	1.083	.833	.916	1.0	1.083	.833	.916	1.0	1.083
L. F. OF TREAD PER INCH OF STAIRWAY WIDTH	.916	1.0	.75	.833	.916	1.0	.75	.833	.916	1.0
FRAMING SQUARE SETTINGS FOR CARRIAGE CUTS — TONGUE	6 5/8"	6 1/8"	8"	7 1/4"	6 11/16"	6 1/8"	8"	7 1/4"	6 11/16"	6 1/8"
FRAMING SQUARE SETTINGS FOR CARRIAGE CUTS — BODY	10 7/8"	11 3/8"	9 1/2"	10 1/4"	10 13/16"	11 3/8"	9 1/2"	10 1/4"	10 13/16"	11 3/8"

FLOOR TO FLOOR RISE	6'8 1/8"	6'8 1/8"	6'8 1/8"	6'8 1/8"	6'8 1/4"	6'8 1/4"	6'8 1/4"	6'8 1/4"	6'8 3/8"	6'8 3/8"
NUMBER OF RISERS	10	11	12	13	10	11	12	13	10	11
HEIGHT OF EACH RISER	8"	7 5/16"	6 11/16"	6 3/16"	8"	7 5/16"	6 11/16"	6 3/16"	8 1/16"	7 5/16"
TOTAL HEIGHT OF RISERS	6'8"	6'8 7/8"	6'8 1/4"	6'8 7/16"	6'8"	6'8 7/16"	6'8 1/4"	6'8 7/16"	6'8 5/8"	6'8 7/16"
RISERS OVERAGE (+) OR UNDERAGE (—)	−1/8"	+5/16"	+1/8"	+5/16"	−1/4"	+3/16"	0	+3/16"	+1/4"	+1/16"
NUMBER OF TREADS	9	10	11	12	9	10	11	12	9	10
WIDTH OF EACH TREAD	9 1/2"	10 3/16"	10 13/16"	11 5/16"	9 1/2"	10 3/16"	10 13/16"	11 5/16"	9 7/16"	10 3/16"
TOTAL RUN OF STAIRWAY	7'1 1/2"	8'5 7/8"	9'10 15/16"	11'3 3/4"	7'1 1/2"	8'5 7/8"	9'10 15/16"	11'3 3/4"	7' 15/16"	8'5 7/8"
MAXIMUM RUN WITH 6'-8" HEADROOM	9 1/2"	1'8 3/8"	1'9 5/8"	1'10 5/8"	9 1/2"	1'8 3/8"	1'9 5/8"	1'10 5/8"	9 7/16"	1'8 3/8"
ANGLE OF INCLINE	40°6'	35°40'	31°44'	28°41'	40°6'	35°40'	31°44'	28°41'	40°30'	35°40'
LENGTH OF CARRIAGE	9'3 25/32"	10'5 13/32"	11'7 27/32"	12'10 23/32"	9'3 25/32"	10'5 13/32"	11'7 27/32"	12'10 23/32"	9'3 23/32"	10'5 13/32"
L. F. OF RISER PER INCH OF STAIRWAY WIDTH	.833	.916	1.0	1.083	.833	.916	1.0	1.083	.833	.916
L. F. OF TREAD PER INCH OF STAIRWAY WIDTH	.75	.833	.916	1.0	.75	.833	.916	1.0	.75	.833
FRAMING SQUARE SETTINGS FOR CARRIAGE CUTS — TONGUE	8"	7 5/16"	6 11/16"	6 3/16"	8"	7 5/16"	6 11/16"	6 3/16"	8 1/16"	7 5/16"
FRAMING SQUARE SETTINGS FOR CARRIAGE CUTS — BODY	9 1/2"	10 3/16"	10 13/16"	11 5/16"	9 1/2"	10 3/16"	10 13/16"	11 5/16"	9 7/16"	10 3/16"

FLOOR TO FLOOR RISE	6'8 3/8"	6'8 3/8"	6'8 1/2"	6'8 1/2"	6'8 1/2"	6'8 1/2"	6'8 5/8"	6'8 5/8"	6'8 5/8"	6'8 5/8"
NUMBER OF RISERS	12	13	10	11	12	13	10	11	12	13
HEIGHT OF EACH RISER	6 11/16"	6 3/16"	8 1/16"	7 5/16"	6 11/16"	6 3/16"	8 1/16"	7 5/16"	6 3/4"	6 3/16"
TOTAL HEIGHT OF RISERS	6'8 1/4"	6'8 7/16"	6'8 5/8"	6'8 7/16"	6'8 1/4"	6'8 7/16"	6'8 5/8"	6'8 7/16"	6'9"	6'8 7/16"
RISERS OVERAGE (+) OR UNDERAGE (—)	−1/8"	+1/16"	+1/8"	−1/16"	−1/4"	−1/16"	0	−3/16"	+3/8"	−3/16"
NUMBER OF TREADS	11	12	9	10	11	12	9	10	11	12
WIDTH OF EACH TREAD	10 13/16"	11 5/16"	9 7/16"	10 3/16"	10 13/16"	11 5/16"	9 7/16"	10 3/16"	10 3/4"	11 5/16"
TOTAL RUN OF STAIRWAY	9'10 15/16"	11'3 3/4"	7'15/16"	8'5 7/8"	9'10 15/16"	11'3 3/4"	7'15/16"	8'5 7/8"	9'10 1/4"	11'3 3/4"
MAXIMUM RUN WITH 6'-8" HEADROOM	1'9 5/8"	1'10 5/8"	9 7/16"	1'8 3/8"	1'9 5/8"	1'10 5/8"	9 7/16"	1'8 3/8"	1'9 1/2"	1'10 5/8"
ANGLE OF INCLINE	31°44'	28°41'	40°30'	35°40'	31°44'	28°41'	40°30'	35°40'	32°7'	28°41'
LENGTH OF CARRIAGE	11'7 27/32"	12'10 23/32"	9'3 23/32"	10'5 13/32"	11'7 27/32"	12'10 23/32"	9'3 23/32"	10'5 13/32"	11'7 5/8"	12'10 23/32"
L. F. OF RISER PER INCH OF STAIRWAY WIDTH	1.0	1.083	.833	.916	1.0	1.083	.833	.916	1.0	1.083
L. F. OF TREAD PER INCH OF STAIRWAY WIDTH	.916	1.0	.75	.833	.916	1.0	.75	.833	.916	1.0
FRAMING SQUARE SETTINGS FOR CARRIAGE CUTS — TONGUE	6 11/16"	6 3/16"	8 1/16"	7 5/16"	6 11/16"	6 3/16"	8 1/16"	7 5/16"	6 3/4"	6 3/16"
FRAMING SQUARE SETTINGS FOR CARRIAGE CUTS — BODY	10 13/16"	11 5/16"	9 7/16"	10 3/16"	10 13/16"	11 5/16"	9 7/16"	10 3/16"	10 3/4"	11 5/16"

FLOOR TO FLOOR RISE	$6'8\frac{3}{4}''$	$6'8\frac{3}{4}''$	$6'8\frac{3}{4}''$	$6'8\frac{3}{4}''$	$6'8\frac{7}{8}''$	$6'8\frac{7}{8}''$	$6'8\frac{7}{8}''$	$6'8\frac{7}{8}''$	6'9"	6'9"
NUMBER OF RISERS	10	11	12	13	10	11	12	13	10	11
HEIGHT OF EACH RISER	$8\frac{1}{16}''$	$7\frac{5}{16}''$	$6\frac{3}{4}''$	$6\frac{3}{16}''$	$8\frac{1}{16}''$	$7\frac{3}{8}''$	$6\frac{3}{4}''$	$6\frac{1}{4}''$	$8\frac{1}{8}''$	$7\frac{3}{8}''$
TOTAL HEIGHT OF RISERS	$6'8\frac{5}{8}''$	$6'8\frac{7}{16}''$	6'9"	$6'8\frac{7}{16}''$	$6'8\frac{5}{8}''$	$6'9\frac{1}{8}''$	6'9"	$6'9\frac{1}{4}''$	$6'9\frac{1}{4}''$	$6'9\frac{1}{8}''$
RISERS OVERAGE (+) OR UNDERAGE (—)	$-\frac{1}{8}''$	$-\frac{5}{16}''$	$+\frac{1}{4}''$	$-\frac{5}{16}''$	$-\frac{1}{4}''$	$+\frac{1}{4}''$	$+\frac{1}{8}''$	$+\frac{3}{8}''$	$+\frac{1}{4}''$	$+\frac{1}{8}''$
NUMBER OF TREADS	9	10	11	12	9	10	11	12	9	10
WIDTH OF EACH TREAD	$9\frac{7}{16}''$	$10\frac{3}{16}''$	$10\frac{3}{4}''$	$11\frac{5}{16}''$	$9\frac{7}{16}''$	$10\frac{1}{8}''$	$10\frac{3}{4}''$	$11\frac{1}{4}''$	$9\frac{3}{8}''$	$10\frac{1}{8}''$
TOTAL RUN OF STAIRWAY	$7'\frac{15}{16}''$	$8'5\frac{7}{8}''$	$9'10\frac{1}{4}''$	$11'3\frac{3}{4}''$	$7'\frac{15}{16}''$	$8'5\frac{1}{4}''$	$9'10\frac{1}{4}''$	11'3"	$7'\frac{3}{8}''$	$8'5\frac{1}{4}''$
MAXIMUM RUN WITH 6'-8" HEADROOM	$9\frac{7}{16}''$	$1'8\frac{3}{8}''$	$1'9\frac{1}{2}''$	$1'10\frac{5}{8}''$	$9\frac{7}{16}''$	$1'8\frac{1}{4}''$	$1'9\frac{1}{2}''$	$1'10\frac{1}{2}''$	$9\frac{3}{8}''$	$1'8\frac{1}{4}''$
ANGLE OF INCLINE	40°30'	35°40'	32°7'	28°41'	40°30'	36°4'	32°7'	29°3'	40°55'	36°4'
LENGTH OF CARRIAGE	$9'3\frac{23}{32}''$	$10'5\frac{13}{32}''$	$11'7\frac{5}{8}''$	$12'10\frac{23}{32}''$	$9'3\frac{23}{32}''$	$10'5\frac{1}{4}''$	$11'7\frac{5}{8}''$	$12'10\frac{7}{16}''$	$9'3\frac{21}{32}''$	$10'5\frac{1}{4}''$
L. F. OF RISER PER INCH OF STAIRWAY WIDTH	.833	.916	1.0	1.083	.833	.916	1.0	1.083	.833	.916
L. F. OF TREAD PER INCH OF STAIRWAY WIDTH	.75	.833	.916	1.0	.75	.833	.916	1.0	.75	.833
FRAMING SQUARE SETTINGS FOR CARRIAGE CUTS — TONGUE	$8\frac{1}{16}''$	$7\frac{5}{16}''$	$6\frac{3}{4}''$	$6\frac{3}{16}''$	$8\frac{1}{16}''$	$7\frac{3}{8}''$	$6\frac{3}{4}''$	$6\frac{1}{4}''$	$8\frac{1}{8}''$	$7\frac{3}{8}''$
FRAMING SQUARE SETTINGS FOR CARRIAGE CUTS — BODY	$9\frac{7}{16}''$	$10\frac{3}{16}''$	$10\frac{3}{4}''$	$11\frac{5}{16}''$	$9\frac{7}{16}''$	$10\frac{1}{8}''$	$10\frac{3}{4}''$	$11\frac{1}{4}''$	$9\frac{3}{8}''$	$10\frac{1}{8}''$

FLOOR TO FLOOR RISE	6'9"	6'9"	6'9 1/8"	6'9 1/8"	6'9 1/8"	6'9 1/8"	6'9 1/4"	6'9 1/4"	6'9 1/4"	6'9 1/4"
NUMBER OF RISERS	12	13	10	11	12	13	10	11	12	13
HEIGHT OF EACH RISER	6 3/4"	6 1/4"	8 1/8"	7 3/8"	6 3/4"	6 1/4"	8 1/8"	7 3/8"	6 3/4"	6 1/4"
TOTAL HEIGHT OF RISERS	6'9"	6'9 1/4"	6'9 1/4"	6'9 1/8"	6'9"	6'9 1/4"	6'9 1/4"	6'9 1/8"	6'9"	6'9 1/4"
RISERS OVERAGE (+) OR UNDERAGE (—)	0	+1/4"	+1/8"	0	−1/8"	+1/8"	0	−1/8"	−1/4"	0
NUMBER OF TREADS	11	12	9	10	11	12	9	10	11	12
WIDTH OF EACH TREAD	10 3/4"	11 1/4"	9 3/8"	10 1/8"	10 3/4"	11 1/4"	9 3/8"	10 1/8"	10 3/4"	11 1/4"
TOTAL RUN OF STAIRWAY	9'10 1/4"	11'3"	7' 3/8"	8'5 1/4"	9'10 1/4"	11'3"	7' 3/8"	8'5 1/4"	9'10 1/4"	11'3"
MAXIMUM RUN WITH 6'-8" HEADROOM	1'9 1/2"	1'10 1/2"	9 3/8"	1'8 1/4"	1'9 1/2"	1'10 1/2"	9 3/8"	1'8 1/4"	1'9 1/2"	1'10 1/2"
ANGLE OF INCLINE	32°7'	29°3'	40°55'	36°4'	32°7'	29°3'	40°55'	36°4'	32°7'	29°3'
LENGTH OF CARRIAGE	11'7 5/8"	12'10 7/16"	9'3 21/32"	10'5 1/4"	11'7 5/8"	12'10 7/16"	9'3 21/32"	10'5 1/4"	11'7 5/8"	12'10 7/16"
L. F. OF RISER PER INCH OF STAIRWAY WIDTH	1.0	1.083	.833	.916	1.0	1.083	.833	.916	1.0	1.083
L. F. OF TREAD PER INCH OF STAIRWAY WIDTH	.916	1.0	.75	.833	.916	1.0	.75	.833	.916	1.0
FRAMING SQUARE SETTINGS FOR CARRIAGE CUTS — TONGUE	6 3/4"	6 1/4"	8 1/8"	7 3/8"	6 3/4"	6 1/4"	8 1/8"	7 3/8"	6 3/4"	6 1/4"
FRAMING SQUARE SETTINGS FOR CARRIAGE CUTS — BODY	10 3/4"	11 1/4"	9 3/8"	10 1/8"	10 3/4"	11 1/4"	9 3/8"	10 1/8"	10 3/4"	11 1/4"

FLOOR TO FLOOR RISE	6'9$\frac{3}{8}$"	6'9$\frac{3}{8}$"	6'9$\frac{3}{8}$"	6'9$\frac{3}{8}$"	6'9$\frac{1}{2}$"	6'9$\frac{1}{2}$"	6'9$\frac{1}{2}$"	6'9$\frac{1}{2}$"	6'9$\frac{5}{8}$"	6'9$\frac{5}{8}$"
NUMBER OF RISERS	10	11	12	13	10	11	12	13	10	11
HEIGHT OF EACH RISER	8$\frac{1}{8}$"	7$\frac{3}{8}$"	6$\frac{13}{16}$"	6$\frac{1}{4}$"	8$\frac{1}{8}$"	7$\frac{7}{16}$"	6$\frac{13}{16}$"	6$\frac{1}{4}$"	8$\frac{3}{16}$"	7$\frac{7}{16}$"
TOTAL HEIGHT OF RISERS	6'9$\frac{1}{4}$"	6'9$\frac{1}{8}$"	6'9$\frac{3}{4}$"	6'9$\frac{1}{4}$"	6'9$\frac{1}{4}$"	6'9$\frac{13}{16}$"	6'9$\frac{3}{4}$"	6'9$\frac{1}{4}$"	6'9$\frac{7}{8}$"	6'9$\frac{13}{16}$"
RISERS OVERAGE (+) OR UNDERAGE (—)	$-\frac{1}{8}$"	$-\frac{1}{4}$"	$+\frac{3}{8}$"	$-\frac{1}{8}$"	$-\frac{1}{4}$"	$+\frac{5}{16}$"	$+\frac{1}{4}$"	$-\frac{1}{4}$"	$+\frac{1}{4}$"	$+\frac{3}{16}$"
NUMBER OF TREADS	9	10	11	12	9	10	11	12	9	10
WIDTH OF EACH TREAD	9$\frac{3}{8}$"	10$\frac{1}{8}$"	10$\frac{11}{16}$"	11$\frac{1}{4}$"	9$\frac{3}{8}$"	10$\frac{1}{16}$"	10$\frac{11}{16}$"	11$\frac{1}{4}$"	9$\frac{5}{16}$"	10$\frac{1}{16}$"
TOTAL RUN OF STAIRWAY	7'$\frac{3}{8}$"	8'5$\frac{1}{4}$"	9'9$\frac{9}{16}$"	11'3"	7'$\frac{3}{8}$"	8'4$\frac{5}{8}$"	9'9$\frac{9}{16}$"	11'3"	6'11$\frac{13}{16}$"	8'4$\frac{5}{8}$"
MAXIMUM RUN WITH 6'-8" HEADROOM	9$\frac{3}{8}$"	1'8$\frac{1}{4}$"	1'9$\frac{3}{8}$"	1'10$\frac{1}{2}$"	9$\frac{3}{8}$"	1'8$\frac{1}{8}$"	1'9$\frac{3}{8}$"	1'10$\frac{1}{2}$"	9$\frac{5}{16}$"	1'8$\frac{1}{8}$"
ANGLE OF INCLINE	40°55'	36°4'	32°31'	29°3'	40°55'	36°28'	32°31'	29°3'	41°19'	36°28'
LENGTH OF CARRIAGE	9'3$\frac{21}{32}$"	10'5$\frac{1}{4}$"	11'7$\frac{7}{16}$"	12'10$\frac{7}{16}$"	9'3$\frac{21}{32}$"	10'5$\frac{1}{8}$"	11'7$\frac{7}{16}$"	12'10$\frac{7}{16}$"	9'3$\frac{19}{32}$"	10'5$\frac{1}{8}$"
L. F. OF RISER PER INCH OF STAIRWAY WIDTH	.833	.916	1.0	1.083	.833	.916	1.0	1.083	.833	.916
L. F. OF TREAD PER INCH OF STAIRWAY WIDTH	.75	.833	.916	1.0	.75	.833	.916	1.0	.75	.833
FRAMING SQUARE SETTINGS FOR CARRIAGE CUTS — TONGUE	8$\frac{1}{8}$"	7$\frac{3}{8}$"	6$\frac{13}{16}$"	6$\frac{1}{4}$"	8$\frac{1}{8}$"	7$\frac{7}{16}$"	6$\frac{13}{16}$"	6$\frac{1}{4}$"	8$\frac{3}{16}$"	7$\frac{7}{16}$"
FRAMING SQUARE SETTINGS FOR CARRIAGE CUTS — BODY	9$\frac{3}{8}$"	10$\frac{1}{8}$"	10$\frac{11}{16}$"	11$\frac{1}{4}$"	9$\frac{3}{8}$"	10$\frac{1}{16}$"	10$\frac{11}{16}$"	11$\frac{1}{4}$"	9$\frac{5}{16}$"	10$\frac{1}{16}$"

FLOOR TO FLOOR RISE	6'9 5/8"	6'9 5/8"	6'9 3/4"	6'9 3/4"	6'9 3/4"	6'9 3/4"	6'9 7/8"	6'9 7/8"	6'9 7/8"	6'9 7/8"
NUMBER OF RISERS	12	13	10	11	12	13	10	11	12	13
HEIGHT OF EACH RISER	6 13/16"	6 1/4"	8 3/16"	7 7/16"	6 13/16"	6 5/16"	8 3/16"	7 7/16"	6 13/16"	6 5/16"
TOTAL HEIGHT OF RISERS	6'9 3/4"	6'9 1/4"	6'9 7/8"	6'9 13/16"	6'9 3/4"	6'10 1/16"	6'9 7/8"	6'9 13/16"	6'9 3/4"	6'10 1/16"
RISERS OVERAGE (+) OR UNDERAGE (—)	+1/8"	−3/8"	+1/8"	+1/16"	0	+5/16"	0	−1/16"	−1/8"	+3/16"
NUMBER OF TREADS	11	12	9	10	11	12	9	10	11	12
WIDTH OF EACH TREAD	10 11/16"	11 1/4"	9 5/16"	10 1/16"	10 11/16"	11 3/16"	9 5/16"	10 1/16"	10 11/16"	11 3/16"
TOTAL RUN OF STAIRWAY	9'9 9/16"	11'3"	6'11 13/16"	8'4 5/8"	9'9 9/16"	11'2 1/4"	6'11 13/16"	8'4 5/8"	9'9 9/16"	11'2 1/4"
MAXIMUM RUN WITH 6'-8" HEADROOM	1'9 3/8"	1'10 1/2"	9 5/16"	1'8 1/8"	1'9 3/8"	1'10 3/8"	9 5/16"	1'8 1/8"	1'9 3/8"	1'10 3/8"
ANGLE OF INCLINE	32°31'	29°3'	41°19'	36°28'	32°31'	29°26'	41°19'	36°28'	32°31'	29°26'
LENGTH OF CARRIAGE	11'7 7/16"	12'10 7/16"	9'3 19/32"	10'5 1/8"	11'7 7/16"	12'10 5/32"	9'3 19/32"	10'5 1/8"	11'7 7/16"	12'10 5/32"
L. F. OF RISER PER INCH OF STAIRWAY WIDTH	1.0	1.083	.833	.916	1.0	1.083	.833	.916	1.0	1.083
L. F. OF TREAD PER INCH OF STAIRWAY WIDTH	.916	1.0	.75	.833	.916	1.0	.75	.833	.916	1.0
FRAMING SQUARE SETTINGS FOR CARRIAGE CUTS — TONGUE	6 13/16"	6 1/4"	8 3/16"	7 7/16"	6 13/16"	6 5/16"	8 3/16"	7 7/16"	6 13/16"	6 5/16"
FRAMING SQUARE SETTINGS FOR CARRIAGE CUTS — BODY	10 11/16"	11 1/4"	9 5/16"	10 1/16"	10 11/16"	11 3/16"	9 5/16"	10 1/16"	10 11/16"	11 3/16"

FLOOR TO FLOOR RISE		6'10"	6'10"	6'10"	6'10"	6'10 1/8"	6'10 1/8"	6'10 1/8"	6'10 1/8"	6'10 1/4"	6'10 1/4"
NUMBER OF RISERS		10	11	12	13	10	11	12	13	10	11
HEIGHT OF EACH RISER		8 3/16"	7 7/16"	6 13/16"	6 5/16"	8 3/16"	7 7/16"	6 7/8"	6 5/16"	8 1/4"	7 1/2"
TOTAL HEIGHT OF RISERS		6'9 7/8"	6'9 13/16"	6'9 3/4"	6'10 1/16"	6'9 7/8"	6'9 13/16"	6'10 1/2"	6'10 1/16"	6'10 1/2"	6'10 1/2"
RISERS OVERAGE (+) OR UNDERAGE (—)		−1/8"	−3/16"	−1/4"	+1/16"	−1/4"	−5/16"	+3/8"	−1/16"	+1/4"	+1/4"
NUMBER OF TREADS		9	10	11	12	9	10	11	12	9	10
WIDTH OF EACH TREAD		9 5/16"	10 1/16"	10 11/16"	11 3/16"	9 5/16"	10 1/16"	10 5/8"	11 3/16"	9 1/4"	10"
TOTAL RUN OF STAIRWAY		6'11 13/16"	8'4 5/8"	9'9 9/16"	11'2 1/4"	6'11 13/16"	8'4 5/8"	9'8 7/8"	11'2 1/4"	6'11 1/4"	8'4"
MAXIMUM RUN WITH 6'-8" HEADROOM		9 5/16"	1'8 1/8"	1'9 3/8"	1'10 3/8"	9 5/16"	1'8 1/8"	1'9 1/4"	1'10 3/8"	9 1/4"	1'8"
ANGLE OF INCLINE		41°19'	36°28'	32°31'	29°26'	41°19'	36°28'	32°54'	29°26'	41°44'	36°52'
LENGTH OF CARRIAGE		9'3 19/32"	10'5 1/8"	11'7 7/16"	12'10 5/32"	9'3 19/32"	10'5 1/8"	11'7 7/32"	12'10 5/32"	9'3 9/16"	10'5"
L. F. OF RISER PER INCH OF STAIRWAY WIDTH		.833	.916	1.0	1.083	.833	.916	1.0	1.083	.833	.916
L. F. OF TREAD PER INCH OF STAIRWAY WIDTH		.75	.833	.916	1.0	.75	.833	.916	1.0	.75	.833
FRAMING SQUARE SETTINGS FOR CARRIAGE CUTS	TONGUE	8 3/16"	7 7/16"	6 13/16"	6 5/16"	8 3/16"	7 7/16"	6 7/8"	6 5/16"	8 1/4"	7 1/2"
	BODY	9 5/16"	10 1/16"	10 11/16"	11 3/16"	9 5/16"	10 1/16"	10 5/8"	11 3/16"	9 1/4"	10"

FLOOR TO FLOOR RISE	6'10 1/4"	6'10 1/4"	6'10 3/8"	6'10 3/8"	6'10 3/8"	6'10 3/8"	6'10 1/2"	6'10 1/2"	6'10 1/2"	6'10 1/2"
NUMBER OF RISERS	12	13	10	11	12	13	10	11	12	13
HEIGHT OF EACH RISER	6 7/8"	6 5/16"	8 1/4"	7 1/2"	6 7/8"	6 5/16"	8 1/4"	7 1/2"	6 7/8"	6 3/8"
TOTAL HEIGHT OF RISERS	6'10 1/2"	6'10 1/16"	6'10 1/2"	6'10 1/2"	6'10 1/2"	6'10 1/16"	6'10 1/2"	6'10 1/2"	6'10 1/2"	6'10 7/8"
RISERS OVERAGE (+) OR UNDERAGE (—)	+1/4"	-3/16"	+1/8"	+1/8"	+1/8"	-5/16"	0	0	0	+3/8"
NUMBER OF TREADS	11	12	9	10	11	12	9	10	11	12
WIDTH OF EACH TREAD	10 5/8"	11 3/16"	9 1/4"	10"	10 5/8"	11 3/16"	9 1/4"	10"	10 5/8"	11 1/8"
TOTAL RUN OF STAIRWAY	9'8 7/8"	11'2 1/4"	6'11 1/4"	8'4"	9'8 7/8"	11'2 1/4"	6'11 1/4"	8'4"	9'8 7/8"	11'1 1/2"
MAXIMUM RUN WITH 6'-8" HEADROOM	1'9 1/4"	1'10 3/8"	9 1/4"	1'8"	1'9 1/4"	1'10 3/8"	9 1/4"	1'8"	1'9 1/4"	1'10 1/4"
ANGLE OF INCLINE	32°54'	29°26'	41°44'	36°52'	32°54'	29°26'	41°44'	36°52'	32°54'	29°49'
LENGTH OF CARRIAGE	11'7 7/32"	12'10 5/32"	9'3 9/16"	10'5"	11'7 7/32"	12'10 5/32"	9'3 9/16"	10'5"	11'7 7/32"	12'9 7/8"
L. F. OF RISER PER INCH OF STAIRWAY WIDTH	1.0	1.083	.833	.916	1.0	1.083	.833	.916	1.0	1.083
L. F. OF TREAD PER INCH OF STAIRWAY WIDTH	.916	1.0	.75	.833	.916	1.0	.75	.833	.916	1.0
FRAMING SQUARE SETTINGS FOR CARRIAGE CUTS — TONGUE	6 7/8"	6 5/16"	8 1/4"	7 1/2"	6 7/8"	6 5/16"	8 1/4"	7 1/2"	6 7/8"	6 3/8"
FRAMING SQUARE SETTINGS FOR CARRIAGE CUTS — BODY	10 5/8"	11 3/16"	9 1/4"	10"	10 5/8"	11 3/16"	9 1/4"	10"	10 5/8"	11 1/8"

FLOOR TO FLOOR RISE	6'10 5/8"	6'10 5/8"	6'10 5/8"	6'10 5/8"	6'10 3/4"	6'10 3/4"	6'10 3/4"	6'10 3/4"	6'10 7/8"	6'10 7/8"
NUMBER OF RISERS	10	11	12	13	10	11	12	13	11	12
HEIGHT OF EACH RISER	8 1/4"	7 1/2"	6 7/8"	6 3/8"	8 1/4"	7 1/2"	6 7/8"	6 3/8"	7 9/16"	6 15/16"
TOTAL HEIGHT OF RISERS	6'10 1/2"	6'10 1/2"	6'10 1/2"	6'10 7/8"	6'10 1/2"	6'10 1/2"	6'10 1/2"	6'10 7/8"	6'11 3/16"	6'11 1/4"
RISERS OVERAGE (+) OR UNDERAGE (—)	−1/8"	−1/8"	−1/8"	+1/4"	−1/4"	−1/4"	−1/4"	+1/8"	+5/16"	+3/8"
NUMBER OF TREADS	9	10	11	12	9	10	11	12	10	11
WIDTH OF EACH TREAD	9 1/4"	10"	10 5/8"	11 1/8"	9 1/4"	10"	10 5/8"	11 1/8"	9 15/16"	10 9/16"
TOTAL RUN OF STAIRWAY	6'11 1/4"	8'4"	9'8 7/8"	11'1 1/2"	6'11 1/4"	8'4"	9'8 7/8"	11'1 1/2"	8'3 3/8"	9'8 3/16"
MAXIMUM RUN WITH 6'-8" HEADROOM	9 1/4"	1'8"	1'9 1/4"	1'10 1/4"	9 1/4"	1'8"	1'9 1/4"	1'10 1/4"	1'7 7/8"	1'9 1/8"
ANGLE OF INCLINE	41°44'	36°52'	32°54'	29°49'	41°44'	36°52'	32°54'	29°49'	37°16'	33°18'
LENGTH OF CARRIAGE	9'3 9/16"	10'5"	11'7 7/32"	12'9 7/8"	9'3 9/16"	10'5"	11'7 7/32"	12'9 7/8"	10'4 7/8"	11'7"
L. F. OF RISER PER INCH OF STAIRWAY WIDTH	.833	.916	1.0	1.083	.833	.916	1.0	1.083	.916	1.0
L. F. OF TREAD PER INCH OF STAIRWAY WIDTH	.75	.833	.916	1.0	.75	.833	.916	1.0	.833	.916
FRAMING SQUARE SETTINGS FOR CARRIAGE CUTS — TONGUE	8 1/4"	7 1/2"	6 7/8"	6 3/8"	8 1/4"	7 1/2"	6 7/8"	6 3/8"	7 9/16"	6 15/16"
FRAMING SQUARE SETTINGS FOR CARRIAGE CUTS — BODY	9 1/4"	10"	10 5/8"	11 1/8"	9 1/4"	10"	10 5/8"	11 1/8"	9 15/16"	10 9/16"

FLOOR TO FLOOR RISE	6'10$\frac{7}{8}$"	6'11"	6'11"	6'11"	6'11$\frac{1}{8}$"	6'11$\frac{1}{8}$"	6'11$\frac{1}{8}$"	6'11$\frac{1}{4}$"	6'11$\frac{1}{4}$"	6'11$\frac{1}{4}$"
NUMBER OF RISERS	13	11	12	13	11	12	13	11	12	13
HEIGHT OF EACH RISER	6$\frac{3}{8}$"	7$\frac{9}{16}$"	6$\frac{15}{16}$"	6$\frac{3}{8}$"	7$\frac{9}{16}$"	6$\frac{15}{16}$"	6$\frac{3}{8}$"	7$\frac{9}{16}$"	6$\frac{15}{16}$"	6$\frac{3}{8}$"
TOTAL HEIGHT OF RISERS	6'10$\frac{7}{8}$"	6'11$\frac{3}{16}$"	6'11$\frac{1}{4}$"	6'10$\frac{7}{8}$"	6'11$\frac{3}{16}$"	6'11$\frac{1}{4}$"	6'10$\frac{7}{8}$"	6'11$\frac{3}{16}$"	6'11$\frac{1}{4}$"	6'10$\frac{7}{8}$"
RISERS OVERAGE (+) OR UNDERAGE (—)	0	+$\frac{3}{16}$"	+$\frac{1}{4}$"	−$\frac{1}{8}$"	+$\frac{1}{16}$"	+$\frac{1}{8}$"	−$\frac{1}{4}$"	−$\frac{1}{16}$"	0	−$\frac{3}{8}$"
NUMBER OF TREADS	12	10	11	12	10	11	12	10	11	12
WIDTH OF EACH TREAD	11$\frac{1}{8}$"	9$\frac{15}{16}$"	10$\frac{9}{16}$"	11$\frac{1}{8}$"	9$\frac{15}{16}$"	10$\frac{9}{16}$"	11$\frac{1}{8}$"	9$\frac{15}{16}$"	10$\frac{9}{16}$"	11$\frac{1}{8}$"
TOTAL RUN OF STAIRWAY	11'1$\frac{1}{2}$"	8'3$\frac{3}{8}$"	9'8$\frac{3}{16}$"	11'1$\frac{1}{2}$"	8'3$\frac{3}{8}$"	9'8$\frac{3}{16}$"	11'1$\frac{1}{2}$"	8'3$\frac{3}{8}$"	9'8$\frac{3}{16}$"	11'1$\frac{1}{2}$"
MAXIMUM RUN WITH 6'-8" HEADROOM	1'10$\frac{1}{4}$"	1'7$\frac{7}{8}$"	1'9$\frac{1}{8}$"	1'10$\frac{1}{4}$"	1'7$\frac{7}{8}$"	1'9$\frac{1}{8}$"	1'10$\frac{1}{4}$"	1'7$\frac{7}{8}$"	1'9$\frac{1}{8}$"	1'10$\frac{1}{4}$"
ANGLE OF INCLINE	29°49'	37°16'	33°18'	29°49'	37°16'	33°18'	29°49'	37°16'	33°18'	29°49'
LENGTH OF CARRIAGE	12'9$\frac{7}{8}$"	10'4$\frac{7}{8}$"	11'7"	12'9$\frac{7}{8}$"	10'4$\frac{7}{8}$"	11'7"	12'9$\frac{7}{8}$"	10'4$\frac{7}{8}$"	11'7"	12'9$\frac{7}{8}$"
L. F. OF RISER PER INCH OF STAIRWAY WIDTH	1.083	.916	1.0	1.083	.916	1.0	1.083	.916	1.0	1.083
L. F. OF TREAD PER INCH OF STAIRWAY WIDTH	1.0	.833	.916	1.0	.833	.916	1.0	.833	.916	1.0
FRAMING SQUARE SETTINGS FOR CARRIAGE CUTS — TONGUE	6$\frac{3}{8}$"	7$\frac{9}{16}$"	6$\frac{15}{16}$"	6$\frac{3}{8}$"	7$\frac{9}{16}$"	6$\frac{15}{16}$"	6$\frac{3}{8}$"	7$\frac{9}{16}$"	6$\frac{15}{16}$"	6$\frac{3}{8}$"
FRAMING SQUARE SETTINGS FOR CARRIAGE CUTS — BODY	11$\frac{1}{8}$"	9$\frac{15}{16}$"	10$\frac{9}{16}$"	11$\frac{1}{8}$"	9$\frac{15}{16}$"	10$\frac{9}{16}$"	11$\frac{1}{8}$"	9$\frac{15}{16}$"	10$\frac{9}{16}$"	11$\frac{1}{8}$"

FLOOR TO FLOOR RISE		6'11 3/8"	6'11 3/8"	6'11 3/8"	6'11 1/2"	6'11 1/2"	6'11 1/2"	6'11 5/8"	6'11 5/8"	6'11 5/8"	6'11 3/4"
NUMBER OF RISERS		11	12	13	11	12	13	11	12	13	11
HEIGHT OF EACH RISER		7 9/16"	6 15/16"	6 7/16"	7 9/16"	6 15/16"	6 7/16"	7 5/8"	7"	6 7/16"	7 5/8"
TOTAL HEIGHT OF RISERS		6'11 3/16"	6'11 1/4"	6'11 11/16"	6'11 3/16"	6'11 1/4"	6'11 11/16"	6'11 7/8"	7'	6'11 11/16"	6'11 7/8"
RISERS OVERAGE (+) OR UNDERAGE (—)		−3/16"	−1/8"	+5/16"	−5/16"	−1/4"	+3/16"	+1/4"	+3/8"	+1/16"	+1/8"
NUMBER OF TREADS		10	11	12	10	11	12	10	11	12	10
WIDTH OF EACH TREAD		9 15/16"	10 9/16"	11 1/16"	9 15/16"	10 9/16"	11 1/16"	9 7/8"	10 1/2"	11 1/16"	9 7/8"
TOTAL RUN OF STAIRWAY		8'3 3/8"	9'8 3/16"	11' 3/4"	8'3 3/8"	9'8 3/16"	11' 3/4"	8'2 3/4"	9'7 1/2"	11' 3/4"	8'2 3/4"
MAXIMUM RUN WITH 6'-8" HEADROOM		1'7 7/8"	1'9 1/8"	1'10 1/8"	1'7 7/8"	1'9 1/8"	1'10 1/8"	1'7 3/4"	1'9"	1'10 1/8"	1'7 3/4"
ANGLE OF INCLINE		37°16'	33°18'	30°12'	37°16'	33°18'	30°12'	37°40'	33°41'	30°12'	37°40'
LENGTH OF CARRIAGE		10'4 7/8"	11'7"	12'9 19/32"	10'4 7/8"	11'7"	12'9 19/32"	10'4 3/4"	11'6 13/16"	12'9 19/32"	10'4 3/4"
L. F. OF RISER PER INCH OF STAIRWAY WIDTH		.916	1.0	1.083	.916	1.0	1.083	.916	1.0	1.083	.916
L. F. OF TREAD PER INCH OF STAIRWAY WIDTH		.833	.916	1.0	.833	.916	1.0	.833	.916	1.0	.833
FRAMING SQUARE SETTINGS FOR CARRIAGE CUTS	TONGUE	7 9/16"	6 15/16"	6 7/16"	7 9/16"	6 15/16"	6 7/16"	7 5/8"	7"	6 7/16"	7 5/8"
	BODY	9 15/16"	10 9/16"	11 1/16"	9 15/16"	10 9/16"	11 1/16"	9 7/8"	10 1/2"	11 1/16"	9 7/8"

FLOOR TO FLOOR RISE		6'11$\frac{3}{4}$"	6'11$\frac{3}{4}$"	6'11$\frac{7}{8}$"	6'11$\frac{7}{8}$"	6'11$\frac{7}{8}$"	7'	7'	7'	7'	7'$\frac{1}{8}$"
NUMBER OF RISERS		12	13	11	12	13	11	12	13	14	11
HEIGHT OF EACH RISER		7"	6$\frac{7}{16}$"	7$\frac{5}{8}$"	7"	6$\frac{7}{16}$"	7$\frac{5}{8}$"	7"	6$\frac{7}{16}$"	6"	7$\frac{5}{8}$"
TOTAL HEIGHT OF RISERS		7'	6'11$\frac{11}{16}$"	6'11$\frac{7}{8}$"	7'	6'11$\frac{11}{16}$"	6'11$\frac{7}{8}$"	7'	6'11$\frac{11}{16}$"	7'	6'11$\frac{7}{8}$"
RISERS OVERAGE (+) OR UNDERAGE (—)		+$\frac{1}{4}$"	−$\frac{1}{16}$"	0	+$\frac{1}{8}$"	−$\frac{3}{16}$"	−$\frac{1}{8}$"	0	−$\frac{5}{16}$"	0	−$\frac{1}{4}$"
NUMBER OF TREADS		11	12	10	11	12	10	11	12	13	10
WIDTH OF EACH TREAD		10$\frac{1}{2}$"	11$\frac{1}{16}$"	9$\frac{7}{8}$"	10$\frac{1}{2}$"	11$\frac{1}{16}$"	9$\frac{7}{8}$"	10$\frac{1}{2}$"	11$\frac{1}{16}$"	11$\frac{1}{2}$"	9$\frac{7}{8}$"
TOTAL RUN OF STAIRWAY		9'7$\frac{1}{2}$"	11'$\frac{3}{4}$"	8'2$\frac{3}{4}$"	9'7$\frac{1}{2}$"	11'$\frac{3}{4}$"	8'2$\frac{3}{4}$"	9'7$\frac{1}{2}$"	11'$\frac{3}{4}$"	12'5$\frac{1}{2}$"	8'2$\frac{3}{4}$"
MAXIMUM RUN WITH 6'-8" HEADROOM		1'9"	1'10$\frac{1}{8}$"	1'7$\frac{3}{4}$"	1'9"	1'10$\frac{1}{8}$"	1'7$\frac{3}{4}$"	1'9"	1'10$\frac{1}{8}$"	1'11"	1'7$\frac{3}{4}$"
ANGLE OF INCLINE		33°41'	30°12'	37°40'	33°41'	30°12'	37°40'	33°41'	30°12'	27°33'	37°40'
LENGTH OF CARRIAGE		11'6$\frac{13}{16}$"	12'9$\frac{19}{32}$"	10'4$\frac{3}{4}$"	11'6$\frac{13}{16}$"	12'9$\frac{19}{32}$"	10'4$\frac{3}{4}$"	11'6$\frac{13}{16}$"	12'9$\frac{19}{32}$"	14'$\frac{5}{8}$"	10'4$\frac{3}{4}$"
L. F. OF RISER PER INCH OF STAIRWAY WIDTH		1.0	1.083	.916	1.0	1.083	.916	1.0	1.083	1.166	.916
L. F. OF TREAD PER INCH OF STAIRWAY WIDTH		.916	1.0	.833	.916	1.0	.833	.916	1.0	1.083	.833
FRAMING SQUARE SETTINGS FOR CARRIAGE CUTS	TONGUE	7"	6$\frac{7}{16}$"	7$\frac{5}{8}$"	7"	6$\frac{7}{16}$"	7$\frac{5}{8}$"	7"	6$\frac{7}{16}$"	6"	7$\frac{5}{8}$"
	BODY	10$\frac{1}{2}$"	11$\frac{1}{16}$"	9$\frac{7}{8}$"	10$\frac{1}{2}$"	11$\frac{1}{16}$"	9$\frac{7}{8}$"	10$\frac{1}{2}$"	11$\frac{1}{16}$"	11$\frac{1}{2}$"	9$\frac{7}{8}$"

FLOOR TO FLOOR RISE		7' 1/8"	7' 1/8"	7' 1/8"	7' 1/4"	7' 1/4"	7' 1/4"	7' 1/4"	7' 3/8"	7' 3/8"	7' 3/8"
NUMBER OF RISERS		12	13	14	11	12	13	14	11	12	13
HEIGHT OF EACH RISER		7"	6 1/2"	6"	7 11/16"	7"	6 1/2"	6"	7 11/16"	7 1/16"	6 1/2"
TOTAL HEIGHT OF RISERS		7'	7' 1/2"	7'	7' 9/16"	7'	7' 1/2"	7'	7' 9/16"	7' 3/4"	7' 1/2"
RISERS OVERAGE (+) OR UNDERAGE (—)		−1/8"	+3/8"	−1/8"	+5/16"	−1/4"	+1/4"	−1/4"	+3/16"	+3/8"	+1/8"
NUMBER OF TREADS		11	12	13	10	11	12	13	10	11	12
WIDTH OF EACH TREAD		10 1/2"	11"	11 1/2"	9 13/16"	10 1/2"	11"	11 1/2"	9 13/16"	10 7/16"	11"
TOTAL RUN OF STAIRWAY		9' 7 1/2"	11'	12' 5 1/2"	8' 2 1/8"	9' 7 1/2"	11'	12' 5 1/2"	8' 2 1/8"	9' 6 13/16"	11'
MAXIMUM RUN WITH 6'-8" HEADROOM		1' 9"	1' 10"	1' 11"	1' 7 5/8"	1' 9"	1' 10"	1' 11"	1' 7 5/8"	1' 8 7/8"	1' 10"
ANGLE OF INCLINE		33°41'	30°35'	27°33'	38°5'	33°41'	30°35'	27°33'	38°5'	34°5'	30°35'
LENGTH OF CARRIAGE		11' 6 13/16"	12' 9 5/16"	14' 5/8"	10' 4 21/32"	11' 6 13/16"	12' 9 5/16"	14' 5/8"	10' 4 21/32"	11' 6 5/8"	12' 9 5/16"
L. F. OF RISER PER INCH OF STAIRWAY WIDTH		1.0	1.083	1.166	.916	1.0	1.083	1.166	.916	1.0	1.083
L. F. OF TREAD PER INCH OF STAIRWAY WIDTH		.916	1.0	1.083	.833	.916	1.0	1.083	.833	.916	1.0
FRAMING SQUARE SETTINGS FOR CARRIAGE CUTS	TONGUE	7"	6 1/2"	6"	7 11/16"	7"	6 1/2"	6"	7 11/16"	7 1/16"	6 1/2"
	BODY	10 1/2"	11"	11 1/2"	9 13/16"	10 1/2"	11"	11 1/2"	9 13/16"	10 7/16"	11"

FLOOR TO FLOOR RISE	7' 3/8"	7' 1/2"	7' 1/2"	7' 1/2"	7' 1/2"	7' 5/8"	7' 5/8"	7' 5/8"	7' 5/8"	7' 3/4"
NUMBER OF RISERS	14	11	12	13	14	11	12	13	14	11
HEIGHT OF EACH RISER	6"	7 11/16"	7 1/16"	6 1/2"	6 1/16"	7 11/16"	7 1/16"	6 1/2"	6 1/16"	7 11/16"
TOTAL HEIGHT OF RISERS	7'	7' 9/16"	7' 3/4"	7' 1/2"	7' 7/8"	7' 9/16"	7' 3/4"	7' 1/2"	7' 7/8"	7' 9/16"
RISERS OVERAGE (+) OR UNDERAGE (—)	−3/8"	+1/16"	+1/4"	0	+3/8"	−1/16"	+1/8"	−1/8"	+1/4"	−3/16"
NUMBER OF TREADS	13	10	11	12	13	10	11	12	13	10
WIDTH OF EACH TREAD	11 1/2"	9 13/16"	10 7/16"	11"	11 7/16"	9 13/16"	10 7/16"	11"	11 7/16"	9 13/16"
TOTAL RUN OF STAIRWAY	12'5 1/2"	8'2 1/8"	9'6 13/16"	11'	12'4 11/16"	8'2 1/8"	9'6 13/16"	11'	12'4 11/16"	8'2 1/8"
MAXIMUM RUN WITH 6'-8" HEADROOM	1'11"	1'7 5/8"	1'8 7/8"	1'10"	1'10 7/8"	1'7 5/8"	1'8 7/8"	1'10"	1'10 7/8"	1'7 5/8"
ANGLE OF INCLINE	27°33'	38°5'	34°5'	30°35'	27°56'	38°5'	34°5'	30°35'	27°56'	38°5'
LENGTH OF CARRIAGE	14' 5/8"	10'4 21/32"	11'6 5/8"	12'9 5/16"	14' 9/32"	10'4 21/32"	11'6 5/8"	12'9 5/16"	14' 9/32"	10'4 21/32"
L. F. OF RISER PER INCH OF STAIRWAY WIDTH	1.166	.916	1.0	1.083	1.166	.916	1.0	1.083	1.166	.916
L. F. OF TREAD PER INCH OF STAIRWAY WIDTH	1.083	.833	.916	1.0	1.083	.833	.916	1.0	1.083	.833
FRAMING SQUARE SETTINGS FOR CARRIAGE CUTS — TONGUE	6"	7 11/16"	7 1/16"	6 1/2"	6 1/16"	7 11/16"	7 1/16"	6 1/2"	6 1/16"	7 11/16"
FRAMING SQUARE SETTINGS FOR CARRIAGE CUTS — BODY	11 1/2"	9 13/16"	10 7/16"	11"	11 7/16"	9 13/16"	10 7/16"	11"	11 7/16"	9 13/16"

FLOOR TO FLOOR RISE	7' 3/4"	7' 3/4"	7' 3/4"	7' 7/8"	7' 7/8"	7' 7/8"	7' 7/8"	7' 1"	7' 1"	7' 1"
NUMBER OF RISERS	12	13	14	11	12	13	14	11	12	13
HEIGHT OF EACH RISER	7 1/16"	6 1/2"	6 1/16"	7 11/16"	7 1/16"	6 1/2"	6 1/16"	7 3/4"	7 1/16"	6 9/16"
TOTAL HEIGHT OF RISERS	7' 3/4"	7' 1/2"	7' 7/8"	7' 9/16"	7' 3/4"	7' 1/2"	7' 7/8"	7' 1 1/4"	7' 3/4"	7' 1 5/16"
RISERS OVERAGE (+) OR UNDERAGE (—)	0	−1/4"	+1/8"	−5/16"	−1/8"	−3/8"	0	+1/4"	−1/4"	+5/16"
NUMBER OF TREADS	11	12	13	10	11	12	13	10	11	12
WIDTH OF EACH TREAD	10 7/16"	11"	11 7/16"	9 13/16"	10 7/16"	11"	11 7/16"	9 3/4"	10 7/16"	10 15/16"
TOTAL RUN OF STAIRWAY	9' 6 13/16"	11'	12' 4 11/16"	8' 2 1/8"	9' 6 13/16"	11'	12' 4 11/16"	8' 1 1/2"	9' 6 13/16"	10' 11 1/4"
MAXIMUM RUN WITH 6'-8" HEADROOM	1' 8 7/8"	1' 10"	1' 10 7/8"	1' 7 5/8"	1' 8 7/8"	1' 10"	1' 10 7/8"	1' 7 1/2"	1' 8 7/8"	1' 9 7/8"
ANGLE OF INCLINE	34°5'	30°35'	27°56'	38°5'	34°5'	30°35'	27°56'	38°29'	34°5'	30°58'
LENGTH OF CARRIAGE	11' 6 5/8"	12' 9 5/16"	14' 9/32"	10' 4 21/32"	11' 6 5/8"	12' 9 5/16"	14' 9/32"	10' 4 17/32"	11' 6 5/8"	12' 9 1/16"
L. F. OF RISER PER INCH OF STAIRWAY WIDTH	1.0	1.083	1.166	.916	1.0	1.083	1.166	.916	1.0	1.083
L. F. OF TREAD PER INCH OF STAIRWAY WIDTH	.916	1.0	1.083	.833	.916	1.0	1.083	.833	.916	1.0
FRAMING SQUARE SETTINGS FOR CARRIAGE CUTS — TONGUE	7 1/16"	6 1/2"	6 1/16"	7 11/16"	7 1/16"	6 1/2"	6 1/16"	7 3/4"	7 1/16"	6 9/16"
FRAMING SQUARE SETTINGS FOR CARRIAGE CUTS — BODY	10 7/16"	11"	11 7/16"	9 13/16"	10 7/16"	11"	11 7/16"	9 3/4"	10 7/16"	10 15/16"

FLOOR TO FLOOR RISE	7'1"	7'1 1/8"	7'1 1/8"	7'1 1/8"	7'1 1/8"	7'1 1/4"	7'1 1/4"	7'1 1/4"	7'1 1/4"	7'1 3/8"
NUMBER OF RISERS	14	11	12	13	14	11	12	13	14	11
HEIGHT OF EACH RISER	6 1/16"	7 3/4"	7 1/8"	6 9/16"	6 1/16"	7 3/4"	7 1/8"	6 9/16"	6 1/16"	7 3/4"
TOTAL HEIGHT OF RISERS	7' 7/8"	7'1 1/4"	7'1 1/2"	7'1 5/16"	7' 7/8"	7'1 1/4"	7'1 1/2"	7'1 5/16"	7' 7/8"	7'1 1/4"
RISERS OVERAGE (+) OR UNDERAGE (—)	−1/8"	+1/8"	+3/8"	+3/16"	−1/4"	0	+1/4"	+1/16"	−3/8"	−1/8"
NUMBER OF TREADS	13	10	11	12	13	10	11	12	13	10
WIDTH OF EACH TREAD	11 7/16"	9 3/4"	10 3/8"	10 15/16"	11 7/16"	9 3/4"	10 3/8"	10 15/16"	11 7/16"	9 3/4"
TOTAL RUN OF STAIRWAY	12'4 11/16"	8'1 1/2"	9'6 1/8"	10'11 1/4"	12'4 11/16"	8'1 1/2"	9'6 1/8"	10'11 1/4"	12'4 11/16"	8'1 1/2"
MAXIMUM RUN WITH 6'-8" HEADROOM	1'10 7/8"	1'7 1/2"	1'8 3/4"	1'9 7/8"	1'10 7/8"	1'7 1/2"	1'8 3/4"	1'9 7/8"	1'10 7/8"	1'7 1/2"
ANGLE OF INCLINE	27°56'	38°29'	34°29'	30°58'	27°56'	38°29'	34°29'	30°58'	27°56'	38°29'
LENGTH OF CARRIAGE	14' 9/32"	10'4 17/32"	11'6 7/16"	12'9 1/16"	14' 9/32"	10'4 17/32"	11'6 7/16"	12'9 1/16"	14' 9/32"	10'4 17/32"
L. F. OF RISER PER INCH OF STAIRWAY WIDTH	1.166	.916	1.0	1.083	1.166	.916	1.0	1.083	1.166	.916
L. F. OF TREAD PER INCH OF STAIRWAY WIDTH	1.083	.833	.916	1.0	1.083	.833	.916	1.0	1.083	.833
FRAMING SQUARE SETTINGS FOR CARRIAGE CUTS — TONGUE	6 1/16"	7 3/4"	7 1/8"	6 9/16"	6 1/16"	7 3/4"	7 1/8"	6 9/16"	6 1/16"	7 3/4"
FRAMING SQUARE SETTINGS FOR CARRIAGE CUTS — BODY	11 7/16"	9 3/4"	10 3/8"	10 15/16"	11 7/16"	9 3/4"	10 3/8"	10 15/16"	11 7/16"	9 3/4"

FLOOR TO FLOOR RISE	7'1 3/8"	7'1 3/8"	7'1 3/8"	7'1 1/2"	7'1 1/2"	7'1 1/2"	7'1 1/2"	7'1 5/8"	7'1 5/8"	7'1 5/8"
NUMBER OF RISERS	12	13	14	11	12	13	14	11	12	13
HEIGHT OF EACH RISER	7 1/8"	6 9/16"	6 1/8"	7 3/4"	7 1/8"	6 9/16"	6 1/8"	7 13/16"	7 1/8"	6 9/16"
TOTAL HEIGHT OF RISERS	7'1 1/2"	7'1 5/16"	7'1 3/4"	7'1 1/4"	7'1 1/2"	7'1 5/16"	7'1 3/4"	7'1 15/16"	7'1 1/2"	7'1 5/16"
RISERS OVERAGE (+) OR UNDERAGE (—)	+1/8"	−1/16"	+3/8"	−1/4"	0	−3/16"	+1/4"	+5/16"	−1/8"	−5/16"
NUMBER OF TREADS	11	12	13	10	11	12	13	10	11	12
WIDTH OF EACH TREAD	10 3/8"	10 15/16"	11 3/8"	9 3/4"	10 3/8"	10 15/16"	11 3/8"	9 11/16"	10 3/8"	10 15/16"
TOTAL RUN OF STAIRWAY	9'6 1/8"	10'11 1/4"	12'3 7/8"	8'1 1/2"	9'6 1/8"	10'11 1/4"	12'3 7/8"	8'7/8"	9'6 1/8"	10'11 1/4"
MAXIMUM RUN WITH 6'-8" HEADROOM	1'8 3/4"	1'9 7/8"	1'10 3/4"	1'7 1/2"	1'8 3/4"	1'9 7/8"	1'10 3/4"	1'7 3/8"	1'8 3/4"	1'9 7/8"
ANGLE OF INCLINE	34°29'	30°58'	28°18'	38°29'	34°29'	30°58'	28°18'	38°53'	34°29'	30°58'
LENGTH OF CARRIAGE	11'6 7/16"	12'9 1/16"	13'11 15/16"	10'4 17/32"	11'6 7/16"	12'9 1/16"	13'11 15/16"	10'4 7/16"	11'6 7/16"	12'9 1/16"
L. F. OF RISER PER INCH OF STAIRWAY WIDTH	1.0	1.083	1.166	.916	1.0	1.083	1.166	.916	1.0	1.083
L. F. OF TREAD PER INCH OF STAIRWAY WIDTH	.916	1.0	1.083	.833	.916	1.0	1.083	.833	.916	1.0
FRAMING SQUARE SETTINGS FOR CARRIAGE CUTS — TONGUE	7 1/8"	6 9/16"	6 1/8"	7 3/4"	7 1/8"	6 9/16"	6 1/8"	7 13/16"	7 1/8"	6 9/16"
FRAMING SQUARE SETTINGS FOR CARRIAGE CUTS — BODY	10 3/8"	10 15/16"	11 3/8"	9 3/4"	10 3/8"	10 15/16"	11 3/8"	9 11/16"	10 3/8"	10 15/16"

FLOOR TO FLOOR RISE	7'1$\frac{5}{8}$"	7'1$\frac{3}{4}$"	7'1$\frac{3}{4}$"	7'1$\frac{3}{4}$"	7'1$\frac{3}{4}$"	7'1$\frac{7}{8}$"	7'1$\frac{7}{8}$"	7'1$\frac{7}{8}$"	7'1$\frac{7}{8}$"	7'2"
NUMBER OF RISERS	14	11	12	13	14	11	12	13	14	11
HEIGHT OF EACH RISER	6$\frac{1}{8}$"	7$\frac{13}{16}$"	7$\frac{1}{8}$"	6$\frac{5}{8}$"	6$\frac{1}{8}$"	7$\frac{13}{16}$"	7$\frac{3}{16}$"	6$\frac{5}{8}$"	6$\frac{1}{8}$"	7$\frac{13}{16}$"
TOTAL HEIGHT OF RISERS	7'1$\frac{3}{4}$"	7'1$\frac{15}{16}$"	7'1$\frac{1}{2}$"	7'2$\frac{1}{8}$"	7'1$\frac{3}{4}$"	7'1$\frac{15}{16}$"	7'2$\frac{1}{4}$"	7'2$\frac{1}{8}$"	7'1$\frac{3}{4}$"	7'1$\frac{15}{16}$"
RISERS OVERAGE (+) OR UNDERAGE (—)	+$\frac{1}{8}$"	+$\frac{3}{16}$"	−$\frac{1}{4}$"	+$\frac{3}{8}$"	0	+$\frac{1}{16}$"	+$\frac{3}{8}$"	+$\frac{1}{4}$"	−$\frac{1}{8}$"	−$\frac{1}{16}$"
NUMBER OF TREADS	13	10	11	12	13	10	11	12	13	10
WIDTH OF EACH TREAD	11$\frac{3}{8}$"	9$\frac{11}{16}$"	10$\frac{3}{8}$"	10$\frac{7}{8}$"	11$\frac{3}{8}$"	9$\frac{11}{16}$"	10$\frac{5}{16}$"	10$\frac{7}{8}$"	11$\frac{3}{8}$"	9$\frac{11}{16}$"
TOTAL RUN OF STAIRWAY	12'3$\frac{7}{8}$"	8'$\frac{7}{8}$"	9'6$\frac{1}{8}$"	10'10$\frac{1}{2}$"	12'3$\frac{7}{8}$"	8'$\frac{7}{8}$"	9'5$\frac{7}{8}$"	10'10$\frac{1}{2}$"	12'3$\frac{7}{8}$"	8'$\frac{7}{8}$"
MAXIMUM RUN WITH 6'-8" HEADROOM	1'10$\frac{3}{4}$"	1'7$\frac{3}{8}$"	1'8$\frac{3}{4}$"	1'9$\frac{3}{4}$"	1'10$\frac{3}{4}$"	1'7$\frac{3}{8}$"	1'8$\frac{5}{8}$"	1'9$\frac{3}{4}$"	1'10$\frac{3}{4}$"	1'7$\frac{3}{8}$"
ANGLE OF INCLINE	28°18'	38°53'	34°29'	31°21'	28°18'	38°53'	34°53'	31°21'	28°18'	38°53'
LENGTH OF CARRIAGE	13'11$\frac{15}{16}$"	10'4$\frac{7}{16}$"	11'6$\frac{7}{16}$"	12'8$\frac{13}{16}$"	13'11$\frac{15}{16}$"	10'4$\frac{7}{16}$"	11'6$\frac{9}{32}$"	12'8$\frac{13}{16}$"	13'11$\frac{15}{16}$"	10'4$\frac{7}{16}$"
L. F. OF RISER PER INCH OF STAIRWAY WIDTH	1.166	.916	1.0	1.083	1.166	.916	1.0	1.083	1.166	.916
L. F. OF TREAD PER INCH OF STAIRWAY WIDTH	1.083	.833	.916	1.0	1.083	.833	.916	1.0	1.083	.833
FRAMING SQUARE SETTINGS FOR CARRIAGE CUTS — TONGUE	6$\frac{1}{8}$"	7$\frac{13}{16}$"	7$\frac{1}{8}$"	6$\frac{5}{8}$"	6$\frac{1}{8}$"	7$\frac{13}{16}$"	7$\frac{3}{16}$"	6$\frac{5}{8}$"	6$\frac{1}{8}$"	7$\frac{13}{16}$"
FRAMING SQUARE SETTINGS FOR CARRIAGE CUTS — BODY	11$\frac{3}{8}$"	9$\frac{11}{16}$"	10$\frac{3}{8}$"	10$\frac{7}{8}$"	11$\frac{3}{8}$"	9$\frac{11}{16}$"	10$\frac{5}{16}$"	10$\frac{7}{8}$"	11$\frac{3}{8}$"	9$\frac{11}{16}$"

FLOOR TO FLOOR RISE	7'2"	7'2"	7'2"	7'2 1/8"	7'2 1/8"	7'2 1/8"	7'2 1/8"	7'2 1/4"	7'2 1/4"	7'2 1/4"
NUMBER OF RISERS	12	13	14	11	12	13	14	11	12	13
HEIGHT OF EACH RISER	7 3/16"	6 5/8"	6 1/8"	7 13/16"	7 3/16"	6 5/8"	6 1/8"	7 13/16"	7 3/16"	6 5/8"
TOTAL HEIGHT OF RISERS	7'2 1/4"	7'2 1/8"	7'1 3/4"	7'1 15/16"	7'2 1/4"	7'2 1/8"	7'1 3/4"	7'1 15/16"	7'2 1/4"	7'2 1/8"
RISERS OVERAGE (+) OR UNDERAGE (—)	+1/4"	+1/8"	−1/4"	−3/16"	+1/8"	0	−3/8"	−5/16"	0	−1/8"
NUMBER OF TREADS	11	12	13	10	11	12	13	10	11	12
WIDTH OF EACH TREAD	10 5/16"	10 7/8"	11 3/8"	9 11/16"	10 5/16"	10 7/8"	11 3/8"	9 11/16"	10 5/16"	10 7/8"
TOTAL RUN OF STAIRWAY	9'5 7/16"	10'10 1/2"	12'3 7/8"	8' 7/8"	9'5 7/16"	10'10 1/2"	12'3 7/8"	8' 7/8"	9'5 7/16"	10'10 1/2"
MAXIMUM RUN WITH 6'-8" HEADROOM	1'8 5/8"	1'9 3/4"	1'10 3/4"	1'7 3/8"	1'8 5/8"	1'9 3/4"	1'10 3/4"	1'7 3/8"	1'8 5/8"	1'9 3/4"
ANGLE OF INCLINE	34°53'	31°21'	28°18'	38°53'	34°53'	31°21'	28°18'	38°53'	34°53'	31°21'
LENGTH OF CARRIAGE	11'6 9/32"	12'8 13/16"	13'11 15/16"	10'4 7/16"	11'6 9/32"	12'8 13/16"	13'11 15/16"	10'4 7/16"	11'6 9/32"	12'8 13/16"
L. F. OF RISER PER INCH OF STAIRWAY WIDTH	1.0	1.083	1.166	.916	1.0	1.083	1.166	.916	1.0	1.083
L. F. OF TREAD PER INCH OF STAIRWAY WIDTH	.916	1.0	1.083	.833	.916	1.0	1.083	.833	.916	1.0
FRAMING SQUARE SETTINGS FOR CARRIAGE CUTS — TONGUE	7 3/16"	6 5/8"	6 1/8"	7 13/16"	7 3/16"	6 5/8"	6 1/8"	7 13/16"	7 3/16"	6 5/8"
BODY	10 5/16"	10 7/8"	11 3/8"	9 11/16"	10 5/16"	10 7/8"	11 3/8"	9 11/16"	10 5/16"	10 7/8"

FLOOR TO FLOOR RISE	7'2 1/4"	7'2 3/8"	7'2 3/8"	7'2 3/8"	7'2 3/8"	7'2 1/2"	7'2 1/2"	7'2 1/2"	7'2 1/2"	7'2 5/8"
NUMBER OF RISERS	14	11	12	13	14	11	12	13	14	11
HEIGHT OF EACH RISER	6 3/16"	7 7/8"	7 3/16"	6 5/8"	6 3/16"	7 7/8"	7 3/16"	6 5/8"	6 3/16"	7 7/8"
TOTAL HEIGHT OF RISERS	7'2 5/8"	7'2 5/8"	7'2 1/4"	7'2 1/8"	7'2 5/8"	7'2 5/8"	7'2 1/4"	7'2 1/8"	7'2 5/8"	7'2 5/8"
RISERS OVERAGE (+) OR UNDERAGE (—)	+3/8"	+1/4"	−1/8"	−1/4"	+1/4"	+1/8"	−1/4"	−3/8"	+1/8"	0
NUMBER OF TREADS	13	10	11	12	13	10	11	12	13	10
WIDTH OF EACH TREAD	11 5/8"	9 5/8"	10 5/16"	10 7/8"	11 5/16"	9 5/8"	10 5/16"	10 7/8"	11 5/16"	9 5/8"
TOTAL RUN OF STAIRWAY	12'3 1/16"	8'1/4"	9'5 7/16"	10'10 1/2"	12'3 1/16"	8'1/4"	9'5 7/16"	10'10 1/2"	12'3 1/16"	8'1/4"
MAXIMUM RUN WITH 6'-8" HEADROOM	1'10 5/8"	1'7 1/4"	1'8 5/8"	1'9 3/4"	1'10 5/8"	1'7 1/4"	1'8 5/8"	1'9 3/4"	1'10 5/8"	1'7 1/4"
ANGLE OF INCLINE	28°41'	39°17'	34°53'	31°21'	28°41'	39°17'	34°53'	31°21'	28°41'	39°17'
LENGTH OF CARRIAGE	13'11 5/8"	10'4 3/8"	11'6 9/32"	12'8 13/16"	13'11 5/8"	10'4 3/8"	11'6 9/32"	12'8 13/16"	13'11 5/8"	10'4 3/8"
L. F. OF RISER PER INCH OF STAIRWAY WIDTH	1.166	.916	1.0	1.083	1.166	.916	1.0	1.083	1.166	.916
L. F. OF TREAD PER INCH OF STAIRWAY WIDTH	1.083	.833	.916	1.0	1.083	.833	.916	1.0	1.083	.833
FRAMING SQUARE SETTINGS FOR CARRIAGE CUTS — TONGUE	6 3/16"	7 7/8"	7 3/16"	6 5/8"	6 3/16"	7 7/8"	7 3/16"	6 5/8"	6 3/16"	7 7/8"
FRAMING SQUARE SETTINGS FOR CARRIAGE CUTS — BODY	11 5/16"	9 5/8"	10 5/16"	10 7/8"	11 5/16"	9 5/8"	10 5/16"	10 7/8"	11 5/16"	9 5/8"

FLOOR TO FLOOR RISE		7'2 5/8"	7'2 5/8"	7'2 5/8"	7'2 3/4"	7'2 3/4"	7'2 3/4"	7'2 3/4"	7'2 7/8"	7'2 7/8"	7'2 7/8"
NUMBER OF RISERS		12	13	14	11	12	13	14	11	12	13
HEIGHT OF EACH RISER		7 1/4"	6 11/16"	6 3/16"	7 7/8"	7 1/4"	6 11/16"	6 3/16"	7 7/8"	7 1/4"	6 11/16"
TOTAL HEIGHT OF RISERS		7'3"	7'2 15/16"	7'2 5/8"	7'2 5/8"	7'3"	7'2 15/16"	7'2 5/8"	7'2 5/8"	7'3"	7'2 15/16"
RISERS OVERAGE (+) OR UNDERAGE (—)		+3/8"	+5/16"	0	−1/8"	+1/4"	+3/16"	−1/8"	−1/4"	+1/8"	+1/16"
NUMBER OF TREADS		11	12	13	10	11	12	13	10	11	12
WIDTH OF EACH TREAD		10 1/4"	10 13/16"	11 5/16"	9 5/8"	10 1/4"	10 13/16"	11 5/16"	9 5/8"	10 1/4"	10 13/16"
TOTAL RUN OF STAIRWAY		9'4 3/4"	10'9 3/4"	12'3 1/16"	8'1/4"	9'4 3/4"	10'9 3/4"	12'3 1/16"	8'1/4"	9'4 3/4"	10'9 3/4"
MAXIMUM RUN WITH 6'-8" HEADROOM		1'8 1/2"	1'9 5/8"	1'10 5/8"	1'7 1/4"	1'8 1/2"	1'9 5/8"	1'10 5/8"	1'7 1/4"	1'8 1/2"	1'9 5/8"
ANGLE OF INCLINE		35°16'	31°44'	28°41'	39°17'	35°16'	31°44'	28°41'	39° 17'	35°16'	31°44'
LENGTH OF CARRIAGE		11'6 3/32"	12'8 9/16"	13'11 5/8"	10'4 3/8"	11'6 3/32"	12'8 9/16"	13'11 5/8"	10'4 3/8"	11'6 3/32"	12'8 9/16"
L. F. OF RISER PER INCH OF STAIRWAY WIDTH		1.0	1.083	1.166	.916	1.0	1.083	1.166	.916	1.0	1.083
L. F. OF TREAD PER INCH OF STAIRWAY WIDTH		.916	1.0	1.083	.833	.916	1.0	1.083	.833	.916	1.0
FRAMING SQUARE SETTINGS FOR CARRIAGE CUTS	TONGUE	7 1/4"	6 11/16"	6 3/16"	7 7/8"	7 1/4"	6 11/16"	6 3/16"	7 7/8"	7 1/4"	6 11/16"
	BODY	10 1/4"	10 13/16"	11 5/16"	9 5/8"	10 1/4"	10 13/16"	11 5/16"	9 5/8"	10 1/4"	10 13/16"

FLOOR TO FLOOR RISE		7'2 7/8"	7'3"	7'3"	7'3"	7'3"	7'3 1/8"	7'3 1/8"	7'3 1/8"	7'3 1/8"	7'3 1/4"
NUMBER OF RISERS		14	11	12	13	14	11	12	13	14	11
HEIGHT OF EACH RISER		6 3/16"	7 15/16"	7 1/4"	6 11/16"	6 3/16"	7 15/16"	7 1/4"	6 11/16"	6 1/4"	7 15/16"
TOTAL HEIGHT OF RISERS		7'2 5/8"	7'3 5/16"	7'3"	7'2 15/16"	7'2 5/8"	7'3 5/16"	7'3"	7'2 15/16"	7'3 1/2"	7'3 5/16"
RISERS OVERAGE (+) OR UNDERAGE (—)		−1/4"	+5/16"	0	−1/16"	−3/8"	+3/16"	−1/8"	−3/16"	+3/8"	+1/16"
NUMBER OF TREADS		13	10	11	12	13	10	11	12	13	10
WIDTH OF EACH TREAD		11 5/16"	9 9/16"	10 1/4"	10 13/16"	11 5/16"	9 9/16"	10 1/4"	10 13/16"	11 1/4"	9 9/16"
TOTAL RUN OF STAIRWAY		12'3 1/16"	7'11 5/8"	9'4 3/4"	10'9 3/4"	12'3 1/16"	7'11 5/8"	9'4 3/4"	10'9 3/4"	12'2 1/4"	7'11 5/8"
MAXIMUM RUN WITH 6'-8" HEADROOM		1'10 5/8"	1'7 1/8"	1'8 1/2"	1'9 5/8"	1'10 5/8"	1'7 1/8"	1'8 1/2"	1'9 5/8"	1'10 1/2"	1'7 1/8"
ANGLE OF INCLINE		28°41'	39°42'	35°16'	31°44'	28°41'	39°42'	35°16'	31°44'	29°3'	39°42'
LENGTH OF CARRIAGE		13'11 5/8"	10'4 9/32"	11'6 3/32"	12'8 9/16"	13'11 5/8"	10'4 9/32"	11'6 3/32"	12'8 9/16"	13'11 5/16"	10'4 9/32"
L. F. OF RISER PER INCH OF STAIRWAY WIDTH		1.166	.916	1.0	1.083	1.166	.916	1.0	1.083	1.166	.916
L. F. OF TREAD PER INCH OF STAIRWAY WIDTH		1.083	.833	.916	1.0	1.083	.833	.916	1.0	1.083	.833
FRAMING SQUARE SETTINGS FOR CARRIAGE CUTS	TONGUE	6 3/16"	7 15/16"	7 1/4"	6 11/16"	6 3/16"	7 15/16"	7 1/4"	6 11/16"	6 1/4"	7 15/16"
	BODY	11 5/16"	9 9/16"	10 1/4"	10 13/16"	11 5/16"	9 9/16"	10 1/4"	10 13/16"	11 1/4"	9 9/16"

FLOOR TO FLOOR RISE		7'3 1/4"	7'3 1/4"	7'3 1/4"	7'3 3/8"	7'3 3/8"	7'3 3/8"	7'3 3/8"	7'3 1/2"	7'3 1/2"	7'3 1/2"
NUMBER OF RISERS		12	13	14	11	12	13	14	11	12	13
HEIGHT OF EACH RISER		7 1/4"	6 11/16"	6 1/4"	7 15/16"	7 5/16"	6 3/4"	6 1/4"	7 15/16"	7 5/16"	6 3/4"
TOTAL HEIGHT OF RISERS		7'3"	7'2 15/16"	7'3 1/2"	7'3 5/16"	7'3 3/4"	7'3 3/4"	7'3 1/2"	7'3 5/16"	7'3 3/4"	7'3 3/4"
RISERS OVERAGE (+) OR UNDERAGE (—)		−1/4"	−5/16"	+1/4"	−1/16"	+3/8"	+3/8"	+1/8"	−3/16"	+1/4"	+1/4"
NUMBER OF TREADS		11	12	13	10	11	12	13	10	11	12
WIDTH OF EACH TREAD		10 1/4"	10 13/16"	11 1/4"	9 9/16"	10 3/16"	10 3/4"	11 1/4"	9 9/16"	10 3/16"	10 3/4"
TOTAL RUN OF STAIRWAY		9'4 3/4"	10'9 3/4"	12'2 1/4"	7'11 5/8"	9'4 1/16"	10'9"	12'2 1/4"	7'11 5/8"	9'4 1/16"	10'9"
MAXIMUM RUN WITH 6'-8" HEADROOM		1'8 1/2"	1'9 5/8"	1'10 1/2"	1'7 1/8"	1'8 3/8"	1'9 1/2"	1'10 1/2"	1'7 1/8"	1'8 3/8"	1'9 1/2"
ANGLE OF INCLINE		35°16'	31°44'	29°3'	39°42'	35°40'	32°7'	29°3'	39°42'	35°40'	32°7'
LENGTH OF CARRIAGE		11'6 3/32"	12'8 9/16"	13'11 5/16"	10'4 9/32"	11'5 15/16"	12'8 5/16"	13'11 5/16"	10'4 9/32"	11'5 15/16"	12'8 5/16"
L. F. OF RISER PER INCH OF STAIRWAY WIDTH		1.0	1.083	1.116	.916	1.0	1.083	1.166	.916	1.0	1.083
L. F. OF TREAD PER INCH OF STAIRWAY WIDTH		.916	1.0	1.083	.833	.916	1.0	1.083	.833	.916	1.0
FRAMING SQUARE SETTINGS FOR CARRIAGE CUTS	TONGUE	7 1/4"	6 11/16"	6 1/4"	7 15/16"	7 5/16"	6 3/4"	6 1/4"	7 15/16"	7 5/16"	6 3/4"
	BODY	10 1/4"	10 13/16"	11 1/4"	9 9/16"	10 3/16"	10 3/4"	11 1/4"	9 9/16"	10 3/16"	10 3/4"

FLOOR TO FLOOR RISE	7'3$\frac{1}{2}$"	7'3$\frac{5}{8}$"	7'3$\frac{5}{8}$"	7'3$\frac{5}{8}$"	7'3$\frac{5}{8}$"	7'3$\frac{3}{4}$"	7'3$\frac{3}{4}$"	7'3$\frac{3}{4}$"	7'3$\frac{3}{4}$"	7'3$\frac{7}{8}$"
NUMBER OF RISERS	14	11	12	13	14	11	12	13	14	11
HEIGHT OF EACH RISER	6$\frac{1}{4}$"	7$\frac{15}{16}$"	7$\frac{5}{16}$"	6$\frac{3}{4}$"	6$\frac{1}{4}$"	8"	7$\frac{5}{16}$"	6$\frac{3}{4}$"	6$\frac{1}{4}$"	8"
TOTAL HEIGHT OF RISERS	7'3$\frac{1}{2}$"	7'3$\frac{5}{16}$"	7'3$\frac{3}{4}$"	7'3$\frac{3}{4}$"	7'3$\frac{1}{2}$"	7'4"	7'3$\frac{3}{4}$"	7'3$\frac{3}{4}$"	7'3$\frac{1}{2}$"	7'4"
RISERS OVERAGE (+) OR UNDERAGE (—)	0	−$\frac{5}{16}$"	+$\frac{1}{8}$"	+$\frac{1}{8}$"	−$\frac{1}{8}$"	+$\frac{1}{4}$"	0	0	−$\frac{1}{4}$"	+$\frac{1}{8}$"
NUMBER OF TREADS	13	10	11	12	13	10	11	12	13	10
WIDTH OF EACH TREAD	11$\frac{1}{4}$"	9$\frac{9}{16}$"	10$\frac{3}{16}$"	10$\frac{3}{4}$"	11$\frac{1}{4}$"	9$\frac{1}{2}$"	10$\frac{3}{16}$"	10$\frac{3}{4}$"	11$\frac{1}{4}$"	9$\frac{1}{2}$"
TOTAL RUN OF STAIRWAY	12'2$\frac{1}{4}$"	7'11$\frac{5}{8}$"	9'4$\frac{1}{16}$"	10'9"	12'2$\frac{1}{4}$"	7'11"	9'4$\frac{1}{16}$"	10'9"	12'2$\frac{1}{4}$"	7'11"
MAXIMUM RUN WITH 6'-8" HEADROOM	1'10$\frac{1}{2}$"	1'7$\frac{1}{8}$"	1'8$\frac{3}{8}$"	1'9$\frac{1}{2}$"	1'10$\frac{1}{2}$"	9$\frac{1}{2}$"	1'8$\frac{3}{8}$"	1'9$\frac{1}{2}$"	1'10$\frac{1}{2}$"	9$\frac{1}{2}$"
ANGLE OF INCLINE	29°3'	39°42'	35°40'	32°7'	29°3'	40°6'	35°40'	32°7'	29°3'	40°6'
LENGTH OF CARRIAGE	13'11$\frac{5}{16}$"	10'4$\frac{9}{32}$"	11'5$\frac{15}{16}$"	12'8$\frac{5}{16}$"	13'11$\frac{5}{16}$"	10'4$\frac{3}{16}$"	11'5$\frac{15}{16}$"	12'8$\frac{5}{16}$"	13'11$\frac{5}{16}$"	10'4$\frac{3}{16}$"
L. F. OF RISER PER INCH OF STAIRWAY WIDTH	1.166	.916	1.0	1.083	1.166	.916	1.0	1.083	1.166	.916
L. F. OF TREAD PER INCH OF STAIRWAY WIDTH	1.083	.833	.916	1.0	1.083	.833	.916	1.0	1.083	.833
FRAMING SQUARE SETTINGS FOR CARRIAGE CUTS — TONGUE	6$\frac{1}{4}$"	7$\frac{15}{16}$"	7$\frac{5}{16}$"	6$\frac{3}{4}$"	6$\frac{1}{4}$"	8"	7$\frac{5}{16}$"	6$\frac{3}{4}$"	6$\frac{1}{4}$"	8"
BODY	11$\frac{1}{4}$"	9$\frac{9}{16}$"	10$\frac{3}{16}$"	10$\frac{3}{4}$"	11$\frac{1}{4}$"	9$\frac{1}{2}$"	10$\frac{3}{16}$"	10$\frac{3}{4}$"	11$\frac{1}{4}$"	9$\frac{1}{2}$"

FLOOR TO FLOOR RISE		7'3 7/8"	7'3 7/8"	7'3 7/8"	7'4"	7'4"	7'4"	7'4"	7'4 1/8"	7'4 1/8"	7'4 1/8"
NUMBER OF RISERS		12	13	14	11	12	13	14	11	12	13
HEIGHT OF EACH RISER		7 5/16"	6 3/4"	6 1/4"	8"	7 5/16"	6 3/4"	6 5/16"	8"	7 3/8"	6 3/4"
TOTAL HEIGHT OF RISERS		7'3 3/4"	7'3 3/4"	7'3 1/2"	7'4"	7'3 3/4"	7'3 3/4"	7'4 3/8"	7'4"	7'4 1/2"	7'3 3/4"
RISERS OVERAGE (+) OR UNDERAGE (—)		−1/8"	−1/8"	−3/8"	0	−1/4"	−1/4"	+3/8"	−1/8"	+3/8"	−3/8"
NUMBER OF TREADS		11	12	13	10	11	12	13	10	11	12
WIDTH OF EACH TREAD		10 3/16"	10 3/4"	11 1/4"	9 1/2"	10 3/16"	10 3/4"	11 3/16"	9 1/2"	10 1/8"	10 3/4"
TOTAL RUN OF STAIRWAY		9'4 1/16"	10'9"	12'2 1/4"	7'11"	9'4 1/16"	10'9"	12'1 7/16"	7'11"	9'3 3/8"	10'9"
MAXIMUM RUN WITH 6'-8" HEADROOM		1'8 3/8"	1'9 1/2"	1'10 1/2"	9 1/2"	1'8 3/8"	1'9 1/2"	1'10 3/8"	9 1/2"	1'8 1/4"	1'9 1/2"
ANGLE OF INCLINE		35°40'	32°7'	29°3'	40°6'	35°40'	32°7'	29°26'	40°6'	36°4'	32°7'
LENGTH OF CARRIAGE		11'5 15/16"	12'8 5/16"	13'11 5/16"	10'4 3/16"	11'5 15/16"	12'8 5/16"	13'11"	10'4 3/16"	11'5 25/32"	12'8 5/16"
L. F. OF RISER PER INCH OF STAIRWAY WIDTH		1.0	1.083	1.166	.916	1.0	1.083	1.166	.916	1.0	1.083
L. F. OF TREAD PER INCH OF STAIRWAY WIDTH		.916	1.0	1.083	.833	.916	1.0	1.083	.833	.916	1.0
FRAMING SQUARE SETTINGS FOR CARRIAGE CUTS	TONGUE	7 5/16"	6 3/4"	6 1/4"	8"	7 5/16"	6 3/4"	6 5/16"	8"	7 3/8"	6 3/4"
	BODY	10 3/16"	10 3/4"	11 1/4"	9 1/2"	10 3/16"	10 3/4"	11 3/16"	9 1/2"	10 1/8"	10 3/4"

FLOOR TO FLOOR RISE	7'4$\frac{1}{8}$"	7'4$\frac{1}{4}$"	7'4$\frac{1}{4}$"	7'4$\frac{1}{4}$"	7'4$\frac{1}{4}$"	7'4$\frac{3}{8}$"	7'4$\frac{3}{8}$"	7'4$\frac{3}{8}$"	7'4$\frac{3}{8}$"	7'4$\frac{1}{2}$"
NUMBER OF RISERS	14	11	12	13	14	11	12	13	14	11
HEIGHT OF EACH RISER	6$\frac{5}{16}$"	8"	7$\frac{3}{8}$"	6$\frac{13}{16}$"	6$\frac{5}{16}$"	8$\frac{1}{16}$"	7$\frac{3}{8}$"	6$\frac{13}{16}$"	6$\frac{5}{16}$"	8$\frac{1}{16}$"
TOTAL HEIGHT OF RISERS	7'4$\frac{3}{8}$"	7'4"	7'4$\frac{1}{2}$"	7'4$\frac{9}{16}$"	7'4$\frac{3}{8}$"	7'4$\frac{11}{16}$"	7'4$\frac{1}{2}$"	7'4$\frac{9}{16}$"	7'4$\frac{3}{8}$"	7'4$\frac{11}{16}$"
RISERS OVERAGE (+) OR UNDERAGE (—)	+$\frac{1}{4}$"	−$\frac{1}{4}$"	+$\frac{1}{4}$"	+$\frac{5}{16}$"	+$\frac{1}{8}$"	+$\frac{5}{16}$"	+$\frac{1}{8}$"	+$\frac{3}{16}$"	0	+$\frac{3}{16}$"
NUMBER OF TREADS	13	10	11	12	13	10	11	12	13	10
WIDTH OF EACH TREAD	11$\frac{3}{16}$"	9$\frac{1}{2}$"	10$\frac{1}{8}$"	10$\frac{11}{16}$"	11$\frac{3}{16}$"	9$\frac{7}{16}$"	10$\frac{1}{8}$"	10$\frac{11}{16}$"	11$\frac{3}{16}$"	9$\frac{7}{16}$"
TOTAL RUN OF STAIRWAY	12'1$\frac{7}{16}$"	7'11"	9'3$\frac{3}{8}$"	10'8$\frac{1}{4}$"	12'1$\frac{7}{16}$"	7'10$\frac{3}{8}$"	9'3$\frac{3}{8}$"	10'8$\frac{1}{4}$"	12'1$\frac{7}{16}$"	7'10$\frac{3}{8}$"
MAXIMUM RUN WITH 6'-8" HEADROOM	1'10$\frac{3}{8}$"	9$\frac{1}{2}$"	1'8$\frac{1}{4}$"	1'9$\frac{3}{8}$"	1'10$\frac{3}{8}$"	9$\frac{7}{16}$"	1'8$\frac{1}{4}$"	1'9$\frac{3}{8}$"	1'10$\frac{3}{8}$"	9$\frac{7}{16}$"
ANGLE OF INCLINE	29°26'	40°6'	36°4'	32°31'	29°26'	40°30'	36°4'	32°31'	29°26'	40°30'
LENGTH OF CARRIAGE	13'11"	10'4$\frac{3}{16}$"	11'5$\frac{25}{32}$"	12'8$\frac{3}{32}$"	13'11"	10'4$\frac{1}{8}$"	11'5$\frac{25}{32}$"	12'8$\frac{3}{32}$"	13'11"	10'4$\frac{1}{8}$"
L. F. OF RISER PER INCH OF STAIRWAY WIDTH	1.166	.916	1.0	1.083	1.166	.916	1.0	1.083	1.166	.916
L. F. OF TREAD PER INCH OF STAIRWAY WIDTH	1.083	.833	.916	1.0	1.083	.833	.916	1.0	1.083	.833
FRAMING SQUARE SETTINGS FOR CARRIAGE CUTS — TONGUE	6$\frac{5}{16}$"	8"	7$\frac{3}{8}$"	6$\frac{13}{16}$"	6$\frac{5}{16}$"	8$\frac{1}{16}$"	7$\frac{3}{8}$"	6$\frac{13}{16}$"	6$\frac{5}{16}$"	8$\frac{1}{16}$"
FRAMING SQUARE SETTINGS FOR CARRIAGE CUTS — BODY	11$\frac{3}{16}$"	9$\frac{1}{2}$"	10$\frac{1}{8}$"	10$\frac{11}{16}$"	11$\frac{3}{16}$"	9$\frac{7}{16}$"	10$\frac{1}{8}$"	10$\frac{11}{16}$"	11$\frac{3}{16}$"	9$\frac{7}{16}$"

FLOOR TO FLOOR RISE		7'4-1/2"	7'4-1/2"	7'4-1/2"	7'4-5/8"	7'4-5/8"	7'4-5/8"	7'4-5/8"	7'4-3/4"	7'4-3/4"	7'4-3/4"
NUMBER OF RISERS		12	13	14	11	12	13	14	11	12	13
HEIGHT OF EACH RISER		7-3/8"	6-13/16"	6-5/16"	8-1/16"	7-3/8"	6-13/16"	6-5/16"	8-1/16"	7-3/8"	6-13/16"
TOTAL HEIGHT OF RISERS		7'4-1/2"	7'4-9/16"	7'4-3/8"	7'4-11/16"	7'4-1/2"	7'4-9/16"	7'4-3/8"	7'4-11/16"	7'4-1/2"	7'4-9/16"
RISERS OVERAGE (+) OR UNDERAGE (—)		0	+1/16"	−1/8"	+1/16"	−1/8"	−1/16"	−1/4"	−1/16"	−1/4"	−3/16"
NUMBER OF TREADS		11	12	13	10	11	12	13	10	11	12
WIDTH OF EACH TREAD		10-1/8"	10-11/16"	11-3/16"	9-7/16"	10-1/8"	10-11/16"	11-3/16"	9-7/16"	10-1/8"	10-11/16"
TOTAL RUN OF STAIRWAY		9'3-3/8"	10'8-1/4"	12'1-7/16"	7'10-3/8"	9'3-3/8"	10'8-1/4"	12'1-7/16"	7'10-3/8"	9'3-3/8"	10'8-1/4"
MAXIMUM RUN WITH 6'-8" HEADROOM		1'8-1/4"	1'9-3/8"	1'10-3/8"	9-7/16"	1'8-1/4"	1'9-3/8"	1'10-3/8"	9-7/16"	1'8-1/4"	1'9-3/8"
ANGLE OF INCLINE		36°4'	32°31'	29°26'	40°30'	36°4'	32°31'	29°26'	40°30'	36°4'	32°31'
LENGTH OF CARRIAGE		11'5-25/32"	12'8-3/32"	13'11"	10'4-1/8"	11'5-25/32"	12'8-3/32"	13'11"	10'4-1/8"	11'5-25/32"	12'8-3/32"
L. F. OF RISER PER INCH OF STAIRWAY WIDTH		1.0	1.083	1.166	.916	1.0	1.083	1.166	.916	1.0	1.083
L. F. OF TREAD PER INCH OF STAIRWAY WIDTH		.916	1.0	1.083	.833	.916	1.0	1.083	.833	.916	1.0
FRAMING SQUARE SETTINGS FOR CARRIAGE CUTS	TONGUE	7-3/8"	6-13/16"	6-5/16"	8-1/16"	7-3/8"	6-13/16"	6-5/16"	8-1/16"	7-3/8"	6-13/16"
	BODY	10-1/8"	10-11/16"	11-3/16"	9-7/16"	10-1/8"	10-11/16"	11-3/16"	9-7/16"	10-1/8"	10-11/16"

FLOOR TO FLOOR RISE	7'4 3/4"	7'4 7/8"	7'4 7/8"	7'4 7/8"	7'4 7/8"	7'5"	7'5"	7'5"	7'5"	7'5 1/8"
NUMBER OF RISERS	14	11	12	13	14	11	12	13	14	11
HEIGHT OF EACH RISER	6 5/16"	8 1/16"	7 7/16"	6 13/16"	6 3/8"	8 1/16"	7 7/16"	6 7/8"	6 3/8"	8 1/8"
TOTAL HEIGHT OF RISERS	7'4 3/8"	7'4 11/16"	7'5 1/4"	7'4 9/16"	7'5 1/4"	7'4 11/16"	7'5 1/4"	7'5 3/8"	7'5 1/4"	7'5 3/8"
RISERS OVERAGE (+) OR UNDERAGE (—)	−3/8"	−3/16"	+3/8"	−5/16"	+3/8"	−5/16"	+1/4"	+3/8"	+1/4"	+1/4"
NUMBER OF TREADS	13	10	11	12	13	10	11	12	13	10
WIDTH OF EACH TREAD	11 3/16"	9 7/16"	10 1/16"	10 11/16"	11 1/8"	9 7/16"	10 1/16"	10 5/8"	11 1/8"	9 3/8"
TOTAL RUN OF STAIRWAY	12'1 7/16"	7'10 3/8"	9'2 11/16"	10'8 1/4"	12' 5/8"	7'10 3/8"	9'2 11/16"	10'7 1/2"	12' 5/8"	7'9 3/4"
MAXIMUM RUN WITH 6'-8" HEADROOM	1'10 3/8"	9 7/16"	1'8 1/8"	1'9 3/8"	1'10 1/4"	9 7/16"	1'8 1/8"	1'9 1/4"	1'10 1/4"	9 3/8"
ANGLE OF INCLINE	29°26'	40°30'	36°28'	32°31'	29°49'	40°30'	36°28'	32°54'	29°49'	40°55'
LENGTH OF CARRIAGE	13'11"	10'4 1/8"	11'5 21/32"	12'8 3/32"	13'10 11/16"	10'4 1/8"	11'5 21/32"	12'7 29/32"	13'10 11/16"	10'4 1/16"
L. F. OF RISER PER INCH OF STAIRWAY WIDTH	1.166	.916	1.0	1.083	1.166	.916	1.0	1.083	1.166	.916
L. F. OF TREAD PER INCH OF STAIRWAY WIDTH	1.083	.833	.916	1.0	1.038	.833	.916	1.0	1.083	.833
FRAMING SQUARE SETTINGS FOR CARRIAGE CUTS — TONGUE	6 5/16"	8 1/16"	7 7/16"	6 13/16"	6 3/8"	8 1/16"	7 7/16"	6 7/8"	6 3/8"	8 1/8"
FRAMING SQUARE SETTINGS FOR CARRIAGE CUTS — BODY	11 3/16"	9 7/16"	10 1/16"	10 11/16"	11 1/8"	9 7/16"	10 1/16"	10 5/8"	11 1/8"	9 3/8"

FLOOR TO FLOOR RISE		7'5$\frac{1}{8}$"	7'5$\frac{1}{8}$"	7'5$\frac{1}{8}$"	7'5$\frac{1}{4}$"	7'5$\frac{1}{4}$"	7'5$\frac{1}{4}$"	7'5$\frac{1}{4}$"	7'5$\frac{3}{8}$"	7'5$\frac{3}{8}$"	7'5$\frac{3}{8}$"
NUMBER OF RISERS		12	13	14	11	12	13	14	11	12	13
HEIGHT OF EACH RISER		7$\frac{7}{16}$"	6$\frac{7}{8}$"	6$\frac{3}{8}$"	8$\frac{1}{8}$"	7$\frac{7}{16}$"	6$\frac{7}{8}$"	6$\frac{3}{8}$"	8$\frac{1}{8}$"	7$\frac{7}{16}$"	6$\frac{7}{8}$"
TOTAL HEIGHT OF RISERS		7'5$\frac{1}{4}$"	7'5$\frac{3}{8}$"	7'5$\frac{1}{4}$"	7'5$\frac{3}{8}$"	7'5$\frac{1}{4}$"	7'5$\frac{3}{8}$"	7'5$\frac{1}{4}$"	7'5$\frac{3}{8}$"	7'5$\frac{1}{4}$"	7'5$\frac{3}{8}$"
RISERS OVERAGE (+) OR UNDERAGE (—)		+$\frac{1}{8}$"	+$\frac{1}{4}$"	+$\frac{1}{8}$"	+$\frac{1}{8}$"	0	+$\frac{1}{8}$"	0	0	−$\frac{1}{8}$"	0
NUMBER OF TREADS		11	12	13	10	11	12	13	10	11	12
WIDTH OF EACH TREAD		10$\frac{1}{16}$"	10$\frac{5}{8}$"	11$\frac{1}{8}$"	9$\frac{3}{8}$"	10$\frac{1}{16}$"	10$\frac{5}{8}$"	11$\frac{1}{8}$"	9$\frac{3}{8}$"	10$\frac{1}{16}$"	10$\frac{5}{8}$"
TOTAL RUN OF STAIRWAY		9'2$\frac{11}{16}$"	10'7$\frac{1}{2}$"	12'$\frac{5}{8}$"	7'9$\frac{3}{4}$"	9'2$\frac{11}{16}$"	10'7$\frac{1}{2}$"	12'$\frac{5}{8}$"	7'9$\frac{3}{4}$"	9'2$\frac{11}{16}$"	10'7$\frac{1}{2}$"
MAXIMUM RUN WITH 6'-8" HEADROOM		1'8$\frac{1}{8}$"	1'9$\frac{1}{4}$"	1'10$\frac{1}{4}$"	9$\frac{3}{8}$"	1'8$\frac{1}{8}$"	1'9$\frac{1}{4}$"	1'10$\frac{1}{4}$"	9$\frac{3}{8}$"	1'8$\frac{1}{8}$"	1'9$\frac{1}{4}$"
ANGLE OF INCLINE		36°28'	32°54'	29°49'	40°55'	36°28'	32°54'	29°49'	40°55'	36°28'	32°54'
LENGTH OF CARRIAGE		11'5$\frac{21}{32}$"	12'7$\frac{29}{32}$"	13'10$\frac{11}{16}$"	10'4$\frac{1}{16}$"	11'5$\frac{21}{32}$"	12'7$\frac{29}{32}$"	13'10$\frac{11}{16}$"	10'4$\frac{1}{16}$"	11'5$\frac{21}{32}$"	12'7$\frac{29}{32}$"
L. F. OF RISER PER INCH OF STAIRWAY WIDTH		1.0	1.083	1.166	.916	1.0	1.083	1.166	.916	1.0	1.083
L. F. OF TREAD PER INCH OF STAIRWAY WIDTH		.916	1.0	1.083	.833	.916	1.0	1.083	.833	.916	1.0
FRAMING SQUARE SETTINGS FOR CARRIAGE CUTS	TONGUE	7$\frac{7}{16}$"	6$\frac{7}{8}$"	6$\frac{3}{8}$"	8$\frac{1}{8}$"	7$\frac{7}{16}$"	6$\frac{7}{8}$"	6$\frac{3}{8}$"	8$\frac{1}{8}$"	7$\frac{7}{16}$"	6$\frac{7}{8}$"
	BODY	10$\frac{1}{16}$"	10$\frac{5}{8}$"	11$\frac{1}{8}$"	9$\frac{3}{8}$"	10$\frac{1}{16}$"	10$\frac{5}{8}$"	11$\frac{1}{8}$"	9$\frac{3}{8}$"	10$\frac{1}{16}$"	10$\frac{5}{8}$"

FLOOR TO FLOOR RISE	7'5 3/8"	7'5 1/2"	7'5 1/2"	7'5 1/2"	7'5 1/2"	7'5 5/8"	7'5 5/8"	7'5 5/8"	7'5 5/8"	7'5 3/4"
NUMBER OF RISERS	14	11	12	13	14	11	12	13	14	11
HEIGHT OF EACH RISER	6 3/8"	8 1/8"	7 7/16"	6 7/8"	6 3/8"	8 1/8"	7 1/2"	6 7/8"	6 3/8"	8 3/16"
TOTAL HEIGHT OF RISERS	7'5 1/4"	7'5 3/8"	7'5 1/4"	7'5 3/8"	7'5 1/4"	7'5 3/8"	7'6"	7'5 3/8"	7'5 1/4"	7'6 1/16"
RISERS OVERAGE (+) OR UNDERAGE (—)	−1/8"	−1/8"	−1/4"	−1/8"	−1/4"	−1/4"	+3/8"	−1/4"	−3/8"	+5/16"
NUMBER OF TREADS	13	10	11	12	13	10	11	12	13	10
WIDTH OF EACH TREAD	11 1/8"	9 3/8"	10 1/16"	10 5/8"	11 1/8"	9 3/8"	10"	10 5/8"	11 1/8"	9 5/16"
TOTAL RUN OF STAIRWAY	12' 5/8"	7'9 3/4"	9'2 11/16"	10'7 1/2"	12' 5/8"	7'9 3/4"	9'2"	10'7 1/2"	12' 5/8"	7'9 1/8"
MAXIMUM RUN WITH 6'-8" HEADROOM	1'10 1/4"	9 3/8"	1'8 1/8"	1'9 1/4"	1'10 1/4"	9 3/8"	1'8"	1'9 1/4"	1'10 1/4"	9 5/16"
ANGLE OF INCLINE	29°49'	40°55'	36°28'	32°54'	29°49'	40°55'	36°52'	32°54'	29°49'	41°19'
LENGTH OF CARRIAGE	13'10 11/16"	10'4 1/16"	11'5 21/32"	12'7 29/32"	13'10 11/16"	10'4 1/16"	11'5 1/2"	12'7 29/32"	13'10 11/16"	10'4"
L. F. OF RISER PER INCH OF STAIRWAY WIDTH	1.166	.916	1.0	1.083	1.166	.916	1.0	1.083	1.166	.916
L. F. OF TREAD PER INCH OF STAIRWAY WIDTH	1.083	.833	.916	1.0	1.083	.833	.916	1.0	1.083	.833
FRAMING SQUARE SETTINGS FOR CARRIAGE CUTS — TONGUE	6 3/8"	8 1/8"	7 7/16"	6 7/8"	6 3/8"	8 1/8"	7 1/2"	6 7/8"	6 3/8"	8 3/16"
FRAMING SQUARE SETTINGS FOR CARRIAGE CUTS — BODY	11 1/8"	9 3/8"	10 1/16"	10 5/8"	11 1/8"	9 3/8"	10"	10 5/8"	11 1/8"	9 5/16"

FLOOR TO FLOOR RISE	7'5 3/4"	7'5 3/4"	7'5 3/4"	7'5 7/8"	7'5 7/8"	7'5 7/8"	7'5 7/8"	7'6"	7'6"	7'6"
NUMBER OF RISERS	12	13	14	11	12	13	14	11	12	13
HEIGHT OF EACH RISER	7 1/2"	6 7/8"	6 7/16"	8 3/16"	7 1/2"	6 15/16"	6 7/16"	8 3/16"	7 1/2"	6 15/16"
TOTAL HEIGHT OF RISERS	7'6"	7'5 3/8"	7'6 1/8"	7'6 1/16"	7'6"	7'6 3/16"	7'6 1/8"	7'6 1/16"	7'6"	7'6 3/16"
RISERS OVERAGE (+) OR UNDERAGE (—)	+1/4"	−3/8"	+3/8"	+3/16"	+1/8"	+5/16"	+1/4"	+1/16"	0	+3/16"
NUMBER OF TREADS	11	12	13	10	11	12	13	10	11	12
WIDTH OF EACH TREAD	10"	10 5/8"	11 1/16"	9 5/16"	10"	10 9/16"	11 1/16"	9 5/16"	10"	10 9/16"
TOTAL RUN OF STAIRWAY	9'2"	10'7 1/2"	11'11 13/16"	7'9 1/8"	9'2"	10'6 3/4"	11'11 13/16"	7'9 1/8"	9'2"	10'6 3/4"
MAXIMUM RUN WITH 6'-8" HEADROOM	1'8"	1'9 1/4"	1'10 1/8"	9 5/16"	1'8"	1'9 1/8"	1'10 1/8"	9 5/16"	1'8"	1'9 1/8"
ANGLE OF INCLINE	36°52'	32°54'	30°12'	41°19'	36°52'	33°18'	30°12'	41°19'	36°52'	33°18'
LENGTH OF CARRIAGE	11'5 1/2"	12'7 29/32"	13'10 13/32"	10'4"	11'5 1/2"	12'7 21/32"	13'10 13/32"	10'4"	11'5 1/2"	12'7 21/32"
L. F. OF RISER PER INCH OF STAIRWAY WIDTH	1.0	1.083	1.166	.916	1.0	1.083	1.166	.916	1.0	1.083
L. F. OF TREAD PER INCH OF STAIRWAY WIDTH	.916	1.0	1.083	.833	.916	1.0	1.083	.833	.916	1.0
FRAMING SQUARE SETTINGS FOR CARRIAGE CUTS — TONGUE	7 1/2"	6 7/8"	6 7/16"	8 3/16"	7 1/2"	6 15/16"	6 7/16"	8 3/16"	7 1/2"	6 15/16"
FRAMING SQUARE SETTINGS FOR CARRIAGE CUTS — BODY	10"	10 5/8"	11 1/16"	9 5/16"	10"	10 9/16"	11 1/16"	9 5/16"	10"	10 9/16"

FLOOR TO FLOOR RISE	7'6"	7'6"	7'6 1/8"	7'6 1/8"	7'6 1/8"	7'6 1/8"	7'6 1/8"	7'6 1/4"	7'6 1/4"	7'6 1/4"
NUMBER OF RISERS	14	15	11	12	13	14	15	11	12	13
HEIGHT OF EACH RISER	6 7/16"	6"	8 3/16"	7 1/2"	6 15/16"	6 7/16"	6"	8 3/16"	7 1/2"	6 15/16"
TOTAL HEIGHT OF RISERS	7'6 1/8"	7'6"	7'6 1/16"	7'6"	7'6 3/16"	7'6 1/8"	7'6"	7'6 1/16"	7'6"	7'6 3/16"
RISERS OVERAGE (+) OR UNDERAGE (—)	+1/8"	0	−1/16"	−1/8"	+1/16"	0	−1/8"	−3/16"	−1/4"	−1/16"
NUMBER OF TREADS	13	14	10	11	12	13	14	10	11	12
WIDTH OF EACH TREAD	11 1/16"	11 1/2"	9 5/16"	10"	10 9/16"	11 1/16"	11 1/2"	9 5/16"	10"	10 9/16"
TOTAL RUN OF STAIRWAY	11'11 13/16"	13'5"	7'9 1/8"	9'2"	10'6 3/4"	11'11 13/16"	13'5"	7'9 1/8"	9'2"	10'6 3/4"
MAXIMUM RUN WITH 6'-8" HEADROOM	1'10 1/8"	1'11"	9 5/16"	1'8"	1'9 1/8"	1'10 1/8"	1'11"	9 5/16"	1'8"	1'9 1/8"
ANGLE OF INCLINE	30°12'	27°33'	41°19'	36°52'	33°18'	30°12'	27°33'	41°19'	36°52'	33°18'
LENGTH OF CARRIAGE	13'10 13/32"	15'1 19/32"	10'4"	11'5 1/2"	12'7 21/32"	13'10 13/32"	15'1 19/32"	10'4"	11'5 1/2"	12'7 21/32"
L. F. OF RISER PER INCH OF STAIRWAY WIDTH	1.166	1.25	.916	1.0	1.083	1.166	1.25	.916	1.0	1.083
L. F. OF TREAD PER INCH OF STAIRWAY WIDTH	1.083	1.166	.833	.916	1.0	1.083	1.166	.833	.916	1.0
FRAMING SQUARE SETTINGS FOR CARRIAGE CUTS — TONGUE	6 7/16"	6"	8 3/16"	7 1/2"	6 15/16"	6 7/16"	6"	8 3/16"	7 1/2"	6 15/16"
FRAMING SQUARE SETTINGS FOR CARRIAGE CUTS — BODY	11 1/16"	11 1/2"	9 5/16"	10"	10 9/16"	11 1/16"	11 1/2"	9 5/16"	10"	10 9/16"

FLOOR TO FLOOR RISE		7'6 1/4"	7'6 1/4"	7'6 3/8"	7'6 3/8"	7'6 3/8"	7'6 3/8"	7'6 3/8"	7'6 1/2"	7'6 1/2"	7'6 1/2"
NUMBER OF RISERS		14	15	11	12	13	14	15	11	12	13
HEIGHT OF EACH RISER		6 7/16"	6"	8 3/16"	7 9/16"	6 15/16"	6 7/16"	6"	8 1/4"	7 9/16"	6 15/16"
TOTAL HEIGHT OF RISERS		7'6 1/8"	7'6"	7'6 1/16"	7'6 3/4"	7'6 3/16"	7'6 1/8"	7'6"	7'6 3/4"	7'6 3/4"	7'6 3/16"
RISERS OVERAGE (+) OR UNDERAGE (—)		−1/8"	−1/4"	−5/16"	+3/8"	−3/16"	−1/4"	−3/8"	+1/4"	+1/4"	−5/16"
NUMBER OF TREADS		13	14	10	11	12	13	14	10	11	12
WIDTH OF EACH TREAD		11 1/16"	11 1/2"	9 5/16"	9 15/16"	10 9/16"	11 1/16"	11 1/2"	9 1/4"	9 15/16"	10 9/16"
TOTAL RUN OF STAIRWAY		11'11 13/16"	13'5"	7'9 1/8"	9'1 5/16"	10'6 3/4"	11'11 13/16"	13'5"	7'8 1/2"	9'1 5/16"	10'6 3/4"
MAXIMUM RUN WITH 6'-8" HEADROOM		1'10 1/8"	1'11"	9 5/16"	1'7 7/8"	1'9 1/8"	1'10 1/8"	1'11"	9 1/4"	1'7 7/8"	1'9 1/8"
ANGLE OF INCLINE		30°12'	27°33'	41°19'	37°16'	33°18'	30°12'	27°33'	41°44'	37°16'	33°18'
LENGTH OF CARRIAGE		13'10 13/32"	15'1 19/32"	10'4"	11'5 3/8"	12'7 21/32"	13'10 13/32"	15'1 19/32"	10'3 15/16"	11'5 3/8"	12'7 21/32"
L. F. OF RISER PER INCH OF STAIRWAY WIDTH		1.166	1.25	.916	1.0	1.083	1.166	1.25	.916	1.0	1.083
L. F. OF TREAD PER INCH OF STAIRWAY WIDTH		1.083	1.166	.833	.916	1.0	1.083	1.166	.833	.916	1.0
FRAMING SQUARE SETTINGS FOR CARRIAGE CUTS	TONGUE	6 7/16"	6"	8 3/16"	7 9/16"	6 15/16"	6 7/16"	6"	8 1/4"	7 9/16"	6 15/16"
	BODY	11 1/16"	11 1/2"	9 5/16"	9 15/16"	10 9/16"	11 1/16"	11 1/2"	9 1/4"	9 15/16"	10 9/16"

FLOOR TO FLOOR RISE	7'6 1/2"	7'6 1/2"	7'6 5/8"	7'6 5/8"	7'6 5/8"	7'6 5/8"	7'6 5/8"	7'6 3/4"	7'6 3/4"	7'6 3/4"
NUMBER OF RISERS	14	15	11	12	13	14	15	11	12	13
HEIGHT OF EACH RISER	6 7/16"	6 1/16"	8 1/4"	7 9/16"	7"	6 1/2"	6 1/16"	8 1/4"	7 9/16"	7"
TOTAL HEIGHT OF RISERS	7'6 1/8"	7'6 15/16"	7'6 3/4"	7'6 3/4"	7'7"	7'7"	7'6 15/16"	7'6 3/4"	7'6 3/4"	7'7"
RISERS OVERAGE (+) OR UNDERAGE (—)	−3/8"	+7/16"	+1/8"	+1/8"	+3/8"	+3/8"	+5/16"	0	0	+1/4"
NUMBER OF TREADS	13	14	10	11	12	13	14	10	11	12
WIDTH OF EACH TREAD	11 1/16"	11 7/16"	9 1/4"	9 15/16"	10 1/2"	11"	11 7/16"	9 1/4"	9 15/16"	10 1/2"
TOTAL RUN OF STAIRWAY	11'11 13/16"	13'4 1/8"	7'8 1/2"	9'1 5/16"	10'6"	11'11"	13'4 1/8"	7'8 1/2"	9'1 5/16"	10'6"
MAXIMUM RUN WITH 6'-8" HEADROOM	1'10 1/8"	1'10 7/8"	9 1/4"	1'7 7/8"	1'9"	1'10"	1'10 7/8"	9 1/4"	1'7 7/8"	1'9"
ANGLE OF INCLINE	30°12'	27°56'	41°44'	37°16'	33°41'	30°35'	27°56'	41°44'	37°16'	33°41'
LENGTH OF CARRIAGE	13'10 13/32"	15'1 7/32"	10'3 15/16"	11'5 3/8"	12'7 7/16"	13'10 3/32"	15'1 7/32"	10'3 15/16"	11'5 3/8"	12'7 7/16"
L. F. OF RISER PER INCH OF STAIRWAY WIDTH	1.166	1.25	.916	1.0	1.083	1.166	1.25	.916	1.0	1.083
L. F. OF TREAD PER INCH OF STAIRWAY WIDTH	1.083	1.166	.833	.916	1.0	1.083	1.166	.833	.916	1.0
FRAMING SQUARE SETTINGS FOR CARRIAGE CUTS — TONGUE	6 7/16"	6 1/16"	8 1/4"	7 9/16"	7"	6 1/2"	6 1/16"	8 1/4"	7 9/16"	7"
FRAMING SQUARE SETTINGS FOR CARRIAGE CUTS — BODY	11 1/16"	11 7/16"	9 1/4"	9 15/16"	10 1/2"	11"	11 7/16"	9 1/4"	9 15/16"	10 1/2"

FLOOR TO FLOOR RISE	7'6$\frac{3}{4}$"	7'6$\frac{3}{4}$"	7'6$\frac{7}{8}$"	7'6$\frac{7}{8}$"	7'6$\frac{7}{8}$"	7'6$\frac{7}{8}$"	7'6$\frac{7}{8}$"	7'7"	7'7"	7'7"
NUMBER OF RISERS	14	15	11	12	13	14	15	11	12	13
HEIGHT OF EACH RISER	6$\frac{1}{2}$"	6$\frac{1}{16}$"	8$\frac{1}{4}$"	7$\frac{9}{16}$"	7"	6$\frac{1}{2}$"	6$\frac{1}{16}$"	8$\frac{1}{4}$"	7$\frac{9}{16}$"	7"
TOTAL HEIGHT OF RISERS	7'7"	7'6$\frac{15}{16}$"	7'6$\frac{3}{4}$"	7'6$\frac{3}{4}$"	7'7"	7'7"	7'6$\frac{15}{16}$"	7'6$\frac{3}{4}$"	7'6$\frac{3}{4}$"	7'7"
RISERS OVERAGE (+) OR UNDERAGE (—)	+$\frac{1}{4}$"	+$\frac{3}{16}$"	−$\frac{1}{8}$"	−$\frac{1}{8}$"	+$\frac{1}{8}$"	+$\frac{1}{8}$"	+$\frac{1}{16}$"	−$\frac{1}{4}$"	−$\frac{1}{4}$"	0
NUMBER OF TREADS	13	14	10	11	12	13	14	10	11	12
WIDTH OF EACH TREAD	11"	11$\frac{7}{16}$"	9$\frac{1}{4}$"	9$\frac{15}{16}$"	10$\frac{1}{2}$"	11"	11$\frac{7}{16}$"	9$\frac{1}{4}$"	9$\frac{15}{16}$"	10$\frac{1}{2}$"
TOTAL RUN OF STAIRWAY	11'11"	13'4$\frac{1}{8}$"	7'8$\frac{1}{2}$"	9'1$\frac{5}{16}$"	10'6"	11'11"	13'4$\frac{1}{8}$"	7'8$\frac{1}{2}$"	9'1$\frac{5}{16}$"	10'6"
MAXIMUM RUN WITH 6'-8" HEADROOM	1'10"	1'10$\frac{7}{8}$"	9$\frac{1}{4}$"	1'7$\frac{7}{8}$"	1'9"	1'10"	1'10$\frac{7}{8}$"	9$\frac{1}{4}$"	1'7$\frac{7}{8}$"	1'9"
ANGLE OF INCLINE	30°35'	27°56'	41°44'	37°16'	33°41'	30°35'	27°56'	41°44'	37°16'	33°41'
LENGTH OF CARRIAGE	13'10$\frac{3}{32}$"	15'1$\frac{7}{32}$"	10'3$\frac{15}{16}$"	11'5$\frac{3}{8}$"	12'7$\frac{7}{16}$"	13'10$\frac{3}{32}$"	15'1$\frac{7}{32}$"	10'3$\frac{15}{16}$"	11'5$\frac{3}{8}$"	12'7$\frac{7}{16}$"
L. F. OF RISER PER INCH OF STAIRWAY WIDTH	1.166	1.25	.916	1.0	1.083	1.166	1.25	.916	1.0	1.083
L. F. OF TREAD PER INCH OF STAIRWAY WIDTH	1.083	1.166	.833	.916	1.0	1.083	1.166	.833	.916	1.0
FRAMING SQUARE SETTINGS FOR CARRIAGE CUTS — TONGUE	6$\frac{1}{2}$"	6$\frac{1}{16}$"	8$\frac{1}{4}$"	7$\frac{9}{16}$"	7"	6$\frac{1}{2}$"	6$\frac{1}{16}$"	8$\frac{1}{4}$"	7$\frac{9}{16}$"	7"
FRAMING SQUARE SETTINGS FOR CARRIAGE CUTS — BODY	11"	11$\frac{7}{16}$"	9$\frac{1}{4}$"	9$\frac{15}{16}$"	10$\frac{1}{2}$"	11"	11$\frac{7}{16}$"	9$\frac{1}{4}$"	9$\frac{15}{16}$"	10$\frac{1}{2}$"

FLOOR TO FLOOR RISE	7'7"	7'7"	7'7$\frac{1}{8}$"	7'7$\frac{1}{8}$"	7'7$\frac{1}{8}$"	7'7$\frac{1}{8}$"	7'7$\frac{1}{4}$"	7'7$\frac{1}{4}$"	7'7$\frac{1}{4}$"	7'7$\frac{1}{4}$"
NUMBER OF RISERS	14	15	12	13	14	15	12	13	14	15
HEIGHT OF EACH RISER	6$\frac{1}{2}$"	6$\frac{1}{16}$"	7$\frac{5}{8}$"	7"	6$\frac{1}{2}$"	6$\frac{1}{16}$"	7$\frac{5}{8}$"	7"	6$\frac{1}{2}$"	6$\frac{1}{16}$"
TOTAL HEIGHT OF RISERS	7'7"	7'6$\frac{15}{16}$"	7'7$\frac{1}{2}$"	7'7"	7'7"	7'6$\frac{15}{16}$"	7'7$\frac{1}{2}$"	7'7"	7'7"	7'6$\frac{15}{16}$"
RISERS OVERAGE (+) OR UNDERAGE (—)	0	$-\frac{1}{16}$"	$+\frac{3}{8}$"	$-\frac{1}{8}$"	$-\frac{1}{8}$"	$-\frac{3}{16}$"	$+\frac{1}{4}$"	$-\frac{1}{4}$"	$-\frac{1}{4}$"	$-\frac{5}{16}$"
NUMBER OF TREADS	13	14	11	12	13	14	11	12	13	14
WIDTH OF EACH TREAD	11"	11$\frac{7}{16}$"	9$\frac{7}{8}$"	10$\frac{1}{2}$"	11"	11$\frac{7}{16}$"	9$\frac{7}{8}$"	10$\frac{1}{2}$"	11"	11$\frac{7}{16}$"
TOTAL RUN OF STAIRWAY	11'11"	13'4$\frac{1}{8}$"	9'$\frac{5}{8}$"	10'6"	11'11"	13'4$\frac{1}{8}$"	9'$\frac{5}{8}$"	10'6"	11'11"	13'4$\frac{1}{8}$"
MAXIMUM RUN WITH 6'-8" HEADROOM	1'10"	1'10$\frac{7}{8}$"	1'7$\frac{3}{4}$"	1'9"	1'10"	1'10$\frac{7}{8}$"	1'7$\frac{3}{4}$"	1'9"	1'10"	1'10$\frac{7}{8}$"
ANGLE OF INCLINE	30°35'	27°56'	37°40'	33°41'	30°35'	27°56'	37°40'	33°41'	30°35'	27°56'
LENGTH OF CARRIAGE	13'10$\frac{3}{32}$"	15'1$\frac{7}{32}$"	11'5$\frac{7}{32}$"	12'7$\frac{7}{16}$"	13'10$\frac{3}{32}$"	15'1$\frac{7}{32}$"	11'5$\frac{7}{32}$"	12'7$\frac{7}{16}$"	13'10$\frac{3}{32}$"	15'1$\frac{7}{32}$"
L. F. OF RISER PER INCH OF STAIRWAY WIDTH	1.166	1.25	1.0	1.083	1.166	1.25	1.0	1.083	1.166	1.25
L. F. OF TREAD PER INCH OF STAIRWAY WIDTH	1.083	1.166	.916	1.0	1.083	1.166	.916	1.0	1.083	1.166
FRAMING SQUARE SETTINGS FOR CARRIAGE CUTS — TONGUE	6$\frac{1}{2}$"	6$\frac{1}{16}$"	7$\frac{5}{8}$"	7"	6$\frac{1}{2}$"	6$\frac{1}{16}$"	7$\frac{5}{8}$"	7"	6$\frac{1}{2}$"	6$\frac{1}{16}$"
BODY	11"	11$\frac{7}{16}$"	9$\frac{7}{8}$"	10$\frac{1}{2}$"	11"	11$\frac{7}{16}$"	9$\frac{7}{8}$"	10$\frac{1}{2}$"	11"	11$\frac{7}{16}$"

FLOOR TO FLOOR RISE	7'7 3/8"	7'7 3/8"	7'7 3/8"	7'7 3/8"	7'7 1/2"	7'7 1/2"	7'7 1/2"	7'7 1/2"	7'7 5/8"	7'7 5/8"
NUMBER OF RISERS	12	13	14	15	12	13	14	15	12	13
HEIGHT OF EACH RISER	7 5/8"	7"	6 1/2"	6 1/16"	7 5/8"	7 1/16"	6 9/16"	6 1/8"	7 5/8"	7 1/16"
TOTAL HEIGHT OF RISERS	7'7 1/2"	7'7"	7'7"	7'6 15/16"	7'7 1/2"	7'7 13/16"	7'7 7/8"	7'7 7/8"	7'7 1/2"	7'7 13/16"
RISERS OVERAGE (+) OR UNDERAGE (—)	+1/8"	−3/8"	−3/8"	−7/16"	0	+5/16"	+3/8"	+3/8"	−1/8"	+3/16"
NUMBER OF TREADS	11	12	13	14	11	12	13	14	11	12
WIDTH OF EACH TREAD	9 7/8"	10 1/2"	11"	11 7/16"	9 7/8"	10 7/16"	10 15/16"	11 3/8"	9 7/8"	10 7/16"
TOTAL RUN OF STAIRWAY	9'5/8"	10'6"	11'11"	13'4 1/8"	9'5/8"	10'5 1/4"	11'10 3/16"	13'3 1/4'	9'5/8"	10'5 1/4"
MAXIMUM RUN WITH 6'-8" HEADROOM	1'7 3/4"	1'9"	1'10"	1'10 7/8"	1'7 3/4"	1'8 7/8"	1'9 7/8"	1'10 3/4"	1'7 3/4"	1'8 7/8"
ANGLE OF INCLINE	37°40'	33°41'	30°35'	27°56'	37°40'	34°5'	30°58'	28°18'	37°40'	34°5'
LENGTH OF CARRIAGE	11'5 7/32"	12'7 7/16"	13'10 3/32"	15'1 7/32"	11'5 7/32"	12'7 7/32"	13'9 13/16"	15'7/8"	11'5 7/32"	12'7 7/32"
L. F. OF RISER PER INCH OF STAIRWAY WIDTH	1.0	1.083	1.166	1.25	1.0	1.083	1.166	1.25	1.0	1.083
L. F. OF TREAD PER INCH OF STAIRWAY WIDTH	.916	1.0	1.083	1.166	.916	1.0	1.083	1.166	.916	1.0
FRAMING SQUARE SETTINGS FOR CARRIAGE CUTS — TONGUE	7 5/8"	7"	6 1/2"	6 1/16"	7 5/8"	7 1/16"	6 9/16"	6 1/8"	7 5/8"	7 1/16"
FRAMING SQUARE SETTINGS FOR CARRIAGE CUTS — BODY	9 7/8"	10 1/2"	11"	11 7/16"	9 7/8"	10 7/16"	10 15/16"	11 3/8"	9 7/8"	10 7/16"

FLOOR TO FLOOR RISE	7'7 5/8"	7'7 5/8"	7'7 3/4"	7'7 3/4"	7'7 3/4"	7'7 3/4"	7'7 7/8"	7'7 7/8"	7'7 7/8"	7'7 7/8"
NUMBER OF RISERS	14	15	12	13	14	15	12	13	14	15
HEIGHT OF EACH RISER	6 9/16"	6 1/8"	7 5/8"	7 1/16"	6 9/16"	6 1/8"	7 11/16"	7 1/16"	6 9/16"	6 1/8"
TOTAL HEIGHT OF RISERS	7'7 7/8"	7'7 7/8"	7'7 1/2"	7'7 13/16"	7'7 7/8"	7'7 7/8"	7'8 1/4"	7'7 13/16"	7'7 7/8"	7'7 7/8"
RISERS OVERAGE (+) OR UNDERAGE (—)	+1/4"	+1/4"	−1/4"	+1/16"	+1/8"	+1/8"	+3/8"	−1/16"	0	0
NUMBER OF TREADS	13	14	11	12	13	14	11	12	13	14
WIDTH OF EACH TREAD	10 15/16"	11 3/8"	9 7/8"	10 7/16"	10 15/16"	11 3/8"	9 13/16"	10 7/16"	10 15/16"	11 3/8"
TOTAL RUN OF STAIRWAY	11'10 3/16"	13'3 1/4"	9' 5/8"	10'5 1/4"	11'10 3/16"	13'3 1/4"	8'11 15/16"	10'5 1/4"	11'10 3/16"	13'3 1/4"
MAXIMUM RUN WITH 6'-8" HEADROOM	1'9 7/8"	1'10 3/4"	1'7 3/4"	1'8 7/8"	1'9 7/8"	1'10 3/4"	1'7 5/8"	1'8 7/8"	1'9 7/8"	1'10 3/4"
ANGLE OF INCLINE	30°58'	28°18'	37°40'	34°5'	30°58'	28°18'	38°5'	34°5'	30°58'	28°18'
LENGTH OF CARRIAGE	13'9 13/16"	15' 7/8"	11'5 7/32"	12'7 7/32"	13'9 13/16"	15' 7/8"	11'5 1/8"	12'7 7/32"	13'9 13/16"	15' 7/8"
L. F. OF RISER PER INCH OF STAIRWAY WIDTH	1.166	1.25	1.0	1.083	1.166	1.25	1.0	1.083	1.166	1.25
L. F. OF TREAD PER INCH OF STAIRWAY WIDTH	1.083	1.166	.916	1.0	1.083	1.166	.916	1.0	1.083	1.166
FRAMING SQUARE SETTINGS FOR CARRIAGE CUTS — TONGUE	6 9/16"	6 1/8"	7 5/8"	7 1/16"	6 9/16"	6 1/8"	7 11/16"	7 1/16"	6 9/16"	6 1/8"
FRAMING SQUARE SETTINGS FOR CARRIAGE CUTS — BODY	10 15/16"	11 3/8"	9 7/8"	10 7/16"	10 15/16"	11 3/8"	9 13/16"	10 7/16"	10 15/16"	11 3/8"

FLOOR TO FLOOR RISE	7'8"	7'8"	7'8"	7'8"	7'8 1/8"	7'8 1/8"	7'8 1/8"	7'8 1/8"	7'8 1/4"	7'8 1/4"
NUMBER OF RISERS	12	13	14	15	12	13	14	15	12	13
HEIGHT OF EACH RISER	7 11/16"	7 1/16"	6 9/16"	6 1/8"	7 11/16"	7 1/16"	6 9/16"	6 1/8"	7 11/16"	7 1/8"
TOTAL HEIGHT OF RISERS	7'8 1/4"	7'7 13/16"	7'7 7/8"	7'7 7/8"	7'8 1/4"	7'7 13/16"	7'7 7/8"	7'7 7/8"	7'8 1/4"	7'8 5/8"
RISERS OVERAGE (+) OR UNDERAGE (—)	+1/4"	−3/16"	−1/8"	−1/8"	+1/8"	−5/16"	−1/4"	−1/4"	0	+3/8"
NUMBER OF TREADS	11	12	13	14	11	12	13	14	11	12
WIDTH OF EACH TREAD	9 13/16"	10 7/16"	10 15/16"	11 3/8"	9 13/16"	10 7/16"	10 15/16"	11 3/8"	9 13/16"	10 3/8"
TOTAL RUN OF STAIRWAY	8'11 15/16"	10'5 1/4"	11'10 3/16"	13'3 1/4"	8'11 15/16"	10'5 1/4"	11'10 3/16"	13'3 1/4"	8'11 15/16"	10'4 1/2"
MAXIMUM RUN WITH 6'-8" HEADROOM	1'7 5/8"	1'8 7/8"	1'9 7/8"	1'10 3/4"	1'7 5/8"	1'8 7/8"	1'9 7/8"	1'10 3/4"	1'7 5/8"	1'8 3/4"
ANGLE OF INCLINE	38°5'	34°5'	30°58'	28°18'	38°5'	34°5'	30°58'	28°18'	38°5'	34°29'
LENGTH OF CARRIAGE	11'5 1/8"	12'7 7/32"	13'9 13/16"	15' 7/8"	11'5 1/8"	12'7 7/32"	13'9 13/16"	15' 7/8"	11'5 1/8"	12'7 1/32"
L. F. OF RISER PER INCH OF STAIRWAY WIDTH	1.0	1.083	1.166	1.25	1.0	1.083	1.166	1.25	1.0	1.083
L. F. OF TREAD PER INCH OF STAIRWAY WIDTH	.916	1.0	1.083	1.166	.916	1.0	1.083	1.166	.916	1.0
FRAMING SQUARE SETTINGS FOR CARRIAGE CUTS — TONGUE	7 11/16"	7 1/16"	6 9/16"	6 1/8"	7 11/16"	7 1/16"	6 9/16"	6 1/8"	7 11/16"	7 1/8"
FRAMING SQUARE SETTINGS FOR CARRIAGE CUTS — BODY	9 13/16"	10 7/16"	10 15/16"	11 3/8"	9 13/16"	10 7/16"	10 15/16"	11 3/8"	9 13/16"	10 3/8"

FLOOR TO FLOOR RISE		7'8 1/4"	7'8 1/4"	7'8 3/8"	7'8 3/8"	7'8 3/8"	7'8 3/8"	7'8 1/2"	7'8 1/2"	7'8 1/2"	7'8 1/2"
NUMBER OF RISERS		14	15	12	13	14	15	12	13	14	15
HEIGHT OF EACH RISER		6 9/16"	6 1/8"	7 11/16"	7 1/8"	6 5/8"	6 3/16"	7 11/16"	7 1/8"	6 5/8"	6 3/16"
TOTAL HEIGHT OF RISERS		7'7 7/8"	7'7 7/8"	7'8 1/4"	7'8 5/8"	7'8 3/4"	7'8 13/16"	7'8 1/4"	7'8 5/8"	7'8 3/4"	7'8 13/16"
RISERS OVERAGE (+) OR UNDERAGE (—)		−3/8"	−3/8"	−1/8"	+1/4"	+3/8"	+7/16"	−1/4"	+1/8"	+1/4"	+5/16"
NUMBER OF TREADS		13	14	11	12	13	14	11	12	13	14
WIDTH OF EACH TREAD		10 15/16"	11 3/8"	9 13/16"	10 3/8"	10 7/8"	11 5/16"	9 13/16"	10 3/8"	10 7/8"	11 5/16"
TOTAL RUN OF STAIRWAY		11'10 3/16"	13'3 1/4"	8'11 15/16"	10'4 1/2"	11'9 3/8"	13'2 3/8"	8'1 15/16"	10'4 1/2"	11'9 3/8"	13'2 3/8"
MAXIMUM RUN WITH 6'-8" HEADROOM		1'9 7/8"	1'10 3/4"	1'7 5/8"	1'8 3/4"	1'9 3/4"	1'10 5/8"	1'7 5/8"	1'8 3/4"	1'9 3/4"	1'10 5/8"
ANGLE OF INCLINE		30°58'	28°18'	38°5'	34°29'	31°21'	28°41'	38°5'	34°29'	31°21'	28°41'
LENGTH OF CARRIAGE		13'9 13/16"	15'7/8"	11'5 1/8"	12'7 1/32"	13'9 17/32"	15'17/32"	11'5 1/8"	12'7 1/32"	13'9 17/32"	15'17/32"
L. F. OF RISER PER INCH OF STAIRWAY WIDTH		1.166	1.25	1.0	1.083	1.166	1.25	1.0	1.083	1.166	1.25
L. F. OF TREAD PER INCH OF STAIRWAY WIDTH		1.083	1.166	.916	1.0	1.083	1.166	.916	1.0	1.083	1.166
FRAMING SQUARE SETTINGS FOR CARRIAGE CUTS	TONGUE	6 9/16"	6 1/8"	7 11/16"	7 1/8"	6 5/8"	6 3/16"	7 11/16"	7 1/8"	6 5/8"	6 3/16"
	BODY	10 15/16"	11 3/8"	9 13/16"	10 3/8"	10 7/8"	11 5/16"	9 13/16"	10 3/8"	10 7/8"	11 5/16"

FLOOR TO FLOOR RISE	7'8 5/8"	7'8 5/8"	7'8 5/8"	7'8 5/8"	7'8 3/4"	7'8 3/4"	7'8 3/4"	7'8 3/4"	7'8 7/8"	7'8 7/8"
NUMBER OF RISERS	12	13	14	15	12	13	14	15	12	13
HEIGHT OF EACH RISER	7 3/4"	7 1/8"	6 5/8"	6 3/16"	7 3/4"	7 1/8"	6 5/8"	6 3/16"	7 3/4"	7 1/8"
TOTAL HEIGHT OF RISERS	7'9"	7'8 5/8"	7'8 3/4"	7'8 13/16"	7'9"	7'8 5/8"	7'8 3/4"	7'8 13/16"	7'9"	7'8 5/8"
RISERS OVERAGE (+) OR UNDERAGE (—)	+3/8"	0	+1/8"	+3/16"	+1/4"	−1/8"	0	+1/16"	+1/8"	−1/4"
NUMBER OF TREADS	11	12	13	14	11	12	13	14	11	12
WIDTH OF EACH TREAD	9 3/4"	10 3/8"	10 7/8"	11 5/16"	9 3/4"	10 3/8"	10 7/8"	11 5/16"	9 3/4"	10 3/8"
TOTAL RUN OF STAIRWAY	8'11 1/4"	10'4 1/2"	11'9 3/8"	13'2 3/8"	8'11 1/4"	10'4 1/2"	11'9 3/8"	13'2 3/8"	8'11 1/4"	10'4 1/2"
MAXIMUM RUN WITH 6'-8" HEADROOM	1'7 1/2"	1'8 3/4"	1'9 3/4"	1'10 5/8"	1'7 1/2"	1'8 3/4"	1'9 3/4"	1'10 5/8"	1'7 1/2"	1'8 3/4"
ANGLE OF INCLINE	38°29'	34°29'	31°21'	28°41'	38°29'	34°29'	31°21'	28°41'	38°29'	34°29'
LENGTH OF CARRIAGE	11'5"	12'7 1/32"	13'9 17/32"	15' 17/32"	11'5"	12'7 1/32"	13'9 17/32"	15' 17/32"	11'5"	12'7 1/32"
L. F. OF RISER PER INCH OF STAIRWAY WIDTH	1.0	1.083	1.166	1.25	1.0	1.083	1.166	1.25	1.0	1.083
L. F. OF TREAD PER INCH OF STAIRWAY WIDTH	.916	1.0	1.083	1.166	.916	1.0	1.083	1.166	.916	1.0
FRAMING SQUARE SETTINGS FOR CARRIAGE CUTS — TONGUE	7 3/4"	7 1/8"	6 5/8"	6 3/16"	7 3/4"	7 1/8"	6 5/8"	6 3/16"	7 3/4"	7 1/8"
FRAMING SQUARE SETTINGS FOR CARRIAGE CUTS — BODY	9 3/4"	10 3/8"	10 7/8"	11 5/16"	9 3/4"	10 3/8"	10 7/8"	11 5/16"	9 3/4"	10 3/8"

FLOOR TO FLOOR RISE		7'8 7/8"	7'8 7/8"	7'9"	7'9"	7'9"	7'9"	7'9 1/8"	7'9 1/8"	7'9 1/8"	7'9 1/8"
NUMBER OF RISERS		14	15	12	13	14	15	12	13	14	15
HEIGHT OF EACH RISER		6 5/8"	6 3/16"	7 3/4"	7 1/8"	6 5/8"	6 3/16"	7 3/4"	7 3/16"	6 5/8"	6 3/16"
TOTAL HEIGHT OF RISERS		7'8 3/4"	7'8 13/16"	7'9"	7'8 5/8"	7'8 3/4"	7'8 13/16"	7'9"	7'9 7/16"	7'8 3/4"	7'8 13/16"
RISERS OVERAGE (+) OR UNDERAGE (—)		−1/8"	−1/16"	0	−3/8"	−1/4"	−3/16"	−1/8"	+5/16"	−3/8"	−5/16"
NUMBER OF TREADS		13	14	11	12	13	14	11	12	13	14
WIDTH OF EACH TREAD		10 7/8"	11 5/16"	9 3/4"	10 3/8"	10 7/8"	11 5/16"	9 3/4"	10 5/16"	10 7/8"	11 5/16"
TOTAL RUN OF STAIRWAY		11'9 3/8"	13'2 3/8"	8'11 1/4"	10'4 1/2"	11'9 3/8"	13'2 3/8"	8'11 1/4"	10'3 3/4"	11'9 3/8"	13'2 3/8"
MAXIMUM RUN WITH 6'-8" HEADROOM		1'9 3/4"	1'10 5/8"	1'7 1/2"	1'8 3/4"	1'9 3/4"	1'10 5/8"	1'7 1/2"	1'8 5/8"	1'9 3/4"	1'10 5/8
ANGLE OF INCLINE		31°21'	28°41'	38°29'	34°29'	31°21'	28°41'	38°29'	34°53'	31°21'	28°41'
LENGTH OF CARRIAGE		13'9 17/32"	15' 17/32"	11'5"	12'7 1/32"	13'9 17/32"	15' 17/32"	11'5"	12'6 27/32"	13'9 17/32"	15' 17/32"
L. F. OF RISER PER INCH OF STAIRWAY WIDTH		1.166	1.25	1.0	1.083	1.166	1.25	1.0	1.083	1.166	1.25
L. F. OF TREAD PER INCH OF STAIRWAY WIDTH		1.083	1.166	.916	1.0	1.083	1.166	.916	1.0	1.083	1.166
FRAMING SQUARE SETTINGS FOR CARRIAGE CUTS	TONGUE	6 5/8"	6 3/16"	7 3/4"	7 1/8"	6 5/8"	6 3/16"	7 3/4"	7 3/16"	6 5/8"	6 3/16"
	BODY	10 7/8"	11 5/16"	9 3/4"	10 3/8"	10 7/8"	11 5/16"	9 3/4"	10 5/16"	10 7/8"	11 5/16"

FLOOR TO FLOOR RISE	7'9 1/4"	7'9 1/4"	7'9 1/4"	7'9 1/4"	7'9 3/8"	7'9 3/8"	7'9 3/8"	7'9 3/8"	7'9 1/2"	7'9 1/2"
NUMBER OF RISERS	12	13	14	15	12	13	14	15	12	13
HEIGHT OF EACH RISER	7 3/4"	7 3/16"	6 11/16"	6 3/16"	7 13/16"	7 3/16"	6 11/16"	6 1/4"	7 13/16"	7 3/16"
TOTAL HEIGHT OF RISERS	7'9"	7'9 7/16"	7'9 5/8"	7'8 13/16"	7'9 3/4"	7'9 7/16"	7'9 5/8"	7'9 3/4"	7'9 3/4"	7'9 7/16"
RISERS OVERAGE (+) OR UNDERAGE (—)	−1/4"	+3/16"	+3/8"	−7/16"	+3/8"	+1/16"	+1/4"	+3/8"	+1/4"	−1/16"
NUMBER OF TREADS	11	12	13	14	11	12	13	14	11	12
WIDTH OF EACH TREAD	9 3/4"	10 5/16"	10 13/16"	11 5/16"	9 11/16"	10 5/16"	10 13/16"	11 1/4"	9 11/16"	10 5/16"
TOTAL RUN OF STAIRWAY	8'11 1/4"	10'3 3/4"	11'8 9/16"	13'2 3/8"	8'10 9/16"	10'3 3/4"	11'8 9/16"	13'1 1/2"	8'10 9/16"	10'3 3/4"
MAXIMUM RUN WITH 6'-8" HEADROOM	1'7 1/2"	1'8 5/8"	1'9 5/8"	1'10 5/8"	1'7 3/8"	1'8 5/8"	1'9 5/8"	1'10 1/2"	1'7 3/8"	1'8 5/8"
ANGLE OF INCLINE	38°29'	34°53'	31°44'	28°41'	38°53'	34°53'	31°44'	29°3'	38°53'	34°53'
LENGTH OF CARRIAGE	11'5"	12'6 27/32"	13'9 9/32"	15' 17/32"	11'4 29/32"	12'6 27/32"	13'9 9/32"	15' 3/16"	11'4 29/32"	12'6 27/32"
L. F. OF RISER PER INCH OF STAIRWAY WIDTH	1.0	1.083	1.166	1.25	1.0	1.083	1.166	1.25	1.0	1.083
L. F. OF TREAD PER INCH OF STAIRWAY WIDTH	.916	1.0	1.083	1.166	.916	1.0	1.083	1.166	.916	1.0
FRAMING SQUARE SETTINGS FOR CARRIAGE CUTS — TONGUE	7 3/4"	7 3/16"	6 11/16"	6 3/16"	7 13/16"	7 3/16"	6 11/16"	6 1/4"	7 13/16"	7 3/16"
BODY	9 3/4"	10 5/16"	10 13/16"	11 5/16"	9 11/16"	10 5/16"	10 13/16"	11 1/4"	9 11/16"	10 5/16"

FLOOR TO FLOOR RISE		7'9 1/2"	7'9 1/2"	7'9 5/8"	7'9 5/8"	7'9 5/8"	7'9 5/8"	7'9 3/4"	7'9 3/4"	7'9 3/4"	7'9 3/4"
NUMBER OF RISERS		14	15	12	13	14	15	12	13	14	15
HEIGHT OF EACH RISER		6 11/16"	6 1/4"	7 13/16"	7 3/16"	6 11/16"	6 1/4"	7 13/16"	7 3/16"	6 11/16"	6 1/4"
TOTAL HEIGHT OF RISERS		7'9 5/8"	7'9 3/4"	7'9 3/4"	7'9 7/16"	7'9 5/8"	7'9 3/4"	7'9 3/4"	7'9 7/16"	7'9 5/8"	7'9 3/4"
RISERS OVERAGE (+) OR UNDERAGE (—)		+1/8"	+1/4"	+1/8"	−3/16"	0	+1/8"	0	−5/16"	−1/8"	0
NUMBER OF TREADS		13	14	11	12	13	14	11	12	13	14
WIDTH OF EACH TREAD		10 13/16"	11 1/4"	9 11/16"	10 5/16"	10 13/16"	11 1/4"	9 11/16"	10 5/16"	10 13/16"	11 1/4"
TOTAL RUN OF STAIRWAY		11'8 9/16"	13'1 1/2"	8'10 9/16"	10'3 3/4"	11'8 9/16"	13'1 1/2"	8'10 9/16"	10'3 3/4"	11'8 9/16"	13'1 1/2"
MAXIMUM RUN WITH 6'-8" HEADROOM		1'9 5/8"	1'10 1/2"	1'7 3/8"	1'8 5/8"	1'9 5/8"	1'10 1/2"	1'7 3/8"	1'8 5/8"	1'9 5/8"	1'10 1/2"
ANGLE OF INCLINE		31°44'	29°3'	38°53'	34°53'	31°44'	29°3'	38°53'	34°53'	31°44'	29°3'
LENGTH OF CARRIAGE		13'9 9/32"	15'3/16"	11'4 29/32"	12'6 27/32"	13'9 9/32"	15'3/16"	11'4 29/32"	12'6 27/32"	13'9 9/32"	15'3/16"
L. F. OF RISER PER INCH OF STAIRWAY WIDTH		1.166	1.25	1.0	1.083	1.166	1.25	1.0	1.083	1.166	1.25
L. F. OF TREAD PER INCH OF STAIRWAY WIDTH		1.083	1.166	.916	1.0	1.083	1.166	.916	1.0	1.083	1.166
FRAMING SQUARE SETTINGS FOR CARRIAGE CUTS	TONGUE	6 11/16"	6 1/4"	7 13/16"	7 3/16"	6 11/16"	6 1/4"	7 13/16"	7 3/16"	6 11/16"	6 1/4"
	BODY	10 13/16"	11 1/4"	9 11/16"	10 5/16"	10 13/16"	11 1/4"	9 11/16"	10 5/16"	10 13/16"	11 1/4"

FLOOR TO FLOOR RISE	7'9 7/8"	7'9 7/8"	7'9 7/8"	7'9 7/8"	7'10	7'10	7'10	7'10"	7'10 1/8"	7'10 1/8"
NUMBER OF RISERS	12	13	14	15	12	13	14	15	12	13
HEIGHT OF EACH RISER	7 13/16"	7 1/4"	6 11/16"	6 1/4"	7 13/16"	7 1/4"	6 11/16"	6 1/4"	7 7/8"	7 1/4"
TOTAL HEIGHT OF RISERS	7'9 3/4"	7'10 1/4"	7'9 5/8"	7'9 3/4"	7'9 3/4"	7'10 1/4"	7'9 5/8"	7'9 3/4"	7'10 1/2"	7'10 1/4"
RISERS OVERAGE (+) OR UNDERAGE (—)	−1/8"	+3/8"	−1/4"	−1/8"	−1/4"	+1/4"	−3/8"	−1/4"	+3/8"	+1/8"
NUMBER OF TREADS	11	12	13	14	11	12	13	14	11	12
WIDTH OF EACH TREAD	9 11/16"	10 1/4"	10 13/16"	11 1/4"	9 11/16"	10 1/4"	10 13/16"	11 1/4"	9 5/8"	10 1/4"
TOTAL RUN OF STAIRWAY	8'10 9/16"	10'3"	11'8 9/16"	13'1 1/2"	8'10 9/16"	10'3"	11'8 9/16"	13'1 1/2"	8'9 7/8"	10'3"
MAXIMUM RUN WITH 6'-8" HEADROOM	1'7 3/8"	1'8 1/2"	1'9 5/8"	1'10 1/2"	1'7 3/8"	1'8 1/2"	1'9 5/8"	1'10 1/2"	1'7 1/4"	1'8 1/2"
ANGLE OF INCLINE	38°53'	35°16'	31°44'	29°3'	38°53'	35°16'	31°44'	29°3'	39°17'	35°16'
LENGTH OF CARRIAGE	11'4 29/32"	12'6 21/32"	13'9 9/32"	15' 3/16"	11'4 29/32"	12'6 21/32"	13'9 9/32"	15' 3/16"	11'4 25/32"	12'6 21/32"
L. F. OF RISER PER INCH OF STAIRWAY WIDTH	1.0	1.083	1.166	1.25	1.0	1.083	1.166	1.25	1.0	1.083
L. F. OF TREAD PER INCH OF STAIRWAY WIDTH	.916	1.0	1.083	1.166	.916	1.0	1.083	1.166	.916	1.0
FRAMING SQUARE SETTINGS FOR CARRIAGE CUTS — TONGUE	7 13/16"	7 1/4"	6 11/16"	6 1/4"	7 13/16"	7 1/4"	6 11/16"	6 1/4"	7 7/8"	7 1/4"
FRAMING SQUARE SETTINGS FOR CARRIAGE CUTS — BODY	9 11/16"	10 1/4"	10 13/16"	11 1/4"	9 11/16"	10 1/4"	10 13/16"	11 1/4"	9 5/8"	10 1/4"

FLOOR TO FLOOR RISE	7'10$\frac{1}{8}$"	7'10$\frac{1}{8}$"	7'10$\frac{1}{4}$"	7'10$\frac{1}{4}$"	7'10$\frac{1}{4}$"	7'10$\frac{1}{4}$"	7'10$\frac{3}{8}$"	7'10$\frac{3}{8}$"	7'10$\frac{3}{8}$"	7'10$\frac{3}{8}$"
NUMBER OF RISERS	14	15	12	13	14	15	12	13	14	15
HEIGHT OF EACH RISER	6$\frac{3}{4}$"	6$\frac{1}{4}$"	7$\frac{7}{8}$"	7$\frac{1}{4}$"	6$\frac{3}{4}$"	6$\frac{5}{16}$"	7$\frac{7}{8}$"	7$\frac{1}{4}$"	6$\frac{3}{4}$"	6$\frac{5}{16}$"
TOTAL HEIGHT OF RISERS	7'10$\frac{1}{2}$"	7'9$\frac{3}{4}$"	7'10$\frac{1}{2}$"	7'10$\frac{1}{4}$"	7'10$\frac{1}{2}$"	7'10$\frac{11}{16}$"	7'10$\frac{1}{2}$"	7'10$\frac{1}{4}$"	7'10$\frac{1}{2}$"	7'10$\frac{11}{16}$"
RISERS OVERAGE (+) OR UNDERAGE (—)	+$\frac{3}{8}$"	−$\frac{3}{8}$"	+$\frac{1}{4}$"	0	+$\frac{1}{4}$"	+$\frac{7}{16}$"	+$\frac{1}{8}$"	−$\frac{1}{8}$"	+$\frac{1}{8}$"	+$\frac{5}{16}$"
NUMBER OF TREADS	13	14	11	12	13	14	11	12	13	14
WIDTH OF EACH TREAD	10$\frac{3}{4}$"	11$\frac{1}{4}$"	9$\frac{5}{8}$"	10$\frac{1}{4}$"	10$\frac{3}{4}$"	11$\frac{3}{16}$"	9$\frac{5}{8}$"	10$\frac{1}{4}$"	10$\frac{3}{4}$"	11$\frac{3}{16}$"
TOTAL RUN OF STAIRWAY	11'7$\frac{3}{4}$"	13'1$\frac{1}{2}$"	8'9$\frac{7}{8}$"	10'3"	11'7$\frac{3}{4}$"	13'$\frac{5}{8}$"	8'9$\frac{7}{8}$"	10'3"	11'7$\frac{3}{4}$"	13'$\frac{5}{8}$"
MAXIMUM RUN WITH 6'-8" HEADROOM	1'9$\frac{1}{2}$"	1'10$\frac{1}{2}$"	1'7$\frac{1}{4}$"	1'8$\frac{1}{2}$"	1'9$\frac{1}{2}$"	1'10$\frac{3}{8}$"	1'7$\frac{1}{4}$"	1'8$\frac{1}{2}$"	1'9$\frac{1}{2}$"	1'10$\frac{3}{8}$"
ANGLE OF INCLINE	32°7'	29°3'	39°17'	35°16'	32°7'	29°26'	39°17'	35°16'	32°7'	29°26'
LENGTH OF CARRIAGE	13'9"	15'$\frac{3}{16}$"	11'4$\frac{25}{32}$"	12'6$\frac{21}{32}$"	13'9"	14'11$\frac{27}{32}$"	11'4$\frac{25}{32}$"	12'6$\frac{21}{32}$"	13'9"	14'11$\frac{27}{32}$"
L. F. OF RISER PER INCH OF STAIRWAY WIDTH	1.166	1.25	1.0	1.083	1.166	1.25	1.0	1.083	1.166	1.25
L. F. OF TREAD PER INCH OF STAIRWAY WIDTH	1.083	1.166	.916	1.0	1.083	1.166	.916	1.0	1.083	1.166
FRAMING SQUARE SETTINGS FOR CARRIAGE CUTS — TONGUE	6$\frac{3}{4}$"	6$\frac{1}{4}$"	7$\frac{7}{8}$"	7$\frac{1}{4}$"	6$\frac{3}{4}$"	6$\frac{5}{16}$"	7$\frac{7}{8}$"	7$\frac{1}{4}$"	6$\frac{3}{4}$"	6$\frac{5}{16}$"
FRAMING SQUARE SETTINGS FOR CARRIAGE CUTS — BODY	10$\frac{3}{4}$"	11$\frac{1}{4}$"	9$\frac{5}{8}$"	10$\frac{1}{4}$"	10$\frac{3}{4}$"	11$\frac{3}{16}$"	9$\frac{5}{8}$"	10$\frac{1}{4}$"	10$\frac{3}{4}$"	11$\frac{3}{16}$"

FLOOR TO FLOOR RISE	7'10 1/2"	7'10 1/2"	7'10 1/2"	7'10 1/2"	7'10 5/8"	7'10 5/8"	7'10 5/8"	7'10 5/8"	7'10 3/4"	7'10 3/4"
NUMBER OF RISERS	12	13	14	15	12	13	14	15	12	13
HEIGHT OF EACH RISER	7 7/8"	7 1/4"	6 3/4"	6 5/16"	7 7/8"	7 1/4"	6 3/4"	6 5/16"	7 7/8"	7 5/16"
TOTAL HEIGHT OF RISERS	7'10 1/2"	7'10 1/4"	7'10 1/2"	7'10 11/16"	7'10 1/2"	7'10 1/4"	7'10 1/2"	7'10 11/16"	7'10 1/2"	7'11 1/16"
RISERS OVERAGE (+) OR UNDERAGE (—)	0	−1/4"	0	+3/16"	−1/8"	−3/8"	−1/8"	+1/16"	−1/4"	+5/16"
NUMBER OF TREADS	11	12	13	14	11	12	13	14	11	12
WIDTH OF EACH TREAD	9 5/8"	10 1/4"	10 3/4"	11 3/16"	9 5/8"	10 1/4"	10 3/4"	11 3/16"	9 5/8"	10 3/16"
TOTAL RUN OF STAIRWAY	8'9 7/8"	10'3"	11'7 3/4"	13' 5/8"	8'9 7/8"	10'3"	11'7 3/4"	13' 5/8"	8'9 7/8"	10'2 1/4"
MAXIMUM RUN WITH 6'-8" HEADROOM	1'7 1/4"	1'8 1/2"	1'9 1/2"	1'10 3/8"	1'7 1/4"	1'8 1/2"	1'9 1/2"	1'10 3/8"	1'7 1/4"	1'8 3/8"
ANGLE OF INCLINE	39°17'	35°16'	32°7'	29°26'	39°17'	35°16'	32°7'	29°26'	39°17'	35°40'
LENGTH OF CARRIAGE	11'4 25/32"	12'6 21/32"	13'9"	14'11 27/32"	11'4 25/32"	12'6 21/32"	13'9"	14'11 27/32"	11'4 25/32"	12'6 15/32"
L. F. OF RISER PER INCH OF STAIRWAY WIDTH	1.0	1.083	1.166	1.25	1.0	1.083	1.166	1.25	1.0	1.083
L. F. OF TREAD PER INCH OF STAIRWAY WIDTH	.916	1.0	1.083	1.166	.916	1.0	1.083	1.166	.916	1.0
FRAMING SQUARE SETTINGS FOR CARRIAGE CUTS — TONGUE	7 7/8"	7 1/4"	6 3/4"	6 5/16"	7 7/8"	7 1/4"	6 3/4"	6 5/16"	7 7/8"	7 5/16"
FRAMING SQUARE SETTINGS FOR CARRIAGE CUTS — BODY	9 5/8"	10 1/4"	10 3/4"	11 3/16"	9 5/8"	10 1/4"	10 3/4"	11 3/16"	9 5/8"	10 3/16"

FLOOR TO FLOOR RISE	7'10 3/4"	7'10 3/4"	7'10 7/8"	7'10 7/8"	7'10 7/8"	7'10 7/8"	7'11"	7'11"	7'11"	7'11"
NUMBER OF RISERS	14	15	12	13	14	15	12	13	14	15
HEIGHT OF EACH RISER	6 3/4"	6 5/16"	7 15/16"	7 5/16"	6 3/4"	6 5/16"	7 15/16"	7 5/16"	6 13/16"	6 5/16"
TOTAL HEIGHT OF RISERS	7'10 1/2"	7'10 11/16"	7'11 1/4"	7'11 1/16"	7'10 1/2"	7'10 11/16"	7'11 1/4"	7'11 1/16"	7'11 3/8"	7'10 11/16"
RISERS OVERAGE (+) OR UNDERAGE (—)	−1/4"	−1/16"	+3/8"	+3/16"	−3/8"	−3/16"	+1/4"	+1/16"	+3/8"	−5/16"
NUMBER OF TREADS	13	14	11	12	13	14	11	12	13	14
WIDTH OF EACH TREAD	10 3/4"	11 3/16"	9 9/16"	10 3/16"	10 3/4"	11 3/16"	9 9/16"	10 3/16"	10 11/16"	11 3/16"
TOTAL RUN OF STAIRWAY	11'7 3/4"	13'5/8"	8'9 3/16"	10'2 1/4"	11'7 3/4"	13'5/8"	8'9 6/16"	10'2 1/4"	11'6 15/16"	13'5/8"
MAXIMUM RUN WITH 6'-8" HEADROOM	1'9 1/2"	1'10 3/8"	1'7 1/8"	1'8 3/8"	1'9 1/2"	1'10 3/8"	1'7 1/8"	1'8 3/8"	1'9 3/8"	1'10 3/8"
ANGLE OF INCLINE	32°7'	29°26'	39°42'	35°40'	32°7'	29°26'	39°42'	35°40'	32°31'	29°26'
LENGTH OF CARRIAGE	13'9"	14'11 27/32"	11'4 11/16"	12'6 15/32"	13'9"	14'11 27/32"	11'4 11/16"	12'6 15/32"	13'8 3/4"	14'11 27/32"
L. F. OF RISER PER INCH OF STAIRWAY WIDTH	1.166	1.25	1.0	1.083	1.166	1.25	1.0	1.083	1.166	1.25
L. F. OF TREAD PER INCH OF STAIRWAY WIDTH	1.083	1.166	.916	1.0	1.083	1.166	.916	1.0	1.083	1.166
FRAMING SQUARE SETTINGS FOR CARRIAGE CUTS — TONGUE	6 3/4"	6 5/16"	7 15/16"	7 5/16"	6 3/4"	6 5/16"	7 15/16"	7 5/16"	6 13/16"	6 5/16"
FRAMING SQUARE SETTINGS FOR CARRIAGE CUTS — BODY	10 3/4"	11 3/16"	9 9/16"	10 3/16"	10 3/4"	11 3/16"	9 9/16"	10 3/16"	10 11/16"	11 3/16"

FLOOR TO FLOOR RISE	7'11 1/8"	7'11 1/8"	7'11 1/8"	7'11 1/8"	7'11 1/4"	7'11 1/4"	7'11 1/4"	7'11 1/4"	7'11 3/8"	7'11 3/8"
NUMBER OF RISERS	12	13	14	15	12	13	14	15	12	13
HEIGHT OF EACH RISER	7 15/16"	7 5/16"	6 13/16"	6 5/16"	7 15/16"	7 5/16"	6 13/16"	6 3/8"	7 15/16"	7 5/16"
TOTAL HEIGHT OF RISERS	7'11 1/4"	7'11 1/16"	7'11 3/8"	7'10 11/16"	7'11 1/4"	7'11 1/16"	7'11 3/8"	7'11 5/8"	7'11 1/4"	7'11 1/16"
RISERS OVERAGE (+) OR UNDERAGE (—)	+1/8"	−1/16"	+1/4"	−7/16"	0	−3/16"	+1/8"	+3/8"	−1/8"	−5/16"
NUMBER OF TREADS	11	12	13	14	11	12	13	14	11	12
WIDTH OF EACH TREAD	9 9/16"	10 3/16"	10 11/16"	11 3/16"	9 9/16"	10 3/16"	10 11/16"	11 1/8"	9 9/16"	10 3/16"
TOTAL RUN OF STAIRWAY	8'9 3/16"	10'2 1/4"	11'6 15/16"	13' 5/8"	8'9 3/16"	10'2 1/4"	11'6 15/16"	12'11 3/4"	8'9 3/16"	10'2 1/4"
MAXIMUM RUN WITH 6'-8" HEADROOM	1'7 1/8"	1'8 3/8"	1'9 3/8"	1'10 3/8"	1'7 1/8"	1'8 3/8"	1'9 3/8"	1'10 1/4"	1'7 1/8"	1'8 3/8"
ANGLE OF INCLINE	39°42'	35°40'	32°31'	29°26'	39°42'	35°40'	32°31'	29°49'	39°42'	35°40'
LENGTH OF CARRIAGE	11'4 11/16"	12'6 15/32"	13'8 3/4"	14'11 27/32"	11'4 11/16"	12'6 15/32"	13'8 3/4"	14'11 1/2"	11'4 11/16"	12'6 15/32"
L. F. OF RISER PER INCH OF STAIRWAY WIDTH	1.0	1.083	1.166	1.25	1.0	1.083	1.166	1.25	1.0	1.083
L. F. OF TREAD PER INCH OF STAIRWAY WIDTH	.916	1.0	1.083	1.166	.916	1.0	1.083	1.166	.916	1.0
FRAMING SQUARE SETTINGS FOR CARRIAGE CUTS — TONGUE	7 15/16"	7 5/16"	6 13/16"	6 5/16"	7 15/16"	7 5/16"	6 13/16"	6 3/8"	7 15/16"	7 5/16"
FRAMING SQUARE SETTINGS FOR CARRIAGE CUTS — BODY	9 9/16"	10 3/16"	10 11/16"	11 3/16"	9 9/16"	10 3/16"	10 11/16"	11 1/8"	9 9/16"	10 3/16"

FLOOR TO FLOOR RISE	$7'11\frac{3}{8}''$	$7'11\frac{3}{8}''$	$7'11\frac{1}{2}''$	$7'11\frac{1}{2}''$	$7'11\frac{1}{2}''$	$7'11\frac{1}{2}''$	$7'11\frac{5}{8}''$	$7'11\frac{5}{8}''$	$7'11\frac{5}{8}''$	$7'11\frac{5}{8}''$
NUMBER OF RISERS	14	15	12	13	14	15	12	13	14	15
HEIGHT OF EACH RISER	$6\frac{13}{16}''$	$6\frac{3}{8}''$	$7\frac{15}{16}''$	$7\frac{3}{8}''$	$6\frac{13}{16}''$	$6\frac{3}{8}''$	8"	$7\frac{3}{8}''$	$6\frac{13}{16}''$	$6\frac{3}{8}''$
TOTAL HEIGHT OF RISERS	$7'11\frac{3}{8}''$	$7'11\frac{5}{8}''$	$7'11\frac{1}{4}''$	$7'11\frac{7}{8}''$	$7'11\frac{3}{8}''$	$7'11\frac{5}{8}''$	8'	$7'11\frac{7}{8}''$	$7'11\frac{3}{8}''$	$7'11\frac{5}{8}''$
RISERS OVERAGE (+) OR UNDERAGE (—)	0	$+\frac{1}{4}''$	$-\frac{1}{4}''$	$+\frac{3}{8}''$	$-\frac{1}{8}''$	$+\frac{1}{8}''$	$+\frac{3}{8}''$	$+\frac{1}{4}''$	$-\frac{1}{4}''$	0
NUMBER OF TREADS	13	14	11	12	13	14	11	12	13	14
WIDTH OF EACH TREAD	$10\frac{11}{16}''$	$11\frac{1}{8}''$	$9\frac{9}{16}''$	$10\frac{1}{8}''$	$10\frac{11}{16}''$	$11\frac{1}{8}''$	$9\frac{1}{2}''$	$10\frac{1}{8}''$	$10\frac{11}{16}''$	$11\frac{1}{8}''$
TOTAL RUN OF STAIRWAY	$11'6\frac{15}{16}''$	$12'11\frac{3}{4}''$	$8'9\frac{3}{16}''$	$10'1\frac{1}{2}''$	$11'6\frac{15}{16}''$	$12'11\frac{3}{4}''$	$8'8\frac{1}{2}''$	$10'1\frac{1}{2}''$	$11'6\frac{15}{16}''$	$12'11\frac{3}{4}''$
MAXIMUM RUN WITH 6'-8" HEADROOM	$1'9\frac{3}{8}''$	$1'10\frac{1}{4}''$	$1'7\frac{1}{8}''$	$1'8\frac{1}{4}''$	$1'9\frac{3}{8}''$	$1'10\frac{1}{4}''$	$9\frac{1}{2}''$	$1'8\frac{1}{4}''$	$1'9\frac{3}{8}''$	$1'10\frac{1}{4}''$
ANGLE OF INCLINE	32°31'	29°49'	39°42'	36°4'	32°31'	29°49'	40°6'	36°4'	32°31'	29°49'
LENGTH OF CARRIAGE	$13'8\frac{3}{4}''$	$14'11\frac{1}{2}''$	$11'4\frac{11}{16}''$	$12'6\frac{5}{16}''$	$13'8\frac{3}{4}''$	$14'11\frac{1}{2}''$	$11'4\frac{5}{8}''$	$12'6\frac{5}{16}''$	$13'8\frac{3}{4}''$	$14'11\frac{1}{2}''$
L. F. OF RISER PER INCH OF STAIRWAY WIDTH	1.166	1.25	1.0	1.083	1.166	1.25	1.0	1.083	1.166	1.25
L. F. OF TREAD PER INCH OF STAIRWAY WIDTH	1.083	1.166	.916	1.0	1.083	1.166	.916	1.0	1.083	1.166
FRAMING SQUARE SETTINGS FOR CARRIAGE CUTS — TONGUE	$6\frac{13}{16}''$	$6\frac{3}{8}''$	$7\frac{15}{16}''$	$7\frac{3}{8}''$	$6\frac{13}{16}''$	$6\frac{3}{8}''$	8"	$7\frac{3}{8}''$	$6\frac{13}{16}''$	$6\frac{3}{8}''$
FRAMING SQUARE SETTINGS FOR CARRIAGE CUTS — BODY	$10\frac{11}{16}''$	$11\frac{1}{8}''$	$9\frac{9}{16}''$	$10\frac{1}{8}''$	$10\frac{11}{16}''$	$11\frac{1}{8}''$	$9\frac{1}{2}''$	$10\frac{1}{8}''$	$10\frac{11}{16}''$	$11\frac{1}{8}''$

FLOOR TO FLOOR RISE	7'11 3/4"	7'11 3/4"	7'11 3/4"	7'11 3/4"	7'11 7/8"	7'11 7/8"	7'11 7/8"	7'11 7/8"
NUMBER OF RISERS	12	13	14	15	12	13	14	15
HEIGHT OF EACH RISER	8"	7 3/8"	6 13/16"	6 3/8"	8"	7 3/8"	6 7/8"	6 3/8"
TOTAL HEIGHT OF RISERS	8'	7'11 7/8"	7'11 3/8"	7'11 5/8"	8'	7'11 7/8"	8' 1/4"	7'11 5/8"
RISERS OVERAGE (+) OR UNDERAGE (—)	+1/4"	+1/8"	−3/8"	−1/8"	+1/8"	0	+3/8"	−1/4"
NUMBER OF TREADS	11	12	13	14	11	12	13	14
WIDTH OF EACH TREAD	9 1/2"	10 1/8"	10 11/16"	11 1/8"	9 1/2"	10 1/8"	10 5/8"	11 1/8"
TOTAL RUN OF STAIRWAY	8'8 1/2"	10'1 1/2"	11'6 15/16"	12'11 3/4"	8'8 1/2"	10'1 1/2"	11'6 1/8"	12'11 3/4"
MAXIMUM RUN WITH 6'-8" HEADROOM	9 1/2"	1'8 1/4"	1'9 3/8"	1'10 1/4"	9 1/2"	1'8 1/4"	1'9 1/4"	1'10 1/4"
ANGLE OF INCLINE	40°6'	36°4'	32°31'	29°49'	40°6'	36°4'	32°54'	29°49'
LENGTH OF CARRIAGE	11'4 5/8"	12'6 5/16"	13'8 3/4"	14'11 1/2"	11'4 5/8"	12'6 5/16"	13'8 17/32"	14'11 1/2"
L. F. OF RISER PER INCH OF STAIRWAY WIDTH	1.0	1.083	1.166	1.25	1.0	1.083	1.166	1.25
L. F. OF TREAD PER INCH OF STAIRWAY WIDTH	.916	1.0	1.083	1.166	.916	1.0	1.083	1.166
FRAMING SQUARE SETTINGS FOR CARRIAGE CUTS — TONGUE	8"	7 3/8"	6 13/16"	6 3/8"	8"	7 3/8"	6 7/8"	6 3/8"
FRAMING SQUARE SETTINGS FOR CARRIAGE CUTS — BODY	9 1/2"	10 1/8"	10 11/16"	11 1/8"	9 1/2"	10 1/8"	10 5/8"	11 1/8"

FLOOR TO FLOOR RISE	8'	8'	8'	8'	8'	8' 1/8"	8' 1/8"	8' 1/8"	8' 1/8"	8' 1/8"
NUMBER OF RISERS	12	13	14	15	16	12	13	14	15	16
HEIGHT OF EACH RISER	8"	7 3/8"	6 7/8"	6 3/8"	6"	8"	7 3/8"	6 7/8"	6 7/16"	6"
TOTAL HEIGHT OF RISERS	8'	7' 11 7/8"	8' 1/4"	7' 11 5/8"	8'	8'	7' 11 7/8"	8' 1/4"	8' 9/16"	8'
RISERS OVERAGE (+) OR UNDERAGE (—)	0	−1/8"	+1/4"	−3/8"	0	−1/8"	−1/4"	+1/8"	+7/16"	−1/8"
NUMBER OF TREADS	11	12	13	14	15	11	12	13	14	15
WIDTH OF EACH TREAD	9 1/2"	10 1/8"	10 5/8"	11 1/8"	11 1/2"	9 1/2"	10 1/8"	10 5/8"	11 1/16"	11 1/2"
TOTAL RUN OF STAIRWAY	8' 8 1/2"	10' 1 1/2"	11' 6 1/8"	12' 11 3/4"	14' 4 1/2"	8' 8 1/2"	10' 1 1/2"	11' 6 1/8"	12' 10 7/8"	14' 4 1/2"
WELL OPENING FOR 6'-8" HEADROOM	8' 10 7/8"	10' 5 3/32"	11' 10 31/32"	13' 6 1/16"	15' 5/32"	8' 10 7/8"	10' 5 3/32"	11' 10 31/32"	13' 5"	15' 5/32"
ANGLE OF INCLINE	40°6'	36°4'	32°54'	29°49'	27°33'	40°6'	36°4'	32°54'	30°12'	27°33'
LENGTH OF CARRIAGE	11' 4 5/8"	12' 6 5/16"	13' 8 17/32"	14' 11 1/2"	16' 2 9/16"	11' 4 5/8"	12' 6 5/16"	13' 8 17/32"	14' 11 3/16"	16' 2 9/16"
L. F. OF RISER PER INCH OF STAIRWAY WIDTH	1.0	1.083	1.166	1.25	1.333	1.0	1.083	1.166	1.25	1.333
L. F. OF TREAD PER INCH OF STAIRWAY WIDTH	.916	1.0	1.083	1.166	1.25	.916	1.0	1.083	1.166	1.25
FRAMING SQUARE SETTINGS FOR CARRIAGE CUTS — TONGUE	8"	7 3/8"	6 7/8"	6 3/8"	6"	8"	7 3/8"	6 7/8"	6 7/16"	6"
FRAMING SQUARE SETTINGS FOR CARRIAGE CUTS — BODY	9 1/2"	10 1/8"	10 5/8"	11 1/8"	11 1/2"	9 1/2"	10 1/8"	10 5/8"	11 1/16"	11 1/2"

FLOOR TO FLOOR RISE	8' 1/4"	8' 1/4"	8' 1/4"	8' 1/4"	8' 1/4"	8' 3/8"	8' 3/8"	8' 3/8"	8' 3/8"	8' 3/8"
NUMBER OF RISERS	12	13	14	15	16	12	13	14	15	16
HEIGHT OF EACH RISER	8"	7 3/8"	6 7/8"	6 7/16"	6"	8 1/16"	7 7/16"	6 7/8"	6 7/16"	6"
TOTAL HEIGHT OF RISERS	8'	7' 11 7/8"	8' 1/4"	8' 9/16"	8'	8' 3/4"	8' 11/16"	8' 1/4"	8' 9/16"	8'
RISERS OVERAGE (+) OR UNDERAGE (—)	−1/4"	−3/8"	0	+5/16"	−1/4"	+3/8"	+5/16"	−1/8"	+3/16"	−3/8"
NUMBER OF TREADS	11	12	13	14	15	11	12	13	14	15
WIDTH OF EACH TREAD	9 1/2"	10 1/8"	10 5/8"	11 1/16"	11 1/2"	9 7/16"	10 1/16"	10 5/8"	11 1/16"	11 1/2"
TOTAL RUN OF STAIRWAY	8' 8 1/2"	10' 1 1/2"	11' 6 1/8"	12' 10 7/8"	14' 4 1/2"	8' 7 13/16"	10' 3/4"	11' 6 1/8"	12' 10 7/8"	14' 4 1/2"
WELL OPENING FOR 6'-8" HEADROOM	8' 10 7/8"	10' 5 3/32"	11' 10 31/32"	13' 5"	15' 5/32"	8' 10 3/32"	10' 4 7/32"	11' 10 31/32"	13' 5"	15' 5/32"
ANGLE OF INCLINE	40°6'	36°4'	32°54'	30°12'	27°33'	40°30'	36°28'	32°54'	30°12'	27°33'
LENGTH OF CARRIAGE	11' 4 5/8"	12' 6 5/16"	13' 8 17/32"	14' 11 3/16"	16' 2 9/16"	11' 4 17/32"	12' 6 5/32"	13' 8 17/32"	14' 11 3/16"	16' 2 9/16"
L. F. OF RISER PER INCH OF STAIRWAY WIDTH	1.0	1.083	1.166	1.25	1.333	1.0	1.083	1.166	1.25	1.333
L. F. OF TREAD PER INCH OF STAIRWAY WIDTH	.916	1.0	1.083	1.166	1.25	.916	1.0	1.083	1.166	1.25
FRAMING SQUARE SETTINGS FOR CARRIAGE CUTS — TONGUE	8"	7 3/8"	6 7/8"	6 7/16"	6"	8 1/16"	7 7/16"	6 7/8"	6 7/16"	6"
FRAMING SQUARE SETTINGS FOR CARRIAGE CUTS — BODY	9 1/2"	10 1/8"	10 5/8"	11 1/16"	11 1/2"	9 7/16"	10 1/16"	10 5/8"	11 1/16"	11 1/2"

FLOOR TO FLOOR RISE	8' 1/2"	8' 1/2"	8' 1/2"	8' 1/2"	8' 1/2"	8' 5/8"	8' 5/8"	8' 5/8"	8' 5/8"	8' 5/8"
NUMBER OF RISERS	12	13	14	15	16	12	13	14	15	16
HEIGHT OF EACH RISER	8 1/16"	7 7/16"	6 7/8"	6 7/16"	6 1/16"	8 1/16"	7 7/16"	6 7/8"	6 7/16"	6 1/16"
TOTAL HEIGHT OF RISERS	8' 3/4"	8' 11/16"	8' 1/4"	8' 9/16"	8' 1"	8' 3/4"	8' 11/16"	8' 1/4"	8' 9/16"	8' 1"
RISERS OVERAGE (+) OR UNDERAGE (—)	+1/4"	+3/16"	−1/4"	+1/16"	+1/2"	+1/8"	+1/16"	−3/8"	−1/16"	+3/8"
NUMBER OF TREADS	11	12	13	14	15	11	12	13	14	15
WIDTH OF EACH TREAD	9 7/16"	10 1/16"	10 5/8"	11 1/16"	11 7/16"	9 7/16"	10 1/16"	10 5/8"	11 1/16"	11 7/16"
TOTAL RUN OF STAIRWAY	8' 7 13/16"	10' 3/4"	11' 6 1/8"	12' 10 7/8"	14' 3 9/16"	8' 7 13/16"	10' 3/4"	11' 6 1/8"	12' 10 7/8"	14' 3 9/16"
WELL OPENING FOR 6'-8" HEADROOM	8' 10 3/32"	10' 4 7/32"	11' 10 31/32"	13' 5"	14' 11"	8' 10 3/32"	10' 4 7/32"	11' 10 31/32"	13' 5"	14' 11"
ANGLE OF INCLINE	40°30'	36°28'	32°54'	30°12'	27°56'	40°30'	36°28'	32°54'	30°12'	27°56'
LENGTH OF CARRIAGE	11' 4 17/32"	12' 6 5/32"	13' 8 17/32"	14' 11 3/16"	16' 2 3/16"	11' 4 17/32"	12' 6 5/32"	13' 8 17/32"	14' 11 3/16"	16' 2 3/16"
L. F. OF RISER PER INCH OF STAIRWAY WIDTH	1.0	1.083	1.166	1.25	1.333	1.0	1.083	1.166	1.25	1.333
L. F. OF TREAD PER INCH OF STAIRWAY WIDTH	.916	1.0	1.083	1.166	1.25	.916	1.0	1.083	1.166	1.25
FRAMING SQUARE SETTINGS FOR CARRIAGE CUTS — TONGUE	8 1/16"	7 7/16"	6 7/8"	6 7/16"	6 1/16"	8 1/16"	7 7/16"	6 7/8"	6 7/16"	6 1/16"
FRAMING SQUARE SETTINGS FOR CARRIAGE CUTS — BODY	9 7/16"	10 1/16"	10 5/8"	11 1/16"	11 7/16"	9 7/16"	10 1/16"	10 5/8"	11 1/16"	11 7/16"

FLOOR TO FLOOR RISE		8'3/4"	8'3/4"	8'3/4"	8'3/4"	8'3/4"	8'7/8"	8'7/8"	8'7/8"	8'7/8"	8'7/8"
NUMBER OF RISERS		12	13	14	15	16	12	13	14	15	16
HEIGHT OF EACH RISER		8 1/16"	7 7/16"	6 15/16"	6 7/16"	6 1/16"	8 1/16"	7 7/16"	6 15/16"	6 7/16"	6 1/16"
TOTAL HEIGHT OF RISERS		8'3/4"	8'11/16"	8'1 1/8"	8'9/16"	8'1"	8'3/4"	8'11/16"	8'1 1/8"	8'9/16"	8'1"
RISERS OVERAGE (+) OR UNDERAGE (—)		0	−1/16"	+3/8"	−3/16"	+1/4"	−1/8"	−3/16"	+1/4"	−5/16"	+1/8"
NUMBER OF TREADS		11	12	13	14	15	11	12	13	14	15
WIDTH OF EACH TREAD		9 7/16"	10 1/16"	10 9/16"	11 1/16"	11 7/16"	9 7/16"	10 1/16"	10 9/16"	11 1/16"	11 7/16"
TOTAL RUN OF STAIRWAY		8'7 13/16"	10'3/4"	11'5 5/16"	12'10 7/8"	14'3 9/16"	8'7 13/16"	10'3/4"	11'5 5/16"	12'10 7/8"	14'3 9/16"
WELL OPENING FOR 6'-8" HEADROOM		8'10 3/32"	10'4 7/32"	11'9 31/32"	13'5"	14'11"	8'10 3/32"	10'4 7/32"	11'9 31/32"	13'5"	14'11"
ANGLE OF INCLINE		40°30'	36°28'	33°18'	30°12'	27°56'	40°30'	36°28'	33°18'	30°12'	27°56'
LENGTH OF CARRIAGE		11'4 17/32"	12'6 5/32"	13'8 9/32"	14'11 3/16"	16'2 3/16"	11'4 17/32"	12'6 5/32"	13'8 9/32"	14'11 3/16"	16'2 3/16"
L. F. OF RISER PER INCH OF STAIRWAY WIDTH		1.0	1.083	1.166	1.25	1.333	1.0	1.083	1.166	1.25	1.333
L. F. OF TREAD PER INCH OF STAIRWAY WIDTH		.916	1.0	1.083	1.166	1.25	.916	1.0	1.083	1.166	1.25
FRAMING SQUARE SETTINGS FOR CARRIAGE CUTS	TONGUE	8 1/16"	7 7/16"	6 15/16"	6 7/16"	6 1/16"	8 1/16"	7 7/16"	6 15/16"	6 7/16"	6 1/16"
	BODY	9 7/16"	10 1/16"	10 9/16"	11 1/16"	11 7/16"	9 7/16"	10 1/16"	10 9/16"	11 1/16"	11 7/16"

FLOOR TO FLOOR RISE	8'1"	8'1"	8'1"	8'1"	8'1"	8'$1\frac{1}{8}$"	8'$1\frac{1}{8}$"	8'$1\frac{1}{8}$"	8'$1\frac{1}{8}$"	8'$1\frac{1}{8}$"
NUMBER OF RISERS	12	13	14	15	16	12	13	14	15	16
HEIGHT OF EACH RISER	$8\frac{1}{16}$"	$7\frac{7}{16}$"	$6\frac{15}{16}$"	$6\frac{7}{16}$"	$6\frac{1}{16}$"	$8\frac{1}{8}$"	$7\frac{1}{2}$"	$6\frac{15}{16}$"	$6\frac{1}{2}$"	$6\frac{1}{16}$"
TOTAL HEIGHT OF RISERS	8'$\frac{3}{4}$"	8'$\frac{11}{16}$"	8'$1\frac{1}{8}$"	8'$\frac{9}{16}$"	8'1"	8'$1\frac{1}{2}$"	8'$1\frac{1}{2}$"	8'$1\frac{1}{8}$"	8'$1\frac{1}{2}$"	8'1"
RISERS OVERAGE (+) OR UNDERAGE (—)	$-\frac{1}{4}$"	$-\frac{5}{16}$"	$+\frac{1}{8}$"	$-\frac{7}{16}$"	0	$+\frac{3}{8}$"	$+\frac{3}{8}$"	0	$+\frac{3}{8}$"	$-\frac{1}{8}$"
NUMBER OF TREADS	11	12	13	14	15	11	12	13	14	15
WIDTH OF EACH TREAD	$9\frac{7}{16}$"	$10\frac{1}{16}$"	$10\frac{9}{16}$"	$11\frac{1}{16}$"	$11\frac{7}{16}$"	$9\frac{3}{8}$"	10"	$10\frac{9}{16}$"	11"	$11\frac{7}{16}$"
TOTAL RUN OF STAIRWAY	8'$7\frac{13}{16}$"	10'$\frac{3}{4}$"	11'$5\frac{5}{16}$"	12'$10\frac{7}{8}$"	14'$3\frac{9}{16}$"	8'$7\frac{1}{8}$"	10'	11'$5\frac{5}{16}$"	12'10"	14'$3\frac{9}{16}$"
WELL OPENING FOR 6'-8" HEADROOM	8'$10\frac{3}{32}$"	10'$4\frac{7}{32}$"	11'$9\frac{31}{32}$"	13'5"	14'11"	8'$9\frac{9}{32}$"	10'$3\frac{11}{32}$"	11'$9\frac{31}{32}$"	13'$3\frac{15}{16}$"	14'11"
ANGLE OF INCLINE	40°30'	36°28'	33°18'	30°12'	27°56'	40°55'	36°52'	33°18'	30°35'	27°56'
LENGTH OF CARRIAGE	11'$4\frac{17}{32}$"	12'$6\frac{5}{32}$"	13'$8\frac{9}{32}$"	14'$11\frac{3}{16}$"	16'$2\frac{3}{16}$"	11'$4\frac{15}{32}$"	12'6"	13'$8\frac{9}{32}$"	14'$10\frac{7}{8}$"	16'$2\frac{3}{16}$"
L. F. OF RISER PER INCH OF STAIRWAY WIDTH	1.0	1.083	1.166	1.25	1.333	1.0	1.083	1.166	1.25	1.333
L. F. OF TREAD PER INCH OF STAIRWAY WIDTH	.916	1.0	1.083	1.166	1.25	.916	1.0	1.083	1.166	1.25
FRAMING SQUARE SETTINGS FOR CARRIAGE CUTS — TONGUE	$8\frac{1}{16}$"	$7\frac{7}{16}$"	$6\frac{15}{16}$"	$6\frac{7}{16}$"	$6\frac{1}{16}$"	$8\frac{1}{8}$"	$7\frac{1}{2}$"	$6\frac{15}{16}$"	$6\frac{1}{2}$"	$6\frac{1}{16}$"
FRAMING SQUARE SETTINGS FOR CARRIAGE CUTS — BODY	$9\frac{7}{16}$"	$10\frac{1}{16}$"	$10\frac{9}{16}$"	$11\frac{1}{16}$"	$11\frac{7}{16}$"	$9\frac{3}{8}$"	10"	$10\frac{9}{16}$"	11"	$11\frac{7}{16}$"

FLOOR TO FLOOR RISE		8'1 1/4"	8'1 1/4"	8'1 1/4"	8'1 1/4"	8'1 1/4"	8'1 3/8"	8'1 3/8"	8'1 3/8"	8'1 3/8"	8'1 3/8"
NUMBER OF RISERS		12	13	14	15	16	12	13	14	15	16
HEIGHT OF EACH RISER		8 1/8"	7 1/2"	6 15/16"	6 1/2"	6 1/16"	8 1/8"	7 1/2"	6 15/16"	6 1/2"	6 1/16"
TOTAL HEIGHT OF RISERS		8'1 1/2"	8'1 1/2"	8'1 1/8"	8'1 1/2"	8'1"	8'1 1/2"	8'1 1/2"	8'1 1/8"	8'1 1/2"	8'1"
RISERS OVERAGE (+) OR UNDERAGE (—)		+1/4"	+1/4"	−1/8"	+1/4"	−1/4"	+1/8"	+1/8"	−1/4"	+1/8"	−3/8"
NUMBER OF TREADS		11	12	13	14	15	11	12	13	14	15
WIDTH OF EACH TREAD		9 3/8"	10"	10 9/16"	11"	11 7/16"	9 3/8"	10"	10 9/16"	11"	11 7/16"
TOTAL RUN OF STAIRWAY		8'7 1/8"	10'	11'5 5/16"	12'10"	14'3 9/16"	8'7 1/8"	10'	11'5 5/16"	12'10"	14'3 9/16"
WELL OPENING FOR 6'-8" HEADROOM		8'9 9/32"	10'3 11/32"	11'9 31/32"	13'3 15/16"	14'11"	8'9 9/32"	10'3 11/32"	11'9 31/32"	13'3 15/16"	14'11"
ANGLE OF INCLINE		40°55'	36°52'	33°18'	30°35'	27°56'	40°55'	36°52'	33°18'	30°35'	27°56'
LENGTH OF CARRIAGE		11'4 15/32"	12'6"	13'8 9/32"	14'10 7/8"	16'2 3/16"	11'4 15/32"	12'6"	13'8 9/32"	14'10 7/8"	16'2 3/16"
L. F. OF RISER PER INCH OF STAIRWAY WIDTH		1.0	1.083	1.166	1.25	1.333	1.0	1.083	1.166	1.25	1.333
L. F. OF TREAD PER INCH OF STAIRWAY WIDTH		.916	1.0	1.083	1.166	1.25	.916	1.0	1.083	1.166	1.25
FRAMING SQUARE SETTINGS FOR CARRIAGE CUTS	**TONGUE**	8 1/8"	7 1/2"	6 15/16"	6 1/2"	6 1/16"	8 1/8"	7 1/2"	6 15/16"	6 1/2"	6 1/16"
	BODY	9 3/8"	10"	10 9/16"	11"	11 7/16"	9 3/8"	10"	10 9/16"	11"	11 7/16"

FLOOR TO FLOOR RISE	8'1 1/2"	8'1 1/2"	8'1 1/2"	8'1 1/2"	8'1 1/2"	8'1 5/8"	8'1 5/8"	8'1 5/8"	8'1 5/8"	8'1 5/8"
NUMBER OF RISERS	12	13	14	15	16	12	13	14	15	16
HEIGHT OF EACH RISER	8 1/8"	7 1/2"	6 15/16"	6 1/2"	6 1/8"	8 1/8"	7 1/2"	7"	6 1/2"	6 1/8"
TOTAL HEIGHT OF RISERS	8'1 1/2"	8'1 1/2"	8'1 1/8"	8'1 1/2"	8'2"	8'1 1/2"	8'1 1/2"	8'2"	8'1 1/2"	8'2"
RISERS OVERAGE (+) OR UNDERAGE (—)	0	0	−3/8"	0	+1/2"	−1/8"	−1/8"	+3/8"	−1/8"	+3/8"
NUMBER OF TREADS	11	12	13	14	15	11	12	13	14	15
WIDTH OF EACH TREAD	9 3/8"	10"	10 9/16"	11"	11 3/8"	9 3/8"	10"	10 1/2"	11"	11 3/8"
TOTAL RUN OF STAIRWAY	8'7 1/8"	10'	11'5 5/16"	12'10"	14'2 5/8"	8'7 1/8"	10'	11'4 1/2"	12'10"	14'2 5/8"
WELL OPENING FOR 6'-8" HEADROOM	8'9 9/32"	10'3 11/32"	11'9 3/32"	13'3 15/16"	14'9 13/16"	8'9 9/32"	10'3 11/32"	11'9"	13'3 15/16"	14'9 13/16"
ANGLE OF INCLINE	40°55'	36°52'	33°18'	30°35'	28°18'	40°55'	36°52'	33°41'	30°35'	28°18'
LENGTH OF CARRIAGE	11'4 15/32"	12'6"	13'8 9/32"	14'10 7/8"	16'1 25/32"	11'4 15/32"	12'6"	13'8 1/16"	14'10 7/8"	16'1 25/32"
L. F. OF RISER PER INCH OF STAIRWAY WIDTH	1.0	1.083	1.166	1.25	1.333	1.0	1.083	1.166	1.25	1.333
L. F. OF TREAD PER INCH OF STAIRWAY WIDTH	.916	1.0	1.083	1.166	1.25	.916	1.0	1.083	1.166	1.25
FRAMING SQUARE SETTINGS FOR CARRIAGE CUTS — TONGUE	8 1/8"	7 1/2"	6 15/16"	6 1/2"	6 1/8"	8 1/8"	7 1/2"	7"	6 1/2"	6 1/8"
FRAMING SQUARE SETTINGS FOR CARRIAGE CUTS — BODY	9 3/8"	10"	10 9/16"	11"	11 3/8"	9 3/8"	10"	10 1/2"	11"	11 3/8"

FLOOR TO FLOOR RISE	$8'1\frac{3}{4}''$	$8'1\frac{3}{4}''$	$8'1\frac{3}{4}''$	$8'1\frac{3}{4}''$	$8'1\frac{3}{4}''$	$8'1\frac{7}{8}''$	$8'1\frac{7}{8}''$	$8'1\frac{7}{8}''$	$8'1\frac{7}{8}''$	$8'1\frac{7}{8}''$
NUMBER OF RISERS	12	13	14	15	16	12	13	14	15	16
HEIGHT OF EACH RISER	$8\frac{1}{8}''$	$7\frac{1}{2}''$	$7''$	$6\frac{1}{2}''$	$6\frac{1}{8}''$	$8\frac{3}{16}''$	$7\frac{1}{2}''$	$7''$	$6\frac{1}{2}''$	$6\frac{1}{8}''$
TOTAL HEIGHT OF RISERS	$8'1\frac{1}{2}''$	$8'1\frac{1}{2}''$	$8'2''$	$8'1\frac{1}{2}''$	$8'2''$	$8'2\frac{1}{4}''$	$8'1\frac{1}{2}''$	$8'2''$	$8'1\frac{1}{2}''$	$8'2''$
RISERS OVERAGE (+) OR UNDERAGE (—)	$-\frac{1}{4}''$	$-\frac{1}{4}''$	$+\frac{1}{4}''$	$-\frac{1}{4}''$	$+\frac{1}{4}''$	$+\frac{3}{8}''$	$-\frac{3}{8}''$	$+\frac{1}{8}''$	$-\frac{3}{8}''$	$+\frac{1}{8}''$
NUMBER OF TREADS	11	12	13	14	15	11	12	13	14	15
WIDTH OF EACH TREAD	$9\frac{3}{8}''$	$10''$	$10\frac{1}{2}''$	$11''$	$11\frac{3}{8}''$	$9\frac{5}{16}''$	$10''$	$10\frac{1}{2}''$	$11''$	$11\frac{3}{8}''$
TOTAL RUN OF STAIRWAY	$8'7\frac{1}{8}''$	$10'$	$11'4\frac{1}{2}''$	$12'10''$	$14'2\frac{5}{8}''$	$8'6\frac{7}{8}''$	$10'$	$11'4\frac{1}{2}''$	$12'10''$	$14'2\frac{5}{8}''$
WELL OPENING FOR 6′-8″ HEADROOM	$8'9\frac{9}{32}''$	$10'3\frac{11}{32}''$	$11'9''$	$13'3\frac{15}{16}''$	$14'9\frac{13}{16}''$	$8'8\frac{1}{2}''$	$10'3\frac{11}{32}''$	$11'9''$	$13'3\frac{15}{16}''$	$14'9\frac{13}{16}''$
ANGLE OF INCLINE	$40°55'$	$36°52'$	$33°41'$	$30°35'$	$28°18'$	$41°19'$	$36°52'$	$33°41'$	$30°35'$	$28°18'$
LENGTH OF CARRIAGE	$11'4\frac{15}{32}''$	$12'6''$	$13'8\frac{1}{16}''$	$14'10\frac{7}{8}''$	$16'1\frac{25}{32}''$	$11'4\frac{13}{32}''$	$12'6''$	$13'8\frac{1}{16}''$	$14'10\frac{7}{8}''$	$16'1\frac{25}{32}''$
L. F. OF RISER PER INCH OF STAIRWAY WIDTH	1.0	1.083	1.166	1.25	1.333	1.0	1.083	1.166	1.25	1.333
L. F. OF TREAD PER INCH OF STAIRWAY WIDTH	.916	1.0	1.083	1.166	1.25	.916	1.0	1.083	1.166	1.25
FRAMING SQUARE SETTINGS FOR CARRIAGE CUTS — TONGUE	$8\frac{1}{8}''$	$7\frac{1}{2}''$	$7''$	$6\frac{1}{2}''$	$6\frac{1}{8}''$	$8\frac{3}{16}''$	$7\frac{1}{2}''$	$7''$	$6\frac{1}{2}''$	$6\frac{1}{8}''$
FRAMING SQUARE SETTINGS FOR CARRIAGE CUTS — BODY	$9\frac{3}{8}''$	$10''$	$10\frac{1}{2}''$	$11''$	$11\frac{3}{8}''$	$9\frac{5}{16}''$	$10''$	$10\frac{1}{2}''$	$11''$	$11\frac{3}{8}''$

FLOOR TO FLOOR RISE	8'2"	8'2"	8'2"	8'2"	8'2"	8'2 1/8"	8'2 1/8"	8'2 1/8"	8'2 1/8"	8'2 1/8"
NUMBER OF RISERS	12	13	14	15	16	12	13	14	15	16
HEIGHT OF EACH RISER	8 3/16"	7 9/16"	7"	6 9/16"	6 1/8"	8 3/16"	7 9/16"	7"	6 9/16"	6 1/8"
TOTAL HEIGHT OF RISERS	8'2 1/4"	8'2 5/16"	8'2"	8'2 7/16"	8'2"	8'2 1/4"	8'2 5/16"	8'2"	8'2 7/16"	8'2"
RISERS OVERAGE (+) OR UNDERAGE (—)	+1/4"	+5/16"	0	+7/16"	0	+1/8"	+3/16"	−1/8"	+5/16"	−1/8"
NUMBER OF TREADS	11	12	13	14	15	11	12	13	14	15
WIDTH OF EACH TREAD	9 5/16"	9 15/16"	10 1/2"	10 15/16"	11 3/8"	9 5/16"	9 15/16"	10 1/2"	10 15/16"	11 3/8"
TOTAL RUN OF STAIRWAY	8'6 7/16"	9'11 1/4"	11'4 1/2"	12'9 1/8"	14'2 5/8"	8'6 7/16"	9'11 1/4"	11'4 1/2"	12'9 1/8"	14'2 5/8"
WELL OPENING FOR 6'-8" HEADROOM	8'8 1/2"	10'2 7/16"	11'9"	13'2 27/32"	14'9 13/16"	8'8 1/2"	10'2 7/16"	11'9"	13'2 27/32"	14'9 13/16"
ANGLE OF INCLINE	41°19'	37°16'	33°41'	30°58'	28°18'	41°19'	37°16'	33°41'	30°58'	28°18'
LENGTH OF CARRIAGE	11'4 13/32"	12'5 27/32"	13'8 1/16"	14'10 9/16"	16'1 25/32"	11'4 13/32"	12'5 27/32"	13'8 1/16"	14'10 9/16"	16'1 25/32"
L. F. OF RISER PER INCH OF STAIRWAY WIDTH	1.0	1.083	1.166	1.25	1.333	1.0	1.083	1.166	1.25	1.333
L. F. OF TREAD PER INCH OF STAIRWAY WIDTH	.916	1.0	1.083	1.166	1.25	.916	1.0	1.083	1.166	1.25
FRAMING SQUARE SETTINGS FOR CARRIAGE CUTS — TONGUE	8 3/16"	7 9/16"	7"	6 9/16"	6 1/8"	8 3/16"	7 9/16"	7"	6 9/16"	6 1/8"
FRAMING SQUARE SETTINGS FOR CARRIAGE CUTS — BODY	9 5/16"	9 15/16"	10 1/2"	10 15/16"	11 3/8"	9 5/16"	9 15/16"	10 1/2"	10 15/16"	11 3/8"

FLOOR TO FLOOR RISE	8'2$\frac{1}{4}$"	8'2$\frac{1}{4}$"	8'2$\frac{1}{4}$"	8'2$\frac{1}{4}$"	8'2$\frac{1}{4}$"	8'2$\frac{3}{8}$"	8'2$\frac{3}{8}$"	8'2$\frac{3}{8}$"	8'2$\frac{3}{8}$"	8'2$\frac{3}{8}$"
NUMBER OF RISERS	12	13	14	15	16	12	13	14	15	16
HEIGHT OF EACH RISER	8$\frac{3}{16}$"	7$\frac{9}{16}$"	7"	6$\frac{9}{16}$"	6$\frac{1}{8}$"	8$\frac{3}{16}$"	7$\frac{9}{16}$"	7"	6$\frac{9}{16}$"	6$\frac{1}{8}$"
TOTAL HEIGHT OF RISERS	8'2$\frac{1}{4}$"	8'2$\frac{5}{16}$"	8'2"	8'2$\frac{7}{16}$"	8'2"	8'2$\frac{1}{4}$"	8'2$\frac{5}{16}$"	8'2"	8'2$\frac{7}{16}$"	8'2"
RISERS OVERAGE (+) OR UNDERAGE (—)	0	+$\frac{1}{16}$"	−$\frac{1}{4}$"	+$\frac{3}{16}$"	−$\frac{1}{4}$"	−$\frac{1}{8}$"	−$\frac{1}{16}$"	−$\frac{3}{8}$"	+$\frac{1}{16}$"	−$\frac{3}{8}$"
NUMBER OF TREADS	11	12	13	14	15	11	12	13	14	15
WIDTH OF EACH TREAD	9$\frac{5}{16}$"	9$\frac{15}{16}$"	10$\frac{1}{2}$"	10$\frac{15}{16}$"	11$\frac{3}{8}$"	9$\frac{5}{16}$"	9$\frac{15}{16}$"	10$\frac{1}{2}$"	10$\frac{15}{16}$"	11$\frac{3}{8}$"
TOTAL RUN OF STAIRWAY	8'6$\frac{7}{16}$"	9'11$\frac{1}{4}$"	11'4$\frac{1}{2}$"	12'9$\frac{1}{8}$"	14'2$\frac{5}{8}$"	8'6$\frac{7}{16}$"	9'11$\frac{1}{4}$"	11'4$\frac{1}{2}$"	12'9$\frac{1}{8}$"	14'2$\frac{5}{8}$"
WELL OPENING FOR 6'-8" HEADROOM	8'8$\frac{1}{2}$"	10'2$\frac{7}{16}$"	11'9"	13'2$\frac{27}{32}$"	14'9$\frac{13}{16}$"	8'8$\frac{1}{2}$"	10'2$\frac{7}{16}$"	11'9"	13'2$\frac{27}{32}$"	14'9$\frac{13}{16}$"
ANGLE OF INCLINE	41°19'	37°16'	33°41'	30°58'	28°18'	41°19'	37°16'	33°41'	30°58'	28°18'
LENGTH OF CARRIAGE	11'4$\frac{13}{32}$"	12'5$\frac{27}{32}$"	13'8$\frac{1}{16}$"	14'10$\frac{9}{16}$"	16'1$\frac{25}{32}$"	11'4$\frac{13}{32}$"	12'5$\frac{27}{32}$"	13'8$\frac{1}{16}$"	14'10$\frac{9}{16}$"	16'1$\frac{25}{32}$"
L. F. OF RISER PER INCH OF STAIRWAY WIDTH	1.0	1.083	1.166	1.25	1.333	1.0	1.083	1.166	1.25	1.333
L. F. OF TREAD PER INCH OF STAIRWAY WIDTH	.916	1.0	1.083	1.166	1.25	.916	1.0	1.083	1.166	1.25
FRAMING SQUARE SETTINGS FOR CARRIAGE CUTS — TONGUE	8$\frac{3}{16}$"	7$\frac{9}{16}$"	7"	6$\frac{9}{16}$"	6$\frac{1}{8}$"	8$\frac{3}{16}$"	7$\frac{9}{16}$"	7"	6$\frac{9}{16}$"	6$\frac{1}{8}$"
BODY	9$\frac{5}{16}$"	9$\frac{15}{16}$"	10$\frac{1}{2}$"	10$\frac{15}{16}$"	11$\frac{3}{8}$"	9$\frac{5}{16}$"	9$\frac{15}{16}$"	10$\frac{1}{2}$"	10$\frac{15}{16}$"	11$\frac{3}{8}$"

FLOOR TO FLOOR RISE	8'2 1/2"	8'2 1/2"	8'2 1/2"	8'2 1/2"	8'2 1/2"	8'2 5/8"	8'2 5/8"	8'2 5/8"	8'2 5/8"	8'2 5/8"
NUMBER OF RISERS	12	13	14	15	16	12	13	14	15	16
HEIGHT OF EACH RISER	8 3/16"	7 9/16"	7 1/16"	6 9/16"	6 3/16"	8 1/4"	7 9/16"	7 1/16"	6 9/16"	6 3/16"
TOTAL HEIGHT OF RISERS	8'2 1/4"	8'2 5/16"	8'2 7/8"	8'2 7/16"	8'3"	8'3"	8'2 5/16"	8'2 7/8"	8'2 7/16"	8'3"
RISERS OVERAGE (+) OR UNDERAGE (—)	−1/4"	−3/16"	+3/8"	−1/16"	+1/2"	+3/8"	−5/16"	+1/4"	−3/16"	+3/8"
NUMBER OF TREADS	11	12	13	14	15	11	12	13	14	15
WIDTH OF EACH TREAD	9 1/4"	9 15/16"	10 7/16"	10 15/16"	11 5/16"	9 1/4"	9 15/16"	10 7/16"	10 15/16"	11 5/16"
TOTAL RUN OF STAIRWAY	8'5 3/4"	9'11 1/4"	11'3 11/16"	12'9 1/8"	14'1 11/16"	8'5 3/4"	9'11 1/4"	11'3 11/16"	12'9 1/8"	14'1 11/16"
WELL OPENING FOR 6'-8" HEADROOM	8'8 1/2"	10'2 7/16"	11'8 1/32"	13'2 27/32"	14'8 21/32"	8'7 23/32"	10'2 7/16"	11'8 1/32"	13'2 27/32"	14'8 21/32"
ANGLE OF INCLINE	41°44'	37°16'	34°5'	30°58'	28°41'	41°44'	37°16'	34°5'	30°58'	28°41'
LENGTH OF CARRIAGE	11'4 11/32"	12'5 27/32"	13'7 27/32"	14'10 9/16"	16'1 13/32"	11'4 11/32"	12'5 27/32"	13'7 27/32"	14'10 9/16"	16'1 13/32"
L. F. OF RISER PER INCH OF STAIRWAY WIDTH	1.0	1.083	1.166	1.25	1.333	1.0	1.083	1.166	1.25	1.333
L. F. OF TREAD PER INCH OF STAIRWAY WIDTH	.916	1.0	1.033	1.166	1.25	.916	1.0	1.083	1.166	1.25
FRAMING SQUARE SETTINGS FOR CARRIAGE CUTS — TONGUE	8 3/16"	7 9/16"	7 1/16"	6 9/16"	6 3/16"	8 1/4"	7 9/16"	7 1/16"	6 9/16"	6 3/16"
FRAMING SQUARE SETTINGS FOR CARRIAGE CUTS — BODY	9 1/4"	9 15/16"	10 7/16"	10 15/16"	11 5/16"	9 1/4"	9 15/16"	10 7/16"	10 15/16"	11 5/16"

FLOOR TO FLOOR RISE		8'2 3/4"	8'2 3/4"	8'2 3/4"	8'2 3/4"	8'2 3/4"	8'2 7/8"	8'2 7/8"	8'2 7/8"	8'2 7/8"	8'2 7/8"
NUMBER OF RISERS		12	13	14	15	16	12	13	14	15	16
HEIGHT OF EACH RISER		8 1/4"	7 5/8"	7 1/16"	6 9/16"	6 3/16"	8 1/4"	7 5/8"	7 1/16"	6 9/16"	6 3/16"
TOTAL HEIGHT OF RISERS		8'3"	8'3 1/8"	8'2 7/8"	8'2 7/16"	8'3"	8'3"	8'3 1/8"	8'2 7/8"	8'2 7/16"	8'3"
RISERS OVERAGE (+) OR UNDERAGE (—)		+1/4"	+3/8"	+1/8"	−5/16"	+1/4"	+1/8"	+1/4"	0	−7/16"	+1/8"
NUMBER OF TREADS		11	12	13	14	15	11	12	13	14	15
WIDTH OF EACH TREAD		9 1/4"	9 7/8"	10 7/16"	10 15/16"	11 5/16"	9 1/4"	9 7/8"	10 7/16"	10 15/16"	11 5/16"
TOTAL RUN OF STAIRWAY		8'5 3/4"	9'10 1/2"	11'3 11/16"	12'9 1/8"	14'1 11/16"	8'5 3/4"	9'10 1/2"	11'3 11/16"	12'9 1/8"	14'1 11/16"
WELL OPENING FOR 6'-8" HEADROOM		8'7 23/32"	10'1 9/16"	11'8 1/32"	13'2 27/32"	14'8 21/32"	8'7 23/32"	10'1 9/16"	11'8 1/32"	13'2 27/32"	14'8 21/32"
ANGLE OF INCLINE		41°44'	37°40'	34°5'	30°58'	28°41'	41°44'	37°40'	34°5'	30°58'	28°41'
LENGTH OF CARRIAGE		11'4 11/32"	12'5 23/32"	13'7 27/32"	14'10 9/16"	16'1 13/32"	11'4 11/32"	12'5 23/32"	13'7 27/32"	14'10 9/16"	16'1 13/32"
L. F. OF RISER PER INCH OF STAIRWAY WIDTH		1.0	1.083	1.166	1.25	1.333	1.0	1.083	1.166	1.25	1.333
L. F. OF TREAD PER INCH OF STAIRWAY WIDTH		.916	1.0	1.083	1.166	1.25	.916	1.0	1.083	1.166	1.25
FRAMING SQUARE SETTINGS FOR CARRIAGE CUTS	TONGUE	8 1/4"	7 5/8"	7 1/16"	6 9/16"	6 3/16"	8 1/4"	7 5/8"	7 1/16"	6 9/16"	6 3/16"
	BODY	9 1/4"	9 7/8"	10 7/16"	10 15/16"	11 5/16"	9 1/4"	9 7/8"	10 7/16"	10 15/16"	11 5/16"

FLOOR TO FLOOR RISE	8'3"	8'3"	8'3"	8'3"	8'3"	8'3$\frac{1}{8}$"	8'3$\frac{1}{8}$"	8'3$\frac{1}{8}$"	8'3$\frac{1}{8}$"	8'3$\frac{1}{8}$"
NUMBER OF RISERS	12	13	14	15	16	12	13	14	15	16
HEIGHT OF EACH RISER	8$\frac{1}{4}$"	7$\frac{5}{8}$"	7$\frac{1}{16}$"	6$\frac{5}{8}$"	6$\frac{3}{16}$"	8$\frac{1}{4}$"	7$\frac{5}{8}$"	7$\frac{1}{16}$"	6$\frac{5}{8}$"	6$\frac{3}{16}$"
TOTAL HEIGHT OF RISERS	8'3"	8'3$\frac{1}{8}$"	8'2$\frac{7}{8}$"	8'3$\frac{3}{8}$"	8'3"	8'3"	8'3$\frac{1}{8}$"	8'2$\frac{7}{8}$"	8'3$\frac{3}{8}$"	8'3"
RISERS OVERAGE (+) OR UNDERAGE (—)	0	+$\frac{1}{8}$"	−$\frac{1}{8}$"	+$\frac{3}{8}$"	0	−$\frac{1}{8}$"	0	−$\frac{1}{4}$"	+$\frac{1}{4}$"	−$\frac{1}{8}$"
NUMBER OF TREADS	11	12	13	14	15	11	12	13	14	15
WIDTH OF EACH TREAD	9$\frac{1}{4}$"	9$\frac{7}{8}$"	10$\frac{7}{16}$"	10$\frac{7}{8}$"	11$\frac{5}{16}$"	9$\frac{1}{4}$"	9$\frac{7}{8}$"	10$\frac{7}{16}$"	10$\frac{7}{8}$"	11$\frac{5}{16}$"
TOTAL RUN OF STAIRWAY	8'5$\frac{3}{4}$"	9'10$\frac{1}{2}$"	11'3$\frac{11}{16}$"	12'8$\frac{1}{4}$"	14'1$\frac{11}{16}$"	8'5$\frac{3}{4}$"	9'10$\frac{1}{2}$"	11'3$\frac{11}{16}$"	12'8$\frac{1}{4}$"	14'1$\frac{11}{16}$"
WELL OPENING FOR 6'-8" HEADROOM	8'7$\frac{23}{32}$"	10'1$\frac{9}{16}$"	11'8$\frac{1}{32}$"	13'1$\frac{25}{32}$"	14'8$\frac{21}{32}$"	8'7$\frac{23}{32}$"	10'1$\frac{9}{16}$"	11'8$\frac{1}{32}$"	13'1$\frac{25}{32}$"	14'8$\frac{21}{32}$"
ANGLE OF INCLINE	41°44'	37°40'	34°5'	31°21'	28°41'	41°44'	37°40'	34°5'	31°21'	28°41'
LENGTH OF CARRIAGE	11'4$\frac{11}{32}$"	12'5$\frac{23}{32}$"	13'7$\frac{27}{32}$"	14'10$\frac{9}{32}$"	16'1$\frac{13}{32}$"	11'4$\frac{11}{32}$"	12'5$\frac{23}{32}$"	13'7$\frac{27}{32}$"	14'10$\frac{9}{32}$"	16'1$\frac{13}{32}$"
L. F. OF RISER PER INCH OF STAIRWAY WIDTH	1.0	1.083	1.166	1.25	1.333	1.0	1.083	1.166	1.25	1.333
L. F. OF TREAD PER INCH OF STAIRWAY WIDTH	.916	1.0	1.083	1.166	1.25	.916	1.0	1.083	1.166	1.25
FRAMING SQUARE SETTINGS FOR CARRIAGE CUTS — TONGUE	8$\frac{1}{4}$"	7$\frac{5}{8}$"	7$\frac{1}{16}$"	6$\frac{5}{8}$"	6$\frac{3}{16}$"	8$\frac{1}{4}$"	7$\frac{5}{8}$"	7$\frac{1}{16}$"	6$\frac{5}{8}$"	6$\frac{3}{16}$"
BODY	9$\frac{1}{4}$"	9$\frac{7}{8}$"	10$\frac{7}{16}$"	10$\frac{7}{8}$"	11$\frac{5}{16}$"	9$\frac{1}{4}$"	9$\frac{7}{8}$"	10$\frac{7}{16}$"	10$\frac{7}{8}$"	11$\frac{5}{16}$"

FLOOR TO FLOOR RISE	8'3 1/4"	8'3 1/4"	8'3 1/4"	8'3 1/4"	8'3 1/4"	8'3 3/8"	8'3 3/8"	8'3 3/8"	8'3 3/8"	8'3 1/2"
NUMBER OF RISERS	12	13	14	15	16	13	14	15	16	13
HEIGHT OF EACH RISER	8 1/4"	7 5/8"	7 1/16"	6 5/8"	6 3/16"	7 5/8"	7 1/8"	6 5/8"	6 3/16"	7 5/8"
TOTAL HEIGHT OF RISERS	8'3"	8'3 1/8"	8'2 7/8"	8'3 3/8"	8'3"	8'3 1/8"	8'3 3/4"	8'3 3/8"	8'3"	8'3 1/8"
RISERS OVERAGE (+) OR UNDERAGE (—)	−1/4"	−1/8"	−3/8"	+1/8"	−1/4"	−1/4"	+3/8"	0	−3/8"	−3/8"
NUMBER OF TREADS	11	12	13	14	15	12	13	14	15	12
WIDTH OF EACH TREAD	9 1/4"	9 7/8"	10 7/16"	10 7/8"	11 5/16"	9 7/8"	10 3/8"	10 7/8"	11 5/16"	9 7/8"
TOTAL RUN OF STAIRWAY	8'5 3/4"	9'10 1/2"	11'3 11/16"	12'8 1/4"	14'1 11/16"	9'10 1/2"	11'2 7/8"	12'8 1/4"	14'1 11/16"	9'10 1/2"
WELL OPENING FOR 6'-8" HEADROOM	8'7 23/32"	10'1 9/16"	11'8 1/32"	13'1 25/32"	14'8 21/32"	10'1 9/16"	11'7 1/16"	13'1 25/32"	14'8 21/32"	10'1 9/16"
ANGLE OF INCLINE	41°44'	37°40'	34°5'	31°21'	28°41'	37°40'	34°29'	31°21'	28°41'	37°40'
LENGTH OF CARRIAGE	11'4 11/32"	12'5 23/32"	13'7 27/32"	14'10 9/32"	16'1 13/32"	12'5 23/32"	13'7 5/8"	14'10 9/32"	16'1 13/32"	12'5 23/32"
L. F. OF RISER PER INCH OF STAIRWAY WIDTH	1.0	1.083	1.166	1.25	1.333	1.083	1.166	1.25	1.333	1.083
L. F. OF TREAD PER INCH OF STAIRWAY WIDTH	.916	1.0	1.083	1.166	1.25	1.0	1.083	1.166	1.25	1.0
FRAMING SQUARE SETTINGS FOR CARRIAGE CUTS — TONGUE	8 1/4"	7 5/8"	7 1/16"	6 5/8"	6 3/16"	7 5/8"	7 1/8"	6 5/8"	6 3/16"	7 5/8"
FRAMING SQUARE SETTINGS FOR CARRIAGE CUTS — BODY	9 1/4"	9 7/8"	10 7/16"	10 7/8"	11 5/16"	9 7/8"	10 3/8"	10 7/8"	11 5/16"	9 7/8"

FLOOR TO FLOOR RISE	8'3$\frac{1}{2}$"	8'3$\frac{1}{2}$"	8'3$\frac{1}{2}$"	8'3$\frac{5}{8}$"	8'3$\frac{5}{8}$"	8'3$\frac{5}{8}$"	8'3$\frac{5}{8}$"	8'3$\frac{3}{4}$"	8'3$\frac{3}{4}$"	8'3$\frac{3}{4}$"
NUMBER OF RISERS	14	15	16	13	14	15	16	13	14	15
HEIGHT OF EACH RISER	7$\frac{1}{8}$"	6$\frac{5}{8}$"	6$\frac{1}{4}$"	7$\frac{11}{16}$"	7$\frac{1}{8}$"	6$\frac{5}{8}$"	6$\frac{1}{4}$"	7$\frac{11}{16}$"	7$\frac{1}{8}$"	6$\frac{5}{8}$"
TOTAL HEIGHT OF RISERS	8'3$\frac{3}{4}$"	8'3$\frac{3}{8}$"	8'4"	8'3$\frac{15}{16}$"	8'3$\frac{3}{4}$"	8'3$\frac{3}{8}$"	8'4"	8'3$\frac{15}{16}$"	8'3$\frac{3}{4}$"	8'3$\frac{3}{8}$"
RISERS OVERAGE (+) OR UNDERAGE (—)	+$\frac{1}{4}$"	−$\frac{1}{8}$"	+$\frac{1}{2}$"	+$\frac{5}{16}$"	+$\frac{1}{8}$"	−$\frac{1}{4}$"	+$\frac{3}{8}$"	+$\frac{3}{16}$"	0	−$\frac{3}{8}$"
NUMBER OF TREADS	13	14	15	12	13	14	15	12	13	14
WIDTH OF EACH TREAD	10$\frac{3}{8}$"	10$\frac{7}{8}$"	11$\frac{1}{4}$"	9$\frac{13}{16}$"	10$\frac{3}{8}$"	10$\frac{7}{8}$"	11$\frac{1}{4}$"	9$\frac{13}{16}$"	10$\frac{3}{8}$"	10$\frac{7}{8}$"
TOTAL RUN OF STAIRWAY	11'2$\frac{7}{8}$"	12'8$\frac{1}{4}$"	14'$\frac{3}{4}$"	9'9$\frac{3}{4}$"	11'2$\frac{7}{8}$"	12'8$\frac{1}{4}$"	14'$\frac{3}{4}$"	9'9$\frac{3}{4}$"	11'2$\frac{7}{8}$"	12'8$\frac{1}{4}$"
WELL OPENING FOR 6'-8" HEADROOM	11'7$\frac{1}{16}$"	13'1$\frac{25}{32}$"	14'7$\frac{1}{2}$"	10'$\frac{23}{32}$"	11'7$\frac{1}{16}$"	13'1$\frac{25}{32}$"	14'7$\frac{1}{2}$"	10'$\frac{23}{32}$"	11'7$\frac{1}{16}$"	13'1$\frac{25}{32}$"
ANGLE OF INCLINE	34°29'	31°21'	29°3'	38°5'	34°29'	31°21'	29°3'	38°5'	34°29'	31°21'
LENGTH OF CARRIAGE	13'7$\frac{5}{8}$"	14'10$\frac{9}{32}$"	16'1$\frac{1}{16}$"	12'5$\frac{9}{16}$"	13'7$\frac{5}{8}$"	14'10$\frac{9}{32}$"	16'1$\frac{1}{16}$"	12'5$\frac{9}{16}$"	13'7$\frac{5}{8}$"	14'10$\frac{9}{32}$"
L. F. OF RISER PER INCH OF STAIRWAY WIDTH	1.166	1.25	1.333	1.083	1.166	1.25	1.333	1.083	1.166	1.25
L. F. OF TREAD PER INCH OF STAIRWAY WIDTH	1.083	1.166	1.25	1.0	1.083	1.166	1.25	1.0	1.083	1.166
FRAMING SQUARE SETTINGS FOR CARRIAGE CUTS — TONGUE	7$\frac{1}{8}$"	6$\frac{5}{8}$"	6$\frac{1}{4}$"	7$\frac{11}{16}$"	7$\frac{1}{8}$"	6$\frac{5}{8}$"	6$\frac{1}{4}$"	7$\frac{11}{16}$"	7$\frac{1}{8}$"	6$\frac{5}{8}$"
FRAMING SQUARE SETTINGS FOR CARRIAGE CUTS — BODY	10$\frac{3}{8}$"	10$\frac{7}{8}$"	11$\frac{1}{4}$"	9$\frac{13}{16}$"	10$\frac{3}{8}$"	10$\frac{7}{8}$"	11$\frac{1}{4}$"	9$\frac{13}{16}$"	10$\frac{3}{8}$"	10$\frac{7}{8}$"

FLOOR TO FLOOR RISE	8'3$\frac{3}{4}$"	8'3$\frac{7}{8}$"	8'3$\frac{7}{8}$"	8'3$\frac{7}{8}$"	8'3$\frac{7}{8}$"	8'4"	8'4"	8'4"	8'4"	8'4$\frac{1}{8}$"
NUMBER OF RISERS	16	13	14	15	16	13	14	15	16	13
HEIGHT OF EACH RISER	6$\frac{1}{4}$"	7$\frac{11}{16}$"	7$\frac{1}{8}$"	6$\frac{11}{16}$"	6$\frac{1}{4}$"	7$\frac{11}{16}$"	7$\frac{1}{8}$"	6$\frac{11}{16}$"	6$\frac{1}{4}$"	7$\frac{11}{16}$"
TOTAL HEIGHT OF RISERS	8'4"	8'3$\frac{15}{16}$"	8'3$\frac{3}{4}$"	8'4$\frac{5}{16}$"	8'4"	8'3$\frac{15}{16}$"	8'3$\frac{3}{4}$"	8'4$\frac{5}{16}$"	8'4"	8'3$\frac{15}{16}$"
RISERS OVERAGE (+) OR UNDERAGE (—)	+$\frac{1}{4}$"	+$\frac{1}{16}$"	−$\frac{1}{8}$"	+$\frac{7}{16}$"	+$\frac{1}{8}$"	−$\frac{1}{16}$"	−$\frac{1}{4}$"	+$\frac{5}{16}$"	0	−$\frac{3}{16}$"
NUMBER OF TREADS	15	12	13	14	15	12	13	14	15	12
WIDTH OF EACH TREAD	11$\frac{1}{4}$"	9$\frac{13}{16}$"	10$\frac{3}{8}$"	10$\frac{13}{16}$"	11$\frac{1}{4}$"	9$\frac{13}{16}$"	10$\frac{3}{8}$"	10$\frac{13}{16}$"	11$\frac{1}{4}$"	9$\frac{13}{16}$"
TOTAL RUN OF STAIRWAY	14'$\frac{3}{4}$"	9'9$\frac{3}{4}$"	11'2$\frac{7}{8}$"	12'7$\frac{3}{8}$"	14'$\frac{3}{4}$"	9'9$\frac{3}{4}$"	11'2$\frac{7}{8}$"	12'7$\frac{3}{8}$"	14'$\frac{3}{4}$"	9'9$\frac{3}{4}$"
WELL OPENING FOR 6'-8" HEADROOM	14'7$\frac{1}{2}$"	10'$\frac{23}{32}$"	11'7$\frac{1}{16}$"	13'$\frac{3}{4}$"	14'7$\frac{1}{2}$"	10'$\frac{23}{32}$"	11'7$\frac{1}{16}$"	13'$\frac{3}{4}$"	14'7$\frac{1}{2}$"	10'$\frac{23}{32}$"
ANGLE OF INCLINE	29°3'	38°5'	34°29'	31°44'	29°3'	38°5'	34°29'	31°44'	29°3'	38°5'
LENGTH OF CARRIAGE	16'1$\frac{1}{16}$"	12'5$\frac{9}{16}$"	13'7$\frac{5}{8}$"	14'10"	16'1$\frac{1}{16}$"	12'5$\frac{9}{16}$"	13'7$\frac{5}{8}$"	14'10"	16'1$\frac{1}{16}$"	12'5$\frac{9}{16}$"
L. F. OF RISER PER INCH OF STAIRWAY WIDTH	1.333	1.083	1.166	1.25	1.333	1.083	1.166	1.25	1.333	1.083
L. F. OF TREAD PER INCH OF STAIRWAY WIDTH	1.25	1.0	1.083	1.166	1.25	1.0	1.083	1.166	1.25	1.0
FRAMING SQUARE SETTINGS FOR CARRIAGE CUTS — TONGUE	6$\frac{1}{4}$"	7$\frac{11}{16}$"	7$\frac{1}{8}$"	6$\frac{11}{16}$"	6$\frac{1}{4}$"	7$\frac{11}{16}$"	7$\frac{1}{8}$"	6$\frac{11}{16}$"	6$\frac{1}{4}$"	7$\frac{11}{16}$"
BODY	11$\frac{1}{4}$"	9$\frac{13}{16}$"	10$\frac{3}{8}$"	10$\frac{13}{16}$"	11$\frac{1}{4}$"	9$\frac{13}{16}$"	10$\frac{3}{8}$"	10$\frac{13}{16}$"	11$\frac{1}{4}$"	9$\frac{13}{16}$"

FLOOR TO FLOOR RISE		8'4 1/8"	8'4 1/8"	8'4 1/8"	8'4 1/4"	8'4 1/4"	8'4 1/4"	8'4 1/4"	8'4 3/8"	8'4 3/8"	8'4 3/8"
NUMBER OF RISERS		14	15	16	13	14	15	16	13	14	15
HEIGHT OF EACH RISER		7 1/8"	6 11/16"	6 1/4"	7 11/16"	7 3/16"	6 11/16"	6 1/4"	7 3/4"	7 3/16"	6 11/16"
TOTAL HEIGHT OF RISERS		8'3 3/4"	8'4 5/16"	8'4"	8'3 15/16"	8'4 5/8"	8'4 5/16"	8'4"	8'4 3/4"	8'4 5/8"	8'4 5/16"
RISERS OVERAGE (+) OR UNDERAGE (—)		−3/8"	+3/16"	−1/8"	−5/16"	+3/8"	+1/16"	−1/4"	+3/8"	+1/4"	−1/16"
NUMBER OF TREADS		13	14	15	12	13	14	15	12	13	14
WIDTH OF EACH TREAD		10 3/8"	10 13/16"	11 1/4"	9 13/16"	10 5/16"	10 13/16"	11 1/4"	9 3/4"	10 5/16"	10 13/16"
TOTAL RUN OF STAIRWAY		11'2 7/8"	12'7 3/8"	14' 3/4"	9'9 3/4"	11'2 1/16"	12'7 3/8"	14' 3/4"	9'9"	11'2 1/16"	12'7 3/8"
WELL OPENING FOR 6'-8" HEADROOM		11'7 1/16"	13' 3/4"	14'7 1/2"	10' 23/32"	11'6 3/32"	13' 3/4"	14'7 1/2"	9'11 27/32"	11'6 3/32"	13' 3/4"
ANGLE OF INCLINE		34°29'	31°44'	29°3'	38°5'	34°53'	31°44'	29°3'	38°29'	34°53'	31°44'
LENGTH OF CARRIAGE		13'7 5/8"	14'10"	16'1 1/16"	12'5 9/16"	13'7 13/32"	14'10"	16'1 1/16"	12'5 15/32"	13'7 13/32"	14'10"
L. F. OF RISER PER INCH OF STAIRWAY WIDTH		1.166	1.25	1.333	1.083	1.166	1.25	1.333	1.083	1.166	1.25
L. F. OF TREAD PER INCH OF STAIRWAY WIDTH		1.083	1.166	1.25	1.0	1.083	1.166	1.25	1.0	1.083	1.166
FRAMING SQUARE SETTINGS FOR CARRIAGE CUTS	TONGUE	7 1/8"	6 11/16"	6 1/4"	7 11/16"	7 3/16"	6 11/16"	6 1/4"	7 3/4"	7 3/16"	6 11/16"
	BODY	10 3/8"	10 13/16"	11 1/4"	9 13/16"	10 5/16"	10 13/16"	11 1/4"	9 3/4"	10 5/16"	10 13/16"

FLOOR TO FLOOR RISE	8'4 3/8"	8'4 1/2"	8'4 1/2"	8'4 1/2"	8'4 1/2"	8'4 5/8"	8'4 5/8"	8'4 5/8"	8'4 5/8"	8'4 3/4"
NUMBER OF RISERS	16	13	14	15	16	13	14	15	16	13
HEIGHT OF EACH RISER	6 1/4"	7 3/4"	7 3/16"	6 11/16"	6 5/16"	7 3/4"	7 3/16"	6 11/16"	6 5/16"	7 3/4"
TOTAL HEIGHT OF RISERS	8'4"	8'4 3/4"	8'4 5/8"	8'4 5/16"	8'5"	8'4 3/4"	8'4 5/8"	8'4 5/16"	8'5"	8'4 3/4"
RISERS OVERAGE (+) OR UNDERAGE (—)	−3/8"	+1/4"	+1/8"	−3/16"	+1/2"	+1/8"	0	−5/16"	+3/8"	0
NUMBER OF TREADS	15	12	13	14	15	12	13	14	15	12
WIDTH OF EACH TREAD	11 1/4"	9 3/4"	10 5/16"	10 13/16"	11 3/16"	9 3/4"	10 5/16"	10 13/16"	11 3/16"	9 3/4"
TOTAL RUN OF STAIRWAY	14' 3/4"	9'9"	11'2 1/16"	12'7 3/8"	13'11 13/16"	9'9"	11'2 1/16"	12'7 3/8"	13'11 13/16"	9'9"
WELL OPENING FOR 6'-8" HEADROOM	14'7 1/2"	9'11 27/32"	11'6 3/32"	13' 3/4"	13'7 5/32"	9'11 27/32"	11'6 3/32"	13' 3/4"	13'7 5/32"	9'11 27/32"
ANGLE OF INCLINE	29°3'	38°29'	34°53'	31°44'	29°26'	38°29'	34°53'	31°44'	29°26'	38°29'
LENGTH OF CARRIAGE	16'1 1/16"	12'5 15/32"	13'7 13/32"	14'10"	16' 11/16"	12'5 15/32"	13'7 13/32"	14'10"	16' 11/16"	12'5 15/32"
L. F. OF RISER PER INCH OF STAIRWAY WIDTH	1.333	1.083	1.166	1.25	1.333	1.083	1.166	1.25	1.333	1.083
L. F. OF TREAD PER INCH OF STAIRWAY WIDTH	1.25	1.0	1.083	1.166	1.25	1.0	1.083	1.166	1.25	1.0
FRAMING SQUARE SETTINGS FOR CARRIAGE CUTS — TONGUE	6 1/4"	7 3/4"	7 3/16"	6 11/16"	6 5/16"	7 3/4"	7 3/16"	6 11/16"	6 5/16"	7 3/4"
FRAMING SQUARE SETTINGS FOR CARRIAGE CUTS — BODY	11 1/4"	9 3/4"	10 5/16"	10 13/16"	11 3/16"	9 3/4"	10 5/16"	10 13/16"	11 3/16"	9 3/4"

FLOOR TO FLOOR RISE		8'4$\frac{3}{4}$"	8'4$\frac{3}{4}$"	8'4$\frac{3}{4}$"	8'4$\frac{7}{8}$"	8'4$\frac{7}{8}$"	8'4$\frac{7}{8}$"	8'4$\frac{7}{8}$"	8'5"	8'5"	8'5"
NUMBER OF RISERS		14	15	16	13	14	15	16	13	14	15
HEIGHT OF EACH RISER		7$\frac{3}{16}$"	6$\frac{11}{16}$"	6$\frac{5}{16}$"	7$\frac{3}{4}$"	7$\frac{3}{16}$"	6$\frac{3}{4}$"	6$\frac{5}{16}$"	7$\frac{3}{4}$"	7$\frac{3}{16}$"	6$\frac{3}{4}$"
TOTAL HEIGHT OF RISERS		8'4$\frac{5}{8}$"	8'4$\frac{5}{16}$"	8'5"	8'4$\frac{3}{4}$"	8'4$\frac{5}{8}$"	8'5$\frac{1}{4}$"	8'5"	8'4$\frac{3}{4}$"	8'4$\frac{5}{8}$"	8'5$\frac{1}{4}$"
RISERS OVERAGE (+) OR UNDERAGE (—)		−$\frac{1}{8}$"	−$\frac{7}{16}$"	+$\frac{1}{4}$"	−$\frac{1}{8}$"	−$\frac{1}{4}$"	+$\frac{3}{8}$"	+$\frac{1}{8}$"	−$\frac{1}{4}$"	−$\frac{3}{8}$"	+$\frac{1}{4}$"
NUMBER OF TREADS		13	14	15	12	13	14	15	12	13	14
WIDTH OF EACH TREAD		10$\frac{5}{16}$"	10$\frac{13}{16}$"	11$\frac{3}{16}$"	9$\frac{3}{4}$"	10$\frac{5}{16}$"	10$\frac{3}{4}$"	11$\frac{3}{16}$"	9$\frac{3}{4}$"	10$\frac{5}{16}$"	10$\frac{3}{4}$"
TOTAL RUN OF STAIRWAY		11'2$\frac{1}{16}$"	12'7$\frac{3}{8}$"	13'11$\frac{13}{16}$"	9'9"	11'2$\frac{1}{16}$"	12'6$\frac{1}{2}$"	13'11$\frac{13}{16}$"	9'9"	11'2$\frac{1}{16}$"	12'6$\frac{1}{2}$"
WELL OPENING FOR 6'-8" HEADROOM		11'6$\frac{3}{32}$"	13'$\frac{3}{4}$"	13'7$\frac{5}{32}$"	9'11$\frac{27}{32}$"	11'6$\frac{3}{32}$"	12'11$\frac{11}{16}$"	13'7$\frac{5}{32}$"	9'11$\frac{27}{32}$"	11'6$\frac{3}{32}$"	12'11$\frac{11}{16}$"
ANGLE OF INCLINE		34°53'	31°44'	29°26'	38°29'	34°53'	32°7'	29°26'	38°29'	34°53'	32°7'
LENGTH OF CARRIAGE		13'7$\frac{13}{32}$"	14'10"	16'$\frac{11}{16}$"	12'5$\frac{15}{32}$"	13'7$\frac{13}{16}$"	14'9$\frac{23}{32}$"	16'$\frac{11}{16}$"	12'5$\frac{15}{32}$"	13'7$\frac{13}{16}$"	14'9$\frac{23}{32}$"
L. F. OF RISER PER INCH OF STAIRWAY WIDTH		1.166	1.25	1.333	1.083	1.166	1.25	1.333	1.083	1.166	1.25
L. F. OF TREAD PER INCH OF STAIRWAY WIDTH		1.083	1.166	1.25	1.0	1.083	1.166	1.25	1.0	1.083	1.166
FRAMING SQUARE SETTINGS FOR CARRIAGE CUTS	TONGUE	7$\frac{3}{16}$"	6$\frac{11}{16}$"	6$\frac{5}{16}$"	7$\frac{3}{4}$"	7$\frac{3}{16}$"	6$\frac{3}{4}$"	6$\frac{5}{16}$"	7$\frac{3}{4}$"	7$\frac{3}{16}$"	6$\frac{3}{4}$"
	BODY	10$\frac{5}{16}$"	10$\frac{13}{16}$"	11$\frac{3}{16}$"	9$\frac{3}{4}$"	10$\frac{5}{16}$"	10$\frac{3}{4}$"	11$\frac{3}{16}$"	9$\frac{3}{4}$"	10$\frac{5}{16}$"	10$\frac{3}{4}$"

FLOOR TO FLOOR RISE	8'5"	8'5 1/8"	8'5 1/8"	8'5 1/8"	8'5 1/8"	8'5 1/4"	8'5 1/4"	8'5 1/4"	8'5 1/4"	8'5 3/8"
NUMBER OF RISERS	16	13	14	15	16	13	14	15	16	13
HEIGHT OF EACH RISER	6 5/16"	7 3/4"	7 1/4"	6 3/4"	6 5/16"	7 13/16"	7 1/4"	6 3/4"	6 5/16"	7 13/16"
TOTAL HEIGHT OF RISERS	8'5"	8'4 3/4"	8'5 1/2"	8'5 1/4"	8'5"	8'5 9/16"	8'5 1/2"	8'5 1/4"	8'5"	8'5 9/16"
RISERS OVERAGE (+) OR UNDERAGE (—)	0	−3/8"	+3/8"	+1/8"	−1/8"	+5/16"	+1/4"	0	−1/4"	+3/16"
NUMBER OF TREADS	15	12	13	14	15	12	13	14	15	12
WIDTH OF EACH TREAD	11 3/16"	9 3/4"	10 1/4"	10 3/4"	11 3/16"	9 11/16"	10 1/4"	10 3/4"	11 3/16"	9 11/16"
TOTAL RUN OF STAIRWAY	13'11 13/16"	9'9"	11'1 1/4"	12'6 1/2"	13'11 13/16"	9'8 1/4"	11'1 1/4"	12'6 1/2"	13'11 13/16"	9'8 1/4"
WELL OPENING FOR 6'-8" HEADROOM	13'7 5/32"	9'11 27/32"	11'5 1/8"	12'11 11/16"	13'7 5/32"	9'10 15/16"	11'5 1/8"	12'11 11/16"	13'7 5/32"	9'10 15/16"
ANGLE OF INCLINE	29°26'	38°29'	35°16'	32°7'	29°26'	38°53'	35°16'	32°7'	29°26'	38°53'
LENGTH OF CARRIAGE	16' 11/16"	12'5 15/32"	13'7 7/32"	14'9 23/32"	16' 11/16"	12'5 11/32"	13'7 7/32"	14'9 23/32"	16' 11/16"	12'5 11/32"
L. F. OF RISER PER INCH OF STAIRWAY WIDTH	1.333	1.083	1.166	1.25	1.333	1.083	1.166	1.25	1.333	1.083
L. F. OF TREAD PER INCH OF STAIRWAY WIDTH	1.25	1.0	1.083	1.166	1.25	1.0	1.083	1.166	1.25	1.0
FRAMING SQUARE SETTINGS FOR CARRIAGE CUTS — TONGUE	6 5/16"	7 3/4"	7 1/4"	6 3/4"	6 5/16"	7 13/16"	7 1/4"	6 3/4"	6 5/16"	7 13/16"
FRAMING SQUARE SETTINGS FOR CARRIAGE CUTS — BODY	11 3/16"	9 3/4"	10 1/4"	10 3/4"	11 3/16"	9 11/16"	10 1/4"	10 3/4"	11 3/16"	9 11/16"

FLOOR TO FLOOR RISE	8'5 3/8"	8'5 3/8"	8'5 3/8"	8'5 1/2"	8'5 1/2"	8'5 1/2"	8'5 1/2"	8'5 5/8"	8'5 5/8"	8'5 5/8"
NUMBER OF RISERS	14	15	16	13	14	15	16	13	14	15
HEIGHT OF EACH RISER	7 1/4"	6 3/4"	6 5/16"	7 13/16"	7 1/4"	6 3/4"	6 3/8"	7 13/16"	7 1/4"	6 3/4"
TOTAL HEIGHT OF RISERS	8'5 1/2"	8'5 1/4"	8'5"	8'5 9/16"	8'5 1/2"	8'5 1/4"	8'6"	8'5 9/16"	8'5 1/2"	8'5 1/4"
RISERS OVERAGE (+) OR UNDERAGE (—)	+1/8"	−1/8"	−3/8"	+1/16"	0	−1/4"	+1/2"	−1/16"	−1/8"	−3/8"
NUMBER OF TREADS	13	14	15	12	13	14	15	12	13	14
WIDTH OF EACH TREAD	10 1/4"	10 3/4"	11 3/16"	9 11/16"	10 1/4"	10 3/4"	11 1/8"	9 11/16"	10 1/4"	10 3/4"
TOTAL RUN OF STAIRWAY	11'1 1/4"	12'6 1/2"	13'11 13/16"	9'8 1/4"	11'1 1/4"	12'6 1/2"	13'10 7/8"	9'8 1/4"	11'1 1/4"	12'6 1/2"
WELL OPENING FOR 6'-8" HEADROOM	11'5 1/8"	12'11 11/16"	13'7 5/32"	9'10 15/16"	11'5 1/8"	12'11 11/16"	13'6 1/16"	9'10 15/16"	11'5 1/8"	12'11 11/16"
ANGLE OF INCLINE	35°16'	32°7'	29°26'	38°53'	35°16'	32°7'	29°49'	38°53'	35°16'	32°7'
LENGTH OF CARRIAGE	13'7 7/32"	14'9 23/32"	16' 11/16"	12'5 11/32"	13'7 7/32"	14'9 23/32"	16' 11/32"	12'5 11/32"	13'7 7/32"	14'9 23/32"
L. F. OF RISER PER INCH OF STAIRWAY WIDTH	1.166	1.25	1.333	1.083	1.166	1.25	1.333	1.083	1.166	1.25
L. F. OF TREAD PER INCH OF STAIRWAY WIDTH	1.083	1.166	1.25	1.0	1.083	1.166	1.25	1.0	1.083	1.168
FRAMING SQUARE SETTINGS FOR CARRIAGE CUTS — TONGUE	7 1/4"	6 3/4"	6 5/16"	7 13/16"	7 1/4"	6 3/4"	6 3/8"	7 13/16"	7 1/4"	6 3/4"
FRAMING SQUARE SETTINGS FOR CARRIAGE CUTS — BODY	10 1/4"	10 3/4"	11 3/16"	9 11/16"	10 1/4"	10 3/4"	11 1/8"	9 11/16"	10 1/4"	10 3/4"

FLOOR TO FLOOR RISE	8'5⅝"	8'5¾"	8'5¾"	8'5¾"	8'5¾"	8'5⅞"	8'5⅞"	8'5⅞"	8'5⅞"	8'6"
NUMBER OF RISERS	16	13	14	15	16	13	14	15	16	13
HEIGHT OF EACH RISER	6⅜"	7 13/16"	7¼"	6 13/16"	6⅜"	7 13/16"	7¼"	6 13/16"	6⅜"	7⅞"
TOTAL HEIGHT OF RISERS	8'6"	8'5 9/16"	8'5½"	8'6 3/16"	8'6"	8'5 9/16"	8'5½"	8'6 3/16"	8'6"	8'6⅜"
RISERS OVERAGE (+) OR UNDERAGE (—)	+⅜"	−3/16"	−¼"	+7/16"	+¼"	−5/16"	−⅜"	+5/16"	+⅛"	+⅜"
NUMBER OF TREADS	15	12	13	14	15	12	13	14	15	12
WIDTH OF EACH TREAD	11⅛"	9 11/16"	10¼"	10 11/16"	11⅛"	9 11/16"	10¼"	10 11/16"	11⅛"	9⅝"
TOTAL RUN OF STAIRWAY	13'10⅞"	9'8¼"	11'1¼"	12'5⅝"	13'10⅞"	9'8¼"	11'1¼"	12'5⅝"	13'10⅞"	9'7½"
WELL OPENING FOR 6'-8" HEADROOM	13'6 1/16"	9'10 15/16"	11'5⅛"	11'11 15/16"	13'6 1/16"	9'10 15/16"	11'5⅛"	11'11 15/16"	13'6 1/16"	9'10 3/32"
ANGLE OF INCLINE	29°49'	38°53'	35°16'	32°31'	29°49'	38°53'	35°16'	32°31'	29°49'	39°17'
LENGTH OF CARRIAGE	16' 11/32"	12'5 11/32"	13'7 7/32"	14'9 7/16"	16' 11/32"	12'5 11/32"	13'7 7/32"	14'9 7/16"	16' 11/32"	12'5 7/32"
L. F. OF RISER PER INCH OF STAIRWAY WIDTH	1.333	1.083	1.166	1.25	1.333	1.083	1.166	1.25	1.333	1.083
L. F. OF TREAD PER INCH OF STAIRWAY WIDTH	1.25	1.0	1.083	1.166	1.25	1.0	1.083	1.166	1.25	1.0
FRAMING SQUARE SETTINGS FOR CARRIAGE CUTS — TONGUE	6⅜"	7 13/16"	7¼"	6 13/16"	6⅜"	7 13/16"	7¼"	6 13/16"	6⅜"	7⅞"
FRAMING SQUARE SETTINGS FOR CARRIAGE CUTS — BODY	11⅛"	9 11/16"	10¼"	10 11/16"	11⅛"	9 11/16"	10¼"	10 11/16"	11⅛"	9⅝"

FLOOR TO FLOOR RISE	8'6"	8'6"	8'6"	8'6"	8'6 1/8"	8'6 1/8"	8'6 1/8"	8'6 1/8"	8'6 1/8"	8'6 1/4"
NUMBER OF RISERS	14	15	16	17	13	14	15	16	17	13
HEIGHT OF EACH RISER	7 5/16"	6 13/16"	6 3/8"	6"	7 7/8"	7 5/16"	6 13/16"	6 3/8"	6"	7 7/8"
TOTAL HEIGHT OF RISERS	8'6 3/8"	8'6 3/16"	8'6"	8'6"	8'6 3/8"	8'6 3/8"	8'6 3/16"	8'6"	8'6"	8'6 3/8"
RISERS OVERAGE (+) OR UNDERAGE (—)	+3/8"	+3/16"	0	0	+1/4"	+1/4"	+1/16"	−1/8"	−1/8"	+1/8"
NUMBER OF TREADS	13	14	15	16	12	13	14	15	16	12
WIDTH OF EACH TREAD	10 3/16"	10 11/16"	11 1/8"	11 1/2"	9 5/8"	10 3/16"	10 11/16"	11 1/8"	11 1/2"	9 5/8"
TOTAL RUN OF STAIRWAY	11' 7/16"	12'5 5/8"	13'10 7/8"	15'4"	9'7 1/2"	11' 7/16"	12'5 5/8"	13'10 7/8"	15'4"	9'7 1/2"
WELL OPENING FOR 6'-8" HEADROOM	11'4 3/16"	11'11 15/16"	13'6 1/16"	15' 5/32"	9'10 3/32"	11'4 3/16"	11'11 15/16"	13'6 1/16"	15' 5/32"	9'10 3/32"
ANGLE OF INCLINE	35°40'	32°31'	29°49'	27°33'	39°17'	35°40'	32°31'	29°49'	27°33'	39°17'
LENGTH OF CARRIAGE	13'7 1/32"	14'9 7/16"	16' 11/32"	17'3 17/32"	12'5 7/32"	13'7 1/32"	14'9 7/16"	16' 11/32"	17'3 17/32"	12'5 7/32"
L. F. OF RISER PER INCH OF STAIRWAY WIDTH	1.166	1.25	1.333	1.416	1.083	1.166	1.25	1.333	1.416	1.083
L. F. OF TREAD PER INCH OF STAIRWAY WIDTH	1.083	1.166	1.25	1.333	1.0	1.083	1.166	1.25	1.333	1.0
FRAMING SQUARE SETTINGS FOR CARRIAGE CUTS — TONGUE	7 5/16"	6 13/16"	6 3/8"	6"	7 7/8"	7 5/16"	6 13/16"	6 3/8"	6"	7 7/8"
FRAMING SQUARE SETTINGS FOR CARRIAGE CUTS — BODY	10 3/16"	10 11/16"	11 1/8"	11 1/2"	9 5/8"	10 3/16"	10 11/16"	11 1/8"	11 1/2"	9 5/8"

FLOOR TO FLOOR RISE	8'6 1/4"	8'6 1/4"	8'6 1/4"	8'6 1/4"	8'6 3/8"	8'6 3/8"	8'6 3/8"	8'6 3/8"	8'6 3/8"	8'6 1/2"
NUMBER OF RISERS	14	15	16	17	13	14	15	16	17	13
HEIGHT OF EACH RISER	7 5/16"	6 13/16"	6 3/8"	6"	7 7/8"	7 5/16"	6 13/16"	6 3/8"	6"	7 7/8"
TOTAL HEIGHT OF RISERS	8'6 3/8"	8'6 3/16"	8'6"	8'6"	8'6 3/8"	8'6 3/8"	8'6 3/16"	8'6"	8'6"	8'6 3/8"
RISERS OVERAGE (+) OR UNDERAGE (—)	+1/8"	−1/16"	−1/4"	−1/4"	0	0	−3/16"	−3/8"	−3/8"	−1/8"
NUMBER OF TREADS	13	14	15	16	12	13	14	15	16	12
WIDTH OF EACH TREAD	10 3/16"	10 11/16"	11 1/8"	11 1/2"	9 5/8"	10 3/16"	10 11/16"	11 1/8"	11 1/2"	9 5/8"
TOTAL RUN OF STAIRWAY	11' 7/16"	12'5 5/8"	13'10 7/8"	15'4"	9'7 1/2"	11' 7/16"	12'5 5/8"	13'10 7/8"	15'4"	9'7 1/2"
WELL OPENING FOR 6'-8" HEADROOM	11'4 3/16"	11'11 15/16"	13'6 1/16"	15' 5/32"	9'10 3/32"	11'4 3/16"	11'11 15/16"	13'6 1/16"	15' 5/32"	9'10 3/32"
ANGLE OF INCLINE	35°40'	32°31'	29°49'	27°33'	39°17'	35°40'	32°31'	29°49'	27°33'	39°17'
LENGTH OF CARRIAGE	13'7 1/32"	14'9 7/16"	16' 11/32"	17'3 17/32"	12'5 7/32"	13'7 1/32"	14'9 7/16"	16' 11/32"	17'3 17/32"	12'5 7/32"
L. F. OF RISER PER INCH OF STAIRWAY WIDTH	1.166	1.25	1.333	1.416	1.083	1.166	1.25	1.333	1.416	1.083
L. F. OF TREAD PER INCH OF STAIRWAY WIDTH	1.083	1.166	1.25	1.333	1.0	1.083	1.166	1.25	1.333	1.0
FRAMING SQUARE SETTINGS FOR CARRIAGE CUTS — TONGUE	7 5/16"	6 13/16	6 3/8"	6"	7 7/8"	7 5/16"	6 13/16"	6 3/8"	6"	7 7/8"
FRAMING SQUARE SETTINGS FOR CARRIAGE CUTS — BODY	10 3/16"	10 11/16"	11 1/8"	11 1/2"	9 5/8"	10 3/16"	10 11/16"	11 1/8"	11 1/2"	9 5/8"

FLOOR TO FLOOR RISE		8'6 1/2"	8'6 1/2"	8'6 1/2"	8'6 1/2"	8'6 5/8"	8'6 5/8"	8'6 5/8"	8'6 5/8"	8'6 5/8"	8'6 3/4"
NUMBER OF RISERS		14	15	16	17	13	14	15	16	17	13
HEIGHT OF EACH RISER		7 5/16"	6 13/16"	6 7/16"	6"	7 7/8"	7 5/16"	6 13/16"	6 7/16"	6 1/16"	7 7/8"
TOTAL HEIGHT OF RISERS		8'6 3/8"	8'6 3/16"	8'7"	8'6"	8'6 3/8"	8'6 3/8"	8'6 3/16"	8'7"	8'7 1/16"	8'6 3/8"
RISERS OVERAGE (+) OR UNDERAGE (—)		−1/8"	−5/16"	+1/2"	−1/2"	−1/4"	−1/4"	−7/16"	+3/8"	+7/16"	−3/8"
NUMBER OF TREADS		13	14	15	16	12	13	14	15	16	12
WIDTH OF EACH TREAD		10 3/16"	10 11/16"	11 1/16"	11 1/2"	9 5/8"	10 3/16"	10 11/16"	11 1/16"	11 7/16"	9 5/8"
TOTAL RUN OF STAIRWAY		11' 7/16"	12'5 5/8"	13'9 15/16"	15'4"	9'7 1/2"	11' 7/16"	12'5 5/8"	13'9 15/16"	15'3"	9'7 1/2"
WELL OPENING FOR 6'-8" HEADROOM		11'4 3/16"	11'11 15/16"	13'5"	15' 5/32"	9'10 3/32"	11'4 3/16"	11'11 15/16"	13'5"	14'11"	9'10 3/32"
ANGLE OF INCLINE		35°40'	32°31'	30°12'	27°33'	39°17'	35°40'	32°31'	30°12'	27°56'	39°17'
LENGTH OF CARRIAGE		13'7 1/32"	14'9 7/16"	16'	17'3 17/32"	12'5 7/32"	13'7 1/32"	14'9 7/16"	16'	17'3 1/8"	12'5 7/32"
L. F. OF RISER PER INCH OF STAIRWAY WIDTH		1.166	1.25	1.333	1.416	1.083	1.166	1.25	1.333	1.416	1.083
L. F. OF TREAD PER INCH OF STAIRWAY WIDTH		1.083	1.166	1.25	1.333	1.0	1.083	1.166	1.25	1.333	1.0
FRAMING SQUARE SETTINGS FOR CARRIAGE CUTS	TONGUE	7 5/16"	6 13/16"	6 7/16"	6"	7 7/8"	7 5/16"	6 13/16"	6 7/16"	6 1/16"	7 7/8"
	BODY	10 3/16"	10 11/16"	11 1/16"	11 1/2"	9 5/8"	10 3/16"	10 11/16"	11 1/16"	11 7/16"	9 5/8"

FLOOR TO FLOOR RISE	8'6 3/4"	8'6 3/4"	8'6 3/4"	8'6 3/4"	8'6 7/8"	8'6 7/8"	8'6 7/8"	8'6 7/8"	8'6 7/8"	8'7"
NUMBER OF RISERS	14	15	16	17	13	14	15	16	17	13
HEIGHT OF EACH RISER	7 5/16"	6 7/8"	6 7/16"	6 1/16"	7 15/16"	7 3/8"	6 7/8"	6 7/16"	6 1/16"	7 15/16"
TOTAL HEIGHT OF RISERS	8'6 3/8"	8'7 1/8"	8'7"	8'7 1/16"	8'7 3/16"	8'7 1/4"	8'7 1/8"	8'7"	8'7 1/16"	8'7 3/16"
RISERS OVERAGE (+) OR UNDERAGE (—)	−3/8"	+3/8"	+1/4"	+5/16"	+5/16"	+3/8"	+1/4"	+1/8"	+3/16"	+3/16"
NUMBER OF TREADS	13	14	15	16	12	13	14	15	16	12
WIDTH OF EACH TREAD	10 3/16"	10 5/8"	11 1/16"	11 7/16"	9 9/16"	10 1/8"	10 5/8"	11 1/16"	11 7/16"	9 9/16"
TOTAL RUN OF STAIRWAY	11' 7/16"	12'4 3/4"	13'9 15/16"	15'3"	9'6 3/4"	10'11 5/8"	12'4 3/4"	13'9 15/16"	15'3"	9'6 3/4"
WELL OPENING FOR 6'-8" HEADROOM	11'4 3/16"	11'10 31/32"	13'5"	14'11"	9'9 1/4"	10'5 3/32"	11'10 31/32"	13'5"	14'11"	9'9 1/4"
ANGLE OF INCLINE	35°40'	32°54'	30°12'	27°56'	39°42'	36°4'	32°54'	30°12'	27°56'	39°42'
LENGTH OF CARRIAGE	13'7 1/32"	14'9 3/16"	16'	17'3 1/8"	12'5 1/8"	13'6 27/32"	14'9 3/16"	16'	17'3 1/8"	12'5 1/8"
L. F. OF RISER PER INCH OF STAIRWAY WIDTH	1.166	1.25	1.333	1.416	1.083	1.166	1.25	1.333	1.416	1.083
L. F. OF TREAD PER INCH OF STAIRWAY WIDTH	1.083	1.166	1.25	1.333	1.0	1.083	1.166	1.25	1.333	1.0
FRAMING SQUARE SETTINGS FOR CARRIAGE CUTS — TONGUE	7 5/16"	6 7/8"	6 7/16"	6 1/16"	7 15/16"	7 3/8"	6 7/8"	6 7/16"	6 1/16"	7 15/16"
FRAMING SQUARE SETTINGS FOR CARRIAGE CUTS — BODY	10 3/16"	10 5/8"	11 1/16"	11 7/16"	9 9/16"	10 1/8"	10 5/8"	11 1/16"	11 7/16"	9 9/16"

FLOOR TO FLOOR RISE	8'7"	8'7"	8'7"	8'7"	8'$7\frac{1}{8}$"	8'$7\frac{1}{8}$"	8'$7\frac{1}{8}$"	8'$7\frac{1}{8}$"	8'$7\frac{1}{8}$"	8'$7\frac{1}{4}$"
NUMBER OF RISERS	14	15	16	17	13	14	15	16	17	13
HEIGHT OF EACH RISER	$7\frac{3}{8}$"	$6\frac{7}{8}$"	$6\frac{7}{16}$"	$6\frac{1}{16}$"	$7\frac{15}{16}$"	$7\frac{3}{8}$"	$6\frac{7}{8}$"	$6\frac{7}{16}$"	$6\frac{1}{16}$"	$7\frac{15}{16}$"
TOTAL HEIGHT OF RISERS	8'$7\frac{1}{4}$"	8'$7\frac{1}{8}$"	8'7"	8'$7\frac{1}{16}$"	8'$7\frac{3}{16}$"	8'$7\frac{1}{4}$"	8'$7\frac{1}{8}$"	8'7"	8'$7\frac{1}{16}$"	8'$7\frac{3}{16}$"
RISERS OVERAGE (+) OR UNDERAGE (—)	$+\frac{1}{4}$"	$+\frac{1}{8}$"	0	$+\frac{1}{16}$"	$+\frac{1}{16}$"	$+\frac{1}{8}$"	0	$-\frac{1}{8}$"	$-\frac{1}{16}$"	$-\frac{1}{16}$"
NUMBER OF TREADS	13	14	15	16	12	13	14	15	16	12
WIDTH OF EACH TREAD	$10\frac{1}{8}$"	$10\frac{5}{8}$"	$11\frac{1}{16}$"	$11\frac{7}{16}$"	$9\frac{9}{16}$"	$10\frac{1}{8}$"	$10\frac{5}{8}$"	$11\frac{1}{16}$"	$11\frac{7}{16}$"	$9\frac{9}{16}$"
TOTAL RUN OF STAIRWAY	10'$11\frac{5}{8}$"	12'$4\frac{3}{4}$"	13'$9\frac{15}{16}$"	15'3"	9'$6\frac{3}{4}$"	10'$11\frac{5}{8}$"	12'$4\frac{3}{4}$"	13'$9\frac{15}{16}$"	15'3"	9'$6\frac{3}{4}$"
WELL OPENING FOR 6'-8" HEADROOM	10'$5\frac{3}{32}$"	11'$10\frac{31}{32}$"	13'5"	14'11"	9'$9\frac{1}{4}$"	10'$5\frac{3}{32}$"	11'$10\frac{31}{32}$"	13'5"	14'11"	9'$9\frac{1}{4}$"
ANGLE OF INCLINE	36°4'	32°54'	30°12'	27°56'	39°42'	36°4'	32°54'	30°12'	27°56'	39°42'
LENGTH OF CARRIAGE	13'$6\frac{27}{32}$"	14'$9\frac{3}{16}$"	16'	17'$3\frac{1}{8}$"	12'$5\frac{1}{8}$"	13'$6\frac{27}{32}$"	14'$9\frac{3}{16}$"	16'	17'$3\frac{1}{8}$"	12'$5\frac{1}{8}$"
L. F. OF RISER PER INCH OF STAIRWAY WIDTH	1.166	1.25	1.333	1.416	1.083	1.166	1.25	1.333	1.416	1.083
L. F. OF TREAD PER INCH OF STAIRWAY WIDTH	1.083	1.166	1.25	1.333	1.0	1.083	1.166	1.25	1.333	1.0
FRAMING SQUARE SETTINGS FOR CARRIAGE CUTS — TONGUE	$7\frac{3}{8}$"	$6\frac{7}{8}$"	$6\frac{7}{16}$"	$6\frac{1}{16}$"	$7\frac{15}{16}$"	$7\frac{3}{8}$"	$6\frac{7}{8}$"	$6\frac{7}{16}$"	$6\frac{1}{16}$"	$7\frac{15}{16}$"
FRAMING SQUARE SETTINGS FOR CARRIAGE CUTS — BODY	$10\frac{1}{8}$"	$10\frac{5}{8}$"	$11\frac{1}{16}$"	$11\frac{7}{16}$"	$9\frac{9}{16}$"	$10\frac{1}{8}$"	$10\frac{5}{8}$"	$11\frac{1}{16}$"	$11\frac{7}{16}$"	$9\frac{9}{16}$"

FLOOR TO FLOOR RISE	8'7$\frac{1}{4}$"	8'7$\frac{1}{4}$"	8'7$\frac{1}{4}$"	8'7$\frac{1}{4}$"	8'7$\frac{3}{8}$"	8'7$\frac{3}{8}$"	8'7$\frac{3}{8}$"	8'7$\frac{3}{8}$"	8'7$\frac{3}{8}$"	8'7$\frac{1}{2}$"
NUMBER OF RISERS	14	15	16	17	13	14	15	16	17	13
HEIGHT OF EACH RISER	7$\frac{3}{8}$"	6$\frac{7}{8}$"	6$\frac{7}{16}$"	6$\frac{1}{16}$"	7$\frac{15}{16}$"	7$\frac{3}{8}$"	6$\frac{7}{8}$"	6$\frac{7}{16}$"	6$\frac{1}{16}$"	7$\frac{15}{16}$"
TOTAL HEIGHT OF RISERS	8'7$\frac{1}{4}$"	8'7$\frac{1}{8}$"	8'7"	8'7$\frac{1}{16}$"	8'7$\frac{3}{16}$"	8'7$\frac{1}{4}$"	8'7$\frac{1}{8}$"	8'7"	8'7$\frac{1}{16}$"	8'7$\frac{3}{16}$"
RISERS OVERAGE (+) OR UNDERAGE (—)	0	$-\frac{1}{8}$"	$-\frac{1}{4}$"	$-\frac{3}{16}$"	$-\frac{3}{16}$"	$-\frac{1}{8}$"	$-\frac{1}{4}$"	$-\frac{3}{8}$"	$-\frac{5}{16}$"	$-\frac{5}{16}$"
NUMBER OF TREADS	13	14	15	16	12	13	14	15	16	12
WIDTH OF EACH TREAD	10$\frac{1}{8}$"	10$\frac{5}{8}$"	11$\frac{1}{16}$"	11$\frac{7}{16}$"	9$\frac{9}{16}$"	10$\frac{1}{8}$"	10$\frac{5}{8}$"	11$\frac{1}{16}$"	11$\frac{7}{16}$"	9$\frac{9}{16}$"
TOTAL RUN OF STAIRWAY	10'11$\frac{5}{8}$"	12'4$\frac{3}{4}$"	13'9$\frac{15}{16}$"	15'3"	9'6$\frac{3}{4}$"	10'11$\frac{5}{8}$"	12'4$\frac{3}{4}$"	13'9$\frac{15}{16}$"	15'3"	9'6$\frac{3}{4}$"
WELL OPENING FOR 6'-8" HEADROOM	10'5$\frac{3}{32}$"	11'10$\frac{31}{32}$"	13'5"	14'11"	9'9$\frac{1}{4}$"	10'5$\frac{3}{32}$"	11'10$\frac{31}{32}$"	13'5"	14'11"	9'9$\frac{1}{4}$"
ANGLE OF INCLINE	36°4'	32°54'	30°12'	27°56'	39°42'	36°4'	32°54'	30°12'	27°56'	39°42'
LENGTH OF CARRIAGE	13'6$\frac{27}{32}$"	14'9$\frac{3}{16}$"	16'	17'3$\frac{1}{8}$"	12'5$\frac{1}{8}$"	13'6$\frac{27}{32}$"	14'9$\frac{3}{16}$"	16'	17'3$\frac{1}{8}$"	12'5$\frac{1}{8}$"
L. F. OF RISER PER INCH OF STAIRWAY WIDTH	1.166	1.25	1.333	1.416	1.083	1.166	1.25	1.333	1.416	1.083
L. F. OF TREAD PER INCH OF STAIRWAY WIDTH	1.083	1.166	1.25	1.333	1.0	1.083	1.166	1.25	1.333	1.0
FRAMING SQUARE SETTINGS FOR CARRIAGE CUTS — TONGUE	7$\frac{3}{8}$"	6$\frac{7}{8}$"	6$\frac{7}{16}$"	6$\frac{1}{16}$"	7$\frac{15}{16}$"	7$\frac{3}{8}$"	6$\frac{7}{8}$"	6$\frac{7}{16}$"	6$\frac{1}{16}$"	7$\frac{15}{16}$"
FRAMING SQUARE SETTINGS FOR CARRIAGE CUTS — BODY	10$\frac{1}{8}$"	10$\frac{5}{8}$"	11$\frac{1}{16}$"	11$\frac{7}{16}$"	9$\frac{9}{16}$"	10$\frac{1}{8}$"	10$\frac{5}{8}$"	11$\frac{1}{16}$"	11$\frac{7}{16}$"	9$\frac{9}{16}$"

FLOOR TO FLOOR RISE		8'7 1/2"	8'7 1/2"	8'7 1/2"	8'7 1/2"	8'7 5/8"	8'7 5/8"	8'7 5/8"	8'7 5/8"	8'7 5/8"	8'7 3/4"
NUMBER OF RISERS		14	15	16	17	13	14	15	16	17	13
HEIGHT OF EACH RISER		7 3/8"	6 7/8"	6 1/2"	6 1/16"	8"	7 3/8"	6 15/16"	6 1/2"	6 1/8"	8"
TOTAL HEIGHT OF RISERS		8'7 1/4"	8'7 1/8"	8'8"	8'7 1/16"	8'8"	8'7 1/4"	8'8 1/16"	8'8"	8'8 1/8"	8'8"
RISERS OVERAGE (+) OR UNDERAGE (—)		−1/4"	−3/8"	+1/2"	−7/16"	+3/8"	−3/8"	+7/16"	+3/8"	+1/2"	+1/4"
NUMBER OF TREADS		13	14	15	16	12	13	14	15	16	12
WIDTH OF EACH TREAD		10 1/8"	10 5/8"	11"	11 7/16"	9 1/2"	10 1/8"	10 9/16"	11"	11 3/8"	9 1/2"
TOTAL RUN OF STAIRWAY		10'11 5/8"	12'4 3/4"	13'9"	15'3"	9'6"	10'11 5/8"	12'3 7/8"	13'9"	15'2"	9'6"
WELL OPENING FOR 6'-8" HEADROOM		10'5 3/32"	11'10 31/32"	13'3 15/16"	14'11"	8'10 7/8"	10'5 3/32"	11'9 31/32"	13'3 15/16"	14'9 13/16"	8'10 7/8"
ANGLE OF INCLINE		36°4'	32°54'	30°35'	27°56'	40°6'	36°4'	33°18'	30°35'	28°18'	40°6'
LENGTH OF CARRIAGE		13'6 27/32"	14'9 3/16"	15'11 21/32"	17'3 1/8"	12'5 1/32"	13'6 27/32"	14'8 29/32"	15'11 21/32"	17'2 23/32"	12'5 1/32"
L. F. OF RISER PER INCH OF STAIRWAY WIDTH		1.166	1.25	1.333	1.416	1.083	1.166	1.25	1.333	1.416	1.083
L. F. OF TREAD PER INCH OF STAIRWAY WIDTH		1.083	1.166	1.25	1.333	1.0	1.083	1.166	1.25	1.333	1.0
FRAMING SQUARE SETTINGS FOR CARRIAGE CUTS	TONGUE	7 3/8"	6 7/8"	6 1/2"	6 1/16"	8"	7 3/8"	6 15/16"	6 1/2"	6 1/8"	8"
	BODY	10 1/8"	10 5/8"	11"	11 7/16"	9 1/2"	10 1/8"	10 9/16"	11"	11 3/8"	9 1/2"

FLOOR TO FLOOR RISE	8'7 3/4"	8'7 3/4"	8'7 3/4"	8'7 3/4"	8'7 7/8"	8'7 7/8"	8'7 7/8"	8'7 7/8"	8'7 7/8"	8'8"
NUMBER OF RISERS	14	15	16	17	13	14	15	16	17	13
HEIGHT OF EACH RISER	7 7/16"	6 15/16"	6 1/2"	6 1/8"	8"	7 7/16"	6 15/16"	6 1/2"	6 1/8"	8"
TOTAL HEIGHT OF RISERS	8'8 1/8"	8'8 1/16"	8'8"	8'8 1/8"	8'8"	8'8 1/8"	8'8 1/16"	8'8"	8'8 1/8"	8'8"
RISERS OVERAGE (+) OR UNDERAGE (—)	+3/8"	+5/16"	+1/4"	+3/8"	+1/8"	+1/4"	+3/16"	+1/8"	+1/4"	0
NUMBER OF TREADS	13	14	15	16	12	13	14	15	16	12
WIDTH OF EACH TREAD	10 1/16"	10 9/16"	11"	11 3/8"	9 1/2"	10 1/16"	10 9/16"	11"	11 3/8"	9 1/2"
TOTAL RUN OF STAIRWAY	10'10 13/16"	12'3 7/8"	13'9"	15'2"	9'6"	10'10 13/16"	12'3 7/8"	13'9"	15'2"	9'6"
WELL OPENING FOR 6'-8" HEADROOM	10'4 7/32"	11'9 31/32"	13'3 15/16"	14'9 13/16"	8'10 7/8"	10'4 7/32"	11'9 31/32"	13'3 15/16"	14'9 13/16"	8'10 7/8"
ANGLE OF INCLINE	36°28'	33°18'	30°35'	28°18'	40°6'	36°28'	33°18'	30°35'	28°18'	40°6'
LENGTH OF CARRIAGE	13'6 21/32"	14'8 29/32"	15'11 21/32"	17'2 23/32"	12'5 1/32"	13'6 21/32"	14'8 29/32"	15'11 21/32"	17'2 23/32"	12'5 1/32"
L. F. OF RISER PER INCH OF STAIRWAY WIDTH	1.166	1.25	1.333	1.416	1.083	1.166	1.25	1.333	1.416	1.083
L. F. OF TREAD PER INCH OF STAIRWAY WIDTH	1.083	1.166	1.25	1.333	1.0	1.083	1.166	1.25	1.333	1.0
FRAMING SQUARE SETTINGS FOR CARRIAGE CUTS — TONGUE	7 7/16"	6 15/16"	6 1/2"	6 1/8"	8"	7 7/16"	6 15/16"	6 1/2"	6 1/8"	8"
FRAMING SQUARE SETTINGS FOR CARRIAGE CUTS — BODY	10 1/16"	10 9/16"	11"	11 3/8"	9 1/2"	10 1/16"	10 9/16"	11"	11 3/8"	9 1/2"

FLOOR TO FLOOR RISE	8'8"	8'8"	8'8"	8'8"	8'8 1/8"	8'8 1/8"	8'8 1/8"	8'8 1/8"	8'8 1/8"	8'8 1/4"
NUMBER OF RISERS	14	15	16	17	13	14	15	16	17	13
HEIGHT OF EACH RISER	7 7/16"	6 15/16"	6 1/2"	6 1/8"	8"	7 7/16"	6 15/16"	6 1/2"	6 1/8"	8"
TOTAL HEIGHT OF RISERS	8'8 1/8"	8'8 1/16"	8'8"	8'8 1/8"	8'8"	8'8 1/8"	8'8 1/16"	8'8"	8'8 1/8"	8'8"
RISERS OVERAGE (+) OR UNDERAGE (—)	+1/8"	+1/16"	0	+1/8"	−1/8"	0	−1/16"	−1/8"	0	−1/4"
NUMBER OF TREADS	13	14	15	16	12	13	14	15	16	12
WIDTH OF EACH TREAD	10 1/16"	10 9/16"	11"	11 3/8"	9 1/2"	10 1/16"	10 9/16"	11"	11 3/8"	9 1/2"
TOTAL RUN OF STAIRWAY	10'10 13/16"	12'3 7/8"	13'9"	15'2"	9'6"	10'10 13/16"	12'3 7/8"	13'9"	15'2"	9'6"
WELL OPENING FOR 6'-8" HEADROOM	10'4 7/32"	11'9 31/32"	13'3 15/16"	14'9 13/16"	8'10 7/8"	10'4 7/32"	11'9 31/32"	13'3 15/16"	14'9 13/16"	8'10 7/8"
ANGLE OF INCLINE	36°28'	33°18'	30°35'	28°18'	40°6'	36°28'	33°18'	30°35'	28°18'	40°6'
LENGTH OF CARRIAGE	13'6 21/32"	14'8 29/32"	15'11 21/32"	17'2 23/32"	12'5 1/32"	13'6 21/32"	14'8 29/32"	15'11 21/32"	17'2 23/32"	12'5 1/32"
L. F. OF RISER PER INCH OF STAIRWAY WIDTH	1.166	1.25	1.333	1.416	1.083	1.166	1.25	1.333	1.416	1.083
L. F. OF TREAD PER INCH OF STAIRWAY WIDTH	1.083	1.166	1.25	1.333	1.0	1.083	1.166	1.25	1.333	1.0
FRAMING SQUARE SETTINGS FOR CARRIAGE CUTS — TONGUE	7 7/16"	6 15/16"	6 1/2"	6 1/8"	8"	7 7/16"	6 15/16"	6 1/2"	6 1/8"	8"
FRAMING SQUARE SETTINGS FOR CARRIAGE CUTS — BODY	10 1/16"	10 9/16"	11"	11 3/8"	9 1/2"	10 1/16"	10 9/16"	11"	11 3/8"	9 1/2"

FLOOR TO FLOOR RISE	8'8 1/4"	8'8 1/4"	8'8 1/4"	8'8 1/4"	8'8 3/8"	8'8 3/8"	8'8 3/8"	8'8 3/8"	8'8 3/8"	8'8 1/2"
NUMBER OF RISERS	14	15	16	17	13	14	15	16	17	13
HEIGHT OF EACH RISER	7 7/16"	6 15/16"	6 1/2"	6 1/8"	8"	7 7/16"	6 15/16"	6 1/2"	6 1/8"	8 1/16"
TOTAL HEIGHT OF RISERS	8'8 1/8"	8'8 1/16"	8'8"	8'8 1/8"	8'8"	8'8 1/8"	8'8 1/16"	8'8"	8'8 1/8"	8'8 13/16"
RISERS OVERAGE (+) OR UNDERAGE (—)	−1/8"	−3/16"	−1/4"	−1/8"	−3/8"	−1/4"	−5/16"	−3/8"	−1/4"	+5/16"
NUMBER OF TREADS	13	14	15	16	12	13	14	15	16	12
WIDTH OF EACH TREAD	10 1/16"	10 9/16"	11"	11 3/8"	9 1/2"	10 1/16"	10 9/16"	11"	11 3/8"	9 7/16"
TOTAL RUN OF STAIRWAY	10'10 13/16"	12'3 7/8"	13'9"	15'2"	9'6"	10'10 13/16"	12'3 7/8"	13'9"	15'2"	9'5 1/4"
WELL OPENING FOR 6'-8" HEADROOM	10'4 7/32"	11'9 31/32"	13'3 15/16"	14'9 13/16"	8'10 7/8"	10'4 7/32"	11'9 31/32"	13'3 15/16"	14'9 13/16"	8'10 3/32"
ANGLE OF INCLINE	36°28'	33°18'	30°35'	28°18'	40°6'	36°28'	33°18'	30°35'	28°18'	40°30'
LENGTH OF CARRIAGE	13'6 21/32"	14'8 29/32"	15'11 21/32"	17'2 23/32"	12'5 1/32"	13'6 21/32"	14'8 29/32"	15'11 21/32"	17'2 23/32"	12'4 15/16"
L. F. OF RISER PER INCH OF STAIRWAY WIDTH	1.166	1.25	1.333	1.416	1.083	1.166	1.25	1.333	1.416	1.083
L. F. OF TREAD PER INCH OF STAIRWAY WIDTH	1.083	1.166	1.25	1.333	1.0	1.083	1.166	1.25	1.333	1.416
FRAMING SQUARE SETTINGS FOR CARRIAGE CUTS — TONGUE	7 7/16"	6 15/16"	6 1/2"	6 1/8"	8"	7 7/16"	6 15/16"	6 1/2"	6 1/8"	8 1/16"
FRAMING SQUARE SETTINGS FOR CARRIAGE CUTS — BODY	10 1/16"	10 9/16"	11"	11 3/8"	9 1/2"	10 1/16"	10 9/16"	11"	11 3/8"	9 7/16"

FLOOR TO FLOOR RISE		8'8 1/2"	8'8 1/2"	8'8 1/2"	8'8 1/2"	8'8 5/8"	8'8 5/8"	8'8 5/8"	8'8 5/8"	8'8 5/8"	8'8 3/4"
NUMBER OF RISERS		14	15	16	17	13	14	15	16	17	13
HEIGHT OF EACH RISER		7 7/16"	6 15/16"	6 9/16"	6 1/8"	8 1/16"	7 1/2"	7"	6 9/16"	6 1/8"	8 1/16"
TOTAL HEIGHT OF RISERS		8'8 1/8"	8'8 1/16"	8'9"	8'8 1/8"	8'8 13/16"	8'9"	8'9"	8'9"	8'8 1/8"	8'8 13/16"
RISERS OVERAGE (+) OR UNDERAGE (—)		−3/8"	−7/16"	+1/2"	−3/8"	+3/16"	+3/8"	+3/8"	+3/8"	−1/2"	+1/16"
NUMBER OF TREADS		13	14	15	16	12	13	14	15	16	12
WIDTH OF EACH TREAD		10 1/16"	10 9/16"	10 15/16"	11 3/8"	9 7/16"	10"	10 1/2"	10 15/16"	11 3/8"	9 7/16"
TOTAL RUN OF STAIRWAY		10'10 13/16"	12'3 7/8"	13'8 1/16"	15'2"	9'5 1/4"	10'10"	12'3"	13'8 1/16"	15'2"	9'5 1/4"
WELL OPENING FOR 6'-8" HEADROOM		10'4 7/32"	11'9 31/32"	13'2 27/32"	14'9 13/16"	8'10 3/32"	10'3 11/32"	11'9"	13'2 27/32"	14'9 13/16"	8'10 3/32"
ANGLE OF INCLINE		36°28'	33°18'	30°58'	28°18'	40°30'	36°52'	33°41'	30°58'	28°18'	40°30'
LENGTH OF CARRIAGE		13'6 21/32"	14'8 29/32"	15'11 11/32"	17'2 23/32"	12'4 15/16"	13'6 1/2"	14'8 21/32"	15'11 11/32"	17'2 23/32"	12'4 15/16"
L. F. OF RISER PER INCH OF STAIRWAY WIDTH		1.166	1.25	1.333	1.416	1.083	1.166	1.25	1.333	1.416	1.083
L. F. OF TREAD PER INCH OF STAIRWAY WIDTH		1.083	1.166	1.25	1.333	1.0	1.083	1.166	1.25	1.333	1.0
FRAMING SQUARE SETTINGS FOR CARRIAGE CUTS	TONGUE	7 7/16"	6 15/16"	6 9/16"	6 1/8"	8 1/16"	7 1/2"	7"	6 9/16"	6 1/8"	8 1/16"
	BODY	10 1/16"	10 9/16"	10 15/16"	11 3/8"	9 7/16"	10"	10 1/2"	10 15/16"	11 3/8"	9 7/16"

FLOOR TO FLOOR RISE		8'8 3/4"	8'8 3/4"	8'8 3/4"	8'8 3/4"	8'8 7/8"	8'8 7/8"	8'8 7/8"	8'8 7/8"	8'8 7/8"	8'9"
NUMBER OF RISERS		14	15	16	17	13	14	15	16	17	13
HEIGHT OF EACH RISER		7 1/2"	7"	6 9/16"	6 3/16"	8 1/16"	7 1/2"	7"	6 9/16"	6 3/16"	8 1/16"
TOTAL HEIGHT OF RISERS		8'9"	8'9"	8'9"	8'9 3/16"	8'8 13/16"	8'9"	8'9"	8'9"	8'9 3/16"	8'8 13/16"
RISERS OVERAGE (+) OR UNDERAGE (—)		+1/4"	+1/4"	+1/4"	+7/16"	−1/16"	+1/8"	+1/8"	+1/8"	+5/16"	−3/16"
NUMBER OF TREADS		13	14	15	16	12	13	14	15	16	12
WIDTH OF EACH TREAD		10"	10 1/2"	10 15/16"	11 5/16"	9 7/16"	10"	10 1/2"	10 15/16"	11 5/16"	9 7/16"
TOTAL RUN OF STAIRWAY		10'10"	12'3"	13'8 1/16"	15'1"	9'5 1/4"	10'10"	12'3"	13'8 1/16"	15'1"	9'5 1/4"
WELL OPENING FOR 6'-8" HEADROOM		10'3 11/32"	11'9"	13'2 27/32"	14'8 21/32"	8'10 3/32"	10'3 11/32"	11'9"	13'2 27/32"	14'8 21/32"	8'10 3/32"
ANGLE OF INCLINE		36°52'	33°41'	30°58'	28°41'	40°30'	36°52'	33°41'	30°58'	28°41'	40°30'
LENGTH OF CARRIAGE		13'6 1/2"	14'8 21/32"	15'11 11/32"	17'2 5/16"	12'4 15/16"	13'6 1/2"	14'8 21/32"	15'11 11/32"	17'2 5/16"	12'4 15/16"
L. F. OF RISER PER INCH OF STAIRWAY WIDTH		1.166	1.25	1.333	1.416	1.083	1.166	1.25	1.333	1.416	1.083
L. F. OF TREAD PER INCH OF STAIRWAY WIDTH		1.083	1.166	1.25	1.333	1.0	1.083	1.166	1.25	1.333	1.0
FRAMING SQUARE SETTINGS FOR CARRIAGE CUTS	TONGUE	7 1/2"	7"	6 9/16"	6 3/16"	8 1/16"	7 1/2"	7"	6 9/16"	6 3/16"	8 1/16"
	BODY	10"	10 1/2"	10 15/16"	11 5/16"	9 7/16"	10"	10 1/2"	10 15/16"	11 5/16"	9 7/16"

FLOOR TO FLOOR RISE	8'9"	8'9"	8'9"	8'9"	8'$9\frac{1}{8}$"	8'$9\frac{1}{8}$"	8'$9\frac{1}{8}$"	8'$9\frac{1}{8}$"	8'$9\frac{1}{8}$"	8'$9\frac{1}{4}$"
NUMBER OF RISERS	14	15	16	17	13	14	15	16	17	13
HEIGHT OF EACH RISER	$7\frac{1}{2}$"	7"	$6\frac{9}{16}$"	$6\frac{3}{16}$"	$8\frac{1}{16}$"	$7\frac{1}{2}$"	7"	$6\frac{9}{16}$"	$6\frac{3}{16}$"	$8\frac{1}{8}$"
TOTAL HEIGHT OF RISERS	8'9"	8'9"	8'9"	8'$9\frac{3}{16}$"	8'$8\frac{13}{16}$"	8'9"	8'9"	8'9"	8'$9\frac{3}{16}$"	8'$9\frac{5}{8}$"
RISERS OVERAGE (+) OR UNDERAGE (—)	0	0	0	$+\frac{3}{16}$"	$-\frac{5}{16}$"	$-\frac{1}{8}$"	$-\frac{1}{8}$"	$-\frac{1}{8}$"	$+\frac{1}{16}$"	$+\frac{3}{8}$"
NUMBER OF TREADS	13	14	15	16	12	13	14	15	16	12
WIDTH OF EACH TREAD	10"	$10\frac{1}{2}$"	$10\frac{15}{16}$"	$11\frac{5}{16}$"	$9\frac{7}{16}$"	10"	$10\frac{1}{2}$"	$10\frac{15}{16}$"	$11\frac{5}{16}$"	$9\frac{3}{8}$"
TOTAL RUN OF STAIRWAY	10'10"	12'3"	13'$8\frac{1}{16}$"	15'1"	9'$5\frac{1}{4}$"	10'10"	12'3"	13'$8\frac{1}{16}$"	15'1"	9'$4\frac{1}{2}$"
WELL OPENING FOR 6'-8" HEADROOM	10'$3\frac{11}{32}$"	11'9"	13'$2\frac{27}{32}$"	14'$8\frac{21}{32}$"	8'$10\frac{3}{32}$"	10'$3\frac{11}{32}$"	11'9"	13'$2\frac{27}{32}$"	14'$8\frac{21}{32}$"	8'$9\frac{9}{32}$"
ANGLE OF INCLINE	36°52'	33°41'	30°58'	28°41'	40°30'	36°52'	33°41'	30°58'	28°41'	40°55'
LENGTH OF CARRIAGE	13'$6\frac{1}{2}$"	14'$8\frac{21}{32}$"	15'$11\frac{11}{32}$"	17'$2\frac{5}{16}$"	12'$4\frac{15}{16}$"	13'$6\frac{1}{2}$"	14'$8\frac{21}{32}$"	15'$11\frac{11}{32}$"	17'$2\frac{5}{16}$"	12'$4\frac{7}{8}$"
L. F. OF RISER PER INCH OF STAIRWAY WIDTH	1.166	1.25	1.333	1.416	1.083	1.166	1.25	1.333	1.416	1.083
L. F. OF TREAD PER INCH OF STAIRWAY WIDTH	1.083	1.166	1.25	1.333	1.0	1.083	1.166	1.25	1.333	1.0
FRAMING SQUARE SETTINGS FOR CARRIAGE CUTS — TONGUE	$7\frac{1}{2}$"	7"	$6\frac{9}{16}$"	$6\frac{3}{16}$"	$8\frac{1}{16}$"	$7\frac{1}{2}$"	7"	$6\frac{9}{16}$"	$6\frac{3}{16}$"	$8\frac{1}{8}$"
FRAMING SQUARE SETTINGS FOR CARRIAGE CUTS — BODY	10"	$10\frac{1}{2}$"	$10\frac{15}{16}$"	$11\frac{5}{16}$"	$9\frac{7}{16}$"	10"	$10\frac{1}{2}$"	$10\frac{15}{16}$"	$11\frac{5}{16}$"	$9\frac{3}{8}$"

FLOOR TO FLOOR RISE		8'9 1/4"	8'9 1/4"	8'9 1/4"	8'9 1/4"	8'9 3/8"	8'9 3/8"	8'9 3/8"	8'9 3/8"	8'9 3/8"	8'9 1/2"
NUMBER OF RISERS		14	15	16	17	13	14	15	16	17	13
HEIGHT OF EACH RISER		7 1/2"	7"	6 9/16"	6 3/16"	8 1/8"	7 1/2"	7"	6 9/16"	6 3/16"	8 1/8"
TOTAL HEIGHT OF RISERS		8'9"	8'9"	8'9"	8'9 3/16"	8'9 5/8"	8'9"	8'9"	8'9"	8'9 3/16"	8'9 5/8"
RISERS OVERAGE (+) OR UNDERAGE (—)		−1/4"	−1/4"	−1/4"	−1/16"	+1/4"	−3/8"	−3/8"	−3/8"	−3/16"	+1/8"
NUMBER OF TREADS		13	14	15	16	12	13	14	15	16	12
WIDTH OF EACH TREAD		10"	10 1/2"	10 15/16"	11 5/16"	9 3/8"	10"	10 1/2"	10 15/16"	11 5/16"	9 3/8"
TOTAL RUN OF STAIRWAY		10'10"	12'3"	13'8 1/16"	15'1"	9'4 1/2"	10'10"	12'3"	13'8 1/16"	15'1"	9'4 1/2"
WELL OPENING FOR 6'-8" HEADROOM		10'3 11/32"	11'9"	13'2 27/32"	14'8 21/32"	8'9 9/32"	10'3 11/32"	11'9"	13'2 27/32"	14'8 21/32"	8'9 9/32"
ANGLE OF INCLINE		36°52'	33°41'	30°58'	28°41'	40°55'	36°52'	33°41'	30°58'	28°41'	40°55'
LENGTH OF CARRIAGE		13'6 1/2"	14'8 21/32"	15'11 11/32"	17'2 5/16"	12'4 7/8"	13'6 1/2"	14'8 21/32"	15'11 11/32"	17'2 5/16"	12'4 7/8"
L. F. OF RISER PER INCH OF STAIRWAY WIDTH		1.166	1.25	1.333	1.416	1.083	1.166	1.25	1.333	1.416	1.083
L. F. OF TREAD PER INCH OF STAIRWAY WIDTH		1.083	1.166	1.25	1.333	1.0	1.083	1.166	1.25	1.333	1.0
FRAMING SQUARE SETTINGS FOR CARRIAGE CUTS	TONGUE	7 1/2"	7"	6 9/16"	6 3/16"	8 1/8"	7 1/2"	7"	6 9/16"	6 3/16"	8 1/8"
	BODY	10"	10 1/2"	10 15/16"	11 5/16"	9 3/8"	10"	10 1/2"	10 15/16"	11 5/16"	9 3/8"

FLOOR TO FLOOR RISE	8'9 1/2"	8'9 1/2"	8'9 1/2"	8'9 1/2"	8'9 5/8"	8'9 5/8"	8'9 5/8"	8'9 5/8"	8'9 5/8"	8'9 3/4"
NUMBER OF RISERS	14	15	16	17	13	14	15	16	17	13
HEIGHT OF EACH RISER	7 9/16"	7 1/16"	6 5/8"	6 3/16"	8 1/8"	7 9/16"	7 1/16"	6 5/8"	6 3/16"	8 1/8"
TOTAL HEIGHT OF RISERS	8'9 7/8"	8'9 15/16"	8'10"	8'9 3/16"	8'9 5/8"	8'9 7/8"	8'9 15/16"	8'10"	8'9 3/16"	8'9 5/8"
RISERS OVERAGE (+) OR UNDERAGE (—)	+3/8"	+7/16"	+1/2"	−5/16"	0	+1/4"	+5/16"	+3/8"	−7/16"	−1/8"
NUMBER OF TREADS	13	14	15	16	12	13	14	15	16	12
WIDTH OF EACH TREAD	9 15/16"	10 7/16"	10 7/8"	11 5/16"	9 3/8"	9 15/16"	10 7/16"	10 7/8"	11 5/16"	9 3/8"
TOTAL RUN OF STAIRWAY	10'9 3/16"	12'2 1/8"	13'7 1/8"	15'1"	9'4 1/2"	10'9 3/16"	12'2 1/8"	13'7 1/8"	15'1"	9'4 1/2"
WELL OPENING FOR 6'-8" HEADROOM	10'2 7/16"	11'8 1/32"	13'1 25/32"	14'8 21/32"	8'9 9/32"	10'2 7/16"	11'8 1/32"	13'1 25/32"	14'8 21/32"	8'9 9/32"
ANGLE OF INCLINE	37°16'	34°5'	31°21'	28°41'	40°55'	37°16'	34°5'	31°21'	28°41'	40°55'
LENGTH OF CARRIAGE	13'6 11/32"	14'8 7/16"	15'11"	17'2 5/16"	12'4 7/8"	13'6 11/32"	14'8 7/16"	15'11"	17'2 5/16"	12'4 7/8"
L. F. OF RISER PER INCH OF STAIRWAY WIDTH	1.166	1.25	1.333	1.416	1.083	1.166	1.25	1.333	1.416	1.083
L. F. OF TREAD PER INCH OF STAIRWAY WIDTH	1.083	1.166	1.25	1.333	1.0	1.083	1.166	1.25	1.333	1.0
FRAMING SQUARE SETTINGS FOR CARRIAGE CUTS — TONGUE	7 9/16"	7 1/16"	6 5/8"	6 3/16"	8 1/8"	7 9/16"	7 1/16"	6 5/8"	6 3/16"	8 1/8"
FRAMING SQUARE SETTINGS FOR CARRIAGE CUTS — BODY	9 15/16"	10 7/16"	10 7/8"	11 5/16"	9 3/8"	9 15/16"	10 7/16"	10 7/8"	11 5/16"	9 3/8"

FLOOR TO FLOOR RISE	8'9 3/4"	8'9 3/4"	8'9 3/4"	8'9 3/4"	8'9 7/8"	8'9 7/8"	8'9 7/8"	8'9 7/8"	8'9 7/8"	8'10"
NUMBER OF RISERS	14	15	16	17	13	14	15	16	17	13
HEIGHT OF EACH RISER	7 9/16"	7 1/16"	6 5/8"	6 1/4"	8 1/8"	7 9/16"	7 1/16"	6 5/8"	6 1/4"	8 1/8"
TOTAL HEIGHT OF RISERS	8'9 7/8"	8'9 15/16"	8'10"	8'10 1/4"	8'9 5/8"	8'9 7/8"	8'9 15/16"	8'10"	8'10 1/4"	8'9 5/8"
RISERS OVERAGE (+) OR UNDERAGE (—)	+1/8"	+3/16"	+1/4"	+1/2"	−1/4"	0	+1/16"	+1/8"	+3/8"	−3/8"
NUMBER OF TREADS	13	14	15	16	12	13	14	15	16	12
WIDTH OF EACH TREAD	9 15/16"	10 7/16"	10 7/8"	11 1/4"	9 3/8"	9 15/16"	10 7/16"	10 7/8"	11 1/4"	9 3/8"
TOTAL RUN OF STAIRWAY	10'9 3/16"	12'2 1/8"	13'7 1/8"	15'	9'4 1/2"	10'9 3/16"	12'2 1/8"	13'7 1/8"	15'	9'4 1/2"
WELL OPENING FOR 6'-8" HEADROOM	10'2 7/16"	11'8 1/32"	13'1 25/32"	14'7 1/2"	8'9 9/32"	10'2 7/16"	11'8 1/32"	13'1 25/32"	14'7 1/2"	8'9 9/32"
ANGLE OF INCLINE	37°16'	34°5'	31°21'	29°3'	40°55'	37°16'	34°5'	31°21'	29°3'	40°55'
LENGTH OF CARRIAGE	13'6 11/32"	14'8 7/16"	15'11"	17'1 29/32"	12'4 7/8"	13'6 11/32"	14'8 7/16"	15'11"	17'1 29/32"	12'4 7/8"
L. F. OF RISER PER INCH OF STAIRWAY WIDTH	1.166	1.25	1.333	1.416	1.083	1.166	1.25	1.333	1.416	1.083
L. F. OF TREAD PER INCH OF STAIRWAY WIDTH	1.083	1.166	1.25	1.333	1.0	1.083	1.166	1.25	1.333	1.0
FRAMING SQUARE SETTINGS FOR CARRIAGE CUTS — TONGUE	7 9/16"	7 1/16"	6 5/8"	6 1/4"	8 1/8"	7 9/16"	7 1/16"	6 5/8"	6 1/4"	8 1/8"
FRAMING SQUARE SETTINGS FOR CARRIAGE CUTS — BODY	9 15/16"	10 7/16"	10 7/8"	11 1/4"	9 3/8"	9 15/16"	10 7/16"	10 7/8"	11 1/4"	9 3/8"

FLOOR TO FLOOR RISE		8'10"	8'10"	8'10"	8'10"	8'10 1/8"	8'10 1/8"	8'10 1/8"	8'10 1/8"	8'10 1/8"	8'10 1/4"
NUMBER OF RISERS		14	15	16	17	13	14	15	16	17	13
HEIGHT OF EACH RISER		7 9/16"	7 1/16"	6 5/8"	6 1/4"	8 3/16"	7 9/16"	7 1/16"	6 5/8"	6 1/4"	8 3/16"
TOTAL HEIGHT OF RISERS		8'9 7/8"	8'9 15/16"	8'10"	8'10 1/4"	8'10 7/16"	8'9 7/8"	8'9 15/16"	8'10"	8'10 1/4"	8'10 7/16"
RISERS OVERAGE (+) OR UNDERAGE (—)		−1/8"	−1/16"	0	+1/4"	+5/16"	−1/4"	−3/16"	−1/8"	+1/8"	+3/16"
NUMBER OF TREADS		13	14	15	16	12	13	14	15	16	12
WIDTH OF EACH TREAD		9 15/16"	10 7/16"	10 7/8"	11 1/4"	9 5/16"	9 15/16"	10 7/16"	10 7/8"	11 1/4"	9 5/16"
TOTAL RUN OF STAIRWAY		10'9 3/16"	12'2 1/8"	13'7 1/8"	15'	9'3 3/4"	10'9 3/16"	12'2 1/8"	13'7 1/8"	15'	9'3 3/4"
WELL OPENING FOR 6'-8" HEADROOM		10'2 7/16"	11'8 1/32"	13'1 25/32"	14'7 1/2"	8'8 1/2"	10'2 7/16"	11'8 1/32"	13'1 25/32"	14'7 1/2"	8'8 1/2"
ANGLE OF INCLINE		37°16'	34°5'	31°21'	29°3'	41°19'	37°16'	34°5'	31°21'	29°3'	41°19'
LENGTH OF CARRIAGE		13'6 11/32"	14'8 7/16"	15'11"	17'1 29/32"	12'4 13/16"	13'6 11/32"	14'8 7/16"	15'11"	17'1 29/32"	12'4 13/16"
L. F. OF RISER PER INCH OF STAIRWAY WIDTH		1.166	1.25	1.333	1.416	1.083	1.166	1.25	1.333	1.416	1.083
L. F. OF TREAD PER INCH OF STAIRWAY WIDTH		1.083	1.166	1.25	1.333	1.0	1.083	1.166	1.25	1.333	1.0
FRAMING SQUARE SETTINGS FOR CARRIAGE CUTS	TONGUE	7 9/16"	7 1/16"	6 5/8"	6 1/4"	8 3/16"	7 9/16"	7 1/16"	6 5/8"	6 1/4"	8 3/16"
	BODY	9 15/16"	10 7/16"	10 7/8"	11 1/4"	9 5/16"	9 15/16"	10 7/16"	10 7/8"	11 1/4"	9 5/16"

FLOOR TO FLOOR RISE	8'10 1/4"	8'10 1/4"	8'10 1/4"	8'10 1/4"	8'10 3/8"	8'10 3/8"	8'10 3/8"	8'10 3/8"	8'10 3/8"	8'10 1/2"
NUMBER OF RISERS	14	15	16	17	13	14	15	16	17	13
HEIGHT OF EACH RISER	7 9/16"	7 1/16"	6 5/8"	6 1/4"	8 3/16"	7 5/8"	7 1/16"	6 5/8"	6 1/4"	8 3/16"
TOTAL HEIGHT OF RISERS	8'9 7/8"	8'9 15/16"	8'10"	8'10 1/4"	8'10 7/16"	8'10 3/4"	8'9 15/16"	8'10"	8'10 1/4"	8'10 7/16"
RISERS OVERAGE (+) OR UNDERAGE (—)	−3/8"	−5/16"	−1/4"	0	+1/16"	+3/8"	−7/16"	−3/8"	−1/8"	−1/16"
NUMBER OF TREADS	13	14	15	16	12	13	14	15	16	12
WIDTH OF EACH TREAD	9 15/16"	10 7/16"	10 7/8"	11 1/4"	9 5/16"	9 7/8"	10 7/16"	10 7/8"	11 1/4"	9 5/16"
TOTAL RUN OF STAIRWAY	10'9 3/16"	12'2 1/8"	13'7 1/8"	15'	9'3 3/4"	10'8 3/8"	12'2 1/8"	13'7 1/8"	15'	9'3 3/4"
WELL OPENING FOR 6'-8" HEADROOM	10'2 7/16"	11'8 1/32"	13'1 25/32"	14'7 1/2"	8'8 1/2"	10'1 9/16"	11'8 1/32"	13'1 25/32"	14'7 1/2"	8'8 1/2"
ANGLE OF INCLINE	37°16'	34°5'	31°21'	29°3'	41°19'	37°40'	34°5'	31°21'	29°3'	41°19'
LENGTH OF CARRIAGE	13'6 11/32"	14'8 7/16"	15'11"	17'1 29/32"	12'4 13/16"	13'6 3/16"	14'8 7/16"	15'11"	17'1 29/32"	12'4 13/16"
L. F. OF RISER PER INCH OF STAIRWAY WIDTH	1.166	1.25	1.333	1.416	1.083	1.166	1.25	1.333	1.416	1.083
L. F. OF TREAD PER INCH OF STAIRWAY WIDTH	1.083	1.166	1.25	1.333	1.0	1.083	1.166	1.25	1.333	1.0
FRAMING SQUARE SETTINGS FOR CARRIAGE CUTS — TONGUE	7 9/16"	7 1/16"	6 5/8"	6 1/4"	8 3/16"	7 5/8"	7 1/16"	6 5/8"	6 1/4"	8 3/16"
FRAMING SQUARE SETTINGS FOR CARRIAGE CUTS — BODY	9 15/16"	10 7/16"	10 7/8"	11 1/4"	9 5/16"	9 7/8"	10 7/16"	10 7/8"	11 1/4"	9 5/16"

FLOOR TO FLOOR RISE	8'10 1/2"	8'10 1/2"	8'10 1/2"	8'10 1/2"	8'10 5/8"	8'10 5/8"	8'10 5/8"	8'10 5/8"	8'10 5/8"	8'10 3/4"
NUMBER OF RISERS	14	15	16	17	13	14	15	16	17	13
HEIGHT OF EACH RISER	7 5/8"	7 1/8"	6 11/16"	6 1/4"	8 3/16"	7 5/8"	7 1/8"	6 11/16"	6 1/4"	8 3/16"
TOTAL HEIGHT OF RISERS	8'10 3/4"	8'10 7/8"	8'11"	8'10 1/4"	8'10 7/16"	8'10 3/4"	8'10 7/8"	8'11"	8'10 1/4"	8'10 7/16"
RISERS OVERAGE (+) OR UNDERAGE (—)	+1/4"	+3/8"	+1/2"	−1/4"	−3/16"	+1/8"	+1/4"	+3/8"	−3/8"	−5/16"
NUMBER OF TREADS	13	14	15	16	12	13	14	15	16	12
WIDTH OF EACH TREAD	9 7/8"	10 3/8"	10 13/16"	11 1/4"	9 5/16"	9 7/8"	10 3/8"	10 13/16"	11 1/4"	9 5/16"
TOTAL RUN OF STAIRWAY	10'8 3/8"	12'1 1/4"	13'6 3/16"	15'	9'3 3/4"	10'8 3/8"	12'1 1/4"	13'6 3/16"	15'	9'3 3/4"
WELL OPENING FOR 6'-8" HEADROOM	10'1 9/16"	11'7 1/16"	13' 3/4"	14'7 1/2"	8'8 1/2"	10'1 9/16"	11'7 1/16"	13' 3/4"	14'7 1/2"	8'8 1/2"
ANGLE OF INCLINE	37°40'	34°29'	31°44'	29°3'	41°19'	37°40'	34°29'	31°44'	29°3'	41°19'
LENGTH OF CARRIAGE	13'6 3/16"	14'8 3/16"	15'10 11/16"	17'1 29/32"	12'4 13/16"	13'6 3/16"	14'8 3/16"	15'10 11/16"	17'1 29/32"	12'4 13/16"
L. F. OF RISER PER INCH OF STAIRWAY WIDTH	1.166	1.25	1.333	1.416	1.083	1.166	1.25	1.333	1.416	1.083
L. F. OF TREAD PER INCH OF STAIRWAY WIDTH	1.083	1.166	1.25	1.333	1.0	1.083	1.166	1.25	1.333	1.0
FRAMING SQUARE SETTINGS FOR CARRIAGE CUTS — TONGUE	7 5/8"	7 1/8"	6 11/16"	6 1/4"	8 3/16"	7 5/8"	7 1/8"	6 11/16"	6 1/4"	8 3/16"
FRAMING SQUARE SETTINGS FOR CARRIAGE CUTS — BODY	9 7/8"	10 3/8"	10 13/16"	11 1/4"	9 5/16"	9 7/8"	10 3/8"	10 13/16"	11 1/4"	9 5/16"

FLOOR TO FLOOR RISE	8'10$\frac{3}{4}$"	8'10$\frac{3}{4}$"	8'10$\frac{3}{4}$"	8'10$\frac{3}{4}$"	8'10$\frac{7}{8}$"	8'10$\frac{7}{8}$"	8'10$\frac{7}{8}$"	8'10$\frac{7}{8}$"	8'10$\frac{7}{8}$"	8'11"
NUMBER OF RISERS	14	15	16	17	13	14	15	16	17	13
HEIGHT OF EACH RISER	7$\frac{5}{8}$"	7$\frac{1}{8}$"	6$\frac{11}{16}$"	6$\frac{1}{4}$"	8$\frac{1}{4}$"	7$\frac{5}{8}$"	7$\frac{1}{8}$"	6$\frac{11}{16}$"	6$\frac{5}{16}$"	8$\frac{1}{4}$"
TOTAL HEIGHT OF RISERS	8'10$\frac{3}{4}$"	8'10$\frac{7}{8}$"	8'11"	8'10$\frac{1}{4}$"	8'11$\frac{1}{4}$"	8'10$\frac{3}{4}$"	8'10$\frac{7}{8}$"	8'11"	8'11$\frac{5}{16}$"	8'11$\frac{1}{4}$"
RISERS OVERAGE (+) OR UNDERAGE (—)	0	+$\frac{1}{8}$"	+$\frac{1}{4}$"	−$\frac{1}{2}$"	+$\frac{3}{8}$"	−$\frac{1}{8}$"	0	+$\frac{1}{8}$"	+$\frac{7}{16}$"	+$\frac{1}{4}$"
NUMBER OF TREADS	13	14	15	16	12	13	14	15	16	12
WIDTH OF EACH TREAD	9$\frac{7}{8}$"	10$\frac{3}{8}$"	10$\frac{13}{16}$"	11$\frac{1}{4}$"	9$\frac{1}{4}$"	9$\frac{7}{8}$"	10$\frac{3}{8}$"	10$\frac{13}{16}$"	11$\frac{3}{16}$"	9$\frac{1}{4}$"
TOTAL RUN OF STAIRWAY	10'8$\frac{3}{8}$"	12'1$\frac{1}{4}$"	13'6$\frac{3}{16}$"	15'	9'3"	10'8$\frac{3}{8}$"	12'1$\frac{1}{4}$"	13'6$\frac{3}{16}$"	14'11"	9'3"
WELL OPENING FOR 6'-8" HEADROOM	10'1$\frac{9}{16}$"	11'7$\frac{1}{16}$"	13'$\frac{3}{4}$"	14'7$\frac{1}{2}$"	8'7$\frac{23}{32}$"	10'1$\frac{9}{16}$"	11'7$\frac{1}{16}$"	13'$\frac{3}{4}$"	13'7$\frac{5}{32}$"	8'7$\frac{23}{32}$"
ANGLE OF INCLINE	37°40'	34°29'	31°44'	29°3'	41°44'	37°40'	34°29'	31°44'	29°26'	41°44'
LENGTH OF CARRIAGE	13'6$\frac{3}{16}$"	14'8$\frac{3}{16}$"	15'10$\frac{11}{16}$"	17'1$\frac{29}{32}$"	12'4$\frac{3}{4}$"	13'6$\frac{3}{16}$"	14'8$\frac{3}{16}$"	15'10$\frac{11}{16}$"	17'1$\frac{17}{32}$"	12'4$\frac{3}{4}$"
L. F. OF RISER PER INCH OF STAIRWAY WIDTH	1.166	1.25	1.333	1.416	1.083	1.166	1.25	1.333	1.416	1.083
L. F. OF TREAD PER INCH OF STAIRWAY WIDTH	1.083	1.166	1.25	1.333	1.0	1.083	1.166	1.25	1.333	1.0
FRAMING SQUARE SETTINGS FOR CARRIAGE CUTS — TONGUE	7$\frac{5}{8}$"	7$\frac{1}{8}$"	6$\frac{11}{16}$"	6$\frac{1}{4}$"	8$\frac{1}{4}$"	7$\frac{5}{8}$"	7$\frac{1}{8}$"	6$\frac{11}{16}$"	6$\frac{5}{16}$"	8$\frac{1}{4}$"
FRAMING SQUARE SETTINGS FOR CARRIAGE CUTS — BODY	9$\frac{7}{8}$"	10$\frac{3}{8}$"	10$\frac{13}{16}$"	11$\frac{1}{4}$"	9$\frac{1}{4}$"	9$\frac{7}{8}$"	10$\frac{3}{8}$"	10$\frac{13}{16}$"	11$\frac{3}{16}$"	9$\frac{1}{4}$"

FLOOR TO FLOOR RISE		$8'11''$	$8'11''$	$8'11''$	$8'11''$	$8'11\frac{1}{8}''$	$8'11\frac{1}{8}''$	$8'11\frac{1}{8}''$	$8'11\frac{1}{8}''$	$8'11\frac{1}{8}''$	$8'11\frac{1}{4}''$
NUMBER OF RISERS		14	15	16	17	13	14	15	16	17	13
HEIGHT OF EACH RISER		$7\frac{5}{8}''$	$7\frac{1}{8}''$	$6\frac{11}{16}''$	$6\frac{5}{16}''$	$8\frac{1}{4}''$	$7\frac{5}{8}''$	$7\frac{1}{8}''$	$6\frac{11}{16}''$	$6\frac{5}{16}''$	$8\frac{1}{4}''$
TOTAL HEIGHT OF RISERS		$8'10\frac{3}{4}''$	$8'10\frac{7}{8}''$	$8'11''$	$8'11\frac{5}{16}''$	$8'11\frac{1}{4}''$	$8'10\frac{3}{4}''$	$8'10\frac{7}{8}''$	$8'11''$	$8'11\frac{5}{16}''$	$8'11\frac{1}{4}''$
RISERS OVERAGE (+) OR UNDERAGE (—)		$-\frac{1}{4}''$	$-\frac{1}{8}''$	0	$+\frac{5}{16}''$	$+\frac{1}{8}''$	$-\frac{3}{8}''$	$-\frac{1}{4}''$	$-\frac{1}{8}''$	$+\frac{3}{16}''$	0
NUMBER OF TREADS		13	14	15	16	12	13	14	15	16	12
WIDTH OF EACH TREAD		$9\frac{7}{8}''$	$10\frac{3}{8}''$	$10\frac{13}{16}''$	$11\frac{3}{16}''$	$9\frac{1}{4}''$	$9\frac{7}{8}''$	$10\frac{3}{8}''$	$10\frac{13}{16}''$	$11\frac{3}{16}''$	$9\frac{1}{4}''$
TOTAL RUN OF STAIRWAY		$10'8\frac{3}{8}''$	$12'1\frac{1}{4}''$	$13'6\frac{3}{16}''$	$14'11''$	$9'3''$	$10'8\frac{3}{8}''$	$12'1\frac{1}{4}''$	$13'6\frac{3}{16}''$	$14'11''$	$9'3''$
WELL OPENING FOR 6'-8" HEADROOM		$10'1\frac{9}{16}''$	$11'7\frac{1}{16}''$	$13'\frac{3}{4}''$	$13'7\frac{5}{32}''$	$8'7\frac{23}{32}''$	$10'1\frac{9}{16}''$	$11'7\frac{1}{16}''$	$13'\frac{3}{4}''$	$13'7\frac{5}{32}''$	$8'7\frac{23}{32}''$
ANGLE OF INCLINE		37°40'	34°29'	31°44'	29°26'	41°44'	37°40'	34°29'	31°44'	29°26'	41°44'
LENGTH OF CARRIAGE		$13'6\frac{3}{16}''$	$14'8\frac{3}{16}''$	$15'10\frac{11}{16}''$	$17'1\frac{17}{32}''$	$12'4\frac{3}{4}''$	$13'6\frac{3}{16}''$	$14'8\frac{3}{16}''$	$15'10\frac{11}{16}''$	$17'1\frac{17}{32}''$	$12'4\frac{3}{4}''$
L. F. OF RISER PER INCH OF STAIRWAY WIDTH		1.166	1.25	1.333	1.416	1.083	1.166	1.25	1.333	1.416	1.083
L. F. OF TREAD PER INCH OF STAIRWAY WIDTH		1.083	1.166	1.25	1.333	1.0	1.083	1.166	1.25	1.333	1.0
FRAMING SQUARE SETTINGS FOR CARRIAGE CUTS	TONGUE	$7\frac{5}{8}''$	$7\frac{1}{8}''$	$6\frac{11}{16}''$	$6\frac{5}{16}''$	$8\frac{1}{4}''$	$7\frac{5}{8}''$	$7\frac{1}{8}''$	$6\frac{11}{16}''$	$6\frac{5}{16}''$	$8\frac{1}{4}''$
	BODY	$9\frac{7}{8}''$	$10\frac{3}{8}''$	$10\frac{13}{16}''$	$11\frac{3}{16}''$	$9\frac{1}{4}''$	$9\frac{7}{8}''$	$10\frac{3}{8}''$	$10\frac{13}{16}''$	$11\frac{3}{16}''$	$9\frac{1}{4}''$

FLOOR TO FLOOR RISE	8'11 1/4"	8'11 1/4"	8'11 1/4"	8'11 1/4"	8'11 3/8"	8'11 3/8"	8'11 3/8"	8'11 3/8"	8'11 3/8"	8'11 1/2"
NUMBER OF RISERS	14	15	16	17	13	14	15	16	17	13
HEIGHT OF EACH RISER	7 11/16"	7 1/8"	6 11/16"	6 5/16"	8 1/4"	7 11/16"	7 3/16"	6 11/16"	6 5/16"	8 1/4"
TOTAL HEIGHT OF RISERS	8'11 5/8"	8'10 7/8"	8'11"	8'11 5/16"	8'11 1/4"	8'11 5/8"	8'11 13/16"	8'11"	8'11 5/16"	8'11 1/4"
RISERS OVERAGE (+) OR UNDERAGE (—)	+3/8"	−3/8"	−1/4"	+1/16"	−1/8"	+1/4"	+7/16"	−3/8"	−1/16"	−1/4"
NUMBER OF TREADS	13	14	15	16	12	13	14	15	16	12
WIDTH OF EACH TREAD	9 13/16"	10 3/8"	10 13/16"	11 3/16"	9 1/4"	9 13/16"	10 5/16"	10 13/16"	11 3/16"	9 1/4"
TOTAL RUN OF STAIRWAY	10'7 9/16"	12'1 1/4"	13'6 3/16"	14'11"	9'3"	10'7 9/16"	12' 3/8"	13'6 3/16"	14'11"	9'3"
WELL OPENING FOR 6'-8" HEADROOM	10' 23/32"	11'7 1/16"	13' 3/4"	13'7 5/32"	8'7 23/32"	10' 23/32"	11'6 3/32"	13' 3/4"	13'7 5/32"	8'7 23/32"
ANGLE OF INCLINE	38°5'	34°29'	31°44'	29°26'	41°44'	38°5'	34°53'	31°44'	29°26'	41°44'
LENGTH OF CARRIAGE	13'6 1/32"	14'8 3/16"	15'10 11/16"	17'1 17/32"	12'4 3/4"	13'6 1/32"	14'7 31/32"	15'10 11/16"	17'1 17/32"	12'4 3/4"
L. F. OF RISER PER INCH OF STAIRWAY WIDTH	1.166	1.25	1.333	1.416	1.083	1.166	1.25	1.333	1.416	1.083
L. F. OF TREAD PER INCH OF STAIRWAY WIDTH	1.083	1.166	1.25	1.333	1.0	1.083	1.166	1.25	1.333	1.0
FRAMING SQUARE SETTINGS FOR CARRIAGE CUTS — TONGUE	7 11/16"	7 1/8"	6 11/16"	6 5/16"	8 1/4"	7 11/16"	7 3/16"	6 11/16"	6 5/16"	8 1/4"
BODY	9 13/16"	10 3/8"	10 13/16"	11 3/16"	9 1/4"	9 13/16"	10 5/16"	10 13/16"	11 3/16"	9 1/4"

FLOOR TO FLOOR RISE	8'11 1/2"	8'11 1/2"	8'11 1/2"	8'11 1/2"	8'11 5/8"	8'11 5/8"	8'11 5/8"	8'11 5/8"	8'11 5/8"	8'11 3/4"
NUMBER OF RISERS	14	15	16	17	13	14	15	16	17	14
HEIGHT OF EACH RISER	7 11/16"	7 3/16"	6 3/4"	6 5/16"	8 1/4"	7 11/16"	7 3/16"	6 3/4"	6 5/16"	7 11/16"
TOTAL HEIGHT OF RISERS	8'11 5/8"	8'11 13/16"	9'	8'11 5/16"	8'11 1/4"	8'11 5/8"	8'11 13/16"	9'	8'11 5/16"	8'11 5/8"
RISERS OVERAGE (+) OR UNDERAGE (—)	+1/8"	+5/16"	+1/2"	−3/16"	−3/8"	0	+3/16"	+3/8"	−5/16"	−1/8"
NUMBER OF TREADS	13	14	15	16	12	13	14	15	16	13
WIDTH OF EACH TREAD	9 13/16"	10 5/16"	10 3/4"	11 3/16"	9 1/4"	9 13/16"	10 5/16"	10 3/4"	11 3/16"	9 13/16"
TOTAL RUN OF STAIRWAY	10'7 9/16"	12' 3/8"	13'5 1/4"	14'11"	9'3"	10'7 9/16"	12' 3/8"	13'5 1/4"	14'11"	10'7 9/16"
WELL OPENING FOR 6'-8" HEADROOM	10' 23/32"	11'6 3/32"	12'11 11/16"	13'7 5/32"	8'7 23/32"	10' 23/32"	11'6 3/32"	12'11 11/16"	13'7 5/32"	10' 23/32"
ANGLE OF INCLINE	38°5'	34°53'	32°7'	29°26'	41°44'	38°5'	34°53'	32°7'	29°26'	38°5'
LENGTH OF CARRIAGE	13'6 1/32"	14'7 31/32"	15'10 13/32"	17'1 17/32"	12'4 3/4"	13'6 1/32"	14'7 31/32"	15'10 13/32"	17'1 17/32"	13'6 1/32"
L. F. OF RISER PER INCH OF STAIRWAY WIDTH	1.166	1.25	1.333	1.416	1.083	1.166	1.25	1.333	1.416	1.166
L. F. OF TREAD PER INCH OF STAIRWAY WIDTH	1.083	1.166	1.25	1.333	1.0	1.083	1.166	1.25	1.333	1.083
FRAMING SQUARE SETTINGS FOR CARRIAGE CUTS — TONGUE	7 11/16"	7 3/16"	6 3/4"	6 5/16"	8 1/4"	7 11/16"	7 3/16"	6 3/4"	6 5/16"	7 11/16"
FRAMING SQUARE SETTINGS FOR CARRIAGE CUTS — BODY	9 13/16"	10 5/16"	10 3/4"	11 3/16"	9 1/4"	9 13/16"	10 5/16"	10 3/4"	11 3/16"	9 13/16"

FLOOR TO FLOOR RISE	8'11 3/4"	8'11 3/4"	8'11 3/4"	8'11 7/8"	8'11 7/8"	8'11 7/8"	8'11 7/8"	9'	9'	9'
NUMBER OF RISERS	15	16	17	14	15	16	17	14	15	16
HEIGHT OF EACH RISER	7 3/16"	6 3/4"	6 5/16"	7 11/16"	7 3/16"	6 3/4"	6 3/8"	7 11/16"	7 3/16"	6 3/4"
TOTAL HEIGHT OF RISERS	8'11 13/16"	9'	8'11 5/16"	8'11 5/8"	8'11 13/16"	9'	9' 3/8"	8'11 5/8"	8'11 13/16"	9'
RISERS OVERAGE (+) OR UNDERAGE (—)	+1/16"	+1/4"	−7/16"	−1/4"	−1/16"	+1/8"	+1/2"	−3/8"	−3/16"	0
NUMBER OF TREADS	14	15	16	13	14	15	16	13	14	15
WIDTH OF EACH TREAD	10 5/16"	10 3/4"	11 3/16"	9 13/16"	10 5/16"	10 3/4"	11 1/8"	9 13/16"	10 5/16"	10 3/4"
TOTAL RUN OF STAIRWAY	12' 3/8"	13'5 1/4"	14'11"	10'7 9/16"	12' 3/8"	13'5 1/4"	14'10"	10'7 9/16"	12' 3/8"	13'5 1/4"
WELL OPENING FOR 6'-8" HEADROOM	11'6 3/32"	12'11 11/16"	13'7 5/32"	10' 23/32"	11'6 3/32"	12'11 11/16"	13'6 1/16"	10' 23/32"	11'6 3/32"	12'11 11/16"
ANGLE OF INCLINE	34°53'	32°7'	29°26'	38°5'	34°53'	32°7'	29°49'	38°5'	34°53'	32°7'
LENGTH OF CARRIAGE	14'7 31/32"	15'10 13/32"	17'1 17/32"	13'6 1/32"	14'7 31/32"	15'10 13/32"	17'1 5/32"	13'6 1/32"	14'7 31/32"	15'10 13/32"
L. F. OF RISER PER INCH OF STAIRWAY WIDTH	1.25	1.333	1.416	1.166	1.25	1.333	1.416	1.166	1.25	1.333
L. F. OF TREAD PER INCH OF STAIRWAY WIDTH	1.166	1.25	1.333	1.083	1.166	1.25	1.333	1.083	1.166	1.25
FRAMING SQUARE SETTINGS FOR CARRIAGE CUTS — TONGUE	7 3/16"	6 3/4"	6 5/16"	7 11/16"	7 3/16"	6 3/4"	6 3/8"	7 11/16"	7 3/16"	6 3/4"
FRAMING SQUARE SETTINGS FOR CARRIAGE CUTS — BODY	10 5/16"	10 3/4"	11 3/16"	9 13/16"	10 5/16"	10 3/4"	11 1/8"	9 13/16"	10 5/16"	10 3/4"

FLOOR TO FLOOR RISE	9'	9'	9' 1/8"	9' 1/8"	9' 1/8"	9' 1/8"	9' 1/8"	9' 1/4"	9' 1/4"	9' 1/4"
NUMBER OF RISERS	17	18	14	15	16	17	18	14	15	16
HEIGHT OF EACH RISER	6 3/8"	6"	7 3/4"	7 1/4"	6 3/4"	6 3/8"	6"	7 3/4"	7 1/4"	6 3/4"
TOTAL HEIGHT OF RISERS	9' 3/8"	9'	9' 1/2"	8' 11 13/16"	9'	9' 3/8"	9'	9' 1/2"	8' 11 13/16"	9'
RISERS OVERAGE (+) OR UNDERAGE (—)	+3/8"	0	+3/8"	−5/16"	−1/8"	+1/4"	−1/8"	+1/4"	−7/16"	−1/4"
NUMBER OF TREADS	16	17	13	14	15	16	17	13	14	15
WIDTH OF EACH TREAD	11 1/8"	11 1/2"	9 3/4"	10 1/4"	10 3/4"	11 1/8"	11 1/2"	9 3/4"	10 1/4"	10 3/4"
TOTAL RUN OF STAIRWAY	14' 10"	16' 3 1/2"	10' 6 3/4"	11' 11 1/2"	13' 5 1/4"	14' 10"	16' 3 1/2"	10' 6 3/4"	11' 11 1/2"	13' 5 1/4"
WELL OPENING FOR 6'-8" HEADROOM	13' 6 1/16"	15' 5/32"	9' 11 27/32"	11' 5 1/8"	12' 11 11/16"	13' 6 1/16"	15' 5/32"	9' 11 27/32"	11' 5 1/8"	12' 11 11/16"
ANGLE OF INCLINE	29°49'	27°33'	38°29'	35°16'	32°7'	29°49'	27°33'	38°29'	35°16'	32°7'
LENGTH OF CARRIAGE	17' 1 5/32"	18' 4 1/2"	13' 5 29/32"	14' 7 25/32"	15' 10 13/32"	17' 1 5/32"	18' 4 1/2"	13' 5 29/32"	14' 7 25/32"	15' 10 13/32"
L. F. OF RISER PER INCH OF STAIRWAY WIDTH	1.416	1.5	1.166	1.25	1.333	1.416	1.5	1.166	1.25	1.333
L. F. OF TREAD PER INCH OF STAIRWAY WIDTH	1.333	1.416	1.083	1.166	1.25	1.333	1.416	1.083	1.166	1.25
FRAMING SQUARE SETTINGS FOR CARRIAGE CUTS — TONGUE	6 3/8"	6"	7 3/4"	7 1/4"	6 3/4"	6 3/8"	6"	7 3/4"	7 1/4"	6 3/4"
FRAMING SQUARE SETTINGS FOR CARRIAGE CUTS — BODY	11 1/8"	11 1/2"	9 3/4"	10 1/4"	10 3/4"	11 1/8"	11 1/2"	9 3/4"	10 1/4"	10 3/4"

FLOOR TO FLOOR RISE		9' 1/4"	9' 1/4"	9' 3/8"	9' 3/8"	9' 3/8"	9' 3/8"	9' 3/8"	9' 1/2"	9' 1/2"	9' 1/2"
NUMBER OF RISERS		17	18	14	15	16	17	18	14	15	16
HEIGHT OF EACH RISER		6 3/8"	6"	7 3/4"	7 1/4"	6 3/4"	6 3/8"	6"	7 3/4"	7 1/4"	6 13/16"
TOTAL HEIGHT OF RISERS		9' 3/8"	9'	9' 1/2"	9' 3/4"	9'	9' 3/8"	9'	9' 1/2"	9' 3/4"	9' 1"
RISERS OVERAGE (+) OR UNDERAGE (—)		+ 1/8"	− 1/4"	+ 1/8"	+ 3/8"	− 3/8"	0	− 3/8"	0	+ 1/4"	+ 1/2"
NUMBER OF TREADS		16	17	13	14	15	16	17	13	14	15
WIDTH OF EACH TREAD		11 1/8"	11 1/2"	9 3/4"	10 1/4"	10 3/4"	11 1/8"	11 1/2"	9 3/4"	10 1/4"	10 11/16"
TOTAL RUN OF STAIRWAY		14' 10"	16' 3 1/2"	10' 6 3/4"	11' 11 1/2"	13' 5 1/4"	14' 10"	16' 3 1/2"	10' 6 3/4"	11' 11 1/2"	13' 4 5/16"
WELL OPENING FOR 6'-8" HEADROOM		13' 6 1/16"	15' 5/32"	9' 11 27/32"	11' 5 1/8"	12' 11 11/16"	13' 6 1/16"	15' 5/32"	9' 11 27/32"	11' 5 1/8"	11' 11 15/16"
ANGLE OF INCLINE		29°49'	27°33'	38°29'	35°16'	32°7'	29°49'	27°33'	38°29'	35°16'	32°31'
LENGTH OF CARRIAGE		17' 1 5/32"	18' 4 1/2"	13' 5 29/32"	14' 7 25/32"	15' 10 13/32"	17' 1 5/32"	18' 4 1/2"	13' 5 29/32"	14' 7 25/32"	15' 10 1/8"
L. F. OF RISER PER INCH OF STAIRWAY WIDTH		1.416	1.5	1.166	1.25	1.333	1.416	1.5	1.166	1.25	1.333
L. F. OF TREAD PER INCH OF STAIRWAY WIDTH		1.333	1.416	1.083	1.166	1.25	1.333	1.416	1.083	1.166	1.25
FRAMING SQUARE SETTINGS FOR CARRIAGE CUTS	TONGUE	6 3/8"	6"	7 3/4"	7 1/4"	6 3/4"	6 3/8"	6"	7 3/4"	7 1/4"	6 13/16"
	BODY	11 1/8"	11 1/2"	9 3/4"	10 1/4"	10 3/4"	11 1/8"	11 1/2"	9 3/4"	10 1/4"	10 11/16"

FLOOR TO FLOOR RISE	9' 1/2"	9' 1/2"	9' 5/8"	9' 5/8"	9' 5/8"	9' 5/8"	9' 5/8"	9' 3/4"	9' 3/4"	9' 3/4"
NUMBER OF RISERS	17	18	14	15	16	17	18	14	15	16
HEIGHT OF EACH RISER	6 3/8"	6"	7 3/4"	7 1/4"	6 13/16"	6 3/8"	6 1/16"	7 3/4"	7 1/4"	6 13/16"
TOTAL HEIGHT OF RISERS	9' 3/8"	9'	9' 1/2"	9' 3/4"	9' 1"	9' 3/8"	9' 1 1/8"	9' 1/2"	9' 3/4"	9' 1"
RISERS OVERAGE (+) OR UNDERAGE (—)	−1/8"	−1/2"	−1/8"	+1/8"	+3/8"	−1/4"	+1/2"	−1/4"	0	+1/4"
NUMBER OF TREADS	16	17	13	14	15	16	17	13	14	15
WIDTH OF EACH TREAD	11 1/8"	11 1/2"	9 3/4"	10 1/4"	10 11/16"	11 1/8"	11 7/16"	9 3/4"	10 1/4"	10 11/16"
TOTAL RUN OF STAIRWAY	14' 10"	16' 3 1/2"	10' 6 3/4"	11' 11 1/2"	13' 4 5/16"	14' 10"	16' 2 7/16"	10' 6 3/4"	11' 11 1/2"	13' 4 5/16"
WELL OPENING FOR 6'-8" HEADROOM	13' 6 1/16"	15' 5/32"	9' 11 27/32"	11' 5 1/8"	11' 11 15/16"	13' 6 1/16"	14' 11"	9' 11 27/32"	11' 5 1/8"	11' 11 15/16"
ANGLE OF INCLINE	29°49'	27°33'	38°29'	35°16'	32°31'	29°49'	27°56'	38°29'	35°16'	32°31'
LENGTH OF CARRIAGE	17' 1 5/32"	18' 4 1/2"	13' 5 29/32"	14' 7 25/32"	15' 10 1/8"	17' 1 5/32"	18' 4 1/16"	13' 5 29/32"	14' 7 25/32"	15' 10 1/8"
L. F. OF RISER PER INCH OF STAIRWAY WIDTH	1.416	1.5	1.166	1.25	1.333	1.416	1.5	1.166	1.25	1.333
L. F. OF TREAD PER INCH OF STAIRWAY WIDTH	1.333	1.416	1.083	1.166	1.25	1.333	1.416	1.083	1.166	1.25
FRAMING SQUARE SETTINGS FOR CARRIAGE CUTS — TONGUE	6 3/8"	6"	7 3/4"	7 1/4"	6 13/16"	6 3/8"	6 1/16"	7 3/4"	7 1/4"	6 13/16"
FRAMING SQUARE SETTINGS FOR CARRIAGE CUTS — BODY	11 1/8"	11 1/2"	9 3/4"	10 1/4"	10 11/16"	11 1/8"	11 7/16"	9 3/4"	10 1/4"	10 11/16"

FLOOR TO FLOOR RISE		9'$\frac{3}{4}$"	9'$\frac{3}{4}$"	9'$\frac{7}{8}$"	9'$\frac{7}{8}$"	9'$\frac{7}{8}$"	9'$\frac{7}{8}$"	9'$\frac{7}{8}$"	9'1"	9'1"	9'1"
NUMBER OF RISERS		17	18	14	15	16	17	18	14	15	16
HEIGHT OF EACH RISER		6$\frac{3}{8}$"	6$\frac{1}{16}$"	7$\frac{3}{4}$"	7$\frac{1}{4}$"	6$\frac{13}{16}$"	6$\frac{3}{8}$"	6$\frac{1}{16}$"	7$\frac{13}{16}$"	7$\frac{1}{4}$"	6$\frac{13}{16}$"
TOTAL HEIGHT OF RISERS		9'$\frac{3}{8}$"	9'1$\frac{1}{8}$"	9'$\frac{1}{2}$"	9'$\frac{3}{4}$"	9'1"	9'$\frac{3}{8}$"	9'1$\frac{1}{8}$"	9'1$\frac{3}{8}$"	9'$\frac{3}{4}$"	9'1"
RISERS OVERAGE (+) OR UNDERAGE (—)		$-\frac{3}{8}$"	$+\frac{3}{8}$"	$-\frac{3}{8}$"	$-\frac{1}{8}$"	$+\frac{1}{8}$"	$-\frac{1}{2}$"	$+\frac{1}{4}$"	$+\frac{3}{8}$"	$-\frac{1}{4}$"	0
NUMBER OF TREADS		16	17	13	14	15	16	17	13	14	15
WIDTH OF EACH TREAD		11$\frac{1}{8}$"	11$\frac{7}{16}$"	9$\frac{3}{4}$"	10$\frac{1}{4}$"	10$\frac{11}{16}$"	11$\frac{1}{8}$"	11$\frac{7}{16}$"	9$\frac{11}{16}$"	10$\frac{1}{4}$"	10$\frac{11}{16}$"
TOTAL RUN OF STAIRWAY		14'10"	16'2$\frac{7}{16}$"	10'6$\frac{3}{4}$"	11'11$\frac{1}{2}$"	13'4$\frac{5}{16}$"	14'10"	16'2$\frac{7}{16}$"	10'5$\frac{15}{16}$"	11'11$\frac{1}{2}$"	13'4$\frac{5}{16}$"
WELL OPENING FOR 6'-8" HEADROOM		13'6$\frac{1}{16}$"	14'11"	9'11$\frac{27}{32}$"	11'5$\frac{1}{8}$"	11'11$\frac{15}{16}$"	13'6$\frac{1}{16}$"	14'11"	9'10$\frac{15}{16}$"	11'5$\frac{1}{8}$"	11'11$\frac{15}{16}$"
ANGLE OF INCLINE		29°49'	27°56'	38°29'	35°16'	32°31'	29°49'	27°56'	38°53'	35°16'	32°31'
LENGTH OF CARRIAGE		17'1$\frac{5}{32}$"	18'4$\frac{1}{16}$"	13'5$\frac{29}{32}$"	14'7$\frac{25}{32}$"	15'10$\frac{1}{8}$"	17'1$\frac{5}{32}$"	18'4$\frac{1}{16}$"	13'5$\frac{25}{32}$"	14'7$\frac{25}{32}$"	15'10$\frac{1}{8}$"
L. F. OF RISER PER INCH OF STAIRWAY WIDTH		1.416	1.5	1.166	1.25	1.333	1.416	1.5	1.166	1.25	1.333
L. F. OF TREAD PER INCH OF STAIRWAY WIDTH		1.333	1.416	1.082	1.166	1.25	1.333	1.416	1.083	1.166	1.25
FRAMING SQUARE SETTINGS FOR CARRIAGE CUTS	TONGUE	6$\frac{3}{8}$"	6$\frac{1}{16}$"	7$\frac{3}{4}$"	7$\frac{1}{4}$"	6$\frac{13}{16}$"	6$\frac{3}{8}$"	6$\frac{1}{16}$"	7$\frac{13}{16}$"	7$\frac{1}{4}$"	6$\frac{13}{16}$"
	BODY	11$\frac{1}{8}$"	11$\frac{7}{16}$"	9$\frac{3}{4}$"	10$\frac{1}{4}$"	10$\frac{11}{16}$"	11$\frac{1}{8}$"	11$\frac{7}{16}$"	9$\frac{11}{16}$"	10$\frac{1}{4}$"	10$\frac{11}{16}$"

FLOOR TO FLOOR RISE	9'1"	9'1"	9'1 1/8"	9'1 1/8"	9'1 1/8"	9'1 1/8"	9'1 1/8"	9'1 1/4"	9'1 1/4"	9'1 1/4"
NUMBER OF RISERS	17	18	14	15	16	17	18	14	15	16
HEIGHT OF EACH RISER	6 7/16"	6 1/16"	7 13/16"	7 1/4"	6 13/16"	6 7/16"	6 1/16"	7 13/16"	7 5/16"	6 13/16"
TOTAL HEIGHT OF RISERS	9'1 7/16"	9'1 1/8"	9'1 3/8"	9' 3/4"	9'1"	9'1 7/16"	9'1 1/8"	9'1 3/8"	9'1 11/16"	9'1"
RISERS OVERAGE (+) OR UNDERAGE (—)	+7/16"	+1/8"	+1/4"	−3/8"	−1/8"	+5/16"	0	+1/8"	+7/16"	−1/4"
NUMBER OF TREADS	16	17	13	14	15	16	17	13	14	15
WIDTH OF EACH TREAD	11 1/16"	11 7/16"	9 11/16"	10 1/4"	10 11/16"	11 1/16"	11 7/16"	9 11/16"	10 3/16"	10 11/16"
TOTAL RUN OF STAIRWAY	14'9"	16'2 7/16"	10'5 15/16"	11'11 1/2"	13'4 5/16"	14'9"	16'2 7/16"	10'5 15/16"	11'10 5/8"	13'4 5/16"
WELL OPENING FOR 6'-8" HEADROOM	13'5"	14'11"	9'10 15/16"	11'5 1/8"	11'11 15/16"	13'5"	14'11"	9'10 15/16"	11'4 3/16"	11'11 15/16"
ANGLE OF INCLINE	30°12'	27°56'	38°53'	35°16'	32°31'	30°12'	27°56'	38°53'	35°40'	32°31'
LENGTH OF CARRIAGE	17' 25/32"	18'4 1/16"	13'5 25/32"	14'7 25/32"	15'10 1/8"	17' 25/32"	18'4 1/16"	13'5 25/32"	14'7 9/16"	15'10 1/8"
L. F. OF RISER PER INCH OF STAIRWAY WIDTH	1.416	1.5	1.166	1.25	1.333	1.416	1.5	1.166	1.25	1.333
L. F. OF TREAD PER INCH OF STAIRWAY WIDTH	1.333	1.416	1.083	1.166	1.25	1.333	1.416	1.083	1.166	1.25
FRAMING SQUARE SETTINGS FOR CARRIAGE CUTS — TONGUE	6 7/16"	6 1/16"	7 13/16"	7 1/4"	6 13/16"	6 7/16"	6 1/16"	7 13/16"	7 5/16"	6 13/16"
FRAMING SQUARE SETTINGS FOR CARRIAGE CUTS — BODY	11 1/16"	11 7/16"	9 11/16"	10 1/4"	10 11/16"	11 1/16"	11 7/16"	9 11/16"	10 3/16"	10 11/16"

FLOOR TO FLOOR RISE	9'1 1/4"	9'1 1/4"	9'1 3/8"	9'1 3/8"	9'1 3/8"	9'1 3/8"	9'1 3/8"	9'1 1/2"	9'1 1/2"	9'1 1/2"
NUMBER OF RISERS	17	18	14	15	16	17	18	14	15	16
HEIGHT OF EACH RISER	6 7/16"	6 1/16"	7 13/16"	7 5/16"	6 13/16"	6 7/16"	6 1/16"	7 13/16"	7 5/16"	6 7/8"
TOTAL HEIGHT OF RISERS	9'1 7/16"	9'1 1/8"	9'1 3/8"	9'1 11/16"	9'1"	9'1 7/16"	9'1 1/8"	9'1 3/8"	9'1 11/16"	9'2"
RISERS OVERAGE (+) OR UNDERAGE (—)	+3/16"	−1/8"	0	+5/16"	−3/8"	+1/16"	−1/4"	−1/8"	+3/16"	+1/2"
NUMBER OF TREADS	16	17	13	14	15	16	17	13	14	15
WIDTH OF EACH TREAD	11 1/16"	11 7/16"	9 11/16"	10 3/16"	10 11/16"	11 1/16"	11 7/16"	9 11/16"	10 3/16"	10 5/8"
TOTAL RUN OF STAIRWAY	14'9"	16'2 7/16"	10'5 15/16"	11'10 5/8"	13'4 5/16"	14'9"	16'2 7/16"	10'5 15/16"	11'10 5/8"	13'3 3/8"
WELL OPENING FOR 6'-8" HEADROOM	13'5"	14'11"	9'10 15/16"	11'4 3/16"	11'11 15/16"	13'5"	14'11"	9'10 15/16"	11'4 3/16"	11'10 31/32"
ANGLE OF INCLINE	30°12'	27°56'	38°53'	35°40'	32°31'	30°12'	27°56'	38°53'	35°40'	32°54'
LENGTH OF CARRIAGE	17' 25/32"	18'4 1/16"	13'5 25/32"	14'7 9/16"	15'10 1/8"	17' 25/32"	18'4 1/16"	13'5 25/32"	14'7 9/16"	15'9 27/32"
L. F. OF RISER PER INCH OF STAIRWAY WIDTH	1.416	1.5	1.166	1.25	1.333	1.416	1.5	1.166	1.25	1.333
L. F. OF TREAD PER INCH OF STAIRWAY WIDTH	1.333	1.416	1.083	1.166	1.25	1.333	1.416	1.083	1.166	1.25
FRAMING SQUARE SETTINGS FOR CARRIAGE CUTS — TONGUE	6 7/16"	6 1/16"	7 13/16"	7 5/16"	6 13/16"	6 7/16"	6 1/16"	7 13/16"	7 5/16"	6 7/8"
FRAMING SQUARE SETTINGS FOR CARRIAGE CUTS — BODY	11 1/16"	11 7/16"	9 11/16"	10 3/16"	10 11/16"	11 1/16"	11 7/16"	9 11/16"	10 3/16"	10 5/8"

FLOOR TO FLOOR RISE	9'1 1/2"	9'1 1/2"	9'1 5/8"	9'1 5/8"	9'1 5/8"	9'1 5/8"	9'1 5/8"	9'1 3/4"	9'1 3/4"	9'1 3/4"
NUMBER OF RISERS	17	18	14	15	16	17	18	14	15	16
HEIGHT OF EACH RISER	6 7/16"	6 1/16"	7 13/16"	7 5/16"	6 7/8"	6 7/16"	6 1/16"	7 13/16"	7 5/16"	6 7/8"
TOTAL HEIGHT OF RISERS	9'1 7/16"	9'1 1/8"	9'1 3/8"	9'1 11/16"	9'2"	9'1 7/16"	9'1 1/8"	9'1 3/8"	9'1 11/16"	9'2"
RISERS OVERAGE (+) OR UNDERAGE (—)	−1/16"	−3/8"	−1/4"	+1/16"	+3/8"	−3/16"	−1/2"	−3/8"	−1/16"	+1/4"
NUMBER OF TREADS	16	17	13	14	15	16	17	13	14	15
WIDTH OF EACH TREAD	11 1/16"	11 7/16"	9 11/16"	10 3/16"	10 5/8"	11 1/16"	11 7/16"	9 11/16"	10 3/16"	10 5/8"
TOTAL RUN OF STAIRWAY	14'9"	16'2 7/16"	10'5 15/16"	11'10 5/8"	13'3 3/8"	14'9"	16'2 7/16"	10'5 15/16"	11'10 5/8"	13'3 3/8"
WELL OPENING FOR 6'-8" HEADROOM	13'5"	14'11"	9'10 15/16"	11'4 3/16"	11'10 31/32"	13'5"	14'11"	9'10 15/16"	11'4 3/16"	11'10 31/32"
ANGLE OF INCLINE	30°12'	27°56'	38°53'	35°40'	32°54'	30°12'	27°56'	38°53'	35°40'	32°54'
LENGTH OF CARRIAGE	17' 25/32"	18'4 1/16"	13'5 25/32"	14'7 9/16"	15'9 27/32"	17' 25/32"	18'4 1/16"	13'5 25/32"	14'7 9/16"	15'9 27/32"
L. F. OF RISER PER INCH OF STAIRWAY WIDTH	1.416	1.5	1.166	1.25	1.333	1.416	1.5	1.166	1.25	1.333
L. F. OF TREAD PER INCH OF STAIRWAY WIDTH	1.333	1.416	1.083	1.166	1.25	1.333	1.416	1.083	1.166	1.25
FRAMING SQUARE SETTINGS FOR CARRIAGE CUTS — TONGUE	6 7/16"	6 1/16"	7 13/16"	7 5/16"	6 7/8"	6 7/16"	6 1/16"	7 13/16"	7 5/16"	6 7/8"
FRAMING SQUARE SETTINGS FOR CARRIAGE CUTS — BODY	11 1/16"	11 7/16"	9 11/16"	10 3/16"	10 5/8"	11 1/16"	11 7/16"	9 11/16"	10 3/16"	10 5/8"

FLOOR TO FLOOR RISE		9'1 3/4"	9'1 3/4"	9'1 7/8"	9'1 7/8"	9'1 7/8"	9'1 7/8"	9'1 7/8"	9'2"	9'2"	9'2"
NUMBER OF RISERS		17	18	14	15	16	17	18	14	15	16
HEIGHT OF EACH RISER		6 7/16"	6 1/8"	7 7/8"	7 5/16"	6 7/8"	6 7/16"	6 1/8"	7 7/8"	7 5/16"	6 7/8"
TOTAL HEIGHT OF RISERS		9'1 7/16"	9'2 1/4"	9'2 1/4"	9'1 11/16"	9'2"	9'1 7/16"	9'2 1/4"	9'2 1/4"	9'1 11/16"	9'2"
RISERS OVERAGE (+) OR UNDERAGE (—)		−5/16"	+1/2"	+3/8"	−3/16"	+1/8"	−7/16"	+3/8"	+1/4"	−5/16"	0
NUMBER OF TREADS		16	17	13	14	15	16	17	13	14	15
WIDTH OF EACH TREAD		11 1/16"	11 3/8"	9 5/8"	10 3/16"	10 5/8"	11 1/16"	11 3/8"	9 5/8"	10 3/16"	10 5/8"
TOTAL RUN OF STAIRWAY		14'9"	16'1 3/8"	10'5 1/8"	11'10 5/8"	13'3 3/8"	14'9"	16'1 3/8"	10'5 1/8"	11'10 5/8"	13'3 3/8"
WELL OPENING FOR 6'-8" HEADROOM		13'5"	14'9 13/16"	9'10 3/32"	11'4 3/16"	11'10 31/32"	13'5"	14'9 13/16"	9'10 3/32"	11'4 3/16"	11'10 31/32"
ANGLE OF INCLINE		30°12'	28°18'	39°17'	35°40'	32°54'	30°12'	28°18'	39°17'	35°40'	32°54'
LENGTH OF CARRIAGE		17' 25/32"	18'3 5/8"	13'5 21/32"	14'7 9/16"	15'9 27/32"	17' 25/32"	18'3 5/8"	13'5 21/32"	14'7 9/16"	15'9 27/32"
L. F. OF RISER PER INCH OF STAIRWAY WIDTH		1.416	1.5	1.166	1.25	1.333	1.416	1.5	1.166	1.25	1.333
L. F. OF TREAD PER INCH OF STAIRWAY WIDTH		1.333	1.416	1.083	1.166	1.25	1.333	1.416	1.083	1.166	1.25
FRAMING SQUARE SETTINGS FOR CARRIAGE CUTS	TONGUE	6 7/16"	6 1/8"	7 7/8"	7 5/16"	6 7/8"	6 7/16"	6 1/8"	7 7/8"	7 5/16"	6 7/8"
	BODY	11 1/16"	11 3/8"	9 5/8"	10 3/16"	10 5/8"	11 1/16"	11 3/8"	9 5/8"	10 3/16"	10 5/8"

FLOOR TO FLOOR RISE		9'2"	9'2"	9'2 1/8"	9'2 1/8"	9'2 1/8"	9'2 1/8"	9'2 1/8"	9'2 1/4"	9'2 1/4"	9'2 1/4"
NUMBER OF RISERS		17	18	14	15	16	17	18	14	15	16
HEIGHT OF EACH RISER		6 1/2"	6 1/8"	7 7/8"	7 5/16"	6 7/8"	6 1/2"	6 1/8"	7 7/8"	7 3/8"	6 7/8"
TOTAL HEIGHT OF RISERS		9'2 1/2"	9'2 1/4"	9'2 1/4"	9'1 11/16"	9'2"	9'2 1/2"	9'2 1/4"	9'2 1/4"	9'2 5/8"	9'2"
RISERS OVERAGE (+) OR UNDERAGE (—)		+1/2"	+1/4"	+1/8"	−7/16"	−1/8"	+3/8"	+1/8"	0	+3/8"	−1/4"
NUMBER OF TREADS		16	17	13	14	15	16	17	13	14	15
WIDTH OF EACH TREAD		11"	11 3/8"	9 5/8"	10 3/16"	10 5/8"	11"	11 3/8"	9 5/8"	10 1/8"	10 5/8"
TOTAL RUN OF STAIRWAY		14'8"	16'1 3/8"	10'5 1/8"	11'10 5/8"	13'3 3/8"	14'8"	16'1 3/8"	10'5 1/8"	11'9 3/4"	13'3 3/8"
WELL OPENING FOR 6'-8" HEADROOM		13'3 15/16"	14'9 13/16"	9'10 3/32"	11'4 3/16"	11'10 31/32"	13'3 15/16"	14'9 13/16"	9'10 3/32"	10'5 3/32"	11'10 31/32"
ANGLE OF INCLINE		30°35'	28°18'	39°17'	35°40'	32°54'	30°35'	28°18'	39°17'	36°4'	32°54'
LENGTH OF CARRIAGE		17' 7/16"	18'3 5/8"	13'5 21/32"	14'7 9/16"	15'9 27/32"	17' 7/16"	18'3 5/8"	13'5 21/32"	14'7 3/8"	15'9 27/32"
L. F. OF RISER PER INCH OF STAIRWAY WIDTH		1.416	1.5	1.166	1.25	1.333	1.416	1.5	1.166	1.25	1.333
L. F. OF TREAD PER INCH OF STAIRWAY WIDTH		1.333	1.416	1.083	1.166	1.25	1.333	1.416	1.083	1.166	1.25
FRAMING SQUARE SETTINGS FOR CARRIAGE CUTS	TONGUE	6 1/2"	6 1/8"	7 7/8"	7 5/16"	6 7/8"	6 1/2"	6 1/8"	7 7/8"	7 3/8"	6 7/8"
	BODY	11"	11 3/8"	9 5/8"	10 3/16"	10 5/8"	11"	11 3/8"	9 5/8"	10 1/8"	10 5/8"

FLOOR TO FLOOR RISE	9'2 1/4"	9'2 1/4"	9'2 3/8"	9'2 3/8"	9'2 3/8"	9'2 3/8"	9'2 3/8"	9'2 1/2"	9'2 1/2"	9'2 1/2"
NUMBER OF RISERS	17	18	14	15	16	17	18	14	15	16
HEIGHT OF EACH RISER	6 1/2"	6 1/8"	7 7/8"	7 3/8"	6 7/8"	6 1/2"	6 1/8"	7 7/8"	7 3/8"	6 15/16"
TOTAL HEIGHT OF RISERS	9'2 1/2"	9'2 1/4"	9'2 1/4"	9'2 5/8"	9'2"	9'2 1/2"	9'2 1/4"	9'2 1/4"	9'2 5/8"	9'3"
RISERS OVERAGE (+) OR UNDERAGE (—)	+1/4"	0	−1/8"	+1/4"	−3/8"	+1/8"	−1/8"	−1/4"	+1/8"	+1/2"
NUMBER OF TREADS	16	17	13	14	15	16	17	13	14	15
WIDTH OF EACH TREAD	11"	11 3/8"	9 5/8"	10 1/8"	10 5/8"	11"	11 3/8"	9 5/8"	10 1/8"	10 9/16"
TOTAL RUN OF STAIRWAY	14'8"	16'1 3/8"	10'5 1/8"	11'9 3/4"	13'3 3/8"	14'8"	16'1 3/8"	10'5 1/8"	11'9 3/4"	13'2 7/16"
WELL OPENING FOR 6'-8" HEADROOM	13'3 15/16"	14'9 13/16"	9'10 3/32"	10'5 3/32"	11'10 31/32"	13'3 15/16"	14'9 13/16"	9'10 3/32"	10'5 3/32"	11'9 31/32"
ANGLE OF INCLINE	30°35'	28°18'	39°17'	36°4'	32°54'	30°35'	28°18'	39°17'	36°4'	33°18'
LENGTH OF CARRIAGE	17' 7/16"	18'3 5/8"	13'5 21/32"	14'7 3/8"	15'9 27/32"	17' 7/16"	18'3 5/8"	13'5 21/32"	14'7 3/8"	15'9 9/16"
L. F. OF RISER PER INCH OF STAIRWAY WIDTH	1.416	1.5	1.166	1.25	1.333	1.416	1.5	1.166	1.25	1.333
L. F. OF TREAD PER INCH OF STAIRWAY WIDTH	1.333	1.416	1.083	1.166	1.25	1.333	1.416	1.083	1.166	1.25
FRAMING SQUARE SETTINGS FOR CARRIAGE CUTS — TONGUE	6 1/2"	6 1/8"	7 7/8"	7 3/8"	6 7/8"	6 1/2"	6 1/8"	7 7/8"	7 3/8"	6 15/16"
FRAMING SQUARE SETTINGS FOR CARRIAGE CUTS — BODY	11"	11 3/8"	9 5/8"	10 1/8"	10 5/8"	11"	11 3/8"	9 5/8"	10 1/8"	10 9/16"

FLOOR TO FLOOR RISE		9'2-1/2"	9'2-1/2"	9'2-5/8"	9'2-5/8"	9'2-5/8"	9'2-5/8"	9'2-5/8"	9'2-3/4"	9'2-3/4"	9'2-3/4"
NUMBER OF RISERS		17	18	14	15	16	17	18	14	15	16
HEIGHT OF EACH RISER		6-1/2"	6-1/8"	7-7/8"	7-3/8"	6-15/16"	6-1/2"	6-1/8"	7-15/16"	7-3/8"	6-15/16"
TOTAL HEIGHT OF RISERS		9'2-1/2"	9'2-1/4"	9'2-1/4"	9'2-5/8"	9'3"	9'2-1/2"	9'2-1/4"	9'3-1/8"	9'2-5/8"	9'3"
RISERS OVERAGE (+) OR UNDERAGE (—)		0	-1/4"	-3/8"	0	+3/8"	-1/8"	-3/8"	+3/8"	-1/8"	+1/4"
NUMBER OF TREADS		16	17	13	14	15	16	17	13	14	15
WIDTH OF EACH TREAD		11"	11-3/8"	9-5/8"	10-1/8"	10-9/16"	11"	11-3/8"	9-9/16"	10-1/8"	10-9/16"
TOTAL RUN OF STAIRWAY		14'8"	16'1-3/8"	10'5-1/8"	11'9-3/4"	13'2-7/16"	14'8"	16'1-3/8"	10'4-5/16"	11'9-3/4"	13'2-7/16"
WELL OPENING FOR 6'-8" HEADROOM		13'3-15/16"	14'9-13/16"	9'10-3/32"	10'5-3/32"	11'9-31/32"	13'3-15/16"	14'9-13/16"	9'9-1/4"	10'5-3/32"	11'9-31/32"
ANGLE OF INCLINE		30°35'	28°18'	39°17'	36°4'	33°18'	30°35'	28°18'	39°42'	36°4'	33°18'
LENGTH OF CARRIAGE		17'7/16"	18'3-5/8"	13'5-21/32"	14'7-3/8"	15'9-9/16"	17'7/16"	18'3-5/8"	13'5-9/16"	14'7-3/8"	15'9-9/16"
L. F. OF RISER PER INCH OF STAIRWAY WIDTH		1.416	1.5	1.166	1.25	1.333	1.416	1.5	1.166	1.25	1.333
L. F. OF TREAD PER INCH OF STAIRWAY WIDTH		1.333	1.416	1.083	1.166	1.25	1.333	1.416	1.083	1.166	1.25
FRAMING SQUARE SETTINGS FOR CARRIAGE CUTS	TONGUE	6-1/2"	6-1/8"	7-7/8"	7-3/8"	6-15/16"	6-1/2"	6-1/8"	7-15/16"	7-3/8"	6-15/16"
	BODY	11"	11-3/8"	9-5/8"	10-1/8"	10-9/16"	11"	11-3/8"	9-9/16"	10-1/8"	10-9/16"

FLOOR TO FLOOR RISE		9'2 3/4"	9'2 3/4"	9'2 7/8"	9'2 7/8"	9'2 7/8"	9'2 7/8"	9'2 7/8"	9'3"	9'3"	9'3"
NUMBER OF RISERS		17	18	14	15	16	17	18	14	15	16
HEIGHT OF EACH RISER		6 1/2"	6 1/8"	7 15/16"	7 3/8"	6 15/16"	6 1/2"	6 3/16"	7 15/16"	7 3/8"	6 15/16"
TOTAL HEIGHT OF RISERS		9'2 1/2"	9'2 1/4"	9'3 1/8"	9'2 5/8"	9'3"	9'2 1/2"	9'3 3/8"	9'3 1/8"	9'2 5/8"	9'3"
RISERS OVERAGE (+) OR UNDERAGE (—)		−1/4"	−1/2"	+1/4"	−1/4"	+1/8"	−3/8"	+1/2"	+1/8"	−3/8"	0
NUMBER OF TREADS		16	17	13	14	15	16	17	13	14	15
WIDTH OF EACH TREAD		11"	11 3/8"	9 9/16"	10 1/8"	10 9/16"	11"	11 5/16"	9 9/16"	10 1/8"	10 9/16"
TOTAL RUN OF STAIRWAY		14'8"	16'1 3/8"	10'4 5/16"	11'9 3/4"	13'2 7/16"	14'8"	16' 5/16"	10'4 5/16"	11'9 3/4"	13'2 7/16"
WELL OPENING FOR 6'-8" HEADROOM		13'3 15/16"	14'9 13/16"	9'9 1/4"	10'5 3/32"	11'9 31/32"	13'3 15/16"	14'8 21/32"	9'9 1/4"	10'5 3/32"	11'9 31/32"
ANGLE OF INCLINE		30°35'	28°18'	39°42'	36°4'	33°18'	30°35'	28°41'	39°42'	36°4'	33°18'
LENGTH OF CARRIAGE		17' 7/16"	18'3 5/8"	13'5 9/16"	14'7 3/8"	15'9 9/16"	17' 7/16"	18'3 3/16"	13'5 9/16"	14'7 3/8"	15'9 9/16"
L. F. OF RISER PER INCH OF STAIRWAY WIDTH		1.416	1.5	1.166	1.25	1.333	1.416	1.5	1.166	1.25	1.333
L. F. OF TREAD PER INCH OF STAIRWAY WIDTH		1.333	1.416	1.083	1.166	1.25	1.333	1.416	1.083	1.166	1.25
FRAMING SQUARE SETTINGS FOR CARRIAGE CUTS	TONGUE	6 1/2"	6 1/8"	7 15/16"	7 3/8"	6 15/16"	6 1/2"	6 3/16"	7 15/16"	7 3/8"	6 15/16"
	BODY	11"	11 3/8"	9 9/16"	10 1/8"	10 9/16"	11"	11 5/16"	9 9/16"	10 1/8"	10 9/16"

FLOOR TO FLOOR RISE		9'3"	9'3"	9'3$\frac{1}{8}$"	9'3$\frac{1}{8}$"	9'3$\frac{1}{8}$"	9'3$\frac{1}{8}$"	9'3$\frac{1}{8}$"	9'3$\frac{1}{4}$"	9'3$\frac{1}{4}$"	9'3$\frac{1}{4}$"
NUMBER OF RISERS		17	18	14	15	16	17	18	14	15	16
HEIGHT OF EACH RISER		6$\frac{1}{2}$"	6$\frac{3}{16}$"	7$\frac{15}{16}$"	7$\frac{7}{16}$"	6$\frac{15}{16}$"	6$\frac{9}{16}$"	6$\frac{3}{16}$"	7$\frac{15}{16}$"	7$\frac{7}{16}$"	6$\frac{15}{16}$"
TOTAL HEIGHT OF RISERS		9'2$\frac{1}{2}$"	9'3$\frac{3}{8}$"	9'3$\frac{1}{8}$"	9'3$\frac{9}{16}$"	9'3"	9'3$\frac{9}{16}$"	9'3$\frac{3}{8}$"	9'3$\frac{1}{8}$"	9'3$\frac{9}{16}$"	9'3"
RISERS OVERAGE (+) OR UNDERAGE (—)		−$\frac{1}{2}$"	+$\frac{3}{8}$"	0	+$\frac{7}{16}$"	−$\frac{1}{8}$"	+$\frac{7}{16}$"	+$\frac{1}{4}$"	−$\frac{1}{8}$"	+$\frac{5}{16}$"	−$\frac{1}{4}$"
NUMBER OF TREADS		16	17	13	14	15	16	17	13	14	15
WIDTH OF EACH TREAD		11"	11$\frac{5}{16}$"	9$\frac{9}{16}$"	10$\frac{1}{16}$"	10$\frac{9}{16}$"	10$\frac{15}{16}$"	11$\frac{5}{16}$"	9$\frac{9}{16}$"	10$\frac{1}{16}$"	10$\frac{9}{16}$"
TOTAL RUN OF STAIRWAY		14'8"	16'$\frac{5}{16}$"	10'4$\frac{5}{16}$"	11'8$\frac{7}{8}$"	13'2$\frac{7}{16}$"	14'7"	16'$\frac{5}{16}$"	10'4$\frac{5}{16}$"	11'8$\frac{7}{8}$"	13'2$\frac{7}{16}$"
WELL OPENING FOR 6'-8" HEADROOM		13'3$\frac{15}{16}$"	14'8$\frac{21}{32}$"	9'9$\frac{1}{4}$"	10'4$\frac{7}{32}$"	11'9$\frac{31}{32}$"	13'2$\frac{27}{32}$"	14'8$\frac{21}{32}$"	9'9$\frac{1}{4}$"	10'4$\frac{7}{32}$"	11'9$\frac{31}{32}$"
ANGLE OF INCLINE		30°35'	28°41'	39°42'	36°28'	33°18'	30°58'	28°41'	39°42'	36°28'	33°18'
LENGTH OF CARRIAGE		17'$\frac{7}{16}$"	18'3$\frac{3}{16}$"	13'5$\frac{9}{16}$"	14'7$\frac{3}{16}$"	15'9$\frac{9}{16}$"	17'$\frac{3}{32}$"	18'3$\frac{3}{16}$"	13'5$\frac{9}{16}$"	14'7$\frac{3}{16}$"	15'9$\frac{9}{16}$"
L. F. OF RISER PER INCH OF STAIRWAY WIDTH		1.416	1.5	1.166	1.25	1.333	1.416	1.5	1.166	1.25	1.333
L. F. OF TREAD PER INCH OF STAIRWAY WIDTH		1.333	1.416	1.083	1.166	1.25	1.333	1.416	1.083	1.166	1.25
FRAMING SQUARE SETTINGS FOR CARRIAGE CUTS	TONGUE	6$\frac{1}{2}$"	6$\frac{3}{16}$"	7$\frac{15}{16}$"	7$\frac{7}{16}$"	6$\frac{15}{16}$"	6$\frac{9}{16}$"	6$\frac{3}{16}$"	7$\frac{15}{16}$"	7$\frac{7}{16}$"	6$\frac{15}{16}$"
	BODY	11"	11$\frac{5}{16}$"	9$\frac{9}{16}$"	10$\frac{1}{16}$"	10$\frac{9}{16}$"	10$\frac{15}{16}$"	11$\frac{5}{16}$"	9$\frac{9}{16}$"	10$\frac{1}{16}$"	10$\frac{9}{16}$"

FLOOR TO FLOOR RISE		8'3 1/4"	9'3 1/4"	9'3 3/8"	9'3 3/8"	9'3 3/8"	9'3 3/8"	9'3 3/8"	9'3 1/2"	9'3 1/2"	9'3 1/2"
NUMBER OF RISERS		17	18	14	15	16	17	18	14	15	16
HEIGHT OF EACH RISER		6 9/16"	6 3/16"	7 15/16"	7 7/16"	6 15/16"	6 9/16"	6 3/16"	7 15/16"	7 7/16"	7"
TOTAL HEIGHT OF RISERS		9'3 9/16"	9'3 3/8"	9'3 1/8"	9'3 9/16"	9'3"	9'3 9/16"	9'3 3/8"	9'3 1/8"	9'3 9/16"	9'4"
RISERS OVERAGE (+) OR UNDERAGE (—)		+5/16"	+1/8"	−1/4"	+3/16"	−3/8"	+3/16"	0	−3/8"	+1/16"	+1/2"
NUMBER OF TREADS		16	17	13	14	15	16	17	13	14	15
WIDTH OF EACH TREAD		10 15/16"	11 5/16"	9 9/16"	10 1/16"	10 9/16"	10 15/16"	11 5/16"	9 9/16"	10 1/16"	10 1/2"
TOTAL RUN OF STAIRWAY		14'7"	16' 5/16"	10'4 5/16"	11'8 7/8"	13'2 7/16"	14'7"	16' 5/16"	10'4 5/16"	11'8 7/8"	13'1 1/2"
WELL OPENING FOR 6'-8" HEADROOM		13'2 27/32"	14'8 21/32"	9'9 1/4"	10'4 7/32"	11'9 31/32"	13'2 27/32"	14'8 21/32"	9'9 1/4"	10'4 7/32"	11'9"
ANGLE OF INCLINE		30°58'	28°41'	39°42'	36°28'	33°18'	30°58'	28°41'	39°42'	36°28'	33°41
LENGTH OF CARRIAGE		17' 3/32"	18'3 3/16"	13'5 9/16"	14'7 3/16"	15'9 9/16"	17' 3/32"	18'3 3/16"	13'5 9/16"	14'7 3/16"	15'9 9/32"
L. F. OF RISER PER INCH OF STAIRWAY WIDTH		1.416	1.5	1.166	1.25	1.333	1.416	1.5	1.166	1.25	1.333
L. F. OF TREAD PER INCH OF STAIRWAY WIDTH		1.333	1.416	1.083	1.166	1.25	1.333	1.416	1.083	1.166	1.25
FRAMING SQUARE SETTINGS FOR CARRIAGE CUTS	TONGUE	6 9/16"	6 3/16"	7 15/16"	7 7/16"	6 15/16"	6 9/16"	6 3/16"	7 15/16"	7 7/16"	7"
	BODY	10 15/16"	11 5/16"	9 9/16"	10 1/16"	10 9/16"	10 15/16"	11 5/16"	9 9/16"	10 1/16"	10 1/2"

FLOOR TO FLOOR RISE	9'3 1/2"	9'3 1/2"	9'3 5/8"	9'3 5/8"	9'3 5/8"	9'3 5/8"	9'3 5/8"	9'3 3/4"	9'3 3/4"	9'3 3/4"
NUMBER OF RISERS	17	18	14	15	16	17	18	14	15	16
HEIGHT OF EACH RISER	6 9/16"	6 3/16"	8"	7 7/16"	7"	6 9/16"	6 3/16"	8"	7 7/16"	7"
TOTAL HEIGHT OF RISERS	9'3 9/16"	9'3 3/8"	9'4"	9'3 9/16"	9'4"	9'3 9/16"	9'3 3/8"	9'4"	9'3 9/16"	9'4"
RISERS OVERAGE (+) OR UNDERAGE (—)	+1/16"	−1/8"	+3/8"	−1/16"	+3/8"	−1/16"	−1/4"	+1/4"	−3/16"	+1/4"
NUMBER OF TREADS	16	17	13	14	15	16	17	13	14	15
WIDTH OF EACH TREAD	10 15/16"	11 5/16"	9 1/2"	10 1/16"	10 1/2"	10 15/16"	11 5/16"	9 1/2"	10 1/16"	10 1/2"
TOTAL RUN OF STAIRWAY	14'7"	16' 5/16"	10'3 1/2"	11'8 7/8"	13'1 1/2"	14'7"	16' 5/16"	10'3 1/2"	11'8 7/8"	13'1 1/2"
WELL OPENING FOR 6'-8" HEADROOM	13'2 27/32"	14'8 21/32"	8'10 7/8"	10'4 7/32"	11'9"	13'2 27/32"	14'8 21/32"	8'10 7/8"	10'4 7/32"	11'9"
ANGLE OF INCLINE	30°58'	28°41'	40°6'	36°28'	33°41'	30°58'	28°41'	40°6'	36°28'	33°41'
LENGTH OF CARRIAGE	17' 3/32"	18'3 3/16"	13'5 15/32"	14'7 3/16"	15'9 9/32"	17' 3/32"	18'3 3/16"	13'5 15/32"	14'7 3/16"	15'9 9/32"
L. F. OF RISER PER INCH OF STAIRWAY WIDTH	1.416	1.5	1.166	1.25	1.333	1.416	1.5	1.166	1.25	1.333
L. F. OF TREAD PER INCH OF STAIRWAY WIDTH	1.333	1.416	1.083	1.166	1.25	1.333	1.416	1.083	1.166	1.25
FRAMING SQUARE SETTINGS FOR CARRIAGE CUTS — TONGUE	6 9/16"	6 3/16"	8"	7 7/16"	7"	6 9/16"	6 3/16"	8"	7 7/16"	7"
FRAMING SQUARE SETTINGS FOR CARRIAGE CUTS — BODY	10 15/16"	11 5/16"	9 1/2"	10 1/16"	10 1/2"	10 15/16"	11 5/16"	9 1/2"	10 1/16"	10 1/2"

FLOOR TO FLOOR RISE	9'3 3/4"	9'3 3/4"	9'3 7/8"	9'3 7/8"	9'3 7/8"	9'3 7/8"	9'3 7/8"	9'4"	9'4"	9'4"
NUMBER OF RISERS	17	18	14	15	16	17	18	14	15	16
HEIGHT OF EACH RISER	6 9/16"	6 3/16"	8"	7 7/16"	7"	6 9/16"	6 3/16"	8"	7 7/16"	7"
TOTAL HEIGHT OF RISERS	9'3 9/16"	9'3 3/8"	9'4"	9'3 9/16"	9'4"	9'3 9/16"	9'3 3/8"	9'4"	9'3 9/16"	9'4"
RISERS OVERAGE (+) OR UNDERAGE (—)	−3/16"	−3/8"	+1/8"	−5/16"	+1/8"	−5/16"	−1/2"	0	−7/16"	0
NUMBER OF TREADS	16	17	13	14	15	16	17	13	14	15
WIDTH OF EACH TREAD	10 15/16"	11 5/16"	9 1/2"	10 1/16"	10 1/2"	10 15/16"	11 5/16"	9 1/2"	10 1/16"	10 1/2"
TOTAL RUN OF STAIRWAY	14'7"	16' 5/16"	10'3 1/2"	11'8 7/8"	13'1 1/2"	14'7"	16' 5/16"	10'3 1/2"	11'8 7/8"	13'1 1/2"
WELL OPENING FOR 6'-8" HEADROOM	13'2 27/32"	14'8 21/32"	8'10 7/8"	10'4 7/32"	11'9"	13'2 27/32"	14'8 21/32"	8'10 7/8"	10'4 7/32"	11'9"
ANGLE OF INCLINE	30°58'	28°41'	40°6'	36°28'	33°41'	30°58'	28°41'	40°6'	36°28'	33°41'
LENGTH OF CARRIAGE	17' 3/32"	18'3 3/16"	13'5 15/16"	14'7 3/16"	15'9 9/32"	17' 3/32"	18'3 3/16"	13'5 15/16"	14'7 3/16"	15'9 9/32"
L. F. OF RISER PER INCH OF STAIRWAY WIDTH	1.416	1.5	1.166	1.25	1.333	1.416	1.5	1.166	1.25	1.333
L. F. OF TREAD PER INCH OF STAIRWAY WIDTH	1.333	1.416	1.083	1.166	1.25	1.333	1.416	1.083	1.166	1.25
FRAMING SQUARE SETTINGS FOR CARRIAGE CUTS — TONGUE	6 9/16"	6 3/16"	8"	7 7/16"	7"	6 9/16"	6 3/16"	8"	7 7/16"	7"
FRAMING SQUARE SETTINGS FOR CARRIAGE CUTS — BODY	10 15/16"	11 5/16"	9 1/2"	10 1/16"	10 1/2"	10 15/16"	11 5/16"	9 1/2"	10 1/16"	10 1/2"

FLOOR TO FLOOR RISE		9'4"	9'4"	9'4 1/8"	9'4 1/8"	9'4 1/8"	9'4 1/8"	9'4 1/8"	9'4 1/4"	9'4 1/4"	9'4 1/4"
NUMBER OF RISERS		17	18	14	15	16	17	18	14	15	16
HEIGHT OF EACH RISER		6 9/16"	6 1/4"	8"	7 1/2"	7"	6 5/8"	6 1/4"	8"	7 1/2"	7"
TOTAL HEIGHT OF RISERS		9'3 9/16"	9'4 1/2"	9'4"	9'4 1/2"	9'4"	9'4 5/8"	9'4 1/2"	9'4"	9'4 1/2"	9'4"
RISERS OVERAGE (+) OR UNDERAGE (—)		−7/16"	+1/2"	−1/8"	+3/8"	−1/8"	+1/2"	+3/8"	−1/4"	+1/4"	−1/4"
NUMBER OF TREADS		16	17	13	14	15	16	17	13	14	15
WIDTH OF EACH TREAD		10 15/16"	11 1/4"	9 1/2"	10"	10 1/2"	10 7/8"	11 1/4"	9 1/2"	10"	10 1/2"
TOTAL RUN OF STAIRWAY		14'7"	15'11 1/4"	10'3 1/2"	11'8"	13'1 1/2"	14'6"	15'11 1/4"	10'3 1/2"	11'8"	13'1 1/2"
WELL OPENING FOR 6'-8" HEADROOM		13'2 27/32"	14'7 1/2"	8'10 7/8"	10'3 11/32"	11'9"	13'1 25/32"	14'7 1/2"	8'10 7/8"	10'3 11/32"	11'9"
ANGLE OF INCLINE		30°58'	29°3'	40°6'	36°52'	33°41'	31°21'	29°3'	40°6'	36°52'	33°41'
LENGTH OF CARRIAGE		17'3/32"	18'2 25/32"	13'5 15/16"	14'7"	15'9 9/32"	16'11 3/4"	18'2 25/32"	13'5 15/16"	14'7"	15'9 9/32"
L. F. OF RISER PER INCH OF STAIRWAY WIDTH		1.416	1.5	1.166	1.25	1.333	1.416	1.5	1.166	1.25	1.333
L. F. OF TREAD PER INCH OF STAIRWAY WIDTH		1.333	1.416	1.083	1.166	1.25	1.333	1.416	1.083	1.166	1.25
FRAMING SQUARE SETTINGS FOR CARRIAGE CUTS	TONGUE	6 9/16"	6 1/4"	8"	7 1/2"	7"	6 5/8"	6 1/4"	8"	7 1/2"	7"
	BODY	10 15/16"	11 1/4"	9 1/2"	10"	10 1/2"	10 7/8"	11 1/4"	9 1/2"	10"	10 1/2"

FLOOR TO FLOOR RISE	9'4 1/4"	9'4 1/4"	9'4 3/8"	9'4 3/8"	9'4 3/8"	9'4 3/8"	9'4 3/8"	9'4 1/2"	9'4 1/2"	9'4 1/2"
NUMBER OF RISERS	17	18	14	15	16	17	18	14	15	16
HEIGHT OF EACH RISER	6 5/8"	6 1/4"	8"	7 1/2"	7"	6 5/8"	6 1/4"	8 1/16"	7 1/2"	7 1/16"
TOTAL HEIGHT OF RISERS	9'4 5/8"	9'4 1/2"	9'4"	9'4 1/2"	9'4"	9'4 5/8"	9'4 1/2"	9'4 7/8"	9'4 1/2"	9'5"
RISERS OVERAGE (+) OR UNDERAGE (—)	+3/8"	+1/4"	−3/8"	+1/8"	−3/8"	+1/4"	+1/8"	+3/8"	0	+1/2"
NUMBER OF TREADS	16	17	13	14	15	16	17	13	14	15
WIDTH OF EACH TREAD	10 7/8"	11 1/4"	9 1/2"	10"	10 1/2"	10 7/8"	11 1/4"	9 7/16"	10"	10 7/16"
TOTAL RUN OF STAIRWAY	14'6"	15'11 1/4"	10'3 1/2"	11'8"	13'1 1/2"	14'6"	15'11 1/4'	10'2 11/16"	11'8"	13' 9/16"
WELL OPENING FOR 6'-8" HEADROOM	13'1 25/32"	14'7 1/2"	8'10 7/8"	10'3 11/32"	11'9"	13'1 25/32"	14'7 1/2"	8'10 3/32"	10'3 11/32"	11'8 1/32"
ANGLE OF INCLINE	31°21'	29°3'	40°6'	36°52'	33°41'	31°21'	29°3'	40°30'	36°52'	34°5'
LENGTH OF CARRIAGE	16'11 3/4"	18'2 25/32"	13'5 15/32"	14'7"	15'9 9/32"	16'11 3/4"	18'2 25/32"	13'5 3/8"	14'7"	15'9 1/32"
L. F. OF RISER PER INCH OF STAIRWAY WIDTH	1.416	1.5	1.166	1.25	1.333	1.416	1.5	1.166	1.25	1.333
L. F. OF TREAD PER INCH OF STAIRWAY WIDTH	1.333	1.416	1.083	1.166	1.25	1.333	1.416	1.083	1.166	1.25
FRAMING SQUARE SETTINGS FOR CARRIAGE CUTS — TONGUE	6 5/8"	6 1/4"	8"	7 1/2"	7"	6 5/8"	6 1/4"	8 1/16"	7 1/2"	7 1/16"
FRAMING SQUARE SETTINGS FOR CARRIAGE CUTS — BODY	10 7/8"	11 1/4"	9 1/2"	10"	10 1/2"	10 7/8"	11 1/4"	9 7/16"	10"	10 7/16"

FLOOR TO FLOOR RISE		9'4 1/2"	9'4 1/2"	9'4 5/8"	9'4 5/8"	9'4 5/8"	9'4 5/8"	9'4 5/8"	9'4 3/4"	9'4 3/4"	9'4 3/4"
NUMBER OF RISERS		17	18	14	15	16	17	18	14	15	16
HEIGHT OF EACH RISER		6 5/8"	6 1/4"	8 1/16"	7 1/2"	7 1/16"	6 5/8"	6 1/4"	8 1/16"	7 1/2"	7 1/16"
TOTAL HEIGHT OF RISERS		9'4 5/8"	9'4 1/2"	9'4 7/8"	9'4 1/2"	9'5"	9'4 5/8"	9'4 1/2"	9'4 7/8"	9'4 1/2"	9'5"
RISERS OVERAGE (+) OR UNDERAGE (—)		+1/8"	0	+1/4"	−1/8"	+3/8"	0	−1/8"	+1/8"	−1/4"	+1/4"
NUMBER OF TREADS		16	17	13	14	15	16	17	13	14	15
WIDTH OF EACH TREAD		10 7/8"	11 1/4"	9 7/16"	10"	10 7/16"	10 7/8"	11 1/4"	9 7/16"	10"	10 7/16"
TOTAL RUN OF STAIRWAY		14'6"	15'11 1/4"	10'2 11/16"	11'8"	13'9/16"	14'6"	15'11 1/4"	10'2 11/16"	11'8"	13'9/16"
WELL OPENING FOR 6'-8" HEADROOM		13'1 25/32"	14'7 1/2"	8'10 3/32"	10'3 11/32"	11'8 1/32"	13'1 25/32"	14'7 1/2"	8'10 3/32"	10'3 11/32"	11'8 1/32"
ANGLE OF INCLINE		31°21'	29°3'	40°30'	36°52'	34°5'	31°21'	29°3'	40°30'	36°52'	34°5'
LENGTH OF CARRIAGE		16'11 3/4"	18'2 25/32"	13'5 3/8"	14'7"	15'9 1/32"	16'11 3/4"	18'2 25/32"	13'5 3/8"	14'7"	15'9 1/32"
L. F. OF RISER PER INCH OF STAIRWAY WIDTH		1.416	1.5	1.166	1.25	1.333	1.416	1.5	1.166	1.25	1.333
L. F. OF TREAD PER INCH OF STAIRWAY WIDTH		1.333	1.416	1.083	1.166	1.25	1.333	1.416	1.083	1.166	1.25
FRAMING SQUARE SETTINGS FOR CARRIAGE CUTS	TONGUE	6 5/8"	6 1/4"	8 1/16"	7 1/2"	7 1/16"	6 5/8"	6 1/4"	8 1/16"	7 1/2"	7 1/16"
	BODY	10 7/8"	11 1/4"	9 7/16"	10"	10 7/16"	10 7/8"	11 1/4"	9 7/16"	10"	10 7/16"

FLOOR TO FLOOR RISE	9'4 3/4"	9'4 3/4"	9'4 7/8"	9'4 7/8"	9'4 7/8"	9'4 7/8"	9'4 7/8"	9'5"	9'5"	9'5"
NUMBER OF RISERS	17	18	14	15	16	17	18	14	15	16
HEIGHT OF EACH RISER	6 5/8"	6 1/4"	8 1/16"	7 1/2"	7 1/16"	6 5/8"	6 1/4"	8 1/16"	7 9/16"	7 1/16"
TOTAL HEIGHT OF RISERS	9'4 5/8"	9'4 1/2"	9'4 7/8"	9'4 1/2"	9'5"	9'4 5/8"	9'4 1/2"	9'4 7/8"	9'5 7/16"	9'5"
RISERS OVERAGE (+) OR UNDERAGE (—)	−1/8"	−1/4"	0	−3/8"	+1/8"	−1/4"	−3/8"	−1/8"	+7/16"	0
NUMBER OF TREADS	16	17	13	14	15	16	17	13	14	15
WIDTH OF EACH TREAD	10 7/8"	11 1/4"	9 7/16"	10"	10 7/16"	10 7/8"	11 1/4"	9 7/16"	9 15/16"	10 7/16"
TOTAL RUN OF STAIRWAY	14'6"	15'11 1/4"	10'2 11/16"	11'8"	13' 9/16"	14'6"	15'11 1/4"	10'2 11/16"	11'7 1/8"	13' 9/16"
WELL OPENING FOR 6'-8" HEADROOM	13'1 25/32"	14'7 1/2"	8'10 3/32"	10'3 11/32"	11'8 1/32"	13'1 25/32"	14'7 1/2"	8'10 3/32"	10'2 7/16"	11'8 1/32"
ANGLE OF INCLINE	31°21'	29°3'	40°30'	36°52'	34°5'	31°21'	29°3'	40°30'	37°16'	34°5'
LENGTH OF CARRIAGE	16'11 3/4"	18'2 25/32"	13'5 3/8"	14'7"	15'9 1/32"	16'11 3/4"	18'2 25/32"	13'5 3/8"	14'6 13/16"	15'9 1/32"
L. F. OF RISER PER INCH OF STAIRWAY WIDTH	1.416	1.5	1.166	1.25	1.333	1.416	1.5	1.166	1.25	1.333
L. F. OF TREAD PER INCH OF STAIRWAY WIDTH	1.333	1.416	1.083	1.166	1.25	1.333	1.416	1.083	1.166	1.25
FRAMING SQUARE SETTINGS FOR CARRIAGE CUTS — TONGUE	6 5/8"	6 1/4"	8 1/16"	7 1/2"	7 1/16"	6 5/8"	6 1/4"	8 1/16"	7 9/16"	7 1/16"
FRAMING SQUARE SETTINGS FOR CARRIAGE CUTS — BODY	10 7/8"	11 1/4"	9 7/16"	10"	10 7/16"	10 7/8"	11 1/4"	9 7/16"	9 15/16"	10 7/16"

FLOOR TO FLOOR RISE	9'5"	9'5"	9'5$\frac{1}{8}$"	9'5$\frac{1}{8}$"	9'5$\frac{1}{8}$"	9'5$\frac{1}{8}$"	9'5$\frac{1}{8}$"	9'5$\frac{1}{4}$"	9'5$\frac{1}{4}$"	9'5$\frac{1}{4}$"
NUMBER OF RISERS	17	18	14	15	16	17	18	14	15	16
HEIGHT OF EACH RISER	6$\frac{5}{8}$"	6$\frac{1}{4}$"	8$\frac{1}{16}$"	7$\frac{9}{16}$"	7$\frac{1}{16}$"	6$\frac{5}{8}$"	6$\frac{5}{16}$"	8$\frac{1}{16}$"	7$\frac{9}{16}$"	7$\frac{1}{16}$"
TOTAL HEIGHT OF RISERS	9'4$\frac{5}{8}$"	9'4$\frac{1}{2}$"	9'4$\frac{7}{8}$"	9'5$\frac{7}{16}$"	9'5"	9'4$\frac{5}{8}$"	9'5$\frac{5}{8}$"	9'4$\frac{7}{8}$"	9'5$\frac{7}{16}$"	9'5"
RISERS OVERAGE (+) OR UNDERAGE (—)	−$\frac{3}{8}$"	−$\frac{1}{2}$"	−$\frac{1}{4}$"	+$\frac{5}{16}$"	−$\frac{1}{8}$"	−$\frac{1}{2}$"	+$\frac{1}{2}$"	−$\frac{3}{8}$"	+$\frac{3}{16}$"	−$\frac{1}{4}$"
NUMBER OF TREADS	16	17	13	14	15	16	17	13	14	15
WIDTH OF EACH TREAD	10$\frac{7}{8}$"	11$\frac{1}{4}$"	9$\frac{7}{16}$"	9$\frac{15}{16}$"	10$\frac{7}{16}$"	10$\frac{7}{8}$"	11$\frac{3}{16}$"	9$\frac{7}{16}$"	9$\frac{15}{16}$"	10$\frac{7}{16}$"
TOTAL RUN OF STAIRWAY	14'6"	15'11$\frac{1}{4}$"	10'2$\frac{11}{16}$"	11'7$\frac{1}{8}$"	13'$\frac{9}{16}$"	14'6"	15'10$\frac{3}{16}$"	10'2$\frac{11}{16}$"	11'7$\frac{1}{8}$"	13'$\frac{9}{16}$"
WELL OPENING FOR 6'-8" HEADROOM	13'1$\frac{25}{32}$"	14'7$\frac{1}{2}$"	8'10$\frac{3}{32}$"	10'2$\frac{7}{16}$"	11'8$\frac{1}{32}$"	13'1$\frac{25}{32}$"	13'7$\frac{5}{32}$"	8'10$\frac{3}{32}$"	10'2$\frac{7}{16}$"	11'8$\frac{1}{32}$"
ANGLE OF INCLINE	31°21'	29°3'	40°30'	37°16'	34°5'	31°21'	29°26'	40°30'	37°16'	34°5'
LENGTH OF CARRIAGE	16'11$\frac{3}{4}$"	18'2$\frac{25}{32}$"	13'5$\frac{3}{8}$"	14'6$\frac{13}{16}$"	15'9$\frac{1}{32}$"	16'11$\frac{3}{4}$"	18'2$\frac{3}{8}$"	13'5$\frac{3}{8}$"	14'6$\frac{13}{16}$"	15'9$\frac{1}{32}$"
L. F. OF RISER PER INCH OF STAIRWAY WIDTH	1.416	1.5	1.166	1.25	1.333	1.416	1.5	1.166	1.25	1.333
L. F. OF TREAD PER INCH OF STAIRWAY WIDTH	1.333	1.416	1.083	1.166	1.25	1.333	1.416	1.083	1.166	1.25
FRAMING SQUARE SETTINGS FOR CARRIAGE CUTS — TONGUE	6$\frac{5}{8}$"	6$\frac{1}{4}$"	8$\frac{1}{16}$"	7$\frac{9}{16}$"	7$\frac{1}{16}$"	6$\frac{5}{8}$"	6$\frac{5}{16}$"	8$\frac{1}{16}$"	7$\frac{9}{16}$"	7$\frac{1}{16}$"
FRAMING SQUARE SETTINGS FOR CARRIAGE CUTS — BODY	10$\frac{7}{8}$"	11$\frac{1}{4}$"	9$\frac{7}{16}$"	9$\frac{15}{16}$"	10$\frac{7}{16}$"	10$\frac{7}{8}$"	11$\frac{3}{16}$"	9$\frac{7}{16}$"	9$\frac{15}{16}$"	10$\frac{7}{16}$"

FLOOR TO FLOOR RISE		9'5 1/4"	9'5 1/4"	9'5 3/8"	9'5 3/8"	9'5 3/8"	9'5 3/8"	9'5 3/8"	9'5 1/2"	9'5 1/2"	9'5 1/2"
NUMBER OF RISERS		17	18	14	15	16	17	18	14	15	16
HEIGHT OF EACH RISER		6 11/16"	6 5/16"	8 1/8"	7 9/16"	7 1/16"	6 11/16"	6 5/16"	8 1/8"	7 9/16"	7 1/8"
TOTAL HEIGHT OF RISERS		9'5 11/16"	9'5 5/8"	9'5 3/4"	9'5 7/16"	9'5"	9'5 11/16"	9'5 5/8"	9'5 3/4"	9'5 7/16"	9'6"
RISERS OVERAGE (+) OR UNDERAGE (—)		+7/16"	+3/8"	+3/8"	+1/16"	−3/8"	+5/16"	+1/4"	+1/4"	−1/16"	+1/2"
NUMBER OF TREADS		16	17	13	14	15	16	17	13	14	15
WIDTH OF EACH TREAD		10 13/16"	11 3/16"	9 3/8"	9 15/16"	10 7/16"	10 13/16"	11 3/16"	9 3/8"	9 15/16"	10 3/8"
TOTAL RUN OF STAIRWAY		14'5"	15'10 3/16"	10'1 7/8"	11'7 1/8"	13' 9/16"	14'5"	15'10 3/16"	10'1 7/8"	11'7 1/8"	12'11 5/8"
WELL OPENING FOR 6'-8" HEADROOM		13' 3/4"	13'7 5/32"	8'9 9/32"	10'2 7/16"	11'8 1/32"	13' 3/4"	13'7 5/32"	8'9 9/32"	10'2 7/16"	11'7 1/16"
ANGLE OF INCLINE		31°44'	29°26'	40°55'	37°16'	34°5'	31°44'	29°26'	40°55'	37°16'	34°29'
LENGTH OF CARRIAGE		16'11 13/32"	18'2 3/8"	13'5 9/32"	14'6 13/16"	15'9 1/32"	16'11 13/32"	18'2 3/8"	13'5 9/32"	14'6 13/16"	15'8 25/32"
L. F. OF RISER PER INCH OF STAIRWAY WIDTH		1.416	1.5	1.166	1.25	1.333	1.416	1.5	1.166	1.25	1.333
L. F. OF TREAD PER INCH OF STAIRWAY WIDTH		1.333	1.416	1.083	1.166	1.25	1.333	1.416	1.083	1.166	1.25
FRAMING SQUARE SETTINGS FOR CARRIAGE CUTS	TONGUE	6 11/16"	6 5/16"	8 1/8"	7 9/16"	7 1/16"	6 11/16"	6 5/16"	8 1/8"	7 9/16"	7 1/8"
	BODY	10 13/16"	11 3/16"	9 3/8"	9 15/16"	10 7/16"	10 13/16"	11 3/16"	9 3/8"	9 15/16"	10 3/8"

FLOOR TO FLOOR RISE		9'5$\frac{1}{2}$"	9'5$\frac{1}{2}$"	9'5$\frac{5}{8}$"	9'5$\frac{5}{8}$"	9'5$\frac{5}{8}$"	9'5$\frac{5}{8}$"	9'5$\frac{5}{8}$"	9'5$\frac{3}{4}$"	9'5$\frac{3}{4}$"	9'5$\frac{3}{4}$"
NUMBER OF RISERS		17	18	14	15	16	17	18	14	15	16
HEIGHT OF EACH RISER		6$\frac{11}{16}$"	6$\frac{5}{16}$"	8$\frac{1}{8}$"	7$\frac{9}{16}$"	7$\frac{1}{8}$"	6$\frac{11}{16}$"	6$\frac{5}{16}$"	8$\frac{1}{8}$"	7$\frac{9}{16}$"	7$\frac{1}{8}$"
TOTAL HEIGHT OF RISERS		9'5$\frac{11}{16}$"	9'5$\frac{5}{8}$"	9'5$\frac{3}{4}$"	9'5$\frac{7}{16}$"	9'6"	9'5$\frac{11}{16}$"	9'5$\frac{5}{8}$"	9'5$\frac{3}{4}$"	9'5$\frac{7}{16}$"	9'6"
RISERS OVERAGE (+) OR UNDERAGE (—)		+$\frac{3}{16}$"	+$\frac{1}{8}$"	+$\frac{1}{8}$"	−$\frac{3}{16}$"	+$\frac{3}{8}$"	+$\frac{1}{16}$"	0	0	−$\frac{5}{16}$"	+$\frac{1}{4}$"
NUMBER OF TREADS		16	17	13	14	15	16	17	13	14	15
WIDTH OF EACH TREAD		10$\frac{13}{16}$"	11$\frac{3}{16}$"	9$\frac{3}{8}$"	9$\frac{15}{16}$"	10$\frac{3}{8}$"	10$\frac{13}{16}$"	11$\frac{3}{16}$"	9$\frac{3}{8}$"	9$\frac{15}{16}$"	10$\frac{3}{8}$"
TOTAL RUN OF STAIRWAY		14'5"	15'10$\frac{3}{16}$"	10'1$\frac{7}{8}$"	11'7$\frac{1}{8}$"	12'11$\frac{5}{8}$"	14'5"	15'10$\frac{3}{16}$"	10'1$\frac{7}{8}$"	11'7$\frac{1}{8}$"	12'11$\frac{5}{8}$"
WELL OPENING FOR 6'-8" HEADROOM		13'$\frac{3}{4}$"	13'7$\frac{5}{32}$"	8'9$\frac{9}{32}$"	10'2$\frac{7}{16}$"	11'7$\frac{1}{16}$"	13'$\frac{3}{4}$"	13'7$\frac{5}{32}$"	8'9$\frac{9}{32}$"	10'2$\frac{7}{16}$"	11'7$\frac{1}{16}$"
ANGLE OF INCLINE		31°44'	29°26'	40°55'	37°16'	34°29'	31°44'	29°26'	40°55'	37°16'	34°29'
LENGTH OF CARRIAGE		16'11$\frac{13}{32}$"	18'2$\frac{3}{8}$"	13'5$\frac{9}{32}$"	14'6$\frac{13}{16}$"	15'8$\frac{25}{32}$"	16'11$\frac{13}{32}$"	18'2$\frac{3}{8}$"	13'5$\frac{9}{32}$"	14'6$\frac{13}{16}$"	15'8$\frac{25}{32}$"
L. F. OF RISER PER INCH OF STAIRWAY WIDTH		1.416	1.5	1.166	1.25	1.333	1.416	1.5	1.166	1.25	1.333
L. F. OF TREAD PER INCH OF STAIRWAY WIDTH		1.333	1.416	1.083	1.166	1.25	1.333	1.416	1.083	1.166	1.25
FRAMING SQUARE SETTINGS FOR CARRIAGE CUTS	TONGUE	6$\frac{11}{16}$"	6$\frac{5}{16}$"	8$\frac{1}{8}$"	7$\frac{9}{16}$"	7$\frac{1}{8}$"	6$\frac{11}{16}$"	6$\frac{5}{16}$"	8$\frac{1}{8}$"	7$\frac{9}{16}$"	7$\frac{1}{8}$"
	BODY	10$\frac{13}{16}$"	11$\frac{3}{16}$"	9$\frac{3}{8}$"	9$\frac{15}{16}$"	10$\frac{3}{8}$"	10$\frac{13}{16}$"	11$\frac{3}{16}$"	9$\frac{3}{8}$"	9$\frac{15}{16}$"	10$\frac{3}{8}$"

FLOOR TO FLOOR RISE		9'5 3/4"	9'5 3/4"	9'5 7/8"	9'5 7/8"	9'5 7/8"	9'5 7/8"	9'5 7/8"	9'6"	9'6"	9'6"
NUMBER OF RISERS		17	18	14	15	16	17	18	14	15	16
HEIGHT OF EACH RISER		6 11/16"	6 5/16"	8 1/8"	7 9/16"	7 1/8"	6 11/16"	6 5/16"	8 1/8"	7 5/8"	7 1/8"
TOTAL HEIGHT OF RISERS		9'5 11/16"	9'5 5/8"	9'5 3/4"	9'5 7/16"	9'6"	9'5 11/16"	9'5 5/8"	9'5 3/4"	9'6 3/8"	9'6"
RISERS OVERAGE (+) OR UNDERAGE (—)		−1/16"	−1/8"	−1/8"	−7/16"	+1/8"	−3/16"	−1/4"	−1/4"	+3/8"	0
NUMBER OF TREADS		16	17	13	14	15	16	17	13	14	15
WIDTH OF EACH TREAD		10 13/16"	11 3/16"	9 3/8"	9 15/16"	10 3/8"	10 13/16"	11 3/16"	9 3/8"	9 7/8"	10 3/8"
TOTAL RUN OF STAIRWAY		14'5"	15'10 3/16"	10'1 7/8"	11'7 1/8"	12'11 5/8"	14'5"	15'10 3/16"	10'1 7/8"	11'6 1/4"	12'11 5/8"
WELL OPENING FOR 6'-8" HEADROOM		13' 3/4"	13'7 5/32"	8'9 9/32"	10'2 7/16"	11'7 1/16"	13' 3/4"	13'7 5/32"	8'9 9/32"	10'1 9/16"	11'7 1/16"
ANGLE OF INCLINE		31°44'	29°26'	40°55'	37°16'	34°29'	31°44'	29°26'	40°55'	37°40'	34°29'
LENGTH OF CARRIAGE		16'11 13/32"	18'2 3/8"	13'5 9/32"	14'6 13/16"	15'8 25/32"	16'11 13/32"	18'2 3/8"	13'5 9/32"	14'6 21/32"	15'8 25/32"
L. F. OF RISER PER INCH OF STAIRWAY WIDTH		1.416	1.5	1.166	1.25	1.333	1.416	1.5	1.166	1.25	1.333
L. F. OF TREAD PER INCH OF STAIRWAY WIDTH		1.333	1.416	1.083	1.166	1.25	1.333	1.416	1.083	1.166	1.25
FRAMING SQUARE SETTINGS FOR CARRIAGE CUTS	TONGUE	6 11/16"	6 5/16"	8 1/8"	7 9/16"	7 1/8"	6 11/16"	6 5/16"	8 1/8"	7 5/8"	7 1/8"
	BODY	10 13/16"	11 3/16"	9 3/8"	9 15/16"	10 3/8"	10 13/16"	11 3/16"	9 3/8"	9 7/8"	10 3/8"

FLOOR TO FLOOR RISE	9'6"	9'6"	9'6"	9'6 1/8"	9'6 1/8"	9'6 1/8"	9'6 1/8"	9'6 1/8"	9'6 1/8"	9'6 1/4"
NUMBER OF RISERS	17	18	19	14	15	16	17	18	19	14
HEIGHT OF EACH RISER	6 11/16"	6 5/16"	6"	8 1/8"	7 5/8"	7 1/8"	6 11/16"	6 5/16"	6"	8 3/16"
TOTAL HEIGHT OF RISERS	9'5 11/16"	9'5 5/8"	9'6"	9'5 3/4"	9'6 3/8"	9'6"	9'5 11/16"	9'5 5/8"	9'6"	9'6 5/8"
RISERS OVERAGE (+) OR UNDERAGE (—)	−5/16"	−3/8"	0	−3/8"	+1/4"	−1/8"	−7/16"	−1/2"	−1/8"	+3/8"
NUMBER OF TREADS	16	17	18	13	14	15	16	17	18	13
WIDTH OF EACH TREAD	10 13/16"	11 3/16"	11 1/2"	9 3/8"	9 7/8"	10 3/8"	10 13/16"	11 3/16"	11 1/2"	9 5/16"
TOTAL RUN OF STAIRWAY	14'5"	15'10 3/16"	17'3"	10'1 7/8"	11'6 1/4"	12'11 5/8"	14'5"	15'10 3/16"	17'3"	10'1 1/16"
WELL OPENING FOR 6'-8" HEADROOM	13' 3/4"	13'7 5/32"	15' 5/32"	8'9 9/32"	10'1 9/16"	11'7 1/16"	13' 3/4"	13'7 5/32"	15' 5/32"	8'8 1/2"
ANGLE OF INCLINE	31°44'	29°26'	27°33'	40°55'	37°40'	34°29'	31°44'	29°26'	27°33'	41°19'
LENGTH OF CARRIAGE	16'11 13/32"	18'2 3/8"	19'5 15/32"	13'5 9/32"	14'6 21/32"	15'8 25/32"	16'11 13/32"	18'2 3/8"	19'5 15/32"	13'5 3/16"
L. F. OF RISER PER INCH OF STAIRWAY WIDTH	1.416	1.5	1.583	1.166	1.25	1.333	1.416	1.5	1.583	1.166
L. F. OF TREAD PER INCH OF STAIRWAY WIDTH	1.333	1.416	1.5	1.083	1.166	1.25	1.333	1.416	1.5	1.083
FRAMING SQUARE SETTINGS FOR CARRIAGE CUTS — TONGUE	6 11/16"	6 5/16"	6"	8 1/8"	7 5/8"	7 1/8"	6 11/16"	6 5/16"	6"	8 3/16"
FRAMING SQUARE SETTINGS FOR CARRIAGE CUTS — BODY	10 13/16"	11 3/16"	11 1/2"	9 3/8"	9 7/8"	10 3/8"	10 13/16"	11 3/16"	11 1/2"	9 5/16"

FLOOR TO FLOOR RISE	$9'6\frac{1}{4}''$	$9'6\frac{1}{4}''$	$9'6\frac{1}{4}''$	$9'6\frac{1}{4}''$	$9'6\frac{1}{4}''$	$9'6\frac{3}{8}''$	$9'6\frac{3}{8}''$	$9'6\frac{3}{8}''$	$9'6\frac{3}{8}''$	$9'6\frac{3}{8}''$
NUMBER OF RISERS	15	16	17	18	19	14	15	16	17	18
HEIGHT OF EACH RISER	$7\frac{5}{8}''$	$7\frac{1}{8}''$	$6\frac{3}{4}''$	$6\frac{3}{8}''$	6"	$8\frac{3}{16}''$	$7\frac{5}{8}''$	$7\frac{1}{8}''$	$6\frac{3}{4}''$	$6\frac{3}{8}''$
TOTAL HEIGHT OF RISERS	$9'6\frac{3}{8}''$	9'6"	$9'6\frac{3}{4}''$	$9'6\frac{3}{4}''$	9'6"	$9'6\frac{5}{8}''$	$9'6\frac{3}{8}''$	9'6"	$9'6\frac{3}{4}''$	$9'6\frac{3}{4}''$
RISERS OVERAGE (+) OR UNDERAGE (—)	$+\frac{1}{8}''$	$-\frac{1}{4}''$	$+\frac{1}{2}''$	$+\frac{1}{2}''$	$-\frac{1}{4}''$	$+\frac{1}{4}''$	0	$-\frac{3}{8}''$	$+\frac{3}{8}''$	$+\frac{3}{8}''$
NUMBER OF TREADS	14	15	16	17	18	13	14	15	16	17
WIDTH OF EACH TREAD	$9\frac{7}{8}''$	$10\frac{3}{8}''$	$10\frac{3}{4}''$	$11\frac{1}{8}''$	$11\frac{1}{2}''$	$9\frac{5}{16}''$	$9\frac{7}{8}''$	$10\frac{3}{8}''$	$10\frac{3}{4}''$	$11\frac{1}{8}''$
TOTAL RUN OF STAIRWAY	$11'6\frac{1}{4}''$	$12'11\frac{5}{8}''$	14'4"	$15'9\frac{1}{8}''$	17'3"	$10'1\frac{1}{16}''$	$11'6\frac{1}{4}''$	$12'11\frac{5}{8}''$	14'4"	$15'9\frac{1}{8}''$
WELL OPENING FOR 6'-8" HEADROOM	$10'1\frac{9}{16}''$	$11'7\frac{1}{16}''$	$12'11\frac{11}{16}''$	$13'6\frac{1}{16}''$	$15'\frac{5}{32}''$	$8'8\frac{1}{2}''$	$10'1\frac{9}{16}''$	$11'7\frac{1}{16}''$	$12'11\frac{11}{16}''$	$13'6\frac{1}{16}''$
ANGLE OF INCLINE	37°40'	34°29'	32°7'	29°49'	27°33'	41°19'	37°40'	34°29'	32°7'	29°49'
LENGTH OF CARRIAGE	$14'6\frac{21}{32}''$	$15'8\frac{25}{32}''$	$16'11\frac{3}{32}''$	$18'1\frac{31}{32}''$	$19'5\frac{15}{32}''$	$13'5\frac{3}{16}''$	$14'6\frac{21}{32}''$	$15'8\frac{25}{32}''$	$16'11\frac{3}{32}''$	$18'1\frac{31}{32}''$
L. F. OF RISER PER INCH OF STAIRWAY WIDTH	1.25	1.333	1.416	1.5	1.583	1.166	1.25	1.333	1.416	1.5
L. F. OF TREAD PER INCH OF STAIRWAY WIDTH	1.166	1.25	1.333	1.416	1.5	1.083	1.166	1.25	1.333	1.416
FRAMING SQUARE SETTINGS FOR CARRIAGE CUTS — TONGUE	$7\frac{5}{8}''$	$7\frac{1}{8}''$	$6\frac{3}{4}''$	$6\frac{3}{8}''$	6"	$8\frac{3}{16}''$	$7\frac{5}{8}''$	$7\frac{1}{8}''$	$6\frac{3}{4}''$	$6\frac{3}{8}''$
FRAMING SQUARE SETTINGS FOR CARRIAGE CUTS — BODY	$9\frac{7}{8}''$	$10\frac{3}{8}''$	$10\frac{3}{4}''$	$11\frac{1}{8}''$	$11\frac{1}{2}''$	$9\frac{5}{16}''$	$9\frac{7}{8}''$	$10\frac{3}{8}''$	$10\frac{3}{4}''$	$11\frac{1}{8}''$

FLOOR TO FLOOR RISE	9'6$\frac{3}{8}$"	9'6$\frac{1}{2}$"	9'6$\frac{1}{2}$"	9'6$\frac{1}{2}$"	9'6$\frac{1}{2}$"	9'6$\frac{1}{2}$"	9'6$\frac{1}{2}$"	9'6$\frac{5}{8}$"	9'6$\frac{5}{8}$"	9'6$\frac{5}{8}$"
NUMBER OF RISERS	19	14	15	16	17	18	19	14	15	16
HEIGHT OF EACH RISER	6"	8$\frac{3}{16}$"	7$\frac{5}{8}$"	7$\frac{3}{16}$"	6$\frac{3}{4}$"	6$\frac{3}{8}$"	6"	8$\frac{3}{16}$"	7$\frac{5}{8}$"	7$\frac{3}{16}$"
TOTAL HEIGHT OF RISERS	9'6"	9'6$\frac{5}{8}$"	9'6$\frac{3}{8}$"	9'7"	9'6$\frac{3}{4}$"	9'6$\frac{3}{4}$"	9'6"	9'6$\frac{5}{8}$"	9'6$\frac{3}{8}$"	9'7"
RISERS OVERAGE (+) OR UNDERAGE (—)	$-\frac{3}{8}$"	$+\frac{1}{8}$"	$-\frac{1}{8}$"	$+\frac{1}{2}$"	$+\frac{1}{4}$"	$+\frac{1}{4}$"	$-\frac{1}{2}$"	0	$-\frac{1}{4}$"	$+\frac{3}{8}$"
NUMBER OF TREADS	18	13	14	15	16	17	18	13	14	15
WIDTH OF EACH TREAD	11$\frac{1}{2}$"	9$\frac{5}{16}$"	9$\frac{7}{8}$"	10$\frac{5}{16}$"	10$\frac{3}{4}$"	11$\frac{1}{8}$"	11$\frac{1}{2}$"	9$\frac{5}{16}$"	9$\frac{7}{8}$"	10$\frac{5}{16}$"
TOTAL RUN OF STAIRWAY	17'3"	10'1$\frac{1}{16}$"	11'6$\frac{1}{4}$"	12'10$\frac{11}{16}$"	14'4"	15'9$\frac{1}{8}$"	17'3"	10'1$\frac{1}{16}$"	11'6$\frac{1}{4}$"	12'10$\frac{11}{16}$"
WELL OPENING FOR 6'-8" HEADROOM	15'$\frac{5}{32}$"	8'8$\frac{1}{2}$"	10'1$\frac{9}{16}$"	11'6$\frac{3}{32}$"	12'11$\frac{11}{16}$"	13'6$\frac{1}{16}$"	15'$\frac{5}{32}$"	8'8$\frac{1}{2}$"	10'1$\frac{9}{16}$"	11'6$\frac{3}{32}$"
ANGLE OF INCLINE	27°33'	41°19'	37°40'	34°53'	32°7'	29°49'	27°33'	41°19'	37°40'	34°53'
LENGTH OF CARRIAGE	19'5$\frac{15}{32}$"	13'5$\frac{3}{16}$"	14'6$\frac{21}{32}$"	15'8$\frac{9}{16}$"	16'11$\frac{3}{32}$"	18'1$\frac{31}{32}$"	19'5$\frac{15}{32}$"	13'5$\frac{3}{16}$"	14'6$\frac{21}{32}$"	15'8$\frac{9}{16}$"
L. F. OF RISER PER INCH OF STAIRWAY WIDTH	1.583	1.166	1.25	1.333	1.416	1.5	1.583	1.166	1.25	1.333
L. F. OF TREAD PER INCH OF STAIRWAY WIDTH	1.5	1.083	1.166	1.25	1.333	1.416	1.5	1.083	1.166	1.25
FRAMING SQUARE SETTINGS FOR CARRIAGE CUTS — TONGUE	6"	8$\frac{3}{16}$"	7$\frac{5}{8}$"	7$\frac{3}{16}$"	6$\frac{3}{4}$"	6$\frac{3}{8}$"	6"	8$\frac{3}{16}$"	7$\frac{5}{8}$"	7$\frac{3}{16}$"
FRAMING SQUARE SETTINGS FOR CARRIAGE CUTS — BODY	11$\frac{1}{2}$"	9$\frac{5}{16}$"	9$\frac{7}{8}$"	10$\frac{5}{16}$"	10$\frac{3}{4}$"	11$\frac{1}{8}$"	11$\frac{1}{2}$"	9$\frac{5}{16}$"	9$\frac{7}{8}$"	10$\frac{5}{16}$"

FLOOR TO FLOOR RISE	9'6 5/8"	9'6 5/8"	9'6 5/8"	9'6 3/4"	9'6 3/4"	9'6 3/4"	9'6 3/4"	9'6 3/4"	9'6 3/4"	9'6 7/8"
NUMBER OF RISERS	17	18	19	14	15	16	17	18	19	14
HEIGHT OF EACH RISER	6 3/4"	6 3/8"	6 1/32"	8 3/16"	7 5/8"	7 3/16"	6 3/4"	6 3/8"	6 1/16"	8 3/16"
TOTAL HEIGHT OF RISERS	9'6 3/4"	9'6 3/4"	9'6 19/32"	9'6 5/8"	9'6 3/8"	9'7"	9'6 3/4"	9'6 3/4"	9'7 3/16"	9'6 5/8"
RISERS OVERAGE (+) OR UNDERAGE (—)	+1/8"	+1/8"	−1/32"	−1/8"	−3/8"	+1/4"	0	0	+7/16"	−1/4"
NUMBER OF TREADS	16	17	18	13	14	15	16	17	18	13
WIDTH OF EACH TREAD	10 3/4"	11 1/8"	11 15/32"	9 5/16"	9 7/8"	10 5/16"	10 3/4"	11 1/8"	11 7/16"	9 5/16"
TOTAL RUN OF STAIRWAY	14'4"	15'9 1/8"	17'2 7/16"	10'1 1/16"	11'6 1/4"	12'10 11/16"	14'4"	15'9 1/8"	17'1 7/8"	10'1 1/16"
WELL OPENING FOR 6'-8" HEADROOM	12'11 11/16"	13'6 1/16"	14'11 19/32"	8'8 1/2"	10'1 9/16"	11'6 3/32"	12'11 11/16"	13'6 1/16"	14'11"	8'8 1/2"
ANGLE OF INCLINE	32°7'	29°49'	27°44'	41°19'	37°40'	34°53'	32°7'	29°49'	27°56'	41°19'
LENGTH OF CARRIAGE	16'11 3/32"	18'1 31/32"	19'5 3/16"	13'5 3/16"	14'6 21/32"	15'8 9/16"	16'11 3/32"	18'1 31/32"	19'5"	13'5 3/16"
L. F. OF RISER PER INCH OF STAIRWAY WIDTH	1.416	1.5	1.583	1.166	1.25	1.333	1.416	1.5	1.583	1.166
L. F. OF TREAD PER INCH OF STAIRWAY WIDTH	1.333	1.416	1.5	1.083	1.166	1.25	1.333	1.416	1.5	1.083
FRAMING SQUARE SETTINGS FOR CARRIAGE CUTS — TONGUE	6 3/4"	6 3/8"	6 1/32"	8 3/16"	7 5/8"	7 3/16"	6 3/4"	6 3/8"	6 1/16"	8 3/16"
FRAMING SQUARE SETTINGS FOR CARRIAGE CUTS — BODY	10 3/4"	11 1/8"	11 15/32"	9 5/16"	9 7/8"	10 5/16"	10 3/4"	11 1/8"	11 7/16"	9 5/16"

FLOOR TO FLOOR RISE	9'6 7/8"	9'6 7/8"	9'6 7/8"	9'6 7/8"	9'6 7/8"	9'7"	9'7"	9'7"	9'7"	9'7"
NUMBER OF RISERS	15	16	17	18	19	14	15	16	17	18
HEIGHT OF EACH RISER	7 11/16"	7 3/16"	6 3/4"	6 3/8"	6 1/16"	8 3/16"	7 11/16"	7 3/16"	6 3/4"	6 3/8"
TOTAL HEIGHT OF RISERS	9'7 5/16"	9'7"	9'6 3/4"	9'6 3/4"	9'7 3/16"	9'6 5/8"	9'7 5/16"	9'7"	9'6 3/4"	9'6 3/4"
RISERS OVERAGE (+) OR UNDERAGE (—)	+7/16"	+1/8"	−1/8"	−1/8"	+5/16"	−3/8"	+5/16"	0	−1/4"	−1/4"
NUMBER OF TREADS	14	15	16	17	18	13	14	15	16	17
WIDTH OF EACH TREAD	9 13/16"	10 5/16"	10 3/4"	11 1/8"	11 7/16"	9 5/16"	9 13/16"	10 5/16"	10 3/4"	11 1/8"
TOTAL RUN OF STAIRWAY	11'5 3/8"	12'10 11/16"	14'4"	15'9 1/8"	17'1 7/8"	10'1 1/16"	11'5 3/8"	12'10 11/16"	14'4"	15'9 1/8"
WELL OPENING FOR 6'-8" HEADROOM	10' 23/32"	11'6 3/32"	12'11 11/16"	13'6 1/16"	14'11"	8'8 1/12"	10' 23/32"	11'6 3/32"	12'11 11/16"	13'6 1/16"
ANGLE OF INCLINE	38°5'	34°53'	32°7'	29°49'	27°56'	41°19'	38°5'	34°53'	32°7'	29°49'
LENGTH OF CARRIAGE	14'6 1/2"	15'8 9/16"	16'11 3/32"	18'1 31/32"	19'5"	13'5 3/16"	14'6 1/2"	15'8 9/16"	16'11 3/32"	18'1 31/32"
L. F. OF RISER PER INCH OF STAIRWAY WIDTH	1.25	1.333	1.416	1.5	1.583	1.166	1.25	1.333	1.416	1.5
L. F. OF TREAD PER INCH OF STAIRWAY WIDTH	1.166	1.25	1.333	1.416	1.5	1.083	1.166	1.25	1.333	1.416
FRAMING SQUARE SETTINGS FOR CARRIAGE CUTS — TONGUE	7 11/16"	7 3/16"	6 3/4"	6 3/8"	6 1/16"	8 3/16"	7 11/16"	7 3/16"	6 3/4"	6 3/8"
FRAMING SQUARE SETTINGS FOR CARRIAGE CUTS — BODY	9 13/16"	10 5/16"	10 3/4"	11 1/8"	11 7/16"	9 5/16"	9 13/16"	10 5/16"	10 3/4"	11 1/8"

FLOOR TO FLOOR RISE	9'7"	9'$7\frac{1}{8}$"	9'$7\frac{1}{8}$"	9'$7\frac{1}{8}$"	9'$7\frac{1}{8}$"	9'$7\frac{1}{8}$"	9'$7\frac{1}{8}$"	9'$7\frac{1}{4}$"	9'$7\frac{1}{4}$"	9'$7\frac{1}{4}$"
NUMBER OF RISERS	19	14	15	16	17	18	19	14	15	16
HEIGHT OF EACH RISER	$6\frac{1}{16}$"	$8\frac{1}{4}$"	$7\frac{11}{16}$"	$7\frac{3}{16}$"	$6\frac{3}{4}$"	$6\frac{3}{8}$"	$6\frac{1}{16}$"	$8\frac{1}{4}$"	$7\frac{11}{16}$"	$7\frac{3}{16}$"
TOTAL HEIGHT OF RISERS	9'$7\frac{3}{16}$"	9'$7\frac{1}{2}$"	9'$7\frac{5}{16}$"	9'7"	9'$6\frac{3}{4}$"	9'$6\frac{3}{4}$"	9'$7\frac{3}{16}$"	9'$7\frac{1}{2}$"	9'$7\frac{5}{16}$"	9'7"
RISERS OVERAGE (+) OR UNDERAGE (—)	$+\frac{3}{16}$"	$+\frac{3}{8}$"	$+\frac{3}{16}$"	$-\frac{1}{8}$"	$-\frac{3}{8}$"	$-\frac{3}{8}$"	$+\frac{1}{16}$"	$+\frac{1}{4}$"	$+\frac{1}{16}$"	$-\frac{1}{4}$"
NUMBER OF TREADS	18	13	14	15	16	17	18	13	14	15
WIDTH OF EACH TREAD	$11\frac{7}{16}$"	$9\frac{1}{4}$"	$9\frac{13}{16}$"	$10\frac{5}{16}$"	$10\frac{3}{4}$"	$11\frac{1}{8}$"	$11\frac{7}{16}$"	$9\frac{1}{4}$"	$9\frac{13}{16}$"	$10\frac{5}{16}$"
TOTAL RUN OF STAIRWAY	17'$1\frac{7}{8}$"	10'$\frac{1}{4}$"	11'$5\frac{3}{8}$"	12'$10\frac{11}{16}$"	14'4"	15'$9\frac{1}{8}$"	17'$1\frac{7}{8}$"	10'$\frac{1}{4}$"	11'$5\frac{3}{8}$"	12'$10\frac{11}{16}$"
WELL OPENING FOR 6'-8" HEADROOM	14'11"	8'$7\frac{23}{32}$"	10'$\frac{23}{32}$"	11'$6\frac{3}{32}$"	12'$11\frac{11}{16}$"	13'$6\frac{1}{16}$"	14'11"	8'$7\frac{23}{32}$"	10'$\frac{23}{32}$"	11'$6\frac{3}{32}$"
ANGLE OF INCLINE	27°56'	41°44'	38°5'	34°53'	32°7'	29°49'	27°56'	41°44'	38°5'	34°53'
LENGTH OF CARRIAGE	19'5"	13'$5\frac{1}{8}$"	14'$6\frac{1}{2}$"	15'$8\frac{9}{16}$"	16'$11\frac{3}{32}$"	18'$1\frac{31}{32}$"	19'5"	13'$5\frac{1}{8}$"	14'$6\frac{1}{2}$"	15'$8\frac{9}{16}$"
L. F. OF RISER PER INCH OF STAIRWAY WIDTH	1.583	1.166	1.25	1.333	1.416	1.5	1.583	1.166	1.25	1.333
L. F. OF TREAD PER INCH OF STAIRWAY WIDTH	1.5	1.083	1.166	1.25	1.333	1.416	1.5	1.083	1.166	1.25
FRAMING SQUARE SETTINGS FOR CARRIAGE CUTS — TONGUE	$6\frac{1}{16}$"	$8\frac{1}{4}$"	$7\frac{11}{16}$"	$7\frac{3}{16}$"	$6\frac{3}{4}$"	$6\frac{3}{8}$"	$6\frac{1}{16}$"	$8\frac{1}{4}$"	$7\frac{11}{16}$"	$7\frac{3}{16}$"
FRAMING SQUARE SETTINGS FOR CARRIAGE CUTS — BODY	$11\frac{7}{16}$"	$9\frac{1}{4}$"	$9\frac{13}{16}$"	$10\frac{5}{16}$"	$10\frac{3}{4}$"	$11\frac{1}{8}$"	$11\frac{7}{16}$"	$9\frac{1}{4}$"	$9\frac{13}{16}$"	$10\frac{5}{16}$"

FLOOR TO FLOOR RISE	9'7 1/4"	9'7 1/4"	9'7 1/4"	9'7 3/8"	9'7 3/8"	9'7 3/8"	9'7 3/8"	9'7 3/8"	9'7 3/8"	9'7 1/2"
NUMBER OF RISERS	17	18	19	14	15	16	17	18	19	14
HEIGHT OF EACH RISER	6 3/4"	6 3/8"	6 1/16"	8 1/4"	7 11/16"	7 3/16"	6 13/16"	6 7/16"	6 1/16"	8 1/4"
TOTAL HEIGHT OF RISERS	9'6 3/4"	9'6 3/4"	9'7 3/16"	9'7 1/2"	9'7 5/16"	9'7"	9'7 13/16"	9'7 7/8"	9'7 3/16"	9'7 1/2"
RISERS OVERAGE (+) OR UNDERAGE (—)	−1/2"	−1/2"	−1/16"	+1/8"	−1/16"	−3/8"	+7/16"	+1/2"	−3/16"	0
NUMBER OF TREADS	16	17	18	13	14	15	16	17	18	13
WIDTH OF EACH TREAD	10 3/4"	11 1/8"	11 7/16"	9 1/4"	9 13/16"	10 5/16"	10 11/16"	11 1/16"	11 7/16"	9 1/4"
TOTAL RUN OF STAIRWAY	14'4"	15'9 1/8"	17'1 7/8"	10' 1/4"	11'5 3/8"	12'10 11/16"	14'3"	15'8 1/16"	17'1 7/8"	10' 1/4"
WELL OPENING FOR 6'-8" HEADROOM	12'11 11/16"	13'6 1/16"	14'11'	8'7 23/32"	10' 23/32"	11'6 3/32"	11'11 15/16"	13'5"	14'11"	8'7 23/32"
ANGLE OF INCLINE	32°7'	29°49'	27°56'	41°44'	38°5'	34°53'	32°31'	30°12'	27°56'	41°44'
LENGTH OF CARRIAGE	16'11 3/32"	18'1 19/32"	19'5"	13'5 1/8"	14'6 1/2"	15'8 9/16"	16'10 25/32"	18'1 19/32"	19'5"	13'5 1/8"
L. F. OF RISER PER INCH OF STAIRWAY WIDTH	1.416	1.5	1.583	1.166	1.25	1.333	1.416	1.5	1.583	1.166
L. F. OF TREAD PER INCH OF STAIRWAY WIDTH	1.333	1.416	1.5	1.083	1.166	1.25	1.333	1.416	1.5	1.083
FRAMING SQUARE SETTINGS FOR CARRIAGE CUTS — TONGUE	6 3/4"	6 3/8"	6 1/16"	8 1/4"	7 11/16"	7 3/16"	6 13/16"	6 7/16"	6 1/16"	8 1/4"
FRAMING SQUARE SETTINGS FOR CARRIAGE CUTS — BODY	10 3/4"	11 1/8"	11 7/16"	9 1/4"	9 13/16"	10 5/16"	10 11/16"	11 1/16"	11 7/16"	9 1/4"

FLOOR TO FLOOR RISE	9'7$\frac{1}{2}$"	9'7$\frac{1}{2}$"	9'7$\frac{1}{2}$"	9'7$\frac{1}{2}$"	9'7$\frac{1}{2}$"	9'7$\frac{5}{8}$"	9'7$\frac{5}{8}$"	9'7$\frac{5}{8}$"	9'7$\frac{5}{8}$"	9'7$\frac{5}{8}$"
NUMBER OF RISERS	15	16	17	18	19	14	15	16	17	18
HEIGHT OF EACH RISER	7$\frac{11}{16}$"	7$\frac{1}{4}$"	6$\frac{13}{16}$"	6$\frac{7}{16}$"	6$\frac{1}{16}$"	8$\frac{1}{4}$"	7$\frac{11}{16}$"	7$\frac{1}{4}$"	6$\frac{13}{16}$"	6$\frac{7}{16}$"
TOTAL HEIGHT OF RISERS	9'7$\frac{5}{16}$"	9'8"	9'7$\frac{13}{16}$"	9'7$\frac{7}{8}$"	9'7$\frac{3}{16}$"	9'7$\frac{1}{2}$"	9'7$\frac{5}{16}$"	9'8"	9'7$\frac{13}{16}$"	9'7$\frac{7}{8}$"
RISERS OVERAGE (+) OR UNDERAGE (—)	−$\frac{3}{16}$"	+$\frac{1}{2}$"	+$\frac{5}{16}$"	+$\frac{3}{8}$"	−$\frac{5}{16}$"	−$\frac{1}{8}$"	−$\frac{5}{16}$"	+$\frac{3}{8}$"	+$\frac{3}{16}$"	+$\frac{1}{4}$"
NUMBER OF TREADS	14	15	16	17	18	13	14	15	16	17
WIDTH OF EACH TREAD	9$\frac{13}{16}$"	10$\frac{1}{4}$"	10$\frac{11}{16}$"	11$\frac{1}{16}$"	11$\frac{7}{16}$"	9$\frac{1}{4}$"	9$\frac{13}{16}$"	10$\frac{1}{4}$"	10$\frac{11}{16}$"	11$\frac{1}{16}$"
TOTAL RUN OF STAIRWAY	11'5$\frac{3}{8}$"	12'9$\frac{3}{4}$"	14'3"	15'8$\frac{1}{16}$"	17'1$\frac{7}{8}$"	10'$\frac{1}{4}$"	11'5$\frac{3}{8}$"	12'9$\frac{3}{4}$"	14'3"	15'8$\frac{1}{16}$"
WELL OPENING FOR 6'-8" HEADROOM	10'$\frac{23}{32}$"	11'5$\frac{1}{8}$"	11'11$\frac{15}{16}$"	13'5"	14'11"	8'7$\frac{23}{32}$"	10'$\frac{23}{32}$"	11'5$\frac{1}{8}$"	11'11$\frac{15}{16}$"	13'5"
ANGLE OF INCLINE	38°5'	35°16'	32°31'	30°12'	27°56'	41°44'	38°5'	35°16'	32°31'	30°12'
LENGTH OF CARRIAGE	14'6$\frac{1}{2}$"	15'8$\frac{5}{16}$"	16'10$\frac{25}{32}$"	18'1$\frac{19}{32}$"	19'5"	13'5$\frac{1}{8}$"	14'6$\frac{1}{2}$"	15'8$\frac{5}{16}$"	16'10$\frac{25}{32}$"	18'1$\frac{19}{32}$"
L. F. OF RISER PER INCH OF STAIRWAY WIDTH	1.25	1.333	1.416	1.5	1.583	1.166	1.25	1.333	1.416	1.5
L. F. OF TREAD PER INCH OF STAIRWAY WIDTH	1.166	1.25	1.333	1.416	1.5	1.083	1.166	1.25	1.333	1.416
FRAMING SQUARE SETTINGS FOR CARRIAGE CUTS — TONGUE	7$\frac{11}{16}$"	7$\frac{1}{4}$"	6$\frac{13}{16}$"	6$\frac{7}{16}$"	6$\frac{1}{16}$"	8$\frac{1}{4}$"	7$\frac{11}{16}$"	7$\frac{1}{4}$"	6$\frac{13}{16}$"	6$\frac{7}{16}$"
FRAMING SQUARE SETTINGS FOR CARRIAGE CUTS — BODY	9$\frac{13}{16}$"	10$\frac{1}{4}$"	10$\frac{11}{16}$"	11$\frac{1}{16}$"	11$\frac{7}{16}$"	9$\frac{1}{4}$"	9$\frac{13}{16}$"	10$\frac{1}{4}$"	10$\frac{11}{16}$"	11$\frac{1}{16}$"

FLOOR TO FLOOR RISE	9'7 5/8"	9'7 3/4"	9'7 3/4"	9'7 3/4"	9'7 3/4"	9'7 3/4"	9'7 3/4"	9'7 7/8"	9'7 7/8"	9'7 7/8"
NUMBER OF RISERS	19	14	15	16	17	18	19	14	15	16
HEIGHT OF EACH RISER	6 1/16"	8 1/4"	7 11/16"	7 1/4"	6 13/16"	6 7/16"	6 3/32"	8 1/4"	7 3/4"	7 1/4"
TOTAL HEIGHT OF RISERS	9'7 3/16"	9'7 1/2"	9'7 5/16"	9'8"	9'7 13/16"	9'7 7/8"	9'7 25/32"	9'7 1/2"	9'8 1/4"	9'8"
RISERS OVERAGE (+) OR UNDERAGE (—)	−7/16"	−1/4"	−7/16"	+1/4"	+1/16"	+1/8"	+1/32"	−3/8"	+3/8"	+1/8"
NUMBER OF TREADS	18	13	14	15	16	17	18	13	14	15
WIDTH OF EACH TREAD	11 7/16"	9 1/4"	9 13/16"	10 1/4"	10 11/16"	11 1/16"	11 13/32"	9 1/4"	9 3/4"	10 1/4"
TOTAL RUN OF STAIRWAY	17'1 7/8"	10'1/4"	11'5 3/8"	12'9 3/4"	14'3"	15'8 1/16"	17'1 5/16"	10'1/4"	11'4 1/2"	12'9 3/4"
WELL OPENING FOR 6'-8" HEADROOM	14'11"	8'7 23/32"	10'23/32"	11'5 1/8"	11'11 15/16"	13'5"	14'10 13/32"	8'7 23/32"	9'11 27/32"	11'5 1/8"
ANGLE OF INCLINE	27°56'	41°44'	38°5'	35°16'	32°31'	30°12'	28°7'	41°44'	38°29'	35°16'
LENGTH OF CARRIAGE	19'5"	13'5 1/8"	14'6 1/2"	15'8 5/16"	16'10 25/32"	18'1 19/32"	19'4 3/4"	13'5 1/8"	14'6 3/8"	15'8 5/16"
L. F. OF RISER PER INCH OF STAIRWAY WIDTH	1.583	1.166	1.25	1.333	1.416	1.5	1.583	1.166	1.25	1.333
L. F. OF TREAD PER INCH OF STAIRWAY WIDTH	1.5	1.083	1.166	1.25	1.333	1.416	1.5	1.083	1.166	1.25
FRAMING SQUARE SETTINGS FOR CARRIAGE CUTS — TONGUE	6 1/16"	8 1/4"	7 11/16"	7 1/4"	6 13/16"	6 7/16"	6 3/32"	8 1/4"	7 3/4"	7 1/4"
FRAMING SQUARE SETTINGS FOR CARRIAGE CUTS — BODY	11 7/16"	9 1/4"	9 13/16"	10 1/4"	10 11/16"	11 1/16"	11 13/32"	9 1/4"	9 3/4"	10 1/4"

FLOOR TO FLOOR RISE		9'7 7/8"	9'7 7/8"	9'7 7/8"	9'8"	9'8"	9'8"	9'8"	9'8"	9'8 1/8"	9'8 1/8"
NUMBER OF RISERS		17	18	19	15	16	17	18	19	15	16
HEIGHT OF EACH RISER		6 13/16"	6 7/16"	6 1/8"	7 3/4"	7 1/4"	6 13/16"	6 7/16"	6 1/8"	7 3/4"	7 1/4"
TOTAL HEIGHT OF RISERS		9'7 13/16"	9'7 7/8"	9'8 3/8"	9'8 1/4"	9'8"	9'7 13/16"	9'7 7/8"	9'8 3/8"	9'8 1/4"	9'8"
RISERS OVERAGE (+) OR UNDERAGE (—)		−1/16"	0	+1/2"	+1/4"	0	−3/16"	−1/8"	+3/8"	+1/8"	−1/8"
NUMBER OF TREADS		16	17	18	14	15	16	17	18	14	15
WIDTH OF EACH TREAD		10 11/16"	11 1/16"	11 3/8"	9 3/4"	10 1/4"	10 11/16"	11 1/16"	11 3/8"	9 3/4"	10 1/4"
TOTAL RUN OF STAIRWAY		14'3"	15'8 1/16"	17'3/4"	11'4 1/2"	12'9 3/4"	14'3"	15'8 1/16"	17'3/4"	11'4 1/2"	12'9 3/4"
WELL OPENING FOR 6'-8" HEADROOM		11'11 15/16"	13'5"	14'9 13/16"	9'11 27/32"	11'5 1/8"	11'11 15/16"	13'5"	14'9 13/16"	9'11 27/32"	11'5 1/8"
ANGLE OF INCLINE		32°31'	30°12'	28°18'	38°29'	35°16'	32°31'	30°12'	28°18'	38°29'	35°16'
LENGTH OF CARRIAGE		16'10 25/32"	18'1 19/32"	19'4 17/32"	14'6 3/8"	15'8 5/16"	16'10 25/32"	18'1 19/32"	19'4 17/32"	14'6 3/8"	15'8 5/16"
L. F. OF RISER PER INCH OF STAIRWAY WIDTH		1.416	1.5	1.583	1.25	1.333	1.416	1.5	1.583	1.25	1.333
L. F. OF TREAD PER INCH OF STAIRWAY WIDTH		1.333	1.416	1.5	1.166	1.25	1.333	1.416	1.5	1.166	1.25
FRAMING SQUARE SETTINGS FOR CARRIAGE CUTS	TONGUE	6 13/16"	6 7/16"	6 1/8"	7 3/4"	7 1/4"	6 13/16"	6 7/16"	6 1/8"	7 3/4"	7 1/4"
	BODY	10 11/16"	11 1/16"	11 3/8"	9 3/4"	10 1/4"	10 11/16"	11 1/16"	11 3/8"	9 3/4"	10 1/4"

FLOOR TO FLOOR RISE		9'8 1/8"	9'8 1/8"	9'8 1/8"	9'8 1/4"	9'8 1/4"	9'8 1/4"	9'8 1/4"	9'8 1/4"	9'8 3/8"	9'8 3/8"
NUMBER OF RISERS		17	18	19	15	16	17	18	19	15	16
HEIGHT OF EACH RISER		6 13/16"	6 7/16"	6 1/8"	7 3/4"	7 1/4"	6 13/16"	6 7/16"	6 1/8"	7 3/4"	7 1/4"
TOTAL HEIGHT OF RISERS		9'7 13/16"	9'7 7/8"	9'8 3/8"	9'8 1/4"	9'8"	9'7 13/16"	9'7 7/8"	9'8 3/8"	9'8 1/4"	9'8"
RISERS OVERAGE (+) OR UNDERAGE (—)		−5/16"	−1/4"	+1/4"	0	−1/4"	−7/16"	−3/8"	+1/8"	−1/8"	−3/8"
NUMBER OF TREADS		16	17	18	14	15	16	17	18	14	15
WIDTH OF EACH TREAD		10 11/16"	11 1/16"	11 3/8"	9 3/4"	10 1/4"	10 11/16"	11 1/16"	11 3/8"	9 3/4"	10 1/4"
TOTAL RUN OF STAIRWAY		14'3"	15'8 1/16"	17' 3/4"	11'4 1/2"	12'9 3/4"	14'3"	15'8 1/16"	17' 3/4"	11'4 1/2"	12'9 3/4"
WELL OPENING FOR 6'-8" HEADROOM		11'11 15/16"	13'5"	14'9 13/16"	9'11 27/32"	11'5 1/8"	11'11 15/16"	13'5"	14'9 13/16"	9'11 27/32"	11'5 1/8"
ANGLE OF INCLINE		32°31'	30°12'	28°18'	38°29'	35°16'	32°31'	30°12'	28°18'	38°29'	35°16'
LENGTH OF CARRIAGE		16'10 25/32"	18'1 19/32"	19'4 17/32"	14'6 3/8"	15'8 5/16"	16'10 25/32"	18'1 19/32"	19'4 17/32"	14'6 3/8"	15'8 5/16"
L. F. OF RISER PER INCH OF STAIRWAY WIDTH		1.416	1.5	1.583	1.25	1.333	1.416	1.5	1.583	1.25	1.333
L. F. OF TREAD PER INCH OF STAIRWAY WIDTH		1.333	1.416	1.5	1.166	1.25	1.333	1.416	1.5	1.166	1.25
FRAMING SQUARE SETTINGS FOR CARRIAGE CUTS	TONGUE	6 13/16"	6 7/16"	6 1/8"	7 3/4"	7 1/4"	6 13/16"	6 7/16"	6 1/8"	7 3/4"	7 1/4"
	BODY	10 11/16"	11 1/16"	11 3/8"	9 3/4"	10 1/4"	10 11/16"	11 1/16"	11 3/8"	9 3/4"	10 1/4"

FLOOR TO FLOOR RISE		9'8 3/8"	9'8 3/8"	9'8 3/8"	9'8 1/2"	9'8 1/2"	9'8 1/2"	9'8 1/2"	9'8 1/2"	9'8 5/8"	9'8 5/8"
NUMBER OF RISERS		17	18	19	15	16	17	18	19	15	16
HEIGHT OF EACH RISER		6 7/8"	6 7/16"	6 1/8"	7 3/4"	7 5/16"	6 7/8"	6 1/2"	6 1/8"	7 3/4"	7 5/16"
TOTAL HEIGHT OF RISERS		9'8 7/8"	9'7 7/8"	9'8 3/8"	9'8 1/4"	9'9"	9'8 7/8"	9'9"	9'8 3/8"	9'8 1/4"	9'9"
RISERS OVERAGE (+) OR UNDERAGE (—)		+1/2"	−1/2"	0	−1/4"	+1/2"	+3/8"	+1/2"	−1/8"	−3/8"	+3/8"
NUMBER OF TREADS		16	17	18	14	15	16	17	18	14	15
WIDTH OF EACH TREAD		10 5/8"	11 1/16"	11 3/8"	9 3/4"	10 3/16"	10 5/8"	11"	11 3/8"	9 3/4"	10 3/16"
TOTAL RUN OF STAIRWAY		14'2"	15'8 1/16"	17'3/4"	11'4 1/2"	12'8 13/16"	14'2"	15'7"	17'3/4"	11'4 1/2"	12'8 13/16"
WELL OPENING FOR 6'-8" HEADROOM		11'10 31/32"	13'5"	14'9 13/16"	9'11 27/32"	11'4 3/16"	11'10 31/32"	13'3 15/16"	14'9 13/16"	9'11 27/32"	11'4 3/16"
ANGLE OF INCLINE		32°54'	30°12'	28°18'	38°29'	35°40'	32°54'	30°35'	28°18'	38°29'	35°40'
LENGTH OF CARRIAGE		16'10 1/2"	18'1 19/32"	19'4 17/32"	14'6 3/8"	15'8 3/32"	16'10 1/2"	18'1 7/32"	19'4 17/32"	14'6 3/8"	15'8 3/32"
L. F. OF RISER PER INCH OF STAIRWAY WIDTH		1.416	1.5	1.583	1.25	1.333	1.416	1.5	1.583	1.25	1.333
L. F. OF TREAD PER INCH OF STAIRWAY WIDTH		1.333	1.416	1.5	1.166	1.25	1.333	1.416	1.5	1.166	1.25
FRAMING SQUARE SETTINGS FOR CARRIAGE CUTS	TONGUE	6 7/8"	6 7/16"	6 1/8"	7 3/4"	7 5/16"	6 7/8"	6 1/2"	6 1/8"	7 3/4"	7 5/16"
	BODY	10 5/8"	11 1/16"	11 3/8"	9 3/4"	10 3/16"	10 5/8"	11"	11 3/8"	9 3/4"	10 3/16"

FLOOR TO FLOOR RISE		9'8 5/8"	9'8 5/8"	9'8 5/8"	9'8 3/4"	9'8 3/4"	9'8 3/4"	9'8 3/4"	9'8 3/4"	9'8 7/8"	9'8 7/8"
NUMBER OF RISERS		17	18	19	15	16	17	18	19	15	16
HEIGHT OF EACH RISER		6 7/8"	6 1/2"	6 1/8"	7 13/16"	7 5/16"	6 7/8"	6 1/2"	6 1/8"	7 13/16"	7 5/16"
TOTAL HEIGHT OF RISERS		9'8 7/8"	9'9"	9'8 3/8"	9'9 3/16"	9'9"	9'8 7/8"	9'9"	9'8 3/8"	9'9 3/16"	9'9"
RISERS OVERAGE (+) OR UNDERAGE (—)		+1/4"	+3/8"	−1/4"	+7/16"	+1/4"	+1/8"	+1/4"	−3/8"	+5/16"	+1/8"
NUMBER OF TREADS		16	17	18	14	15	16	17	18	14	15
WIDTH OF EACH TREAD		10 5/8"	11"	11 3/8"	9 11/16"	10 3/16"	10 5/8"	11"	11 3/8"	9 11/16"	10 3/16"
TOTAL RUN OF STAIRWAY		14'2"	15'7"	17' 3/4"	11'3 5/8"	12'8 13/16"	14'2"	15'7"	17' 3/4"	11'3 5/8"	12'8 13/16"
WELL OPENING FOR 6'-8" HEADROOM		11'10 31/32"	13'3 15/16"	14'9 13/16"	9'10 15/16"	11'4 3/16"	11'10 31/32"	13'3 15/16"	14'9 13/16"	9'10 15/16"	11'4 3/16"
ANGLE OF INCLINE		32°54'	30°35'	28°18'	38°53'	35°40'	32°54'	30°35'	28°18'	38°53'	35°40'
LENGTH OF CARRIAGE		16'10 1/2"	18'1 7/32"	19'4 17/32"	14'6 7/32"	15'8 3/32"	16'10 1/2"	18'1 7/32"	19'4 17/32"	14'6 7/32"	15'8 3/32"
L. F. OF RISER PER INCH OF STAIRWAY WIDTH		1.416	1.5	1.583	1.25	1.333	1.416	1.5	1.583	1.25	1.333
L. F. OF TREAD PER INCH OF STAIRWAY WIDTH		1.333	1.416	1.5	1.166	1.25	1.333	1.416	1.5	1.166	1.25
FRAMING SQUARE SETTINGS FOR CARRIAGE CUTS	TONGUE	6 7/8"	6 1/2"	6 1/8"	7 13/16"	7 5/16"	6 7/8"	6 1/2"	6 1/8"	7 13/16"	7 5/16"
	BODY	10 5/8"	11"	11 3/8"	9 11/16"	10 3/16"	10 5/8"	11"	11 3/8"	9 11/16"	10 3/16"

FLOOR TO FLOOR RISE		9'8 7/8"	9'8 7/8"	9'8 7/8"	9'9"	9'9"	9'9"	9'9"	9'9"	9'9 1/8"	9'9 1/8"
NUMBER OF RISERS		17	18	19	15	16	17	18	19	15	16
HEIGHT OF EACH RISER		6 7/8"	6 1/2"	6 1/8"	7 13/16"	7 5/16"	6 7/8"	6 1/2"	6 5/32"	7 13/16"	7 5/16"
TOTAL HEIGHT OF RISERS		9'8 7/8"	9'9"	9'8 3/8"	9'9 3/16"	9'9"	9'8 7/8"	9'9"	9'8 31/32"	9'9 3/16"	9'9"
RISERS OVERAGE (+) OR UNDERAGE (—)		0	+1/8"	−1/2"	+3/16"	0	−1/8"	0	−1/32"	+1/16"	−1/8"
NUMBER OF TREADS		16	17	18	14	15	16	17	18	14	15
WIDTH OF EACH TREAD		10 5/8"	11"	11 3/8"	9 11/16"	10 3/16"	10 5/8"	11"	11 11/32"	9 11/16"	10 3/16"
TOTAL RUN OF STAIRWAY		14'2"	15'7"	17'3/4"	11'3 5/8"	12'8 13/16"	14'2"	15'7"	17'3/16"	11'3 5/8"	12'8 13/16"
WELL OPENING FOR 6'-8" HEADROOM		11'10 31/32"	13'3 15/16"	14'9 13/16"	9'10 15/16"	11'4 3/16"	11'10 31/32"	13'3 15/16"	14'9 1/4"	9'10 15/16"	11'4 3/16"
ANGLE OF INCLINE		32°54'	30°35'	28°18'	38°53'	35°40'	32°54'	30°35'	28°29'	38°53'	35°40'
LENGTH OF CARRIAGE		16'10 1/2"	18'1 7/32"	19'4 17/32"	14'6 7/32"	15'8 3/32"	16'10 1/2"	18'1 7/32"	19'4 5/16"	14'6 7/32"	15'8 3/32"
L. F. OF RISER PER INCH OF STAIRWAY WIDTH		1.416	1.5	1.583	1.25	1.333	1.416	1.5	1.583	1.25	1.333
L. F. OF TREAD PER INCH OF STAIRWAY WIDTH		1.333	1.416	1.5	1.166	1.25	1.333	1.416	1.5	1.166	1.25
FRAMING SQUARE SETTINGS FOR CARRIAGE CUTS	TONGUE	6 7/8"	6 1/2"	6 1/8"	7 13/16"	7 5/16"	6 7/8"	6 1/2"	6 5/32"	7 13/16"	7 5/16"
	BODY	10 5/8"	11"	11 3/8"	9 11/16"	10 3/16"	10 5/8"	11"	11 11/32"	9 11/16"	10 3/16"

FLOOR TO FLOOR RISE	$9'9\frac{1}{8}''$	$9'9\frac{1}{8}''$	$9'9\frac{1}{8}''$	$9'9\frac{1}{4}''$	$9'9\frac{1}{4}''$	$9'9\frac{1}{4}''$	$9'9\frac{1}{4}''$	$9'9\frac{1}{4}''$	$9'9\frac{3}{8}''$	$9'9\frac{3}{8}''$
NUMBER OF RISERS	17	18	19	15	16	17	18	19	15	16
HEIGHT OF EACH RISER	$6\frac{7}{8}''$	$6\frac{1}{2}''$	$6\frac{3}{16}''$	$7\frac{13}{16}''$	$7\frac{5}{16}''$	$6\frac{7}{8}''$	$6\frac{1}{2}''$	$6\frac{3}{16}''$	$7\frac{13}{16}''$	$7\frac{5}{16}''$
TOTAL HEIGHT OF RISERS	$9'8\frac{7}{8}''$	9'9"	$9'9\frac{9}{16}''$	$9'9\frac{3}{16}''$	9'9"	$9'8\frac{7}{8}''$	9'9"	$9'9\frac{9}{16}''$	$9'9\frac{3}{16}''$	9'9"
RISERS OVERAGE (+) OR UNDERAGE (—)	$-\frac{1}{4}''$	$-\frac{1}{8}''$	$+\frac{7}{16}''$	$-\frac{1}{16}''$	$-\frac{1}{4}''$	$-\frac{3}{8}''$	$-\frac{1}{4}''$	$+\frac{5}{16}''$	$-\frac{3}{16}''$	$-\frac{3}{8}''$
NUMBER OF TREADS	16	17	18	14	15	16	17	18	14	15
WIDTH OF EACH TREAD	$10\frac{5}{8}''$	11"	$11\frac{5}{16}''$	$9\frac{11}{16}''$	$10\frac{3}{16}''$	$10\frac{5}{8}''$	11"	$11\frac{5}{16}''$	$9\frac{11}{16}''$	$10\frac{3}{16}''$
TOTAL RUN OF STAIRWAY	14'2"	15'7"	$16'11\frac{5}{8}''$	$11'3\frac{5}{8}''$	$12'8\frac{13}{16}''$	14'2"	15'7"	$16'11\frac{5}{8}''$	$11'3\frac{5}{8}''$	$12'8\frac{13}{16}''$
WELL OPENING FOR 6'-8" HEADROOM	$11'10\frac{31}{32}''$	$13'3\frac{15}{16}''$	$14'8\frac{21}{32}''$	$9'10\frac{15}{16}''$	$11'4\frac{3}{16}''$	$11'10\frac{31}{32}''$	$13'3\frac{15}{16}''$	$14'8\frac{21}{32}''$	$9'10\frac{15}{16}''$	$11'4\frac{3}{16}''$
ANGLE OF INCLINE	32°54'	30°35'	28°41'	38°53'	35°40'	32°54'	30°35'	28°41'	38°53'	35°40'
LENGTH OF CARRIAGE	$16'10\frac{1}{2}''$	$18'1\frac{7}{32}''$	$19'4\frac{3}{32}''$	$14'6\frac{7}{32}''$	$15'8\frac{3}{32}''$	$16'10\frac{1}{2}''$	$18'1\frac{7}{32}''$	$19'4\frac{3}{32}''$	$14'6\frac{7}{32}''$	$15'8\frac{3}{32}''$
L. F. OF RISER PER INCH OF STAIRWAY WIDTH	1.416	1.5	1.583	1.25	1.333	1.416	1.5	1.583	1.25	1.333
L. F. OF TREAD PER INCH OF STAIRWAY WIDTH	1.333	1.416	1.5	1.166	1.25	1.333	1.416	1.5	1.166	1.25
FRAMING SQUARE SETTINGS FOR CARRIAGE CUTS — TONGUE	$6\frac{7}{8}''$	$6\frac{1}{2}''$	$6\frac{3}{16}''$	$7\frac{13}{16}''$	$7\frac{5}{16}''$	$6\frac{7}{8}''$	$6\frac{1}{2}''$	$6\frac{3}{16}''$	$7\frac{13}{16}''$	$7\frac{5}{16}''$
FRAMING SQUARE SETTINGS FOR CARRIAGE CUTS — BODY	$10\frac{5}{8}''$	11"	$11\frac{5}{16}''$	$9\frac{11}{16}''$	$10\frac{3}{16}''$	$10\frac{5}{8}''$	11"	$11\frac{5}{16}''$	$9\frac{11}{16}''$	$10\frac{3}{16}''$

FLOOR TO FLOOR RISE	9'9 3/8"	9'9 3/8"	9'9 3/8"	9'9 1/2"	9'9 1/2"	9'9 1/2"	9'9 1/2"	9'9 1/2"	9'9 5/8"	9'9 5/8"
NUMBER OF RISERS	17	18	19	15	16	17	18	19	15	16
HEIGHT OF EACH RISER	6 7/8"	6 1/2"	6 3/16"	7 13/16"	7 3/8"	6 15/16"	6 1/2"	6 3/16"	7 13/16"	7 3/8"
TOTAL HEIGHT OF RISERS	9'8 7/8"	9'9"	9'9 9/16"	9'9 3/16"	9'10"	9'9 15/16"	9'9"	9'9 9/16"	9'9 3/16"	9'10"
RISERS OVERAGE (+) OR UNDERAGE (—)	−1/2"	−3/8"	+3/16"	−5/16"	+1/2"	+7/16"	−1/2"	+1/16"	−7/16"	+3/8"
NUMBER OF TREADS	16	17	18	14	15	16	17	18	14	15
WIDTH OF EACH TREAD	10 5/8"	11"	11 5/16"	9 11/16"	10 1/8"	10 9/16"	11"	11 5/16"	9 11/16"	10 1/8"
TOTAL RUN OF STAIRWAY	14'2"	15'7"	16'11 5/8"	11'3 5/8"	12'7 7/8"	14'1"	15'7"	16'11 5/8"	11'3 5/8"	12'7 7/8"
WELL OPENING FOR 6'-8" HEADROOM	11'10 31/32"	13'3 15/16"	14'8 21/32"	9'10 15/16"	10'5 3/32"	11'9 31/32"	13'3 15/16"	14'8 21/32"	9'10 15/16"	10'5 3/32"
ANGLE OF INCLINE	32°54'	30°35'	28°41'	38°53'	36°4'	33°18'	30°35'	28°41'	38°53'	36°4'
LENGTH OF CARRIAGE	16'10 1/2"	18'1 7/32"	19'4 3/32"	14'6 7/32"	15'7 29/32"	16'10 3/16"	18'1 7/32"	19'4 3/32"	14'6 7/32"	15'7 29/32"
L. F. OF RISER PER INCH OF STAIRWAY WIDTH	1.416	1.5	1.583	1.25	1.333	1.416	1.5	1.583	1.25	1.333
L. F. OF TREAD PER INCH OF STAIRWAY WIDTH	1.333	1.416	1.5	1.166	1.25	1.333	1.416	1.5	1.166	1.25
FRAMING SQUARE SETTINGS FOR CARRIAGE CUTS — TONGUE	6 7/8"	6 1/2"	6 3/16"	7 13/16"	7 3/8"	6 15/16"	6 1/2"	6 3/16"	7 13/16"	7 3/8"
FRAMING SQUARE SETTINGS FOR CARRIAGE CUTS — BODY	10 5/8"	11"	11 5/16"	9 11/16"	10 1/8"	10 9/16"	11"	11 5/16"	9 11/16"	10 1/8"

FLOOR TO FLOOR RISE		9'9$\frac{5}{8}$"	9'9$\frac{5}{8}$"	9'9$\frac{5}{8}$"	9'9$\frac{3}{4}$"	9'9$\frac{3}{4}$"	9'9$\frac{3}{4}$"	9'9$\frac{3}{4}$"	9'9$\frac{3}{4}$"	9'9$\frac{7}{8}$"	9'9$\frac{7}{8}$"
NUMBER OF RISERS		17	18	19	15	16	17	18	19	15	16
HEIGHT OF EACH RISER		6$\frac{15}{16}$"	6$\frac{9}{16}$"	6$\frac{3}{16}$"	7$\frac{7}{8}$"	7$\frac{3}{8}$"	6$\frac{15}{16}$"	6$\frac{9}{16}$"	6$\frac{3}{16}$"	7$\frac{7}{8}$"	7$\frac{3}{8}$"
TOTAL HEIGHT OF RISERS		9'9$\frac{15}{16}$"	9'10$\frac{1}{8}$"	9'9$\frac{9}{16}$"	9'10$\frac{1}{8}$"	9'10"	9'9$\frac{15}{16}$"	9'10$\frac{1}{8}$"	9'9$\frac{9}{16}$"	9'10$\frac{1}{8}$"	9'10"
RISERS OVERAGE (+) OR UNDERAGE (—)		+$\frac{5}{16}$"	+$\frac{1}{2}$"	−$\frac{1}{16}$"	+$\frac{3}{8}$"	+$\frac{1}{4}$"	+$\frac{3}{16}$"	+$\frac{3}{8}$"	−$\frac{3}{16}$"	+$\frac{1}{4}$"	+$\frac{1}{8}$"
NUMBER OF TREADS		16	17	18	14	15	16	17	18	14	15
WIDTH OF EACH TREAD		10$\frac{9}{16}$"	10$\frac{15}{16}$"	11$\frac{5}{16}$"	9$\frac{5}{8}$"	10$\frac{1}{8}$"	10$\frac{9}{16}$"	10$\frac{15}{16}$"	11$\frac{5}{16}$"	9$\frac{5}{8}$"	10$\frac{1}{8}$"
TOTAL RUN OF STAIRWAY		14'1"	15'5$\frac{15}{16}$"	16'11$\frac{5}{8}$"	11'2$\frac{3}{4}$"	12'7$\frac{7}{8}$"	14'1"	15'5$\frac{15}{16}$"	16'11$\frac{5}{8}$"	11'2$\frac{3}{4}$"	12'7$\frac{7}{8}$"
WELL OPENING FOR 6'-8" HEADROOM		11'9$\frac{31}{32}$"	13'2$\frac{27}{32}$"	14'8$\frac{21}{32}$"	9'10$\frac{3}{32}$"	10'5$\frac{3}{32}$"	11'9$\frac{31}{32}$"	13'2$\frac{27}{32}$"	14'8$\frac{21}{32}$"	9'10$\frac{3}{32}$"	10'5$\frac{3}{32}$"
ANGLE OF INCLINE		33°18'	30°58'	28°41'	39°17'	36°4'	33°18'	30°58'	28°41'	39°17'	36°4'
LENGTH OF CARRIAGE		16'10$\frac{3}{16}$"	18'$\frac{27}{32}$"	19'4$\frac{3}{32}$"	14'6$\frac{3}{32}$"	15'7$\frac{29}{32}$"	16'10$\frac{3}{16}$"	18'$\frac{27}{32}$"	19'4$\frac{3}{32}$"	14'6$\frac{3}{32}$"	15'7$\frac{29}{32}$"
L. F. OF RISER PER INCH OF STAIRWAY WIDTH		1.416	1.5	1.583	1.25	1.333	1.416	1.5	1.583	1.25	1.333
L. F. OF TREAD PER INCH OF STAIRWAY WIDTH		1.333	1.416	1.5	1.166	1.25	1.333	1.416	1.5	1.166	1.25
FRAMING SQUARE SETTINGS FOR CARRIAGE CUTS	TONGUE	6$\frac{15}{16}$"	6$\frac{9}{16}$"	6$\frac{3}{16}$"	7$\frac{7}{8}$"	7$\frac{3}{8}$"	6$\frac{15}{16}$"	6$\frac{9}{16}$"	6$\frac{3}{16}$"	7$\frac{7}{8}$"	7$\frac{3}{8}$"
	BODY	10$\frac{9}{16}$"	10$\frac{15}{16}$"	11$\frac{5}{16}$"	9$\frac{5}{8}$"	10$\frac{1}{8}$"	10$\frac{9}{16}$"	10$\frac{15}{16}$"	11$\frac{5}{16}$"	9$\frac{5}{8}$"	10$\frac{1}{8}$"

FLOOR TO FLOOR RISE	9'9 7/8"	9'9 7/8"	9'9 7/8"	9'10"	9'10"	9'10"	9'10"	9'10"	9'10 1/8"	9'10 1/8"
NUMBER OF RISERS	17	18	19	15	16	17	18	19	15	16
HEIGHT OF EACH RISER	6 15/16"	6 9/16"	6 3/16"	7 7/8"	7 3/8"	6 15/16"	6 9/16"	6 3/16"	7 7/8"	7 3/8"
TOTAL HEIGHT OF RISERS	9'9 15/16"	9'10 1/8"	9'9 9/16"	9'10 1/8"	9'10"	9'9 15/16"	9'10 1/8"	9'9 9/16"	9'10 1/8"	9'10"
RISERS OVERAGE (+) OR UNDERAGE (—)	+1/16"	+1/4"	−5/16"	+1/8"	0	−1/16"	+1/8"	−7/16"	0	−1/8"
NUMBER OF TREADS	16	17	18	14	15	16	17	18	14	15
WIDTH OF EACH TREAD	10 9/16"	10 15/16"	11 5/16"	9 5/8"	10 1/8"	10 9/16"	10 15/16"	11 5/16"	9 5/8"	10 1/8"
TOTAL RUN OF STAIRWAY	14'1"	15'5 15/16"	16'11 5/8"	11'2 3/4"	12'7 7/8"	14'1"	15'5 15/16"	16'11 5/8"	11'2 3/4"	12'7 7/8"
WELL OPENING FOR 6'-8" HEADROOM	11'9 31/32"	13'2 27/32"	14'8 21/32"	9'10 3/32"	10'5 3/32"	11'9 31/32"	13'2 27/32"	14'8 21/32"	9'10 3/32"	10'5 3/32"
ANGLE OF INCLINE	33°18'	30°58'	28°41'	39°17'	36°4'	33°18'	30°58'	28°41'	39°17'	36°4'
LENGTH OF CARRIAGE	16'10 3/16"	18' 27/32"	19'4 3/32"	14'6 3/32"	15'7 29/32"	16'10 3/16"	18' 27/32"	19'4 3/32"	14'6 3/32"	15'7 29/32"
L. F. OF RISER PER INCH OF STAIRWAY WIDTH	1.416	1.5	1.583	1.25	1.333	1.416	1.5	1.583	1.25	1.333
L. F. OF TREAD PER INCH OF STAIRWAY WIDTH	1.333	1.416	1.5	1.166	1.25	1.333	1.416	1.5	1.166	1.25
FRAMING SQUARE SETTINGS FOR CARRIAGE CUTS — TONGUE	6 15/16"	6 9/16"	6 3/16"	7 7/8"	7 3/8"	6 15/16"	6 9/16"	6 3/16"	7 7/8"	7 3/8"
FRAMING SQUARE SETTINGS FOR CARRIAGE CUTS — BODY	10 9/16"	10 15/16"	11 5/16"	9 5/8"	10 1/8"	10 9/16"	10 15/16"	11 5/16"	9 5/8"	10 1/8"

FLOOR TO FLOOR RISE		9'10 1/8"	9'10 1/8"	9'10 1/8"	9'10 1/4"	9'10 1/4"	9'10 1/4"	9'10 1/4"	9'10 1/4"	9'10 3/8"	9'10 3/8"
NUMBER OF RISERS		17	18	19	15	16	17	18	19	15	16
HEIGHT OF EACH RISER		6 15/16"	6 9/16"	6 7/32"	7 7/8"	7 3/8"	6 15/16"	6 9/16"	6 1/4"	7 7/8"	7 3/8"
TOTAL HEIGHT OF RISERS		9'9 15/16"	9'10 1/8"	9'10 5/32"	9'10 1/8"	9'10"	9'9 15/16"	9'10 1/8"	9'10 3/4"	9'10 1/8"	9'10"
RISERS OVERAGE (+) OR UNDERAGE (—)		−3/16"	0	+1/32"	−1/8"	−1/4"	−5/16"	−1/8"	+1/2"	−1/4"	−3/8"
NUMBER OF TREADS		16	17	18	14	15	16	17	18	14	15
WIDTH OF EACH TREAD		10 9/16"	10 15/16"	11 9/32"	9 5/8"	10 1/8"	10 9/16"	10 15/16"	11 1/4"	9 5/8"	10 1/8"
TOTAL RUN OF STAIRWAY		14'1"	15'5 15/16"	16'11 1/16"	11'2 3/4"	12'7 7/8"	14'1"	15'5 15/16"	16'10 1/2"	11'2 3/4"	12'7 7/8"
WELL OPENING FOR 6'-8" HEADROOM		11'9 31/32"	13'2 27/32"	14'8 3/32"	9'10 3/32"	10'5 3/32"	11'9 31/32"	13'2 27/32"	14'7 1/2"	9'10 3/32"	10'5 3/32"
ANGLE OF INCLINE		33°18'	30°58'	28°52'	39°17'	36°4'	33°18'	30°58'	29°3'	39°17'	36°4'
LENGTH OF CARRIAGE		16'10 3/16"	18' 27/32"	19'3 7/8"	14'6 3/32"	15'7 29/32"	16'10 3/16"	18' 27/32"	19'3 21/32"	14'6 3/32"	15'7 29/32"
L. F. OF RISER PER INCH OF STAIRWAY WIDTH		1.416	1.5	1.583	1.25	1.333	1.416	1.5	1.583	1.25	1.333
L. F. OF TREAD PER INCH OF STAIRWAY WIDTH		1.333	1.416	1.5	1.166	1.25	1.333	1.416	1.5	1.166	1.25
FRAMING SQUARE SETTINGS FOR CARRIAGE CUTS	TONGUE	6 15/16"	6 9/16"	6 7/32"	7 7/8"	7 3/8"	6 15/16"	6 9/16"	6 1/4"	7 7/8"	7 3/8"
	BODY	10 9/16"	10 15/16"	11 9/32"	9 5/8"	10 1/8"	10 9/16"	10 15/16"	11 1/4"	9 5/8"	10 1/8"

FLOOR TO FLOOR RISE		9'10 3/8"	9'10 3/8"	9'10 3/8"	9'10 1/2"	9'10 1/2"	9'10 1/2"	9'10 1/2"	9'10 1/2"	9'10 5/8"	9'10 5/8"
NUMBER OF RISERS		17	18	19	15	16	17	18	19	15	16
HEIGHT OF EACH RISER		6 15/16"	6 9/16"	6 1/4"	7 7/8"	7 7/16"	7"	6 9/16"	6 1/4"	7 15/16"	7 7/16"
TOTAL HEIGHT OF RISERS		9'9 15/16"	9'10 1/8"	9'10 3/4"	9'10 1/8"	9'11"	9'11"	9'10 1/8"	9'10 3/4"	9'11 1/16"	9'11"
RISERS OVERAGE (+) OR UNDERAGE (—)		−7/16"	−1/4"	+3/8"	−3/8"	+1/2"	+1/2"	−3/8"	+1/4"	+7/16"	+3/8"
NUMBER OF TREADS		16	17	18	14	15	16	17	18	14	15
WIDTH OF EACH TREAD		10 9/16"	10 15/16"	11 1/4"	9 5/8"	10 1/16"	10 1/2"	10 15/16"	11 1/4"	9 9/16"	10 1/16"
TOTAL RUN OF STAIRWAY		14'1"	15'5 15/16"	16'10 1/2"	11'2 3/4"	12'6 15/16"	14'	15'5 15/16"	16'10 1/2"	11'1 7/8"	12'6 15/16"
WELL OPENING FOR 6'-8" HEADROOM		11'9 31/32"	13'2 27/32"	14'7 1/2"	9'10 3/32"	10'4 7/32"	11'9"	13'2 27/32"	14'7 1/2"	9'9 1/4"	10'4 7/32"
ANGLE OF INCLINE		33°18'	30°58'	29°3'	39°17'	36°28'	33°41'	30°58'	29°3'	39°42'	36°28'
LENGTH OF CARRIAGE		16'10 3/16"	18' 27/32"	19'3 21/32"	14'6 3/32"	15'7 11/16"	16'9 29/32"	18' 27/32"	19'3 21/32"	14'6"	15'7 11/16"
L. F. OF RISER PER INCH OF STAIRWAY WIDTH		1.416	1.5	1.583	1.25	1.333	1.416	1.5	1.583	1.25	1.333
L. F. OF TREAD PER INCH OF STAIRWAY WIDTH		1.333	1.416	1.5	1.166	1.25	1.333	1.416	1.5	1.166	1.25
FRAMING SQUARE SETTINGS FOR CARRIAGE CUTS	TONGUE	6 15/16"	6 9/16"	6 1/4"	7 7/8"	7 7/16"	7"	6 9/16"	6 1/4"	7 15/16"	7 7/16"
	BODY	10 9/16"	10 15/16"	11 1/4"	9 5/8"	10 1/16"	10 1/2"	10 15/16"	11 1/4"	9 9/16"	10 1/16"

FLOOR TO FLOOR RISE		9'10$\frac{5}{8}$"	9'10$\frac{5}{8}$"	9'10$\frac{5}{8}$"	9'10$\frac{3}{4}$"	9'10$\frac{3}{4}$"	9'10$\frac{3}{4}$"	9'10$\frac{3}{4}$"	9'10$\frac{3}{4}$"	9'10$\frac{7}{8}$"	9'10$\frac{7}{8}$"
NUMBER OF RISERS		17	18	19	15	16	17	18	19	15	16
HEIGHT OF EACH RISER		7"	6$\frac{9}{16}$"	6$\frac{1}{4}$"	7$\frac{15}{16}$"	7$\frac{7}{16}$"	7"	6$\frac{5}{8}$"	6$\frac{1}{4}$"	7$\frac{15}{16}$"	7$\frac{7}{16}$"
TOTAL HEIGHT OF RISERS		9'11"	9'10$\frac{1}{8}$"	9'10$\frac{3}{4}$"	9'11$\frac{1}{16}$"	9'11"	9'11"	9'11$\frac{1}{4}$"	9'10$\frac{3}{4}$"	9'11$\frac{1}{16}$"	9'11"
RISERS OVERAGE (+) OR UNDERAGE (—)		+$\frac{3}{8}$"	−$\frac{1}{2}$"	+$\frac{1}{8}$"	+$\frac{5}{16}$"	+$\frac{1}{4}$"	+$\frac{1}{4}$"	+$\frac{1}{2}$"	0	+$\frac{3}{16}$"	+$\frac{1}{8}$"
NUMBER OF TREADS		16	17	18	14	15	16	17	18	14	15
WIDTH OF EACH TREAD		10$\frac{1}{2}$"	10$\frac{15}{16}$"	11$\frac{1}{4}$"	9$\frac{9}{16}$"	10$\frac{1}{16}$"	10$\frac{1}{2}$"	10$\frac{7}{8}$"	11$\frac{1}{4}$"	9$\frac{9}{16}$"	10$\frac{1}{16}$"
TOTAL RUN OF STAIRWAY		14'	15'5$\frac{15}{16}$"	16'10$\frac{1}{2}$"	11'1$\frac{7}{8}$"	12'6$\frac{15}{16}$"	14'	15'4$\frac{7}{8}$"	16'10$\frac{1}{2}$"	11'1$\frac{7}{8}$"	12'6$\frac{15}{16}$"
WELL OPENING FOR 6'-8" HEADROOM		11'9"	13'2$\frac{27}{32}$"	14'7$\frac{1}{2}$"	9'9$\frac{1}{4}$"	10'4$\frac{7}{32}$"	11'9"	13'1$\frac{23}{32}$"	14'7$\frac{1}{2}$"	9'9$\frac{1}{4}$"	10'4$\frac{7}{32}$"
ANGLE OF INCLINE		33°41'	30°58'	29°3'	39°42'	36°28'	33°41'	31°21'	29°3'	39°42'	36°28'
LENGTH OF CARRIAGE		16'9$\frac{29}{32}$"	18'$\frac{27}{32}$"	19'3$\frac{21}{32}$"	14'6"	15'7$\frac{11}{16}$"	16'9$\frac{29}{32}$"	18'$\frac{15}{32}$"	19'3$\frac{21}{32}$"	14'6"	15'7$\frac{11}{16}$"
L. F. OF RISER PER INCH OF STAIRWAY WIDTH		1.416	1.5	1.583	1.25	1.333	1.416	1.5	1.583	1.25	1.333
L. F. OF TREAD PER INCH OF STAIRWAY WIDTH		1.333	1.416	1.5	1.166	1.25	1.333	1.416	1.5	1.166	1.25
FRAMING SQUARE SETTINGS FOR CARRIAGE CUTS	TONGUE	7"	6$\frac{9}{16}$"	6$\frac{1}{4}$"	7$\frac{15}{16}$"	7$\frac{7}{16}$"	7"	6$\frac{5}{8}$"	6$\frac{1}{4}$"	7$\frac{15}{16}$"	7$\frac{7}{16}$"
	BODY	10$\frac{1}{2}$"	10$\frac{15}{16}$"	11$\frac{1}{4}$"	9$\frac{9}{16}$"	10$\frac{1}{16}$"	10$\frac{1}{2}$"	10$\frac{7}{8}$"	11$\frac{1}{4}$"	9$\frac{9}{16}$"	10$\frac{1}{16}$"

FLOOR TO FLOOR RISE	9'10 7/8"	9'10 7/8"	9'10 7/8"	9'11"	9'11"	9'11"	9'11"	9'11"	9'11 1/8"	9'11 1/8"
NUMBER OF RISERS	17	18	19	15	16	17	18	19	15	16
HEIGHT OF EACH RISER	7"	6 5/8"	6 1/4"	7 15/16"	7 7/16"	7"	6 5/8"	6 1/4"	7 15/16"	7 7/16"
TOTAL HEIGHT OF RISERS	9'11"	9'11 1/4"	9'10 3/4"	9'11 1/16"	9'11"	9'11"	9'11 1/4"	9'10 3/4"	9'11 1/16"	9'11"
RISERS OVERAGE (+) OR UNDERAGE (—)	+1/8"	+3/8"	−1/8"	+1/16"	0	0	+1/4"	−1/4"	−1/16"	−1/8"
NUMBER OF TREADS	16	17	18	14	15	16	17	18	14	15
WIDTH OF EACH TREAD	10 1/2"	10 7/8"	11 1/4"	9 9/16"	10 1/16"	10 1/2"	10 7/8"	11 1/4"	9 9/16"	10 1/16"
TOTAL RUN OF STAIRWAY	14'	15'4 7/8"	16'10 1/2"	11'1 7/8"	12'6 15/16"	14'	15'4 7/8"	16'10 1/2"	11'1 7/8"	12'6 15/16"
WELL OPENING FOR 6'-8" HEADROOM	11'9"	13'1 25/32"	14'7 1/2"	9'9 1/4"	10'4 7/32"	11'9"	13'1 25/32"	14'7 1/2"	9'9 1/4"	10'4 7/32"
ANGLE OF INCLINE	33°41'	31°21'	29°3'	39°42'	36°28'	33°41'	31°21'	29°3'	39°42'	36°28'
LENGTH OF CARRIAGE	16'9 29/32"	18' 15/32"	19'3 21/32"	14'6"	15'7 11/16"	16'9 29/32"	18' 15/32"	19'3 21/32"	14'6"	15'7 11/16"
L. F. OF RISER PER INCH OF STAIRWAY WIDTH	1.416	1.5	1.583	1.25	1.333	1.416	1.5	1.583	1.25	1.333
L. F. OF TREAD PER INCH OF STAIRWAY WIDTH	1.333	1.416	1.5	1.166	1.25	1.333	1.416	1.5	1.166	1.25
FRAMING SQUARE SETTINGS FOR CARRIAGE CUTS — TONGUE	7"	6 5/8"	6 1/4"	7 15/16"	7 7/16"	7"	6 5/8"	6 1/4"	7 15/16"	7 7/16"
FRAMING SQUARE SETTINGS FOR CARRIAGE CUTS — BODY	10 1/2"	10 7/8"	11 1/4"	9 9/16"	10 1/16"	10 1/2"	10 7/8"	11 1/4"	9 9/16"	10 1/16"

FLOOR TO FLOOR RISE	9'11 1/8"	9'11 1/8"	9'11 1/8"	9'11 1/4"	9'11 1/4"	9'11 1/4"	9'11 1/4"	9'11 1/4"	9'11 3/8"	9'11 3/8"
NUMBER OF RISERS	17	18	19	15	16	17	18	19	15	16
HEIGHT OF EACH RISER	7"	6 5/8"	6 1/4"	7 15/16"	7 7/16"	7"	6 5/8"	6 1/4"	7 15/16"	7 7/16"
TOTAL HEIGHT OF RISERS	9'11"	9'11 1/4"	9'10 3/4"	9'11 1/16"	9'11"	9'11"	9'11 1/4"	9'10 3/4"	9'11 1/16"	9'11"
RISERS OVERAGE (+) OR UNDERAGE (—)	−1/8"	+1/8"	−3/8"	−3/16"	−1/4"	−1/4"	0	−1/2"	−5/16"	−3/8"
NUMBER OF TREADS	16	17	18	14	15	16	17	18	14	15
WIDTH OF EACH TREAD	10 1/2"	10 7/8"	11 1/4"	9 9/16"	10 1/16"	10 1/2"	10 7/8"	11 1/4"	9 9/16"	10 1/16"
TOTAL RUN OF STAIRWAY	14'	15'4 7/8"	16'10 1/2"	11'1 7/8"	12'6 15/16"	14'	15'4 7/8"	16'10 1/2"	11'1 7/8"	12'6 15/16"
WELL OPENING FOR 6'-8" HEADROOM	11'9"	13'1 25/32"	14'7 1/2"	9'9 1/4"	10'4 7/32"	11'9"	13'1 25/32"	14'7 1/2"	9'9 1/4"	10'4 7/32"
ANGLE OF INCLINE	33°41'	31°21'	29°3'	39°42'	36°28'	33°41'	31°21'	29°3'	39°42'	36°28'
LENGTH OF CARRIAGE	16'9 29/32"	18' 15/32"	19'3 21/32"	14'6"	15'7 11/16"	16'9 29/32"	18' 15/32"	19'3 21/32"	14'6"	15'7 11/16"
L. F. OF RISER PER INCH OF STAIRWAY WIDTH	1.416	1.5	1.583	1.25	1.333	1.416	1.5	1.583	1.25	1.333
L. F. OF TREAD PER INCH OF STAIRWAY WIDTH	1.333	1.416	1.5	1.166	1.25	1.333	1.416	1.5	1.166	1.25
FRAMING SQUARE SETTINGS FOR CARRIAGE CUTS — TONGUE	7"	6 5/8"	6 1/4"	7 15/16"	7 7/16"	7"	6 5/8"	6 1/4"	7 15/16"	7 7/16"
FRAMING SQUARE SETTINGS FOR CARRIAGE CUTS — BODY	10 1/2"	10 7/8"	11 1/4"	9 9/15"	10 1/16"	10 1/2"	10 7/8"	11 1/4"	9 9/16"	10 1/16"

FLOOR TO FLOOR RISE	9'11 3/8"	9'11 3/8"	9'11 3/8"	9'11 1/2"	9'11 1/2"	9'11 1/2"	9'11 1/2"	9'11 1/2"	9'11 5/8"	9'11 5/8"
NUMBER OF RISERS	17	18	19	15	16	17	18	19	15	16
HEIGHT OF EACH RISER	7"	6 5/8"	6 9/32"	7 15/16"	7 1/2"	7"	6 5/8"	6 5/16"	8"	7 1/2"
TOTAL HEIGHT OF RISERS	9'11"	9'11 1/4"	9'11 11/32"	9'11 1/16"	10'	9'11"	9'11 1/4"	9'11 15/16"	10'	10'
RISERS OVERAGE (+) OR UNDERAGE (—)	−3/8"	−1/8"	−1/32"	−7/16"	+1/2"	−1/2"	−1/4"	+7/16"	+3/8"	+3/8"
NUMBER OF TREADS	16	17	18	14	15	16	17	18	14	15
WIDTH OF EACH TREAD	10 1/2"	10 7/8"	11 7/32"	9 9/16"	10"	10 1/2"	10 7/8"	11 3/16"	9 1/2"	10"
TOTAL RUN OF STAIRWAY	14'	15'4 7/8"	16'9 15/16"	11'1 7/8"	12'6"	14'	15'4 7/8"	16'9 3/8"	11'1"	12'6"
WELL OPENING FOR 6'-8" HEADROOM	11'9"	13'1 25/32"	14'6 29/32"	9'9 1/4"	10'3 11/32"	11'9"	13'1 25/32"	13'7 5/32"	8'10 7/8"	10'3 11/32"
ANGLE OF INCLINE	33°41'	31°21'	29°15'	39°42'	36°52'	33°41'	31°21'	29°26'	40°6'	36°52'
LENGTH OF CARRIAGE	16'9 29/32"	18' 15/32"	19'3 7/16"	14'6"	15'7 1/2"	16'9 29/32"	18' 15/32"	19'3 7/32"	14'5 7/8"	15'7 1/2"
L. F. OF RISER PER INCH OF STAIRWAY WIDTH	1.416	1.5	1.583	1.25	1.333	1.416	1.5	1.583	1.25	1.333
L. F. OF TREAD PER INCH OF STAIRWAY WIDTH	1.333	1.416	1.5	1.166	1.25	1.333	1.416	1.5	1.166	1.25
FRAMING SQUARE SETTINGS FOR CARRIAGE CUTS — TONGUE	7"	6 5/8"	6 9/32"	7 15/16"	7 1/2"	7"	6 5/8"	6 5/16"	8"	7 1/2"
FRAMING SQUARE SETTINGS FOR CARRIAGE CUTS — BODY	10 1/2"	10 7/8"	11 7/32"	9 9/16"	10"	10 1/2"	10 7/8"	11 3/16"	9 1/2"	10"

FLOOR TO FLOOR RISE		9'11 5/8"	9'11 5/8"	9'11 5/8"	9'11 3/4"	9'11 3/4"	9'11 3/4"	9'11 3/4"	9'11 3/4"	9'11 7/8"	9'11 7/8"
NUMBER OF RISERS		17	18	19	15	16	17	18	19	15	16
HEIGHT OF EACH RISER		7 1/16"	6 5/8"	6 5/16"	8"	7 1/2"	7 1/16"	6 5/8"	6 5/16"	8"	7 1/2"
TOTAL HEIGHT OF RISERS		10' 1/16"	9'11 1/4"	9'11 15/16"	10'	10'	10' 1/16"	9'11 1/4"	9'11 15/16"	10'	10'
RISERS OVERAGE (+) OR UNDERAGE (—)		+7/16"	−3/8"	+5/16"	+1/4"	+1/4"	+5/16"	−1/2"	+3/16"	+1/8"	+1/8"
NUMBER OF TREADS		16	17	18	14	15	16	17	18	14	15
WIDTH OF EACH TREAD		10 7/16"	10 7/8"	11 3/16"	9 1/2"	10"	10 7/16"	10 7/8"	11 3/16"	9 1/2"	10"
TOTAL RUN OF STAIRWAY		13'11"	15'4 7/8"	16'9 3/8"	11'1"	12'6"	13'11"	15'4 7/8"	16'9 3/8"	11'1"	12'6"
WELL OPENING FOR 6'-8" HEADROOM		11'8 1/32"	13'1 25/32"	13'7 5/32"	8'10 7/8"	10'3 11/32"	11'8 1/32"	13'1 25/32"	13'7 5/32"	8'10 7/8"	10'3 11/32"
ANGLE OF INCLINE		34°5'	31°21'	29°26'	40°6'	36°52'	34°5'	31°21'	29°26'	40°6'	36°52'
LENGTH OF CARRIAGE		16'9 5/8"	18' 15/32"	19'3 7/32"	14'5 7/8"	15'7 1/2"	16'9 5/8"	18' 15/32"	19'3 7/32"	14'5 7/8"	15'7 1/2"
L. F. OF RISER PER INCH OF STAIRWAY WIDTH		1.416	1.5	1.583	1.25	1.333	1.416	1.5	1.583	1.25	1.333
L. F. OF TREAD PER INCH OF STAIRWAY WIDTH		1.333	1.416	1.5	1.166	1.25	1.333	1.416	1.5	1.166	1.25
FRAMING SQUARE SETTINGS FOR CARRIAGE CUTS	TONGUE	7 1/16"	6 5/8"	6 5/16"	8"	7 1/2"	7 1/16"	6 5/8"	6 5/16"	8"	7 1/2"
	BODY	10 7/16"	10 7/8"	11 3/16"	9 1/2"	10"	10 7/16"	10 7/8"	11 3/16"	9 1/2"	10"

FLOOR TO FLOOR RISE		9'11 7/8"	9'11 7/8"	9'11 7/8"	10'	10'	10'	10'	10'	10'	10 1/8"
NUMBER OF RISERS		17	18	19	15	16	17	18	19	20	15
HEIGHT OF EACH RISER		7 1/16"	6 11/16"	6 5/16"	8"	7 1/2"	7 1/16"	6 11/16"	6 5/16"	6"	8"
TOTAL HEIGHT OF RISERS		10' 1/16"	10' 3/8"	9'11 15/16"	10'	10'	10' 1/16"	10' 3/8"	9'11 15/16"	10'	10'
RISERS OVERAGE (+) OR UNDERAGE (—)		+3/16"	+1/2"	+1/16"	0	0	+1/16"	+3/8"	−1/16"	0	−1/8"
NUMBER OF TREADS		16	17	18	14	15	16	17	18	19	14
WIDTH OF EACH TREAD		10 7/16"	10 13/16"	11 3/16"	9 1/2"	10"	10 7/16"	10 13/16"	11 3/16"	11 1/2"	9 1/2"
TOTAL RUN OF STAIRWAY		13'11"	15'3 13/16"	16'9 3/8"	11'1"	12'6"	13'11"	15'3 13/16"	16'9 3/8"	18'2 1/2"	11'1"
WELL OPENING FOR 6'-8" HEADROOM		11'8 1/32"	13' 3/4"	13'7 5/32"	8'10 7/8"	10'3 11/32"	11'8 1/32"	13' 3/4"	13'7 5/32"	15' 5/32"	8'10 7/8"
ANGLE OF INCLINE		34°5'	31°44'	29°26'	40°6'	36°52'	34°5'	31°44'	29°26'	27°33'	40°6'
LENGTH OF CARRIAGE		16'9 5/8"	18' 1/8"	19'3 7/32"	14'5 7/8"	15'7 1/2"	16'9 5/8"	18' 1/8"	19'3 7/32"	20'6 7/16"	14'5 7/8"
L. F. OF RISER PER INCH OF STAIRWAY WIDTH		1.416	1.5	1.583	1.25	1.333	1.416	1.5	1.583	1.666	1.25
L. F. OF TREAD PER INCH OF STAIRWAY WIDTH		1.333	1.416	1.5	1.166	1.25	1.333	1.416	1.5	1.583	1.166
FRAMING SQUARE SETTINGS FOR CARRIAGE CUTS	TONGUE	7 1/16"	6 11/16"	6 5/16"	8"	7 1/2"	7 1/16"	6 11/16"	6 5/16"	6"	8"
	BODY	10 7/16"	10 13/16"	11 3/16"	9 1/2"	10"	10 7/16"	10 13/16"	11 3/16"	11 1/2"	9 1/2"

FLOOR TO FLOOR RISE		10'1/8"	10'1/8"	10'1/8"	10'1/8"	10'1/8"	10'1/4"	10'1/4"	10'1/4"	10'1/4"	10'1/4"
NUMBER OF RISERS		16	17	18	19	20	15	16	17	18	19
HEIGHT OF EACH RISER		7 1/2"	7 1/16"	6 11/16"	6 5/16"	6"	8"	7 1/2"	7 1/16"	6 11/16"	6 5/16"
TOTAL HEIGHT OF RISERS		10'	10'1/16"	10'3/8"	9'11 15/16"	10'	10'	10'	10'1/16"	10'3/8"	9'11 15/16"
RISERS OVERAGE (+) OR UNDERAGE (—)		−1/8"	−1/16"	+1/4"	−3/16"	−1/8"	−1/4"	−1/4"	−3/16"	+1/8"	−5/16"
NUMBER OF TREADS		15	16	17	18	19	14	15	16	17	18
WIDTH OF EACH TREAD		10"	10 7/16"	10 13/16"	11 3/16"	11 1/2"	9 1/2"	10"	10 7/16"	10 13/16"	11 3/16"
TOTAL RUN OF STAIRWAY		12'6"	13'11"	15'3 13/16"	16'9 3/8"	18'2 1/2"	11'1"	12'6"	13'11"	15'3 13/16"	16'9 3/8"
WELL OPENING FOR 6'-8" HEADROOM		10'3 11/32"	11'8 1/32"	13'3/4""	13'7 5/32"	15'5/32"	8'10 7/8"	10'3 11/32"	11'8 1/32"	13'3/4""	13'7 5/32"
ANGLE OF INCLINE		36°52'	34°5'	31°44'	29°26'	27°33'	40°6'	36°52'	34°5'	31°44'	29°26'
LENGTH OF CARRIAGE		15'7 1/2"	16'9 5/8"	18'1/8"	19'3 7/32"	20'6 7/16"	14'5 7/8"	15'7 1/2"	16'9 5/8"	18'1/8"	19'3 7/32"
L. F. OF RISER PER INCH OF STAIRWAY WIDTH		1.333	1.416	1.5	1.583	1.666	1.25	1.333	1.416	1.5	1.583
L. F. OF TREAD PER INCH OF STAIRWAY WIDTH		1.25	1.333	1.416	1.5	1.583	1.166	1.25	1.333	1.416	1.5
FRAMING SQUARE SETTINGS FOR CARRIAGE CUTS	TONGUE	7 1/2"	7 1/16"	6 11/16"	6 5/16"	6"	8"	7 1/2"	7 1/16"	6 11/16"	6 5/16"
	BODY	10"	10 7/16"	10 13/16"	11 3/16"	11 1/2"	9 1/2"	10"	10 7/16"	10 13/16"	11 3/16"

FLOOR TO FLOOR RISE	10' 1/4"	10' 3/8"	10' 3/8"	10' 3/8"	10' 3/8"	10' 3/8"	10' 3/8"	10' 1/2"	10' 1/2"	10' 1/2"
NUMBER OF RISERS	20	15	16	17	18	19	20	15	16	17
HEIGHT OF EACH RISER	6"	8"	7 1/2"	7 1/16"	6 11/16"	6 5/16"	6"	8 1/16"	7 9/16"	7 1/16"
TOTAL HEIGHT OF RISERS	10'	10'	10'	10' 1/16"	10' 3/8"	9' 11 15/16"	10'	10' 15/16"	10' 1"	10' 1/16"
RISERS OVERAGE (+) OR UNDERAGE (—)	−1/4"	−3/8"	−3/8"	−5/16"	0	−7/16"	−3/8"	+7/16"	+1/2"	−7/16"
NUMBER OF TREADS	19	14	15	16	17	18	19	14	15	16
WIDTH OF EACH TREAD	11 1/2"	9 1/2"	10"	10 7/16"	10 13/16"	11 3/16"	11 1/2"	9 7/16"	9 15/16"	10 7/16"
TOTAL RUN OF STAIRWAY	18' 2 1/2"	11' 1"	12' 6"	13' 11"	15' 3 13/16"	16' 9 3/8"	18' 2 1/2"	11' 1/8"	12' 5 1/16"	13' 11"
WELL OPENING FOR 6'-8" HEADROOM	15' 5/32"	8' 10 7/8"	10' 3 11/32"	11' 8 1/32"	13' 3/4""	13' 7 5/32"	15' 5/32"	8' 10 3/32"	10' 2 7/16"	11' 8 1/32"
ANGLE OF INCLINE	27°33'	40°6'	36°52'	34°5'	31°44'	29°26'	27°33'	40°30'	37°16'	34°5'
LENGTH OF CARRIAGE	20' 6 7/16"	14' 5 7/8"	15' 7 1/2"	16' 9 5/8"	18' 1/8"	19' 3 7/32"	20' 6 7/16"	14' 5 25/32"	15' 7 5/16"	16' 9 5/8"
L. F. OF RISER PER INCH OF STAIRWAY WIDTH	1.666	1.25	1.333	1.416	1.5	1.583	1.666	1.25	1.333	1.416
L. F. OF TREAD PER INCH OF STAIRWAY WIDTH	1.583	1.166	1.25	1.333	1.416	1.5	1.583	1.166	1.25	1.333
FRAMING SQUARE SETTINGS FOR CARRIAGE CUTS — TONGUE	6"	8"	7 1/2"	7 1/16"	6 11/16"	6 5/16"	6"	8 1/16"	7 9/16"	7 1/16"
FRAMING SQUARE SETTINGS FOR CARRIAGE CUTS — BODY	11 1/2"	9 1/2"	10"	10 7/16"	10 13/16"	11 3/16"	11 1/2"	9 7/16"	9 15/16"	10 7/16"

FLOOR TO FLOOR RISE	$10'\frac{1}{2}''$	$10'\frac{1}{2}''$	$10'\frac{1}{2}''$	$10'\frac{5}{8}''$	$10'\frac{5}{8}''$	$10'\frac{5}{8}''$	$10'\frac{5}{8}''$	$10'\frac{5}{8}''$	$10'\frac{5}{8}''$	$10'\frac{3}{4}''$
NUMBER OF RISERS	18	19	20	15	16	17	18	19	20	15
HEIGHT OF EACH RISER	$6\frac{11}{16}''$	$6\frac{11}{32}''$	$6''$	$8\frac{1}{16}''$	$7\frac{9}{16}''$	$7\frac{1}{8}''$	$6\frac{11}{16}''$	$6\frac{3}{8}''$	$6\frac{1}{32}''$	$8\frac{1}{16}''$
TOTAL HEIGHT OF RISERS	$10'\frac{3}{8}''$	$10'\frac{17}{32}''$	$10'$	$10'\frac{15}{16}''$	$10'1''$	$10'1\frac{1}{8}''$	$10'\frac{3}{8}''$	$10'1\frac{1}{8}''$	$10'\frac{5}{8}''$	$10'\frac{15}{16}''$
RISERS OVERAGE (+) OR UNDERAGE (—)	$-\frac{1}{8}''$	$+\frac{1}{32}''$	$-\frac{1}{2}''$	$+\frac{5}{16}''$	$+\frac{3}{8}''$	$+\frac{1}{2}''$	$-\frac{1}{4}''$	$+\frac{1}{2}''$	0	$+\frac{3}{16}''$
NUMBER OF TREADS	17	18	19	14	15	16	17	18	19	14
WIDTH OF EACH TREAD	$10\frac{13}{16}''$	$11\frac{5}{32}''$	$11\frac{1}{2}''$	$9\frac{7}{16}''$	$9\frac{15}{16}''$	$10\frac{3}{8}''$	$10\frac{13}{16}''$	$11\frac{1}{8}''$	$11\frac{15}{32}''$	$9\frac{7}{16}''$
TOTAL RUN OF STAIRWAY	$15'3\frac{13}{16}''$	$16'8\frac{13}{16}''$	$18'2\frac{1}{2}''$	$11'\frac{1}{8}''$	$12'5\frac{1}{16}''$	$13'10''$	$15'3\frac{13}{16}''$	$16'8\frac{1}{4}''$	$18'1\frac{29}{32}''$	$11'\frac{1}{8}''$
WELL OPENING FOR 6′-8″ HEADROOM	$13'\frac{3}{4}''$	$13'6\frac{5}{8}''$	$15'\frac{5}{32}''$	$8'10\frac{3}{32}''$	$10'2\frac{7}{16}''$	$11'7\frac{1}{16}''$	$13'\frac{3}{4}''$	$13'6\frac{1}{16}''$	$14'11\frac{19}{32}''$	$8'10\frac{3}{32}''$
ANGLE OF INCLINE	31°44'	29°37'	27°33'	40°30'	37°16'	34°29'	31°44'	29°49'	27°44'	40°30'
LENGTH OF CARRIAGE	$18'\frac{1}{8}''$	$19'3''$	$20'6\frac{7}{16}''$	$14'5\frac{25}{32}''$	$15'7\frac{5}{16}''$	$16'9\frac{3}{8}''$	$18'\frac{1}{8}''$	$19'2\frac{13}{16}''$	$20'6\frac{3}{16}''$	$14'5\frac{25}{32}''$
L. F. OF RISER PER INCH OF STAIRWAY WIDTH	1.5	1.583	1.666	1.25	1.333	1.416	1.5	1.583	1.666	1.25
L. F. OF TREAD PER INCH OF STAIRWAY WIDTH	1.416	1.5	1.583	1.166	1.25	1.333	1.416	1.5	1.583	1.166
FRAMING SQUARE SETTINGS FOR CARRIAGE CUTS — TONGUE	$6\frac{11}{16}''$	$6\frac{11}{32}''$	$6''$	$8\frac{1}{16}''$	$7\frac{9}{16}''$	$7\frac{1}{8}''$	$6\frac{11}{16}''$	$6\frac{3}{8}''$	$6\frac{1}{32}''$	$8\frac{1}{16}''$
FRAMING SQUARE SETTINGS FOR CARRIAGE CUTS — BODY	$10\frac{13}{16}''$	$11\frac{5}{32}''$	$11\frac{1}{2}''$	$9\frac{7}{16}''$	$9\frac{15}{16}''$	$10\frac{3}{8}''$	$10\frac{13}{16}''$	$11\frac{1}{8}''$	$11\frac{15}{32}''$	$9\frac{7}{16}''$

FLOOR TO FLOOR RISE		10' 3/4"	10' 3/4"	10' 3/4"	10' 3/4"	10' 3/4"	10' 7/8"	10' 7/8"	10' 7/8"	10' 7/8"	10' 7/8"
NUMBER OF RISERS		16	17	18	19	20	15	16	17	18	19
HEIGHT OF EACH RISER		7 9/16"	7 1/8"	6 11/16"	6 3/8"	6 1/16"	8 1/16"	7 9/16"	7 1/8"	6 11/16"	6 3/8"
TOTAL HEIGHT OF RISERS		10'1"	10'1 1/8"	10' 3/8"	10'1 1/8"	10'1 1/4"	10' 15/16"	10'1"	10'1 1/8"	10' 3/8"	10'1 1/8"
RISERS OVERAGE (+) OR UNDERAGE (—)		+1/4"	+3/8"	−3/8"	+3/8"	+1/2"	+1/16"	+1/8"	+1/4"	−1/2"	+1/4"
NUMBER OF TREADS		15	16	17	18	19	14	15	16	17	18
WIDTH OF EACH TREAD		9 15/16"	10 3/8"	10 13/16"	11 1/8"	11 7/16"	9 7/16"	9 15/16"	10 3/8"	10 13/16"	11 1/8"
TOTAL RUN OF STAIRWAY		12'5 1/16"	13'10"	15'3 13/16"	16'8 1/4"	18'1 5/16"	11' 1/8"	12'5 1/16"	13'10"	15'3 13/16"	16'8 1/4"
WELL OPENING FOR 6'-8" HEADROOM		10'2 7/16"	11'7 1/16"	13' 3/4""	13'6 1/16"	14'11"	8'10 3/32"	10'2 7/16"	11'7 1/16"	13' 3/4"	13'6 1/16"
ANGLE OF INCLINE		37°16'	34°29'	31°44'	29°49'	27°56'	40°30'	37°16'	34°29'	31°44'	29°49'
LENGTH OF CARRIAGE		15'7 5/16"	16'9 3/8"	18' 1/8"	19'2 13/16"	20'5 15/16"	14'5 25/32"	15'7 5/16"	16'9 3/8"	18' 1/8"	19'2 13/16"
L. F. OF RISER PER INCH OF STAIRWAY WIDTH		1.333	1.416	1.5	1.583	1.666	1.25	1.333	1.416	1.5	1.583
L. F. OF TREAD PER INCH OF STAIRWAY WIDTH		1.25	1.333	1.416	1.5	1.583	1.166	1.25	1.333	1.416	1.5
FRAMING SQUARE SETTINGS FOR CARRIAGE CUTS	TONGUE	7 9/16"	7 1/8"	6 11/16"	6 3/8"	6 1/16"	8 1/16"	7 9/16"	7 1/8"	6 11/16"	6 3/8"
	BODY	9 15/16"	10 3/8"	10 13/16"	11 1/8"	11 7/16"	9 7/16"	9 15/16"	10 3/8"	10 13/16"	11 1/8"

FLOOR TO FLOOR RISE	10' 7/8"	10'1"	10'1"	10'1"	10'1"	10'1"	10'1"	10'1 1/8"	10'1 1/8"	10'1 1/8"
NUMBER OF RISERS	20	15	16	17	18	19	20	15	16	17
HEIGHT OF EACH RISER	6 1/16"	8 1/16"	7 9/16"	7 1/8"	6 3/4"	6 3/8"	6 1/16"	8 1/16"	7 9/16"	7 1/8"
TOTAL HEIGHT OF RISERS	10'1 1/4"	10' 15/16"	10'1"	10'1 1/8"	10'1 1/2"	10'1 1/8"	10'1 1/4"	10' 15/16"	10'1"	10'1 1/8"
RISERS OVERAGE (+) OR UNDERAGE (—)	+3/8"	−1/16"	0	+1/8"	+1/2"	+1/8"	+1/4"	−3/16"	−1/8"	0
NUMBER OF TREADS	19	14	15	16	17	18	19	14	15	16
WIDTH OF EACH TREAD	11 7/16"	9 7/16"	9 15/16"	10 3/8"	10 3/4"	11 1/8"	11 7/16"	9 7/16"	9 15/16"	10 3/8"
TOTAL RUN OF STAIRWAY	18'1 5/16"	11' 1/8"	12'5 1/16"	13'10"	15'2 3/4"	16'8 1/4"	18'1 5/16"	11' 1/8"	12'5 1/16"	13'10"
WELL OPENING FOR 6'-8" HEADROOM	14'11"	8'10 3/32"	10'2 7/16"	11'7 1/16"	12'11 14/6"	13'6 1/16"	14'11"	8'10 3/32"	10'2 7/16"	11'7 1/16"
ANGLE OF INCLINE	27°56'	40°30'	37°16'	34°29'	32°7'	29°49'	27°56'	40°30'	37°16'	34°29'
LENGTH OF CARRIAGE	20'5 15/16"	14'5 25/32"	15'7 5/16"	16'9 3/8"	17'11 25/32"	19'2 13/16"	20'5 15/16"	14'5 25/32"	15'7 5/16"	16'9 3/8"
L. F. OF RISER PER INCH OF STAIRWAY WIDTH	1.166	1.25	1.333	1.416	1.5	1.583	1.666	1.25	1.333	1.416
L. F. OF TREAD PER INCH OF STAIRWAY WIDTH	1.583	1.166	1.25	1.333	1.416	1.5	1.583	1.166	1.25	1.333
FRAMING SQUARE SETTINGS FOR CARRIAGE CUTS — TONGUE	6 1/16"	8 1/16"	7 9/16"	7 1/8"	6 3/4"	6 3/8"	6 1/16"	8 1/16"	7 9/16"	7 1/8"
FRAMING SQUARE SETTINGS FOR CARRIAGE CUTS — BODY	11 7/16"	9 7/16"	9 15/16"	10 3/8"	10 3/4"	11 1/8"	11 7/16"	9 7/16"	9 15/16"	10 3/8"

FLOOR TO FLOOR RISE	10'1 1/8"	10'1 1/8"	10'1 1/8"	10'1 1/4"	10'1 1/4"	10'1 1/4"	10'1 1/4"	10'1 1/4"	10'1 1/4"	10'1 3/8"
NUMBER OF RISERS	18	19	20	15	16	17	18	19	20	15
HEIGHT OF EACH RISER	6 3/4"	6 3/8"	6 1/16"	8 1/16"	7 9/16"	7 1/8"	6 3/4"	6 3/8"	6 1/16"	8 1/16"
TOTAL HEIGHT OF RISERS	10'1 1/2"	10'1 1/8"	10'1 1/4"	10' 15/16"	10'1"	10'1 1/8"	10'1 1/2"	10'1 1/8"	10'1 1/4"	10' 15/16"
RISERS OVERAGE (+) OR UNDERAGE (—)	+3/8"	0	+1/8"	−5/16"	−1/4"	−1/8"	+1/4"	−1/8"	0	−7/16"
NUMBER OF TREADS	17	18	19	14	15	16	17	18	19	14
WIDTH OF EACH TREAD	10 3/4"	11 1/8"	11 7/16"	9 7/16"	9 15/16"	10 3/8"	10 3/4"	11 1/8"	11 7/16"	9 7/16"
TOTAL RUN OF STAIRWAY	15'2 3/4"	16'8 1/4"	18'1 5/16"	11' 1/8"	12'5 1/16"	13'10"	15'2 3/4"	16'8 1/4"	18'1 5/16"	11' 1/8"
WELL OPENING FOR 6'-8" HEADROOM	12'11 11/16"	13'6 1/16"	14'11"	8'10 3/32"	10'2 7/16"	11'7 1/16"	12'11 11/6"	13'6 1/16"	14'11"	8'10 3/32"
ANGLE OF INCLINE	32°7'	29°49'	27°56'	40°30'	37°16'	34°29'	32°7'	29°49'	27°56'	40°30'
LENGTH OF CARRIAGE	17'11 25/32"	19'2 13/16"	20'5 15/16"	14'5 25/32"	15'7 5/16"	16'9 3/8"	17'11 25/32"	19'2 13/16"	20'5 15/16"	14'5 25/32"
L. F. OF RISER PER INCH OF STAIRWAY WIDTH	1.5	1.583	1.666	1.25	1.333	1.416	1.5	1.583	1.666	1.25
L. F. OF TREAD PER INCH OF STAIRWAY WIDTH	1.416	1.5	1.583	1.666	1.25	1.333	1.416	1.5	1.583	1.666
FRAMING SQUARE SETTINGS FOR CARRIAGE CUTS — TONGUE	6 3/4"	6 3/8"	6 1/16"	8 1/16"	7 9/16"	7 1/8"	6 3/4"	6 3/8"	6 1/16"	8 1/16"
FRAMING SQUARE SETTINGS FOR CARRIAGE CUTS — BODY	10 3/4"	11 1/8"	11 7/16"	9 7/16"	9 15/16"	10 3/8"	10 3/4"	11 1/8"	11 7/16"	9 7/16"

FLOOR TO FLOOR RISE		10'1 3/8"	10'1 3/8"	10'1 3/8"	10'1 3/8"	10'1 3/8"	10'1 1/2"	10'1 1/2"	10'1 1/2"	10'1 1/2"	10'1 1/2"
NUMBER OF RISERS		16	17	18	19	20	15	16	17	18	19
HEIGHT OF EACH RISER		7 9/16"	7 1/8"	6 3/4"	6 3/8"	6 1/16"	8 1/8"	7 5/8"	7 1/8"	6 3/4"	6 3/8"
TOTAL HEIGHT OF RISERS		10'1"	10'1 1/8"	10'1 1/2"	10'1 1/8"	10'1 1/4"	10'1 7/8"	10'2"	10'1 1/8"	10'1 1/2"	10'1 1/8"
RISERS OVERAGE (+) OR UNDERAGE (—)		−3/8"	−1/4"	+1/8"	−1/4"	−1/8"	+3/8"	+1/2"	−3/8"	0	−3/8"
NUMBER OF TREADS		15	16	17	18	19	14	15	16	17	18
WIDTH OF EACH TREAD		9 15/16"	10 3/8"	10 3/4"	11 1/8"	11 7/16"	9 3/8"	9 7/8"	10 3/8"	10 3/4"	11 1/8"
TOTAL RUN OF STAIRWAY		12'5 1/16"	13'10"	15'2 3/4"	16'8 1/4"	18'1 5/16"	10'11 1/4"	12'4 1/8"	13'10"	15'2 3/4"	16'8 1/4"
WELL OPENING FOR 6'-8" HEADROOM		10'2 7/16"	11'7 1/16"	12'11 11/16"	13'6 1/16"	14'11"	8'9 9/32"	10'1 9/16"	11'7 1/16"	12'11 11/16"	13'6 1/16"
ANGLE OF INCLINE		37°16'	34°29'	32°7'	29°49'	27°56'	40°55'	37°40'	34°29'	32°7'	29°49'
LENGTH OF CARRIAGE		15'7 5/16"	16'9 3/8"	17'11 25/32"	19'2 13/16"	20'5 15/16"	14'5 11/16"	15'7 1/8"	16'9 3/8"	17'11 25/32"	19'2 13/16"
L. F. OF RISER PER INCH OF STAIRWAY WIDTH		1.333	1.416	1.5	1.583	1.666	1.25	1.333	1.416	1.5	1.583
L. F. OF TREAD PER INCH OF STAIRWAY WIDTH		1.25	1.333	1.416	1.5	1.583	1.166	1.25	1.333	1.416	1.5
FRAMING SQUARE SETTINGS FOR CARRIAGE CUTS	TONGUE	7 9/16"	7 1/8"	6 3/4"	6 3/8"	6 1/16"	8 1/8"	7 5/8"	7 1/8"	6 3/4"	6 3/8"
	BODY	9 15/16"	10 3/8"	10 3/4"	11 1/8"	11 7/16"	9 3/8"	9 7/8"	10 3/8"	10 3/4"	11 1/8"

FLOOR TO FLOOR RISE		10'1 1/2"	10'1 5/8"	10'1 5/8"	10'1 5/8"	10'1 5/8"	10'1 5/8"	10'1 5/8"	10'1 3/4"	10'1 3/4"	10'1 3/4"
NUMBER OF RISERS		20	15	16	17	18	19	20	15	16	17
HEIGHT OF EACH RISER		6 1/16"	8 1/8"	7 5/8"	7 1/8"	6 3/4"	6 3/8"	6 1/16"	8 1/8"	7 5/8"	7 3/16"
TOTAL HEIGHT OF RISERS		10'1 1/4"	10'1 7/8"	10'2"	10'1 1/8"	10'1 1/2"	10'1 1/8"	10'1 1/4"	10'1 7/8"	10'2"	10'2 3/16"
RISERS OVERAGE (+) OR UNDERAGE (—)		−1/4"	+1/4"	+3/8"	−1/2"	−1/8"	−1/2"	−3/8"	+1/8"	+1/4"	+7/16"
NUMBER OF TREADS		19	14	15	16	17	18	19	14	15	16
WIDTH OF EACH TREAD		11 7/16"	9 3/8"	9 7/8"	10 3/8"	10 3/4"	11 1/8"	11 7/16"	9 3/8"	9 7/8"	10 5/16"
TOTAL RUN OF STAIRWAY		18'1 5/16"	10'11 1/4"	12'4 1/8"	13'10"	15'2 3/4"	16'8 1/4"	18'1 5/16"	10'11 1/4"	12'4 1/8"	13'9"
WELL OPENING FOR 6'-8" HEADROOM		14'11"	8'9 9/32"	10'1 9/16"	11'7 1/16"	12'11 11/6"	13'6 1/16"	14'11'	8'9 9/32"	10'1 9/16"	11'6 3/32"
ANGLE OF INCLINE		27°56'	40°55'	37°40'	34°29'	32°7'	29°49'	27°56'	40°55'	37°40'	34°53'
LENGTH OF CARRIAGE		20'5 15/16"	14'5 11/16"	15'7 1/8"	16'9 3/8"	17'11 25/32"	19'2 13/16"	20'5 15/16"	14'5 11/16"	15'7 1/8"	16'9 1/8"
L. F. OF RISER PER INCH OF STAIRWAY WIDTH		1.666	1.25	1.333	1.416	1.5	1.583	1.666	1.25	1.333	1.416
L. F. OF TREAD PER INCH OF STAIRWAY WIDTH		1.583	1.166	1.25	1.333	1.416	1.5	1.583	1.166	1.25	1.333
FRAMING SQUARE SETTINGS FOR CARRIAGE CUTS	TONGUE	6 1/16"	8 1/8"	7 5/8"	7 1/8"	6 3/4"	6 3/8"	6 1/16"	8 1/8"	7 5/8"	7 3/16"
	BODY	11 7/16"	9 3/8"	9 7/8"	10 3/8"	10 3/4"	11 1/8"	11 7/16"	9 3/8"	9 7/8"	10 5/16"

FLOOR TO FLOOR RISE		10'1 3/4"	10'1 3/4"	10'1 3/4"	10'1 7/8"	10'1 7/8"	10'1 7/8"	10'1 7/8"	10'1 7/8"	10'1 7/8"	10'2"
NUMBER OF RISERS		18	19	20	15	16	17	18	19	20	15
HEIGHT OF EACH RISER		6 3/4"	6 13/32"	6 1/16"	8 1/8"	7 5/8"	7 3/16"	6 3/4"	6 7/16"	6 3/32"	8 1/8"
TOTAL HEIGHT OF RISERS		10'1 1/2"	10'1 23/32"	10'1 1/4"	10'1 7/8"	10'2"	10'2 3/16"	10'1 1/2"	10'2 5/16"	10'1 7/8"	10'1 7/8"
RISERS OVERAGE (+) OR UNDERAGE (—)		−1/4"	−1/32"	−1/2"	0	+1/8"	+5/16"	−3/8"	+7/16"	0	−1/8"
NUMBER OF TREADS		17	18	19	14	15	16	17	18	19	14
WIDTH OF EACH TREAD		10 3/4"	11 3/32"	11 7/16"	9 3/8"	9 7/8"	10 5/16"	10 3/4"	11 1/16"	11 13/32"	9 3/8"
TOTAL RUN OF STAIRWAY		15'2 3/4"	16'7 11/16"	18'1 5/16"	10'11 1/4"	12'4 1/8"	13'9"	15'2 3/4"	16'7 1/8"	18' 23/32"	10'11 1/4"
WELL OPENING FOR 6'-8" HEADROOM		12'11 11/16"	13'5 17/32"	14'11"	8'9 9/32"	10'1 9/16"	11'6 3/32"	12'11 11/16"	13'5"	14'10 13/32"	8'9 9/32"
ANGLE OF INCLINE		32°7'	30°0'	27°56'	40°55'	37°40'	34°53'	32°7'	30°12'	28°7'	40°55'
LENGTH OF CARRIAGE		17'11 25/32"	19'2 19/32"	20'5 15/16"	14'5 11/16"	15'7 1/8"	16'9 1/8"	17'11 25/32"	19'2 3/8"	20'5 23/32"	14'5 11/16"
L. F. OF RISER PER INCH OF STAIRWAY WIDTH		1.5	1.583	1.666	1.25	1.333	1.416	1.5	1.583	1.666	1.25
L. F. OF TREAD PER INCH OF STAIRWAY WIDTH		1.416	1.5	1.583	1.166	1.25	1.333	1.416	1.5	1.583	1.166
FRAMING SQUARE SETTINGS FOR CARRIAGE CUTS	TONGUE	6 3/4"	6 13/32"	6 1/16"	8 1/8"	7 5/8"	7 3/16"	6 3/4"	6 7/16"	6 3/32"	8 1/8"
	BODY	10 3/4"	11 3/32"	11 7/16"	9 3/8"	9 7/8"	10 5/16"	10 3/4"	11 1/16"	11 13/32"	9 3/8"

FLOOR TO FLOOR RISE		10'2"	10'2"	10'2"	10'2"	10'2"	10'2 1/8"	10'2 1/8"	10'2 1/8"	10'2 1/8"	10'2 1/8"
NUMBER OF RISERS		16	17	18	19	20	15	16	17	18	19
HEIGHT OF EACH RISER		7 5/8"	7 3/16"	6 3/4"	6 7/16"	6 1/8"	8 1/8"	7 5/8"	7 3/16"	6 13/16"	6 7/16"
TOTAL HEIGHT OF RISERS		10'2"	10'2 3/16"	10'1 1/2"	10'2 5/16"	10'2 1/2"	10'1 7/8"	10'2"	10'2 3/16"	10'2 5/8"	10'2 5/16"
RISERS OVERAGE (+) OR UNDERAGE (—)		0	+3/16"	−1/2"	+5/16"	+1/2"	−1/4"	−1/8"	+1/16"	+1/2"	+3/16"
NUMBER OF TREADS		15	16	17	18	19	14	15	16	17	18
WIDTH OF EACH TREAD		9 7/8"	10 5/16"	10 3/4"	11 1/16"	11 3/8"	9 3/8"	9 7/8"	10 5/16"	10 11/16"	11 1/16"
TOTAL RUN OF STAIRWAY		12'4 1/8"	13'9"	15'2 3/4"	16'7 1/8"	18' 1/8"	10'11 1/4"	12'4 1/8"	13'9"	15'1 11/16"	16'7 1/8"
WELL OPENING FOR 6'-8" HEADROOM		10'1 9/16"	11'6 3/32"	12'11 11/16"	13'5"	14'9 13/16"	8'9 9/32"	10'1 9/16"	11'6 3/32"	11'11 15/16"	13'5"
ANGLE OF INCLINE		37°40'	34°53'	32°7'	30°12'	28°18'	40°55'	37°40'	34°53'	32°31'	30°12'
LENGTH OF CARRIAGE		15'7 1/8"	16'9 1/8"	17'11 25/32"	19'2 3/8"	20'5 15/32"	14'5 11/16"	15'7 1/8"	16'9 1/8"	17'11 15/32"	19'2 3/8"
L. F. OF RISER PER INCH OF STAIRWAY WIDTH		1.333	1.416	1.5	1.583	1.666	1.25	1.333	1.416	1.5	1.583
L. F. OF TREAD PER INCH OF STAIRWAY WIDTH		1.25	1.333	1.416	1.5	1.583	1.166	1.25	1.333	1.416	1.5
FRAMING SQUARE SETTINGS FOR CARRIAGE CUTS	TONGUE	7 5/8"	7 3/16"	6 3/4"	6 7/16"	6 1/8"	8 1/8"	7 5/8"	7 3/16"	6 13/16"	6 7/16"
	BODY	9 7/8"	10 5/16"	10 3/4"	11 1/16"	11 3/8"	9 3/8"	9 7/8"	10 5/16"	10 11/16"	11 1/16"

FLOOR TO FLOOR RISE		10'2 1/8"	10'2 1/4"	10'2 1/4"	10'2 1/4"	10'2 1/4"	10'2 1/4"	10'2 1/4"	10'2 3/8"	10'2 3/8"	10'2 3/8"
NUMBER OF RISERS		20	15	16	17	18	19	20	15	16	17
HEIGHT OF EACH RISER		6 1/8"	8 1/8"	7 5/8"	7 3/16"	6 13/16"	6 7/16"	6 1/8"	8 3/16"	7 5/8"	7 3/16"
TOTAL HEIGHT OF RISERS		10'2 1/2"	10'1 7/8"	10'2"	10'2 3/16"	10'2 5/8"	10'2 5/16"	10'2 1/2"	10'2 13/16"	10'2"	10'2 3/16"
RISERS OVERAGE (+) OR UNDERAGE (—)		+3/8"	−3/8"	−1/4"	−1/16"	+3/8"	+1/16"	+1/4"	+7/16"	−3/8"	−3/16"
NUMBER OF TREADS		19	14	15	16	17	18	19	14	15	16
WIDTH OF EACH TREAD		11 3/8"	9 3/8"	9 7/8"	10 5/16"	10 11/16"	11 1/16"	11 3/8"	9 5/16"	9 7/8"	10 5/16"
TOTAL RUN OF STAIRWAY		18' 1/8"	10'11 1/4"	12'4 1/8"	13'9"	15'1 11/16"	16'7 1/8"	18' 1/8"	10'10 3/8"	12'4 1/8"	13'9"
WELL OPENING FOR 6'-8" HEADROOM		14'9 13/16"	8'9 9/32"	10'1 9/16"	11'6 3/32"	11'11 15/16"	13'5"	14'9 13/16"	8'8 1/2"	10'1 9/16"	11'6 3/32"
ANGLE OF INCLINE		28°18'	40°55'	37°40'	34°53'	32°31'	30°12'	28°18'	41°19'	37°40'	34°53'
LENGTH OF CARRIAGE		20'5 15/32"	14'5 11/16"	15'7 1/8"	16'9 1/8"	17'11 15/32"	19'2 3/8"	20'5 15/32"	14'5 19/32"	15'7 1/8"	16'9 1/8"
L. F. OF RISER PER INCH OF STAIRWAY WIDTH		1.666	1.25	1.333	1.416	1.5	1.583	1.666	1.25	1.333	1.416
L. F. OF TREAD PER INCH OF STAIRWAY WIDTH		1.583	1.166	1.25	1.333	1.416	1.5	1.583	1.166	1.25	1.333
FRAMING SQUARE SETTINGS FOR CARRIAGE CUTS	TONGUE	6 1/8"	8 1/8"	7 5/8"	7 3/16"	6 13/16"	6 7/16"	6 1/8"	8 3/16"	7 5/8"	7 3/16"
	BODY	11 3/8"	9 3/8"	9 7/8"	10 5/16"	10 11/16"	11 1/16"	11 3/8"	9 5/16"	9 7/8"	10 5/16"

FLOOR TO FLOOR RISE	10'2 3/8"	10'2 3/8"	10'2 3/8"	10'2 1/2"	10'2 1/2"	10'2 1/2"	10'2 1/2"	10'2 1/2"	10'2 1/2"	10'2 5/8"
NUMBER OF RISERS	18	19	20	15	16	17	18	19	20	15
HEIGHT OF EACH RISER	6 13/16"	6 7/16"	6 1/8"	8 3/16"	7 11/16"	7 3/16"	6 13/16"	6 7/16"	6 1/8"	8 3/16"
TOTAL HEIGHT OF RISERS	10'2 5/8"	10'2 5/16"	10'2 1/2"	10'2 13/16"	10'3"	10'2 3/16"	10'2 5/8"	10'2 5/16"	10'2 1/2"	10'2 13/16"
RISERS OVERAGE (+) OR UNDERAGE (—)	+1/4"	−1/16"	+1/8"	+5/16"	+1/2"	−5/16"	+1/8"	−3/16"	0	+3/16"
NUMBER OF TREADS	17	18	19	14	15	16	17	18	19	14
WIDTH OF EACH TREAD	10 11/16"	11 1/16"	11 3/8"	9 5/16"	9 13/16"	10 5/16"	10 11/16"	11 1/16"	11 3/8"	9 5/16"
TOTAL RUN OF STAIRWAY	15'1 11/16"	16'7 1/8"	18'1/8"	10'10 3/8"	12'3 3/16"	13'9"	15'1 11/16"	16'7 1/8"	18'1/8"	10'10 3/8"
WELL OPENING FOR 6'-8" HEADROOM	11'11 15/16"	13'5"	14'9 13/16"	8'8 1/2"	10'23/32"	11'6 3/32"	11'11 15/16"	13'5"	14'9 13/16"	8'8 1/2"
ANGLE OF INCLINE	32°31'	30°12'	28°18'	41°19'	38°5'	34°53'	32°31'	30°12'	28°18'	41°19'
LENGTH OF CARRIAGE	17'11 15/32"	19'2 3/8"	20'5 15/32"	14'5 19/32"	15'6 31/32"	16'9 1/8"	17'11 15/32"	19'2 3/8"	20'5 15/32"	14'5 19/32"
L. F. OF RISER PER INCH OF STAIRWAY WIDTH	1.5	1.583	1.666	1.25	1.333	1.416	1.5	1.583	1.666	1.25
L. F. OF TREAD PER INCH OF STAIRWAY WIDTH	1.416	1.5	1.583	1.166	1.25	1.333	1.416	1.5	1.583	1.166
FRAMING SQUARE SETTINGS FOR CARRIAGE CUTS — TONGUE	6 13/16"	6 7/16"	6 1/8"	8 3/16"	7 11/16"	7 3/16"	6 13/16"	6 7/16"	6 1/8"	8 3/16"
FRAMING SQUARE SETTINGS FOR CARRIAGE CUTS — BODY	10 11/16"	11 1/16"	11 3/8"	9 5/16"	9 13/16"	10 5/16"	10 11/16"	11 1/16"	11 3/8"	9 5/16"

FLOOR TO FLOOR RISE		10'2 5/8"	10'2 5/8"	10'2 5/8"	10'2 5/8"	10'2 5/8"	10'2 3/4"	10'2 3/4"	10'2 3/4"	10'2 3/4"	10'2 3/4"
NUMBER OF RISERS		16	17	18	19	20	15	16	17	18	19
HEIGHT OF EACH RISER		7 11/16"	7 3/16"	6 13/16"	6 7/16"	6 1/8"	8 3/16"	7 11/16"	7 1/4"	6 13/16"	6 7/16"
TOTAL HEIGHT OF RISERS		10'3"	10'2 3/16"	10'2 5/8"	10'2 5/16"	10'2 1/2"	10'2 13/16"	10'3"	10'3 1/4"	10'2 5/8"	10'2 5/16"
RISERS OVERAGE (+) OR UNDERAGE (—)		+3/8"	−7/16"	0	−5/16"	−1/8"	+1/16"	+1/4"	+1/2"	−1/8"	−7/16"
NUMBER OF TREADS		15	16	17	18	19	14	15	16	17	18
WIDTH OF EACH TREAD		9 13/16"	10 5/16"	10 11/16"	11 1/16"	11 3/8"	9 5/16"	9 13/16"	10 1/4"	10 11/16"	11 1/16"
TOTAL RUN OF STAIRWAY		12'3 3/16"	13'9"	15'1 11/16"	16'7 1/8"	18' 1/8"	10'10 3/8"	12'3 3/16"	13'8"	15'1 11/16"	16'7 1/8"
WELL OPENING FOR 6'-8" HEADROOM		10' 23/32"	11'6 3/32"	11'11 15/16"	13'5"	14'9 13/16"	8'8 1/2"	10' 23/32"	11'5 1/8"	11'11 15/16"	13'5"
ANGLE OF INCLINE		38°5'	34°53'	32°31'	30°12'	28°18'	41°19'	38°5'	35°16'	32°31'	30°12'
LENGTH OF CARRIAGE		15'6 31/32"	16'9 1/8"	17'11 15/32"	19'2 3/8"	20'5 15/32"	14'5 19/32"	15'6 31/32"	16'8 7/8"	17'11 15/32"	19'2 3/8"
L. F. OF RISER PER INCH OF STAIRWAY WIDTH		1.333	1.416	1.5	1.583	1.666	1.25	1.333	1.416	1.5	1.583
L. F. OF TREAD PER INCH OF STAIRWAY WIDTH		1.25	1.333	1.416	1.5	1.583	1.166	1.25	1.333	1.416	1.5
FRAMING SQUARE SETTINGS FOR CARRIAGE CUTS	TONGUE	7 11/16"	7 3/16"	6 13/16"	6 7/16"	6 1/8"	8 3/16"	7 11/16"	7 1/4"	6 13/16"	6 7/16"
	BODY	9 13/16"	10 5/16"	10 11/16"	11 1/16"	11 3/8"	9 5/16"	9 13/16"	10 1/4"	10 11/16"	11 1/16"

FLOOR TO FLOOR RISE	10'2 3/4"	10'2 7/8"	10'2 7/8"	10'2 7/8"	10'2 7/8"	10'2 7/8"	10'2 7/8"	10'3"	10'3"	10'3"
NUMBER OF RISERS	20	15	16	17	18	19	20	15	16	17
HEIGHT OF EACH RISER	6 1/8"	8 3/16"	7 11/16"	7 1/4"	6 13/16"	6 15/32"	6 1/8"	8 3/16"	7 11/16"	7 1/4"
TOTAL HEIGHT OF RISERS	10'2 1/2"	10'2 13/16"	10'3"	10'3 1/4"	10'2 5/8"	10'2 29/32"	10'2 1/2"	10'2 13/16"	10'3"	10'3 1/4"
RISERS OVERAGE (+) OR UNDERAGE (—)	−1/4"	−1/16"	+1/8"	+3/8"	−1/4"	+1/32"	−3/8"	−3/16"	0	+1/4"
NUMBER OF TREADS	19	14	15	16	17	18	19	14	15	16
WIDTH OF EACH TREAD	11 3/8"	9 5/16"	9 13/16"	10 1/4"	10 11/16"	11 1/32"	11 3/8"	9 5/16"	9 13/16"	10 1/4"
TOTAL RUN OF STAIRWAY	18' 1/8"	10'10 3/8"	12'3 3/16"	13'8"	15'1 11/16"	16'6 9/16"	18' 1/8"	10'10 3/8"	12'3 3/16"	13'8"
WELL OPENING FOR 6'-8" HEADROOM	14'9 13/16"	8'8 1/2"	10' 23/32"	11'5 1/8"	11'11 15/16"	13'4 15/32"	14'9 13/16"	8'8 1/2"	10' 23/32"	11'5 1/8"
ANGLE OF INCLINE	28°18'	41°19'	38°5'	35°16'	32°31'	30°23'	28°18'	41°19'	38°5'	35°16'
LENGTH OF CARRIAGE	20'5 15/32"	14'5 19/32"	15'6 31/32"	16'8 7/8"	17'11 15/32"	19'2 3/16"	20'5 15/32"	14'5 19/32"	15'6 31/32"	16'8 7/8"
L. F. OF RISER PER INCH OF STAIRWAY WIDTH	1.666	1.25	1.333	1.416	1.5	1.583	1.666	1.25	1.333	1.416
L. F. OF TREAD PER INCH OF STAIRWAY WIDTH	1.583	1.166	1.25	1.333	1.416	1.5	1.583	1.166	1.25	1.333
FRAMING SQUARE SETTINGS FOR CARRIAGE CUTS — TONGUE	6 1/8"	8 3/16"	7 11/16"	7 1/4"	6 13/16"	6 15/32"	6 1/8"	8 3/16"	7 11/16"	7 1/4"
FRAMING SQUARE SETTINGS FOR CARRIAGE CUTS — BODY	11 3/8"	9 5/16"	9 13/16"	10 1/4"	10 11/16"	11 1/32"	11 3/8"	9 5/16"	9 13/16"	10 1/4"

FLOOR TO FLOOR RISE		10'3"	10'3"	10'3"	10'3 1/8"	10'3 1/8"	10'3 1/8"	10'3 1/8"	10'3 1/8"	10'3 1/8"	10'3 1/4"
NUMBER OF RISERS		18	19	20	15	16	17	18	19	20	15
HEIGHT OF EACH RISER		6 13/16"	6 1/2"	6 1/8"	8 3/16"	7 11/16"	7 1/4"	6 13/16"	6 1/2"	6 5/32"	8 3/16"
TOTAL HEIGHT OF RISERS		10'2 5/8"	10'3 1/2"	10'2 1/2"	10'2 13/16"	10'3"	10'3 1/4"	10'2 5/8"	10'3 1/2"	10'3 1/8"	10'2 13/16"
RISERS OVERAGE (+) OR UNDERAGE (—)		−3/8"	+1/2"	−1/2"	−5/16"	−1/8"	+1/8"	−1/2"	+3/8"	0	−7/16"
NUMBER OF TREADS		17	18	19	14	15	16	17	18	19	14
WIDTH OF EACH TREAD		10 11/16"	11"	11 3/8"	9 5/16"	9 13/16"	10 1/4"	10 11/16"	11"	11 11/32"	9 5/16"
TOTAL RUN OF STAIRWAY		15'1 11/16"	16'6"	18' 1/8"	10'10 3/8"	12'3 3/16"	13'8"	15'1 11/16"	16'6"	17'11 17/32"	10'10 3/8"
WELL OPENING FOR 6'-8" HEADROOM		11'11 15/16"	13'3 15/16"	14'9 13/16"	8'8 1/2"	10' 23/32"	11'5 1/8"	11'11 15/16"	13'3 15/16"	14'9 1/4"	8'8 1/2"
ANGLE OF INCLINE		32°31'	30°35'	28°18'	41°19'	38°5'	35°16'	32°31'	30°35'	28°29'	41°19'
LENGTH OF CARRIAGE		17'11 15/32"	19'1 31/32"	20'5 15/32"	14'5 19/32"	15'6 31/32"	16'8 7/8"	17'11 15/32"	19'1 31/32"	20'5 7/32"	14'5 19/32"
L. F. OF RISER PER INCH OF STAIRWAY WIDTH		1.5	1.583	1.666	1.25	1.333	1.416	1.5	1.583	1.666	1.25
L. F. OF TREAD PER INCH OF STAIRWAY WIDTH		1.416	1.5	1.583	1.166	1.25	1.333	1.416	1.5	1.583	1.166
FRAMING SQUARE SETTINGS FOR CARRIAGE CUTS	TONGUE	6 13/16"	6 1/2"	6 1/8"	8 3/16"	7 11/16"	7 1/4"	6 13/16"	6 1/2"	6 5/32"	8 3/16"
	BODY	10 11/16"	11"	11 3/8"	9 5/16"	9 13/16"	10 1/4"	10 11/16"	11"	11 11/32"	9 5/16"

FLOOR TO FLOOR RISE	10'3 1/4"	10'3 1/4"	10'3 1/4"	10'3 1/4"	10'3 1/4"	10'3 3/8"	10'3 3/8"	10'3 3/8"	10'3 3/8"	10'3 3/8"
NUMBER OF RISERS	16	17	18	19	20	15	16	17	18	19
HEIGHT OF EACH RISER	7 11/16"	7 1/4"	6 7/8"	6 1/2"	6 3/16"	8 1/4"	7 11/16"	7 1/4"	6 7/8"	6 1/2"
TOTAL HEIGHT OF RISERS	10'3"	10'3 1/4"	10'3 3/4"	10'3 1/2"	10'3 3/4"	10'3 3/4"	10'3"	10'3 1/4"	10'3 3/4"	10'3 1/2"
RISERS OVERAGE (+) OR UNDERAGE (-)	-1/4"	0	+1/2"	+1/4"	+1/2"	+3/8"	-3/8"	-1/8"	+3/8"	+1/8"
NUMBER OF TREADS	15	16	17	18	19	14	15	16	17	18
WIDTH OF EACH TREAD	9 13/16"	10 1/4"	10 5/8"	11"	11 5/16"	9 1/4"	9 13/16"	10 1/4"	10 5/8"	11"
TOTAL RUN OF STAIRWAY	12'3 3/16"	13'8"	15'5/8"	16'6"	17'10 15/16"	10'9 1/2"	12'3 3/16"	13'8"	15'5/8"	16'6"
WELL OPENING FOR 6'-8" HEADROOM	10'23/32"	11'5 1/8"	11'10 31/32"	13'3 15/16"	14'8 21/32"	8'7 23/32"	10'23/32"	11'5 1/8"	11'10 31/32"	13'3 15/16"
ANGLE OF INCLINE	38°5'	35°16'	32°54'	30°35'	28°41'	41°44'	38°5'	35°16'	32°54'	30°35'
LENGTH OF CARRIAGE	15'6 31/32"	16'8 7/8"	17'11 5/32"	19'1 31/32"	20'5"	14'5 17/32"	15'6 31/32"	16'8 7/8"	17'11 5/32"	19'1 31/32"
L. F. OF RISER PER INCH OF STAIRWAY WIDTH	1.333	1.416	1.5	1.583	1.666	1.25	1.333	1.416	1.5	1.583
L. F. OF TREAD PER INCH OF STAIRWAY WIDTH	1.25	1.333	1.416	1.5	1.583	1.166	1.25	1.333	1.416	1.5
FRAMING SQUARE SETTINGS FOR CARRIAGE CUTS — TONGUE	7 11/16"	7 1/4"	6 7/8"	6 1/2"	6 3/16"	8 1/4"	7 11/16"	7 1/4"	6 7/8"	6 1/2"
FRAMING SQUARE SETTINGS FOR CARRIAGE CUTS — BODY	9 13/16"	10 1/4"	10 5/8"	11"	11 5/16"	9 1/4"	9 13/16"	10 1/4"	10 5/8"	11"

FLOOR TO FLOOR RISE		10'3 3/8"	10'3 1/2"	10'3 1/2"	10'3 1/2"	10'3 1/2"	10'3 1/2"	10'3 1/2"	10'3 5/8"	10'3 5/8"	10'3 5/8"
NUMBER OF RISERS		20	15	16	17	18	19	20	15	16	17
HEIGHT OF EACH RISER		6 3/16"	8 1/4"	7 3/4"	7 1/4"	6 7/8"	6 1/2"	6 3/16"	8 1/4"	7 3/4"	7 1/4"
TOTAL HEIGHT OF RISERS		10'3 3/4"	10'3 3/4"	10'4"	10'3 1/4"	10'3 3/4"	10'3 1/2"	10'3 3/4"	10'3 3/4"	10'4"	10'3 1/4"
RISERS OVERAGE (+) OR UNDERAGE (—)		+3/8"	+1/4"	+1/2"	−1/4"	+1/4"	0	+1/4"	+1/8"	+3/8"	−3/8"
NUMBER OF TREADS		19	14	15	16	17	18	19	14	15	16
WIDTH OF EACH TREAD		11 5/16"	9 1/4"	9 3/4"	10 1/4"	10 5/8"	11"	11 5/16"	9 1/4"	9 3/4"	10 1/4"
TOTAL RUN OF STAIRWAY		17'10 15/16"	10'9 1/2"	12'2 1/4"	13'8"	15'5/8"	16'6"	17'10 15/16"	10'9 1/2"	12'2 1/4"	13'8"
WELL OPENING FOR 6'-8" HEADROOM		14'8 21/32"	8'7 23/32"	9'11 27/32"	11'5 1/8"	11'10 31/32"	13'3 15/16"	14'8 21/32"	8'7 23/32"	9'11 27/32"	11'5 1/8"
ANGLE OF INCLINE		28°41'	41°44'	38°29'	35°16'	32°54'	30°35'	28°41'	41°44'	38°29'	35°16'
LENGTH OF CARRIAGE		20'5"	14'5 17/32"	15'6 13/16"	16'8 7/8"	17'11 5/32"	19'1 31/32"	20'5"	14'5 17/32"	15'6 13/16"	16'8 7/8"
L. F. OF RISER PER INCH OF STAIRWAY WIDTH		1.666	1.25	1.333	1.416	1.5	1.583	1.666	1.25	1.333	1.416
L. F. OF TREAD PER INCH OF STAIRWAY WIDTH		1.583	1.166	1.25	1.333	1.416	1.5	1.583	1.166	1.25	1.333
FRAMING SQUARE SETTINGS FOR CARRIAGE CUTS	TONGUE	6 3/16"	8 1/4"	7 3/4"	7 1/4"	6 7/8"	6 1/2"	6 3/16"	8 1/4"	7 3/4"	7 1/4"
	BODY	11 5/16"	9 1/4"	9 3/4"	10 1/4"	10 5/8"	11"	11 5/16"	9 1/4"	9 3/4"	10 1/4"

FLOOR TO FLOOR RISE	10'3 5/8"	10'3 5/8"	10'3 5/8"	10'3 3/4"	10'3 3/4"	10'3 3/4"	10'3 3/4"	10'3 3/4"	10'3 3/4"	10'3 7/8"
NUMBER OF RISERS	18	19	20	15	16	17	18	19	20	15
HEIGHT OF EACH RISER	6 7/8"	6 1/2"	6 3/16"	8 1/4"	7 3/4"	7 1/4"	6 7/8"	6 1/2"	6 3/16"	8 1/4"
TOTAL HEIGHT OF RISERS	10'3 3/4"	10'3 1/2"	10'3 3/4"	10'3 3/4"	10'4"	10'3 1/4"	10'3 3/4"	10'3 1/2"	10'3 3/4"	10'3 3/4"
RISERS OVERAGE (+) OR UNDERAGE (—)	+1/8"	−1/8"	+1/8"	0	+1/4"	−1/2"	0	−1/4"	0	−1/8"
NUMBER OF TREADS	17	18	19	14	15	16	17	18	19	14
WIDTH OF EACH TREAD	10 5/8"	11"	11 5/16"	9 1/4"	9 3/4"	10 1/4"	10 5/8"	11"	11 5/16"	9 1/4"
TOTAL RUN OF STAIRWAY	15' 5/8"	16'6"	17'10 15/16"	10'9 1/2"	12'2 1/4"	13'8"	15' 5/8"	16'6"	17'10 15/16"	10'9 1/2"
WELL OPENING FOR 6'-8" HEADROOM	11'10 31/32"	13'3 15/16"	14'8 21/32"	8'7 23/32"	9'11 27/32"	11'5 1/8"	11'10 31/32"	13'3 15/16	14'8 21/32"	8'7 23/32"
ANGLE OF INCLINE	32°54'	30°35'	28°41'	41°44'	38°29'	35°16'	32°54'	30°35'	28°41'	41°44'
LENGTH OF CARRIAGE	17'11 5/32"	19'1 31/32"	20'5"	14'5 17/32"	15'6 13/16"	16'8 7/8"	17'11 5/32"	19'1 31/32"	20'5"	14'5 17/32"
L. F. OF RISER PER INCH OF STAIRWAY WIDTH	1.5	1.583	1.666	1.25	1.333	1.416	1.5	1.583	1.666	1.25
L. F. OF TREAD PER INCH OF STAIRWAY WIDTH	1.416	1.5	1.583	1.166	1.25	1.333	1.416	1.5	1.583	1.166
FRAMING SQUARE SETTINGS FOR CARRIAGE CUTS — TONGUE	6 7/8"	6 1/2"	6 3/16"	8 1/4"	7 3/4"	7 1/4"	6 7/8"	6 1/2"	6 3/16"	8 1/4"
FRAMING SQUARE SETTINGS FOR CARRIAGE CUTS — BODY	10 5/8"	11"	11 5/16"	9 1/4"	9 3/4"	10 1/4"	10 5/8"	11"	11 5/16"	9 1/4"

FLOOR TO FLOOR RISE	10'3 7/8"	10'3 7/8"	10'3 7/8"	10'3 7/8"	10'3 7/8"	10'4"	10'4"	10'4"	10'4"	10'4"
NUMBER OF RISERS	16	17	18	19	20	15	16	17	18	19
HEIGHT OF EACH RISER	7 3/4"	7 5/16"	6 7/8"	6 1/2"	6 3/16"	8 1/4"	7 3/4"	7 5/16"	6 7/8"	6 1/2"
TOTAL HEIGHT OF RISERS	10'4"	10'4 5/16"	10'3 3/4"	10'3 1/2"	10'3 3/4"	10'3 3/4"	10'4"	10'4 5/16"	10'3 3/4"	10'3 1/2"
RISERS OVERAGE (+) OR UNDERAGE (—)	+1/8"	+7/16"	−1/8"	−3/8"	−1/8"	−1/4"	0	+5/16"	−1/4"	−1/2"
NUMBER OF TREADS	15	16	17	18	19	14	15	16	17	18
WIDTH OF EACH TREAD	9 3/4"	10 3/16"	10 5/8"	11"	11 5/16"	9 1/4"	9 3/4"	10 3/16"	10 5/8"	11"
TOTAL RUN OF STAIRWAY	12'2 1/4"	13'7"	15' 5/8"	16'6"	17'10 15/16"	10'9 1/2"	12'2 1/4"	13'7"	15' 5/8"	16'6"
WELL OPENING FOR 6'-8" HEADROOM	9'11 27/32"	11'4 3/16"	11'10 31/32"	13'3 15/16"	14'8 21/32"	8'7 23/32"	9'11 27/32"	11'4 3/16"	11'10 31/32"	13'3 15/16"
ANGLE OF INCLINE	38°29'	35°40'	32°54'	30°35'	28°41'	41°44'	38°29'	35°40'	32°54'	30°35'
LENGTH OF CARRIAGE	15'6 13/16"	16'8 21/32"	17'11 5/32"	19'1 31/32"	20'5"	14'5 17/32"	15'6 13/16"	16'8 21/32"	17'11 5/32"	19'1 31/32"
L. F. OF RISER PER INCH OF STAIRWAY WIDTH	1.333	1.416	1.5	1.583	1.666	1.25	1.333	1.416	1.5	1.583
L. F. OF TREAD PER INCH OF STAIRWAY WIDTH	1.25	1.333	1.416	1.5	1.583	1.166	1.25	1.333	1.416	1.5
FRAMING SQUARE SETTINGS FOR CARRIAGE CUTS — TONGUE	7 3/4"	7 5/16"	6 7/8"	6 1/2"	6 3/16"	8 1/4"	7 3/4"	7 5/16"	6 7/8"	6 1/2"
FRAMING SQUARE SETTINGS FOR CARRIAGE CUTS — BODY	9 3/4"	10 3/16"	10 5/8"	11"	11 5/16"	9 1/4"	9 3/4"	10 3/16"	10 5/8"	11"

FLOOR TO FLOOR RISE	10'4"	10'4$\frac{1}{8}$"	10'4$\frac{1}{8}$"	10'4$\frac{1}{8}$"	10'4$\frac{1}{8}$"	10'4$\frac{1}{8}$"	10'4$\frac{1}{8}$"	10'4$\frac{1}{4}$"	10'4$\frac{1}{4}$"	10'4$\frac{1}{4}$"
NUMBER OF RISERS	20	15	16	17	18	19	20	16	17	18
HEIGHT OF EACH RISER	6$\frac{3}{16}$"	8$\frac{1}{4}$"	7$\frac{3}{4}$"	7$\frac{5}{16}$"	6$\frac{7}{8}$"	6$\frac{17}{32}$"	6$\frac{3}{16}$"	7$\frac{3}{4}$"	7$\frac{5}{16}$"	6$\frac{7}{8}$"
TOTAL HEIGHT OF RISERS	10'3$\frac{3}{4}$"	10'3$\frac{3}{4}$"	10'4"	10'4$\frac{5}{16}$"	10'3$\frac{3}{4}$"	10'4$\frac{3}{32}$"	10'3$\frac{3}{4}$"	10'4"	10'4$\frac{5}{16}$"	10'3$\frac{3}{4}$"
RISERS OVERAGE (+) OR UNDERAGE (—)	$-\frac{1}{4}$"	$-\frac{3}{8}$"	$-\frac{1}{8}$"	$+\frac{3}{16}$"	$-\frac{3}{8}$"	$-\frac{1}{32}$"	$-\frac{3}{8}$"	$-\frac{1}{4}$"	$+\frac{1}{16}$"	$-\frac{1}{2}$"
NUMBER OF TREADS	19	14	15	16	17	18	19	15	16	17
WIDTH OF EACH TREAD	11$\frac{5}{16}$"	9$\frac{1}{4}$"	9$\frac{3}{4}$"	10$\frac{3}{16}$"	10$\frac{5}{8}$"	10$\frac{31}{32}$"	11$\frac{5}{16}$"	9$\frac{3}{4}$"	10$\frac{3}{16}$"	10$\frac{5}{8}$"
TOTAL RUN OF STAIRWAY	7'10$\frac{15}{16}$"	10'9$\frac{1}{2}$"	12'2$\frac{1}{4}$"	13'7"	15'$\frac{5}{8}$"	16'5$\frac{7}{16}$"	17'10$\frac{15}{16}$"	12'2$\frac{1}{4}$"	13'7"	15'$\frac{5}{8}$"
WELL OPENING FOR 6'-8" HEADROOM	14'8$\frac{21}{32}$"	8'7$\frac{23}{32}$"	9'11$\frac{27}{32}$"	11'4$\frac{3}{16}$"	11'10$\frac{31}{32}$"	13'3$\frac{13}{32}$"	14'8$\frac{21}{32}$"	9'11$\frac{27}{32}$"	11'4$\frac{3}{16}$"	11'10$\frac{31}{32}$"
ANGLE OF INCLINE	28°41'	41°44'	38°29'	35°40'	32°54'	30°46'	28°41'	38°29'	35°40'	32°54'
LENGTH OF CARRIAGE	20'5"	14'5$\frac{17}{32}$"	15'6$\frac{13}{16}$"	16'8$\frac{21}{32}$"	17'11$\frac{5}{32}$"	19'1$\frac{25}{32}$"	20'5"	15'6$\frac{13}{16}$"	16'8$\frac{21}{32}$"	17'11$\frac{5}{32}$"
L. F. OF RISER PER INCH OF STAIRWAY WIDTH	1.666	1.25	1.333	1.416	1.5	1.583	1.666	1.333	1.416	1.5
L. F. OF TREAD PER INCH OF STAIRWAY WIDTH	1.583	1.166	1.25	1.333	1.416	1.5	1.583	1.25	1.333	1.416
FRAMING SQUARE SETTINGS FOR CARRIAGE CUTS — TONGUE	6$\frac{3}{16}$"	8$\frac{1}{4}$"	7$\frac{3}{4}$"	7$\frac{5}{16}$"	6$\frac{7}{8}$"	6$\frac{17}{32}$"	6$\frac{3}{16}$"	7$\frac{3}{4}$"	7$\frac{5}{16}$"	6$\frac{7}{8}$"
FRAMING SQUARE SETTINGS FOR CARRIAGE CUTS — BODY	11$\frac{5}{16}$"	9$\frac{1}{4}$"	9$\frac{3}{4}$"	10$\frac{3}{16}$"	10$\frac{5}{8}$"	10$\frac{31}{32}$"	11$\frac{5}{16}$"	9$\frac{3}{4}$"	10$\frac{3}{16}$"	10$\frac{5}{8}$"

FLOOR TO FLOOR RISE		10'4 1/4"	10'4 1/4"	10'4 3/8"	10'4 3/8"	10'4 3/8"	10'4 3/8"	10'4 3/8"	10'4 1/2"	10'4 1/2"	10'4 1/2"
NUMBER OF RISERS		19	20	16	17	18	19	20	16	17	18
HEIGHT OF EACH RISER		6 9/16"	6 3/16"	7 3/4"	7 5/16"	6 15/16"	6 9/16"	6 7/32"	7 13/16"	7 5/16"	6 15/16"
TOTAL HEIGHT OF RISERS		10'4 11/16"	10'3 3/4"	10'4"	10'4 5/16"	10'4 7/8"	10'4 11/16"	10'4 3/8"	10'5"	10'4 5/16"	10'4 7/8"
RISERS OVERAGE (+) OR UNDERAGE (—)		+7/16"	−1/2"	−3/8"	−1/16"	+1/2"	+5/16"	0	+1/2"	−3/16"	+3/8"
NUMBER OF TREADS		18	19	15	16	17	18	19	15	16	17
WIDTH OF EACH TREAD		10 15/16"	11 5/16"	9 3/4"	10 3/16"	10 9/16"	10 15/16"	11 9/32"	9 11/16"	10 3/16"	10 9/16"
TOTAL RUN OF STAIRWAY		16'4 7/8"	17'10 15/16"	12'2 1/4"	13'7"	14'11 9/16"	16'4 7/8"	17'10 11/32"	12'1 5/16"	13'7"	14'11 9/16"
WELL OPENING FOR 6'-8" HEADROOM		13'2 27/32"	14'8 21/32"	9'11 27/32"	11'4 3/16"	11'9 31/32"	13'2 27/32"	14'8 3/32"	9'10 15/16"	11'4 3/16"	11'9 31/32"
ANGLE OF INCLINE		30°58'	28°41'	38°29'	35°40'	33°18'	30°58'	28°52'	38°53'	35°40'	33°18'
LENGTH OF CARRIAGE		19'1 19/32"	20'5"	15'6 13/16"	16'8 21/32"	17'10 27/32"	19'1 19/32"	20'4 3/4"	15'6 11/16"	16'8 21/32"	17'10 27/32"
L. F. OF RISER PER INCH OF STAIRWAY WIDTH		1.583	1.666	1.333	1.416	1.5	1.583	1.666	1.333	1.416	1.5
L. F. OF TREAD PER INCH OF STAIRWAY WIDTH		1.5	1.583	1.25	1.333	1.416	1.5	1.583	1.25	1.333	1.416
FRAMING SQUARE SETTINGS FOR CARRIAGE CUTS	TONGUE	6 9/16"	6 3/16"	7 3/4"	7 5/16"	6 15/16"	6 9/16"	6 7/32"	7 13/16"	7 5/16"	6 15/16"
	BODY	10 15/16"	11 5/16"	9 3/4"	10 3/16"	10 9/16"	10 15/16"	11 9/32"	9 11/16"	10 3/16"	10 9/16"

FLOOR TO FLOOR RISE	10'4 1/2"	10'4 1/2"	10'4 5/8"	10'4 5/8"	10'4 5/8"	10'4 5/8"	10'4 5/8"	10'4 3/4"	10'4 3/4"	10'4 3/4"
NUMBER OF RISERS	19	20	16	17	18	19	20	16	17	18
HEIGHT OF EACH RISER	6 9/16"	6 1/4"	7 13/16"	7 5/16"	6 15/16"	6 9/16"	6 1/4"	7 13/16"	7 5/16"	6 15/16"
TOTAL HEIGHT OF RISERS	10'4 11/16"	10'5"	10'5"	10'4 5/16"	10'4 7/8"	10'4 11/16"	10'5"	10'5"	10'4 5/16"	10'4 7/8"
RISERS OVERAGE (+) OR UNDERAGE (—)	+3/16"	+1/2"	+3/8"	−5/16"	+1/4"	+1/16"	+3/8"	+1/4"	−7/16"	+1/8"
NUMBER OF TREADS	18	19	15	16	17	18	19	15	16	17
WIDTH OF EACH TREAD	10 15/16"	11 1/4"	9 11/16"	10 3/16"	10 9/16"	10 15/16"	11 1/4"	9 11/16"	10 3/16"	10 9/16"
TOTAL RUN OF STAIRWAY	16'4 7/8"	17'9 3/4"	12'1 5/16"	13'7"	14'11 9/16"	16'4 7/8"	17'9 3/4"	12'1 5/16"	13'7"	14'11 9/16"
WELL OPENING FOR 6'-8" HEADROOM	13'2 27/32"	14'7 1/2"	9'10 15/16"	11'4 3/16"	11'9 31/32"	13'2 27/32"	14'7 1/2"	9'10 15/16"	11'4 3/16"	11'9 31/32"
ANGLE OF INCLINE	30°58'	29°3'	38°53'	35°40'	33°18'	30°58'	29°3'	38°53'	35°40'	33°18'
LENGTH OF CARRIAGE	19'1 19/32"	20'4 17/32"	15'6 11/16"	16'8 21/32"	17'10 27/32"	19'1 19/32"	20'4 17/32"	15'6 11/16"	16'8 21/32"	17'10 27/32"
L. F. OF RISER PER INCH OF STAIRWAY WIDTH	1.583	1.666	1.333	1.416	1.5	1.583	1.666	1.333	1.416	1.5
L. F. OF TREAD PER INCH OF STAIRWAY WIDTH	1.5	1.583	1.25	1.333	1.416	1.5	1.583	1.25	1.333	1.416
FRAMING SQUARE SETTINGS FOR CARRIAGE CUTS — TONGUE	6 9/16"	6 1/4"	7 13/16"	7 5/16"	6 15/16"	6 9/16"	6 1/4"	7 13/16"	7 5/16"	6 15/16"
FRAMING SQUARE SETTINGS FOR CARRIAGE CUTS — BODY	10 15/16"	11 1/4"	9 11/16"	10 3/16"	10 9/16"	10 15/16"	11 1/4"	9 11/16"	10 3/16"	10 9/16"

FLOOR TO FLOOR RISE	10'4 3/4"	10'4 3/4"	10'4 7/8"	10'4 7/8"	10'4 7/8"	10'4 7/8"	10'4 7/8"	10'5"	10'5"	10'5"
NUMBER OF RISERS	19	20	16	17	18	19	20	16	17	18
HEIGHT OF EACH RISER	6 9/16"	6 1/4"	7 13/16"	7 3/8"	6 15/16"	6 9/16"	6 1/4"	7 13/16"	7 3/8"	6 15/16"
TOTAL HEIGHT OF RISERS	10'4 11/16"	10'5"	10'5"	10'5 3/8"	10'4 7/8"	10'4 11/16"	10'5"	10'5"	10'5 3/8"	10'4 7/8"
RISERS OVERAGE (+) OR UNDERAGE (—)	−1/16"	+1/4"	+1/8"	+1/2"	0	−3/16"	+1/8"	0	+3/8"	−1/8"
NUMBER OF TREADS	18	19	15	16	17	18	19	15	16	17
WIDTH OF EACH TREAD	10 15/16"	11 1/4"	9 11/16"	10 1/8"	10 9/16"	10 15/16"	11 1/4"	9 11/16"	10 1/8"	10 9/16"
TOTAL RUN OF STAIRWAY	16'4 7/8"	17'9 3/4"	12'1 5/16"	13'6"	14'11 9/16"	16'4 7/8"	17'9 3/4"	12'1 5/16"	13'6"	14'11 9/16"
WELL OPENING FOR 6'-8" HEADROOM	13'2 27/32"	14'7 1/2"	9'10 15/16"	10'5 3/32"	11'9 31/32"	13'2 27/32"	14'7 1/2"	9'10 15/16"	10'5 3/32"	11'9 31/32"
ANGLE OF INCLINE	30°58'	29°3'	38°53'	36°4'	33°18'	30°58'	29°3'	38°53'	36°4'	33°18'
LENGTH OF CARRIAGE	19'1 19/32"	20'4 17/32"	15'6 11/16"	16'8 13/32"	17'10 27/32"	19'1 19/32"	20'4 17/32"	15'6 11/16"	16'8 13/32"	17'10 27/32"
L. F. OF RISER PER INCH OF STAIRWAY WIDTH	1.583	1.666	1.333	1.416	1.5	1.583	1.666	1.333	1.416	1.5
L. F. OF TREAD PER INCH OF STAIRWAY WIDTH	1.5	1.583	1.25	1.333	1.416	1.5	1.583	1.25	1.333	1.416
FRAMING SQUARE SETTINGS FOR CARRIAGE CUTS — TONGUE	6 9/16"	6 1/4"	7 13/16"	7 3/8"	6 15/16"	6 9/16"	6 1/4"	7 13/16"	7 3/8"	6 15/16"
FRAMING SQUARE SETTINGS FOR CARRIAGE CUTS — BODY	10 15/16"	11 1/4"	9 11/16"	10 1/8"	10 9/16"	10 15/16"	11 1/4"	9 11/16"	10 1/8"	10 9/16"

FLOOR TO FLOOR RISE	10'5"	10'5"	10'5$\frac{1}{8}$"	10'5$\frac{1}{8}$"	10'5$\frac{1}{8}$"	10'5$\frac{1}{8}$"	10'5$\frac{1}{8}$"	10'5$\frac{1}{4}$"	10'5$\frac{1}{4}$"	10'5$\frac{1}{4}$"
NUMBER OF RISERS	19	20	16	17	18	19	20	16	17	18
HEIGHT OF EACH RISER	6$\frac{9}{16}$"	6$\frac{1}{4}$"	7$\frac{13}{16}$"	7$\frac{3}{8}$"	6$\frac{15}{16}$"	6$\frac{9}{16}$"	6$\frac{1}{4}$"	7$\frac{13}{16}$"	7$\frac{3}{8}$"	6$\frac{15}{16}$"
TOTAL HEIGHT OF RISERS	10'4$\frac{11}{16}$"	10'5"	10'5"	10'5$\frac{3}{8}$"	10'4$\frac{7}{8}$"	10'4$\frac{11}{16}$"	10'5"	10'5"	10'5$\frac{3}{8}$"	10'4$\frac{7}{8}$"
RISERS OVERAGE (+) OR UNDERAGE (—)	$-\frac{5}{16}$"	0	$-\frac{1}{8}$"	$+\frac{1}{4}$"	$-\frac{1}{4}$"	$-\frac{7}{16}$"	$-\frac{1}{8}$"	$-\frac{1}{4}$"	$+\frac{1}{8}$"	$-\frac{3}{8}$"
NUMBER OF TREADS	18	19	15	16	17	18	19	15	16	17
WIDTH OF EACH TREAD	10$\frac{15}{16}$"	11$\frac{1}{4}$"	9$\frac{11}{16}$"	10$\frac{1}{8}$"	10$\frac{9}{16}$"	10$\frac{15}{16}$"	11$\frac{1}{4}$"	9$\frac{11}{16}$"	10$\frac{1}{8}$"	10$\frac{9}{16}$"
TOTAL RUN OF STAIRWAY	16'4$\frac{7}{8}$"	17'9$\frac{3}{4}$"	12'1$\frac{5}{16}$"	13'6"	14'11$\frac{9}{16}$"	16'4$\frac{7}{8}$"	17'9$\frac{3}{4}$"	12'1$\frac{5}{16}$"	13'6"	14'11$\frac{9}{16}$"
WELL OPENING FOR 6'-8" HEADROOM	13'2$\frac{27}{32}$"	14'7$\frac{1}{2}$"	9'10$\frac{15}{16}$"	10'5$\frac{3}{32}$"	11'9$\frac{31}{32}$"	13'2$\frac{27}{32}$"	14'7$\frac{1}{2}$"	9'10$\frac{15}{16}$"	10'5$\frac{3}{32}$"	11'9$\frac{31}{32}$"
ANGLE OF INCLINE	30°58'	29°3'	38°53'	36°4'	33°18'	30°58'	29°3'	38°53'	36°4'	33°18'
LENGTH OF CARRIAGE	19'1$\frac{19}{32}$"	20'4$\frac{17}{32}$"	15'6$\frac{11}{16}$"	16'8$\frac{13}{32}$"	17'10$\frac{27}{32}$"	19'1$\frac{19}{32}$"	20'4$\frac{17}{32}$"	15'6$\frac{11}{16}$"	16'8$\frac{13}{32}$"	17'10$\frac{27}{32}$"
L. F. OF RISER PER INCH OF STAIRWAY WIDTH	1.583	1.666	1.333	1.416	1.5	1.583	1.666	1.333	1.416	1.5
L. F. OF TREAD PER INCH OF STAIRWAY WIDTH	1.5	1.583	1.25	1.333	1.416	1.5	1.583	1.25	1.333	1.416
FRAMING SQUARE SETTINGS FOR CARRIAGE CUTS — TONGUE	6$\frac{9}{16}$"	6$\frac{1}{4}$"	7$\frac{13}{16}$"	7$\frac{3}{8}$"	6$\frac{9}{16}$"	6$\frac{9}{16}$"	6$\frac{1}{4}$"	7$\frac{13}{16}$"	7$\frac{3}{8}$"	6$\frac{15}{16}$"
FRAMING SQUARE SETTINGS FOR CARRIAGE CUTS — BODY	10$\frac{15}{16}$"	11$\frac{1}{4}$"	9$\frac{11}{16}$"	10$\frac{1}{8}$"	10$\frac{15}{16}$"	10$\frac{15}{16}$"	11$\frac{1}{4}$"	9$\frac{11}{16}$"	10$\frac{1}{8}$"	10$\frac{9}{16}$"

FLOOR TO FLOOR RISE	10'5 1/4"	10'5 1/4"	10'5 3/8"	10'5 3/8"	10'5 3/8"	10'5 3/8"	10'5 3/8"	10'5 1/2"	10'5 1/2"	10'5 1/2"
NUMBER OF RISERS	19	20	16	17	18	19	20	16	17	18
HEIGHT OF EACH RISER	6 19/32"	6 1/4"	7 13/16"	7 3/8"	6 15/16"	6 5/8"	6 1/4"	7 7/8"	7 3/8"	7"
TOTAL HEIGHT OF RISERS	10'5 9/32"	10'5"	10'5"	10'5 3/8"	10'4 7/8"	10'5 7/8"	10'5"	10'6"	10'5 3/8"	10'6"
RISERS OVERAGE (+) OR UNDERAGE (—)	+1/32"	−1/4"	−3/8"	0	−1/2"	+1/2"	−3/8"	+1/2"	−1/8"	+1/2"
NUMBER OF TREADS	18	19	15	16	17	18	19	15	16	17
WIDTH OF EACH TREAD	10 29/32"	11 1/4"	9 11/16"	10 1/8"	10 9/16"	10 7/8"	11 1/4"	9 5/8"	10 1/8"	10 1/2"
TOTAL RUN OF STAIRWAY	16'4 5/16"	17'9 3/4"	12'1 5/16"	13'6"	14'11 9/16"	16'3 3/4"	17'9 3/4"	12'3/8"	13'6"	14'10 1/2"
WELL OPENING FOR 6'-8" HEADROOM	13'2 5/16"	14'7 1/2"	9'10 15/16"	10'5 3/32"	11'9 31/32"	13'1 25/32"	14'7 1/2"	9'10 3/32"	10'5 3/32"	11'9"
ANGLE OF INCLINE	31°9'	29°3'	38°53'	36°4'	33°18'	31°21'	29°3'	39°17'	36°4'	33°41'
LENGTH OF CARRIAGE	19'1 13/32"	20'4 17/32"	15'6 11/16"	16'8 13/32"	17'10 27/32"	19'1 7/32"	20'4 17/32"	15'6 17/32"	16'8 13/32"	17'10 17/32"
L. F. OF RISER PER INCH OF STAIRWAY WIDTH	1.583	1.666	1.333	1.416	1.5	1.583	1.666	1.333	1.416	1.5
L. F. OF TREAD PER INCH OF STAIRWAY WIDTH	1.5	1.583	1.25	1.333	1.416	1.5	1.583	1.25	1.333	1.416
FRAMING SQUARE SETTINGS FOR CARRIAGE CUTS — TONGUE	6 19/32"	6 1/4"	7 13/16"	7 3/8"	6 15/16"	6 5/8"	6 1/4"	7 7/8"	7 3/8"	7"
FRAMING SQUARE SETTINGS FOR CARRIAGE CUTS — BODY	10 29/32"	11 1/4"	9 11/16"	10 1/8"	10 9/16"	10 7/8"	11 1/4"	9 5/8"	10 1/8"	10 1/2"

FLOOR TO FLOOR RISE	10'5 1/2"	10'5 1/2"	10'5 5/8"	10'5 5/8"	10'5 5/8"	10'5 5/8"	10'5 5/8"	10'5 3/4"	10'5 3/4"	10'5 3/4"
NUMBER OF RISERS	19	20	16	17	18	19	20	16	17	18
HEIGHT OF EACH RISER	6 5/8"	6 1/4"	7 7/8"	7 3/8"	7"	6 5/8"	6 9/32"	7 7/8"	7 3/8"	7"
TOTAL HEIGHT OF RISERS	10'5 7/8"	10'5"	10'6"	10'5 3/8"	10'6"	10'5 7/8"	10'5 5/8"	10'6"	10'5 3/8"	10'6"
RISERS OVERAGE (+) OR UNDERAGE (—)	+3/8"	−1/2"	+3/8"	−1/4"	+3/8"	+1/4"	0	+1/4"	−3/8"	+1/4"
NUMBER OF TREADS	18	19	15	16	17	18	19	15	16	17
WIDTH OF EACH TREAD	10 7/8"	11 1/4"	9 5/8"	10 1/8"	10 1/2"	10 7/8"	11 7/32"	9 5/8"	10 1/8"	10 1/2"
TOTAL RUN OF STAIRWAY	16'3 3/4"	17'9 3/4"	12'3/8"	13'6"	14'10 1/2"	16'3 3/4"	17'9 5/32"	12'3/8"	13'6"	14'10 1/2"
WELL OPENING FOR 6'-8" HEADROOM	13'1 25/32"	14'7 1/2"	9'10 3/32"	10'5 3/32"	11'9"	13'1 25/32"	14'6 29/32"	9'10 3/32"	10'5 3/32"	11'9"
ANGLE OF INCLINE	31°21'	29°3'	39°17'	36°4'	33°41'	31°21'	29°15'	39°17'	36°4'	33°41'
LENGTH OF CARRIAGE	19'1 7/32"	20'4 17/32"	15'6 17/32"	16'8 13/32"	17'10 17/32"	19'1 7/32"	20'4 9/32"	15'6 17/32"	16'8 13/32"	17'10 17/32"
L. F. OF RISER PER INCH OF STAIRWAY WIDTH	1.583	1.666	1.333	1.416	1.5	1.583	1.666	1.333	1.416	1.5
L. F. OF TREAD PER INCH OF STAIRWAY WIDTH	1.5	1.583	1.25	1.333	1.416	1.5	1.583	1.25	1.333	1.416
FRAMING SQUARE SETTINGS FOR CARRIAGE CUTS — TONGUE	6 5/8"	6 1/4"	7 7/8"	7 3/8"	7"	6 5/8"	6 9/32"	7 7/8"	7 3/8"	7"
FRAMING SQUARE SETTINGS FOR CARRIAGE CUTS — BODY	10 7/8"	11 1/4"	9 5/8"	10 1/8"	10 1/2"	10 7/8"	11 7/32"	9 5/8"	10 1/8"	10 1/2"

FLOOR TO FLOOR RISE		10'5$\frac{3}{4}$"	10'5$\frac{3}{4}$"	10'5$\frac{7}{8}$"	10'5$\frac{7}{8}$"	10'5$\frac{7}{8}$"	10'5$\frac{7}{8}$"	10'5$\frac{7}{8}$"	10'6"	10'6"	10'6"
NUMBER OF RISERS		19	20	16	17	18	19	20	16	17	18
HEIGHT OF EACH RISER		6$\frac{5}{8}$"	6$\frac{5}{16}$"	7$\frac{7}{8}$"	7$\frac{3}{8}$"	7"	6$\frac{5}{8}$"	6$\frac{5}{16}$"	7$\frac{7}{8}$"	7$\frac{7}{16}$"	7"
TOTAL HEIGHT OF RISERS		10'5$\frac{7}{8}$"	10'6$\frac{1}{4}$"	10'6"	10'5$\frac{3}{8}$"	10'6"	10'5$\frac{7}{8}$"	10'6$\frac{1}{4}$"	10'6"	10'6$\frac{7}{16}$"	10'6"
RISERS OVERAGE (+) OR UNDERAGE (—)		+$\frac{1}{8}$"	+$\frac{1}{2}$"	+$\frac{1}{8}$"	−$\frac{1}{2}$"	+$\frac{1}{8}$"	0	+$\frac{3}{8}$"	0	+$\frac{7}{16}$"	0
NUMBER OF TREADS		18	19	15	16	17	18	19	15	16	17
WIDTH OF EACH TREAD		10$\frac{7}{8}$"	11$\frac{3}{16}$"	9$\frac{5}{8}$"	10$\frac{1}{8}$"	10$\frac{1}{2}$"	10$\frac{7}{8}$"	11$\frac{3}{16}$"	9$\frac{5}{8}$"	10$\frac{1}{16}$"	10$\frac{1}{2}$"
TOTAL RUN OF STAIRWAY		16'3$\frac{3}{4}$"	17'8$\frac{9}{16}$"	12'$\frac{3}{8}$"	13'6"	14'10$\frac{1}{2}$"	16'3$\frac{3}{4}$"	17'8$\frac{9}{16}$"	12'$\frac{3}{8}$"	13'5"	14'10$\frac{1}{2}$"
WELL OPENING FOR 6'-8" HEADROOM		13'1$\frac{25}{32}$"	13'7$\frac{5}{32}$"	9'10$\frac{3}{32}$"	10'5$\frac{3}{32}$"	11'9"	13'1$\frac{25}{32}$"	13'7$\frac{5}{32}$"	9'10$\frac{3}{32}$"	10'4$\frac{7}{32}$"	11'9"
ANGLE OF INCLINE		31°21'	29°26'	39°17'	36°4'	33°41'	31°21'	29°26'	39°17'	36°28'	33°41'
LENGTH OF CARRIAGE		19'1$\frac{7}{32}$"	20'4$\frac{1}{16}$"	15'6$\frac{17}{32}$"	16'8$\frac{13}{32}$"	17'10$\frac{17}{32}$"	19'1$\frac{7}{32}$"	20'4$\frac{1}{16}$"	15'6$\frac{17}{32}$"	16'8$\frac{7}{32}$"	17'10$\frac{17}{32}$"
L. F. OF RISER PER INCH OF STAIRWAY WIDTH		1.583	1.666	1.333	1.416	1.5	1.583	1.666	1.333	1.416	1.5
L. F. OF TREAD PER INCH OF STAIRWAY WIDTH		1.5	1.583	1.25	1.333	1.416	1.5	1.583	1.25	1.333	1.416
FRAMING SQUARE SETTINGS FOR CARRIAGE CUTS	TONGUE	6$\frac{5}{8}$"	6$\frac{5}{16}$"	7$\frac{7}{8}$"	7$\frac{3}{8}$"	7"	6$\frac{5}{8}$"	6$\frac{5}{16}$"	7$\frac{7}{8}$"	7$\frac{7}{16}$"	7"
	BODY	10$\frac{7}{8}$"	11$\frac{3}{16}$"	9$\frac{5}{8}$"	10$\frac{1}{8}$"	10$\frac{1}{2}$"	10$\frac{7}{8}$"	11$\frac{3}{16}$"	9$\frac{5}{8}$"	10$\frac{1}{16}$"	10$\frac{1}{2}$"

FLOOR TO FLOOR RISE		10'6"	10'6"	10'6"	10'6$\frac{1}{8}$"	10'6$\frac{1}{8}$"	10'6$\frac{1}{8}$"	10'6$\frac{1}{8}$"	10'6$\frac{1}{8}$"	10'6$\frac{1}{8}$"	10'6$\frac{1}{4}$"
NUMBER OF RISERS		19	20	21	16	17	18	19	20	21	16
HEIGHT OF EACH RISER		6$\frac{5}{8}$"	6$\frac{5}{16}$"	6"	7$\frac{7}{8}$"	7$\frac{7}{16}$"	7"	6$\frac{5}{8}$"	6$\frac{5}{16}$"	6"	7$\frac{7}{8}$"
TOTAL HEIGHT OF RISERS		10'5$\frac{7}{8}$"	10'6$\frac{1}{4}$"	10'6"	10'6"	10'6$\frac{7}{16}$"	10'6"	10'5$\frac{7}{8}$"	10'6$\frac{1}{4}$"	10'6"	10'6"
RISERS OVERAGE (+) OR UNDERAGE (—)		$-\frac{1}{8}$"	$+\frac{1}{4}$"	0	$-\frac{1}{8}$"	$+\frac{5}{16}$"	$-\frac{1}{8}$"	$-\frac{1}{4}$"	$+\frac{1}{8}$"	$-\frac{1}{8}$"	$-\frac{1}{4}$"
NUMBER OF TREADS		18	19	20	15	16	17	18	19	20	15
WIDTH OF EACH TREAD		10$\frac{7}{8}$"	11$\frac{3}{16}$"	11$\frac{1}{2}$"	9$\frac{5}{8}$"	10$\frac{1}{16}$"	10$\frac{1}{2}$"	10$\frac{7}{8}$"	11$\frac{3}{16}$"	11$\frac{1}{2}$"	9$\frac{5}{8}$"
TOTAL RUN OF STAIRWAY		16'3$\frac{3}{4}$"	17'8$\frac{9}{16}$"	19'2"	12'$\frac{3}{8}$"	13'5"	14'10$\frac{1}{2}$"	16'3$\frac{3}{4}$"	17'8$\frac{9}{16}$"	19'2"	12'$\frac{3}{8}$"
WELL OPENING FOR 6'-8" HEADROOM		13'1$\frac{25}{32}$"	13'7$\frac{5}{32}$"	15'$\frac{5}{32}$"	9'10$\frac{3}{32}$"	10'4$\frac{7}{32}$"	11'9"	13'1$\frac{25}{32}$"	13'7$\frac{5}{32}$"	15'$\frac{5}{32}$"	9'10$\frac{3}{32}$"
ANGLE OF INCLINE		31°21'	29°26'	27°33'	39°17'	36°28'	33°41'	31°21'	29°26'	27°33'	39°17'
LENGTH OF CARRIAGE		19'1$\frac{7}{32}$"	20'4$\frac{1}{16}$"	21'7$\frac{7}{16}$"	15'6$\frac{17}{32}$"	16'8$\frac{7}{32}$"	17'10$\frac{17}{32}$"	19'1$\frac{7}{32}$"	20'4$\frac{1}{16}$"	21'7$\frac{7}{16}$"	15'6$\frac{17}{32}$"
L. F. OF RISER PER INCH OF STAIRWAY WIDTH		1.583	1.666	1.75	1.333	1.416	1.5	1.583	1.666	1.75	1.333
L. F. OF TREAD PER INCH OF STAIRWAY WIDTH		1.5	1.583	1.666	1.25	1.333	1.416	1.5	1.583	1.666	1.25
FRAMING SQUARE SETTINGS FOR CARRIAGE CUTS	TONGUE	6$\frac{5}{8}$"	6$\frac{5}{16}$"	6"	7$\frac{7}{8}$"	7$\frac{7}{16}$"	7"	6$\frac{5}{8}$"	6$\frac{5}{16}$"	6"	7$\frac{7}{8}$"
	BODY	10$\frac{7}{8}$"	11$\frac{3}{16}$"	11$\frac{1}{2}$"	9$\frac{5}{8}$"	10$\frac{1}{16}$"	10$\frac{1}{2}$"	10$\frac{7}{8}$"	11$\frac{3}{16}$"	11$\frac{1}{2}$"	9$\frac{5}{8}$"

FLOOR TO FLOOR RISE	10'6$\frac{1}{4}$"	10'6$\frac{1}{4}$"	10'6$\frac{1}{4}$"	10'6$\frac{1}{4}$"	10'6$\frac{1}{4}$"	10'6$\frac{3}{8}$"	10'6$\frac{3}{8}$"	10'6$\frac{3}{8}$"	10'6$\frac{3}{8}$"	10'6$\frac{3}{8}$"
NUMBER OF RISERS	17	18	19	20	21	16	17	18	19	20
HEIGHT OF EACH RISER	7$\frac{7}{16}$"	7"	6$\frac{5}{8}$"	6$\frac{5}{16}$"	6"	7$\frac{7}{8}$"	7$\frac{7}{16}$"	7"	6$\frac{5}{8}$"	6$\frac{5}{16}$"
TOTAL HEIGHT OF RISERS	10'6$\frac{7}{16}$"	10'6"	10'5$\frac{7}{8}$"	10'6$\frac{1}{4}$"	10'6"	10'6"	10'6$\frac{7}{16}$"	10'6"	10'5$\frac{7}{8}$"	10'6$\frac{1}{4}$"
RISERS OVERAGE (+) OR UNDERAGE (—)	+$\frac{3}{16}$"	−$\frac{1}{4}$"	−$\frac{3}{8}$"	0	−$\frac{1}{4}$"	−$\frac{3}{8}$"	+$\frac{1}{16}$"	−$\frac{3}{8}$"	−$\frac{1}{2}$"	−$\frac{1}{8}$"
NUMBER OF TREADS	16	17	18	19	20	15	16	17	18	19
WIDTH OF EACH TREAD	10$\frac{1}{16}$"	10$\frac{1}{2}$"	10$\frac{7}{8}$"	11$\frac{3}{16}$"	11$\frac{1}{2}$"	9$\frac{5}{8}$"	10$\frac{1}{16}$"	10$\frac{1}{2}$"	10$\frac{7}{8}$"	11$\frac{3}{16}$"
TOTAL RUN OF STAIRWAY	13'5"	14'10$\frac{1}{2}$"	16'3$\frac{3}{4}$"	17'8$\frac{9}{16}$"	19'2"	12'$\frac{3}{8}$"	13'5"	14'10$\frac{1}{2}$"	16'3$\frac{3}{4}$"	17'8$\frac{9}{16}$"
WELL OPENING FOR 6'-8" HEADROOM	10'4$\frac{7}{32}$"	11'9"	13'1$\frac{25}{32}$"	13'7$\frac{5}{32}$"	15'$\frac{5}{32}$"	9'10$\frac{3}{32}$"	10'4$\frac{7}{32}$"	11'9"	13'1$\frac{25}{32}$"	13'7$\frac{5}{32}$"
ANGLE OF INCLINE	36°28'	33°41'	31°21'	29°26'	27°33'	39°17'	36°28'	33°41'	31°21'	29°26'
LENGTH OF CARRIAGE	16'8$\frac{7}{32}$"	17'10$\frac{17}{32}$"	19'1$\frac{7}{32}$"	20'4$\frac{1}{16}$"	21'7$\frac{7}{16}$"	15'6$\frac{17}{32}$"	16'8$\frac{7}{32}$"	17'10$\frac{17}{32}$"	19'1$\frac{7}{32}$"	20'4$\frac{1}{16}$"
L. F. OF RISER PER INCH OF STAIRWAY WIDTH	1.416	1.5	1.583	1.666	1.75	1.333	1.416	1.5	1.583	1.666
L. F. OF TREAD PER INCH OF STAIRWAY WIDTH	1.333	1.416	1.5	1.583	1.666	1.25	1.333	1.416	1.5	1.583
FRAMING SQUARE SETTINGS FOR CARRIAGE CUTS — TONGUE	7$\frac{7}{16}$"	7"	6$\frac{5}{8}$"	6$\frac{5}{16}$"	6"	7$\frac{7}{8}$"	7$\frac{7}{16}$"	7"	6$\frac{5}{8}$"	6$\frac{5}{16}$"
BODY	10$\frac{1}{16}$"	10$\frac{1}{2}$"	10$\frac{7}{8}$"	11$\frac{3}{16}$"	11$\frac{1}{2}$"	9$\frac{5}{8}$"	10$\frac{1}{16}$"	10$\frac{1}{2}$"	10$\frac{7}{8}$"	11$\frac{3}{16}$"

FLOOR TO FLOOR RISE		10'6 3/8"	10'6 1/2"	10'6 1/2"	10'6 1/2"	10'6 1/2"	10'6 1/2"	10'6 1/2"	10'6 5/8"	10'6 5/8"	10'6 5/8"
NUMBER OF RISERS		21	16	17	18	19	20	21	16	17	18
HEIGHT OF EACH RISER		6"	7 15/16"	7 7/16"	7"	6 21/32"	6 5/16"	6"	7 15/16"	7 7/16"	7 1/16"
TOTAL HEIGHT OF RISERS		10'6"	10'7"	10'6 7/16"	10'6"	10'6 15/32"	10'6 1/4"	10'6"	10'7"	10'6 7/16"	10'7 1/8"
RISERS OVERAGE (+) OR UNDERAGE (—)		−3/8"	+1/2"	−1/16"	−1/2"	−1/32"	−1/4"	−1/2"	+3/8"	−3/16"	+1/2"
NUMBER OF TREADS		20	15	16	17	18	19	20	15	16	17
WIDTH OF EACH TREAD		11 1/2"	9 9/16"	10 1/16"	10 1/2"	10 27/32"	11 3/16"	11 1/2"	9 9/16"	10 1/16"	10 7/16"
TOTAL RUN OF STAIRWAY		19'2"	11'11 7/16"	13'5"	14'10 1/2"	16'3 3/16"	17'8 9/16"	19'2"	11'11 7/16"	13'5"	14'9 7/16"
WELL OPENING FOR 6'-8" HEADROOM		15' 3/32"	9'9 1/4"	10'4 7/32"	11'9"	13'1 9/16"	13'7 5/32"	15' 5/32"	9'9 1/4"	10'4 7/32"	11'8 1/32"
ANGLE OF INCLINE		27°33'	39°42'	36°28'	33°41'	31°33'	29°26'	27°33'	39°42'	36°28'	34°5'
LENGTH OF CARRIAGE		21'7 7/16"	15'6 13/32"	16'8 7/32"	17'10 17/32"	19'1 1/32"	20'4 1/16"	21'7 7/16"	15'6 13/32"	16'8 7/32"	17'10 1/4"
L. F. OF RISER PER INCH OF STAIRWAY WIDTH		1.75	1.333	1.416	1.5	1.583	1.666	1.75	1.333	1.416	1.5
L. F. OF TREAD PER INCH OF STAIRWAY WIDTH		1.666	1.25	1.333	1.416	1.5	1.583	1.666	1.25	1.333	1.416
FRAMING SQUARE SETTINGS FOR CARRIAGE CUTS	TONGUE	6"	7 15/16"	7 7/16"	7"	6 21/32"	6 5/16"	6"	7 15/16"	7 7/16"	7 1/16"
	BODY	11 1/2"	9 9/16"	10 1/16"	10 1/2"	10 27/32"	11 3/16"	11 1/2"	9 9/16"	10 1/16"	10 7/16"

FLOOR TO FLOOR RISE		10'6 5/8"	10'6 5/8"	10'6 5/8"	10'6 3/4"	10'6 3/4"	10'6 3/4"	10'6 3/4"	10'6 3/4"	10'6 3/4"	10'6 7/8"
NUMBER OF RISERS		19	20	21	16	17	18	19	20	21	16
HEIGHT OF EACH RISER		6 11/16"	6 5/16"	6 1/32"	7 15/16"	7 7/16"	7 1/16"	6 11/16"	6 5/16"	6 1/32"	7 15/16"
TOTAL HEIGHT OF RISERS		10'7 1/16"	10'6 1/4"	10'6 21/32"	10'7"	10'6 7/16"	10'7 1/8"	10'7 1/16"	10'6 1/4"	10'6 21/32"	10'7"
RISERS OVERAGE (+) OR UNDERAGE (—)		+7/16"	−3/8"	+1/32"	+1/4"	−5/16"	+3/8"	+5/16"	−1/2"	−3/32"	+1/8"
NUMBER OF TREADS		18	19	20	15	16	17	18	19	20	15
WIDTH OF EACH TREAD		10 13/16"	11 3/16"	11 15/32"	9 9/16"	10 1/16"	10 7/16"	10 13/16"	11 3/16"	11 15/32"	9 9/16"
TOTAL RUN OF STAIRWAY		16'2 5/8"	17'8 9/16"	19'1 3/8"	11'11 7/16"	13'5"	14'9 7/16"	16'2 5/8"	17'8 9/16"	19'1 3/8"	11'11 7/16"
WELL OPENING FOR 6'-8" HEADROOM		13' 3/4"	13'7 5/32"	14'11 19/32"	9'9 1/4"	10'4 7/32"	11'8 1/32"	13' 3/4"	13'7 5/32"	14'11 19/32"	9'9 1/4"
ANGLE OF INCLINE		31°44'	29°26'	27°44'	39°42'	36°28'	34°5'	31°44'	29°26'	27°44'	39°42'
LENGTH OF CARRIAGE		19' 27/32"	20'4 1/16"	21'7 5/32"	15'6 13/16"	16'8 7/32"	17'10 1/4"	19' 27/32"	20'4 1/16"	21'7 5/32"	15'6 13/32"
L. F. OF RISER PER INCH OF STAIRWAY WIDTH		1.583	1.666	1.75	1.333	1.416	1.5	1.583	1.666	1.75	1.333
L. F. OF TREAD PER INCH OF STAIRWAY WIDTH		1.5	1.583	1.666	1.25	1.333	1.416	1.5	1.583	1.666	1.25
FRAMING SQUARE SETTINGS FOR CARRIAGE CUTS	TONGUE	6 11/16"	6 5/16"	6 1/32"	7 15/16"	7 7/16"	7 1/16"	6 11/16"	6 5/16"	6 1/32"	7 15/16"
	BODY	10 13/16"	11 3/16"	11 15/32"	9 9/16"	10 1/16"	10 7/16"	10 13/16"	11 3/16"	11 15/32"	9 9/16"

FLOOR TO FLOOR RISE	10'6 7/8"	10'6 7/8"	10'6 7/8"	10'6 7/8"	10'6 7/8"	10'7"	10'7"	10'7"	10'7"	10'7"
NUMBER OF RISERS	17	18	19	20	21	16	17	18	19	20
HEIGHT OF EACH RISER	7 7/16"	7 1/16"	6 11/16"	6 11/32"	6 1/16"	7 15/16"	7 1/2"	7 1/16"	6 11/16"	6 3/8"
TOTAL HEIGHT OF RISERS	10'6 7/16"	10'7 1/8"	10'7 1/16"	10'6 7/8"	10'7 5/16"	10'7"	10'7 1/2"	10'7 1/8"	10'7 1/16"	10'7 1/2"
RISERS OVERAGE (+) OR UNDERAGE (—)	−7/16"	+1/4"	+3/16"	0	+7/16"	0	+1/2"	+1/8"	+1/16"	+1/2"
NUMBER OF TREADS	16	17	18	19	20	15	16	17	18	19
WIDTH OF EACH TREAD	10 1/16"	10 7/16"	10 13/16"	11 5/32"	11 7/16"	9 9/16"	10"	10 7/16"	10 13/16"	11 1/8"
TOTAL RUN OF STAIRWAY	13'5"	14'9 7/16"	16'2 5/8"	17'7 31/32"	19'3/4"	11'11 7/16"	13'4"	14'9 7/16"	16'2 5/8"	17'7 3/8"
WELL OPENING FOR 6'-8" HEADROOM	10'4 7/32"	11'8 1/32"	13'3/4"	13'6 5/8"	14'11"	9'9 1/4"	10'3 11/32"	11'8 1/32"	13'3/4"	13'6 1/16"
ANGLE OF INCLINE	36°28'	34°5'	31°44'	29°37'	27°56'	39°42'	36°52'	34°5'	31°44'	29°49'
LENGTH OF CARRIAGE	16'8 7/32"	17'10 1/4"	19'27/32"	20'3 27/32"	21'6 29/32"	15'6 13/32"	16'8"	17'10 1/4"	19'27/32"	20'3 5/8"
L. F. OF RISER PER INCH OF STAIRWAY WIDTH	1.416	1.5	1.583	1.666	1.75	1.333	1.416	1.5	1.583	1.666
L. F. OF TREAD PER INCH OF STAIRWAY WIDTH	1.333	1.416	1.5	1.583	1.666	1.25	1.333	1.416	1.5	1.583
FRAMING SQUARE SETTINGS FOR CARRIAGE CUTS — TONGUE	7 7/16"	7 1/16"	6 11/16"	6 11/32"	6 1/16"	7 15/16"	7 1/2"	7 1/16"	6 11/16"	6 3/8"
FRAMING SQUARE SETTINGS FOR CARRIAGE CUTS — BODY	10 1/16"	10 7/16"	10 13/16"	11 5/32"	11 7/16"	9 9/16"	10"	10 7/16"	10 13/16"	11 1/8"

FLOOR TO FLOOR RISE	10'7"	10'7$\frac{1}{8}$"	10'7$\frac{1}{8}$"	10'7$\frac{1}{8}$"	10'7$\frac{1}{8}$"	10'7$\frac{1}{8}$"	10'7$\frac{1}{8}$"	10'7$\frac{1}{4}$"	10'7$\frac{1}{4}$"	10'7$\frac{1}{4}$"
NUMBER OF RISERS	21	16	17	18	19	20	21	16	17	18
HEIGHT OF EACH RISER	6$\frac{1}{16}$"	7$\frac{15}{16}$"	7$\frac{1}{2}$"	7$\frac{1}{16}$"	6$\frac{11}{16}$"	6$\frac{3}{8}$"	6$\frac{1}{16}$"	7$\frac{15}{16}$"	7$\frac{1}{2}$"	7$\frac{1}{16}$"
TOTAL HEIGHT OF RISERS	10'7$\frac{5}{16}$"	10'7"	10'7$\frac{1}{2}$"	10'7$\frac{1}{8}$"	10'7$\frac{1}{16}$"	10'7$\frac{1}{2}$"	10'7$\frac{5}{16}$"	10'7"	10'7$\frac{1}{2}$"	10'7$\frac{1}{8}$"
RISERS OVERAGE (+) OR UNDERAGE (—)	+$\frac{5}{16}$"	−$\frac{1}{8}$"	+$\frac{3}{8}$"	0	−$\frac{1}{16}$"	+$\frac{3}{8}$"	+$\frac{3}{16}$"	−$\frac{1}{4}$"	+$\frac{1}{4}$"	−$\frac{1}{8}$"
NUMBER OF TREADS	20	15	16	17	18	19	20	15	16	17
WIDTH OF EACH TREAD	11$\frac{7}{16}$"	9$\frac{9}{16}$"	10"	10$\frac{7}{16}$"	10$\frac{13}{16}$"	11$\frac{1}{8}$"	11$\frac{7}{16}$"	9$\frac{9}{16}$"	10"	10$\frac{7}{16}$"
TOTAL RUN OF STAIRWAY	19'$\frac{3}{4}$"	11'11$\frac{7}{16}$"	13'4"	14'9$\frac{7}{16}$"	16'2$\frac{5}{8}$"	17'7$\frac{3}{8}$"	19'$\frac{3}{4}$"	11'11$\frac{7}{16}$"	13'4"	14'9$\frac{7}{16}$"
WELL OPENING FOR 6'-8" HEADROOM	14'11"	9'9$\frac{1}{4}$"	10'3$\frac{11}{32}$"	11'8$\frac{1}{32}$"	13'$\frac{3}{4}$"	13'6$\frac{1}{16}$"	14'11"	9'9$\frac{1}{4}$"	10'3$\frac{11}{32}$"	11'8$\frac{1}{32}$"
ANGLE OF INCLINE	27°56'	39°42'	36°52'	34°5'	31°44'	29°49'	27°56'	39°42'	36°52'	34°5'
LENGTH OF CARRIAGE	21'6$\frac{29}{32}$"	15'6$\frac{13}{32}$"	16'8"	17'10$\frac{1}{4}$"	19'$\frac{27}{32}$"	20'3$\frac{5}{8}$"	21'6$\frac{29}{32}$"	15'6$\frac{13}{32}$"	16'8"	17'10$\frac{1}{4}$"
L. F. OF RISER PER INCH OF STAIRWAY WIDTH	1.75	1.333	1.416	1.5	1.583	1.666	1.75	1.333	1.416	1.5
L. F. OF TREAD PER INCH OF STAIRWAY WIDTH	1.666	1.25	1.333	1.416	1.5	1.583	1.666	1.25	1.333	1.416
FRAMING SQUARE SETTINGS FOR CARRIAGE CUTS — TONGUE	6$\frac{1}{16}$"	7$\frac{15}{16}$"	7$\frac{1}{2}$"	7$\frac{1}{16}$"	6$\frac{11}{16}$"	6$\frac{3}{8}$"	6$\frac{1}{16}$"	7$\frac{15}{16}$"	7$\frac{1}{2}$"	7$\frac{1}{16}$"
FRAMING SQUARE SETTINGS FOR CARRIAGE CUTS — BODY	11$\frac{7}{16}$"	9$\frac{9}{16}$"	10"	10$\frac{7}{16}$"	10$\frac{13}{16}$"	11$\frac{1}{8}$"	11$\frac{7}{16}$"	9$\frac{9}{16}$"	10"	10$\frac{7}{16}$"

FLOOR TO FLOOR RISE	10'7 1/4"	10'7 1/4"	10'7 1/4"	10'7 3/8"	10'7 3/8"	10'7 3/8"	10'7 3/8"	10'7 3/8"	10'7 3/8"	10'7 1/2"
NUMBER OF RISERS	19	20	21	16	17	18	19	20	21	16
HEIGHT OF EACH RISER	6 11/16"	6 3/8"	6 1/16"	7 15/16"	7 1/2"	7 1/16"	6 11/16"	6 3/8"	6 1/16"	8"
TOTAL HEIGHT OF RISERS	10'7 1/16"	10'7 1/2"	10'7 5/16"	10'7"	10'7 1/2"	10'7 1/8"	10'7 1/16"	10'7 1/2"	10'7 5/16"	10'8"
RISERS OVERAGE (+) OR UNDERAGE (—)	−3/16"	+1/4"	+1/16"	−3/8"	+1/8"	−1/4"	−5/16"	+1/8"	−1/16"	+1/2"
NUMBER OF TREADS	18	19	20	15	16	17	18	19	20	15
WIDTH OF EACH TREAD	10 13/16"	11 1/8"	11 7/16"	9 9/16"	10"	10 7/16"	10 13/16'	11 1/8"	11 7/16"	9 1/2"
TOTAL RUN OF STAIRWAY	16'2 5/8"	17'7 3/8"	19' 3/4"	11'11 7/16"	13'4"	14'9 7/16"	16'2 5/8"	17'7 3/8"	19' 3/4"	11'10 1/2"
WELL OPENING FOR 6'-8" HEADROOM	13' 3/4"	13'6 1/16"	14'11"	9'9 1/4"	10'3 11/32"	11'8 1/32"	13' 3/4"	13'6 1/16"	14'11"	8'10 7/8"
ANGLE OF INCLINE	31°44'	29°49'	27°56'	39°42'	36°52'	34°5'	31°44'	29°49'	27°56'	40°6'
LENGTH OF CARRIAGE	19' 27/32"	20'3 5/8"	21'6 29/32"	15'6 13/32"	16'8"	17'10 1/4"	19' 27/32"	20'3 5/8"	21'6 29/32"	15'6 9/32"
L. F. OF RISER PER INCH OF STAIRWAY WIDTH	1.583	1.666	1.75	1.333	1.416	1.5	1.583	1.666	1.75	1.333
L. F. OF TREAD PER INCH OF STAIRWAY WIDTH	1.5	1.583	1.666	1.25	1.333	1.416	1.5	1.583	1.666	1.25
FRAMING SQUARE SETTINGS FOR CARRIAGE CUTS — TONGUE	6 11/16"	6 3/8"	6 1/16"	7 15/16"	7 1/2"	7 1/16"	6 11/16"	6 3/8"	6 1/16"	8"
FRAMING SQUARE SETTINGS FOR CARRIAGE CUTS — BODY	10 13/16"	11 1/8"	11 7/16"	9 9/16"	10"	10 7/16"	10 13/16"	11 1/8"	11 7/16"	9 1/2"

FLOOR TO FLOOR RISE	10'7 1/2"	10'7 1/2"	10'7 1/2"	10'7 1/2"	10'7 1/2"	10'7 5/8"	10'7 5/8"	10'7 5/8"	10'7 5/8"	10'7 5/8"
NUMBER OF RISERS	17	18	19	20	21	16	17	18	19	20
HEIGHT OF EACH RISER	7 1/2"	7 1/16"	6 11/16"	6 3/8"	6 1/16"	8"	7 1/2"	7 1/16"	6 23/32"	6 3/8"
TOTAL HEIGHT OF RISERS	10'7 1/2"	10'7 1/8"	10'7 1/16"	10'7 1/2"	10'7 5/16"	10'8"	10'7 1/2"	10'7 1/8"	10'7 21/32"	10'7 1/2"
RISERS OVERAGE (+) OR UNDERAGE (—)	0	−3/8"	−7/16"	0	−3/16"	+3/8"	−1/8"	−1/2"	+1/32"	−1/8"
NUMBER OF TREADS	16	17	18	19	20	15	16	17	18	19
WIDTH OF EACH TREAD	10"	10 7/16"	10 13/16"	11 1/8"	11 7/16"	9 1/2"	10"	10 7/16"	10 25/32"	11 1/8"
TOTAL RUN OF STAIRWAY	13'4"	14'9 7/16"	16'2 5/8"	17'7 3/8"	19'3/4"	11'10 1/2"	13'4"	14'9 7/16"	16'2 1/16"	17'7 3/8"
WELL OPENING FOR 6'-8" HEADROOM	10'3 11/32"	11'8 1/32"	13'3/4"	13'6 1/16"	14'11"	8'10 7/8"	10'3 11/32"	11'8 1/32"	13'7/32"	13'6 1/16"
ANGLE OF INCLINE	36°52'	34°5'	31°44'	29°49'	27°56'	40°6'	36°52'	34°5'	31°56'	29°49'
LENGTH OF CARRIAGE	16'8"	17'10 1/4"	19'27/32"	20'3 5/8"	21'6 29/32"	15'6 9/32"	16'8"	17'10 1/4"	19'21/32"	20'3 5/8"
L. F. OF RISER PER INCH OF STAIRWAY WIDTH	1.416	1.5	1.583	1.666	1.75	1.333	1.416	1.5	1.583	1.666
L. F. OF TREAD PER INCH OF STAIRWAY WIDTH	1.333	1.416	1.5	1.583	1.666	1.25	1.333	1.416	1.5	1.583
FRAMING SQUARE SETTINGS FOR CARRIAGE CUTS — TONGUE	7 1/2"	7 1/16"	6 11/16"	6 3/8"	6 1/16"	8"	7 1/2"	7 1/16"	6 23/32"	6 3/8"
FRAMING SQUARE SETTINGS FOR CARRIAGE CUTS — BODY	10"	10 7/16"	10 13/16"	11 1/8"	11 7/16"	9 1/2"	10"	10 7/16"	10 25/32"	11 1/8"

FLOOR TO FLOOR RISE		10'7 5/8"	10'7 3/4"	10'7 3/4"	10'7 3/4"	10'7 3/4"	10'7 3/4"	10'7 3/4"	10'7 7/8"	10'7 7/8"	10'7 7/8"
NUMBER OF RISERS		21	16	17	18	19	20	21	16	17	18
HEIGHT OF EACH RISER		6 1/16"	8"	7 1/2"	7 1/8"	6 3/4"	6 3/8"	6 1/16"	8"	7 1/2"	7 1/8"
TOTAL HEIGHT OF RISERS		10'7 5/16"	10'8"	10'7 1/2"	10'8 1/4"	10'8 1/4"	10'7 1/2"	10'7 5/16"	10'8"	10'7 1/2"	10'8 1/4"
RISERS OVERAGE (+) OR UNDERAGE (—)		−5/16"	+1/4"	−1/4"	+1/2"	+1/2"	−1/4"	−7/16"	+1/8"	−3/8"	+3/8"
NUMBER OF TREADS		20	15	16	17	18	19	20	15	16	17
WIDTH OF EACH TREAD		11 7/16"	9 1/2"	10"	10 3/8"	10 3/4"	11 1/8"	11 7/16"	9 1/2"	10"	10 3/8"
TOTAL RUN OF STAIRWAY		19' 3/4"	11'10 1/2"	13'4"	14'8 3/8"	16'1 1/2"	17'7 3/8"	19' 3/4"	11'10 1/2"	13'4"	14'8 3/8"
WELL OPENING FOR 6'-8" HEADROOM		14'11"	8'10 7/8"	10'3 11/32"	11'7 1/16"	12'11 11/16"	13'6 1/16"	14'11"	8'10 7/8"	10'3 11/32"	11'7 1/16"
ANGLE OF INCLINE		27°56'	40°6'	36°52'	34°29'	32°7'	29°49'	27°56'	40°6'	36°52'	34°29'
LENGTH OF CARRIAGE		21'6 29/32"	15'6 9/32"	16'8"	17'9 31/32"	19' 15/32"	20'3 5/8"	21'6 29/32"	15'6 9/32"	16'8"	17'9 31/32"
L. F. OF RISER PER INCH OF STAIRWAY WIDTH		1.75	1.333	1.416	1.5	1.583	1.666	1.75	1.333	1.416	1.5
L. F. OF TREAD PER INCH OF STAIRWAY WIDTH		1.666	1.25	1.333	1.416	1.5	1.583	1.666	1.25	1.333	1.416
FRAMING SQUARE SETTINGS FOR CARRIAGE CUTS	TONGUE	6 1/16"	8"	7 1/2"	7 1/8"	6 3/4"	6 3/8"	6 1/16"	8"	7 1/2"	7 1/8"
	BODY	11 7/16"	9 1/2"	10"	10 3/8"	10 3/4"	11 1/8"	11 7/16"	9 1/2"	10"	10 3/8"

FLOOR TO FLOOR RISE		10'7 7/8"	10'7 7/8"	10'7 7/8"	10'8"	10'8"	10'8"	10'8"	10'8"	10'8"	10'8 1/8"
NUMBER OF RISERS		19	20	21	16	17	18	19	20	21	16
HEIGHT OF EACH RISER		6 3/4"	6 3/8"	6 3/32"	8"	7 1/2"	7 1/8"	6 3/4"	6 3/8"	6 3/32"	8"
TOTAL HEIGHT OF RISERS		10'8 1/4"	10'7 1/2"	10'7 31/32"	10'8"	10'7 1/2"	10'8 1/4"	10'8 1/4"	10'7 1/2"	10'7 31/32"	10'8"
RISERS OVERAGE (+) OR UNDERAGE (—)		+3/8"	−3/8"	+3/32"	0	−1/2"	+1/4"	+1/4"	−1/2"	−1/32"	−1/8"
NUMBER OF TREADS		18	19	20	15	16	17	18	19	20	15
WIDTH OF EACH TREAD		10 3/4"	11 1/8"	11 13/32"	9 1/2"	10"	10 3/8"	10 3/4"	11 1/8"	11 13/32"	9 1/2"
TOTAL RUN OF STAIRWAY		16'1 1/2"	17'7 3/8"	19' 1/8"	11'10 1/2"	13'4"	14'8 3/8"	16'1 1/2"	17'7 3/8"	19' 1/8"	11'10 1/2"
WELL OPENING FOR 6'-8" HEADROOM		12'11 11/16"	13'6 1/16"	14'10 13/32"	8'10 7/8"	10'3 11/32"	11'7 1/16"	12'11 11/16"	13'6 1/16"	14'10 13/32"	8'10 7/8"
ANGLE OF INCLINE		32°7'	29°49'	28°7'	40°6'	36°52'	34°29'	32°7'	29°49'	28°7'	40°6'
LENGTH OF CARRIAGE		19' 15/32"	20'3 5/8"	21'6 5/8"	15'6 9/32"	16'8"	17'9 31/32"	19' 15/32"	20'3 5/8"	21'6 5/8"	15'6 9/32"
L. F. OF RISER PER INCH OF STAIRWAY WIDTH		1.583	1.666	1.75	1.333	1.416	1.5	1.583	1.666	1.75	1.333
L. F. OF TREAD PER INCH OF STAIRWAY WIDTH		1.5	1.583	1.666	1.25	1.333	1.416	1.5	1.583	1.666	1.25
FRAMING SQUARE SETTINGS FOR CARRIAGE CUTS	TONGUE	6 3/4"	6 3/8"	6 3/32"	8"	7 1/2"	7 1/8"	6 3/4"	6 3/8"	6 3/32"	8"
	BODY	10 3/4"	11 1/8"	11 13/32"	9 1/2"	10"	10 3/8"	10 3/4"	11 1/8"	11 13/32"	9 1/2"

FLOOR TO FLOOR RISE	10'8 1/8"	10'8 1/8"	10'8 1/8"	10'8 1/8"	10'8 1/8"	10'8 1/4"	10'8 1/4"	10'8 1/4"	10'8 1/4"	10'8 1/4"
NUMBER OF RISERS	17	18	19	20	21	16	17	18	19	20
HEIGHT OF EACH RISER	7 9/16"	7 1/8"	6 3/4"	6 13/32"	6 1/8"	8"	7 9/16"	7 1/8"	6 3/4"	6 7/16"
TOTAL HEIGHT OF RISERS	10'8 9/16"	10'8 1/4"	10'8 1/4"	10'8 1/8"	10'8 5/8"	10'8"	10'8 9/16"	10'8 1/4"	10'8 1/4"	10'8 3/4"
RISERS OVERAGE (+) OR UNDERAGE (—)	+7/16"	+1/8"	+1/8"	0	+1/2"	−1/4"	+5/16"	0	0	+1/2"
NUMBER OF TREADS	16	17	18	19	20	15	16	17	18	19
WIDTH OF EACH TREAD	9 15/16"	10 3/8"	10 3/4"	11 3/32"	11 3/8"	9 1/2"	9 15/16"	10 3/8"	10 3/4"	11 1/16"
TOTAL RUN OF STAIRWAY	13'3"	14'8 3/8"	16'1 1/2"	17'6 25/32"	18'11 1/2"	11'10 1/2"	13'3"	14'8 3/8"	16'1 1/2"	17'6 3/16"
WELL OPENING FOR 6'-8" HEADROOM	10'2 7/16"	11'7 1/16"	12'11 11/16"	13'5 17/32"	14'9 13/16"	8'10 7/8"	10'2 7/16"	11'7 1/16"	12'11 11/16"	13'5"
ANGLE OF INCLINE	37°16'	34°29'	32°7'	30°0'	28°18'	40°6'	37°16'	34°29'	32°7'	30°12'
LENGTH OF CARRIAGE	16'7 13/16"	17'9 31/32"	19' 15/32"	20'3 13/32"	21'6 3/8"	15'6 9/32"	16'7 13/16"	17'9 31/32"	19' 15/32"	20'3 3/16"
L. F. OF RISER PER INCH OF STAIRWAY WIDTH	1.416	1.5	1.583	1.666	1.75	1.333	1.416	1.5	1.583	1.666
L. F. OF TREAD PER INCH OF STAIRWAY WIDTH	1.333	1.416	1.5	1.583	1.666	1.25	1.333	1.416	1.5	1.583
FRAMING SQUARE SETTINGS FOR CARRIAGE CUTS — TONGUE	7 9/16"	7 1/8"	6 3/4"	6 13/32"	6 1/8"	8"	7 9/16"	7 1/8"	6 3/4"	6 7/16"
FRAMING SQUARE SETTINGS FOR CARRIAGE CUTS — BODY	9 15/16"	10 3/8"	10 3/4"	11 3/32"	11 3/8"	9 1/2"	9 15/16"	10 3/8"	10 3/4"	11 1/16"

FLOOR TO FLOOR RISE	10'8 1/4"	10'8 3/8"	10'8 3/8"	10'8 3/8"	10'8 3/8"	10'8 3/8"	10'8 3/8"	10'8 1/2"	10'8 1/2"	10'8 1/2"
NUMBER OF RISERS	21	16	17	18	19	20	21	16	17	18
HEIGHT OF EACH RISER	6 1/8"	8"	7 9/16"	7 1/8"	6 3/4"	6 7/16"	6 1/8"	8 1/16"	7 9/16"	7 1/8"
TOTAL HEIGHT OF RISERS	10'8 5/8"	10'8"	10'8 9/16"	10'8 1/4"	10'8 1/4"	10'8 3/4"	10'8 5/8"	10'9"	10'8 9/16"	10'8 1/4"
RISERS OVERAGE (+) OR UNDERAGE (—)	+3/8"	−3/8"	+3/16"	−1/8"	−1/8"	+3/8"	+1/4"	+1/2"	+1/16"	−1/4"
NUMBER OF TREADS	20	15	16	17	18	19	20	15	16	17
WIDTH OF EACH TREAD	11 3/8"	9 1/2"	9 15/16"	10 3/8"	10 3/4"	11 1/16"	11 3/8"	9 7/16"	9 15/16"	10 3/8"
TOTAL RUN OF STAIRWAY	18'11 1/2"	11'10 1/2"	13'3"	14'8 3/8"	16'1 1/2"	17'6 3/16"	18'11 1/2"	11'9 9/16"	13'3"	14'8 3/8"
WELL OPENING FOR 6'-8" HEADROOM	14'9 13/16"	8'10 7/8"	10'2 7/16"	11'7 1/16"	12'11 11/16"	13'5"	14'9 13/16"	8'10 3/32"	10'2 7/16"	11'7 1/16"
ANGLE OF INCLINE	28°18'	40°6'	37°16'	34°29'	32°7'	30°12'	28°18'	40°30'	37°16'	34°29'
LENGTH OF CARRIAGE	21'6 3/8"	15'6 9/32"	16'7 13/16"	17'9 31/32"	19' 15/32"	20'3 3/16"	21'6 3/8"	15'6 3/16"	16'7 13/16"	17'9 31/32"
L. F. OF RISER PER INCH OF STAIRWAY WIDTH	1.75	1.333	1.416	1.5	1.583	1.666	1.75	1.333	1.416	1.5
L. F. OF TREAD PER INCH OF STAIRWAY WIDTH	1.666	1.25	1.333	1.416	1.5	1.583	1.666	1.25	1.333	1.416
FRAMING SQUARE SETTINGS FOR CARRIAGE CUTS — TONGUE	6 1/8"	8"	7 9/16"	7 1/8"	6 3/4"	6 7/16"	6 1/8"	8 1/16"	7 9/16"	7 1/8"
FRAMING SQUARE SETTINGS FOR CARRIAGE CUTS — BODY	11 3/8"	9 1/2"	9 15/16"	10 3/8"	10 3/4"	11 1/16"	11 3/8"	9 7/16"	9 15/16"	10 3/8"

FLOOR TO FLOOR RISE	10'8 1/2"	10'8 1/2"	10'8 1/2"	10'8 5/8"	10'8 5/8"	10'8 5/8"	10'8 5/8"	10'8 5/8"	10'8 5/8"	10'8 3/4"
NUMBER OF RISERS	19	20	21	16	17	18	19	20	21	16
HEIGHT OF EACH RISER	6 3/4"	6 7/16"	6 1/8"	8 1/16"	7 9/16"	7 1/8"	6 3/4"	6 7/16"	6 1/8"	8 1/16"
TOTAL HEIGHT OF RISERS	10'8 1/4"	10'8 3/4"	10'8 5/8"	10'9"	10'8 9/16"	10'8 1/4"	10'8 1/4"	10'8 3/4"	10'8 5/8"	10'9"
RISERS OVERAGE (+) OR UNDERAGE (—)	−1/4"	+1/4"	+1/8"	+3/8"	−1/16"	−3/8"	−3/8"	+1/8"	0	+1/4"
NUMBER OF TREADS	18	19	20	15	16	17	18	19	20	15
WIDTH OF EACH TREAD	10 3/4"	11 1/16"	11 3/8"	9 7/16"	9 15/16"	10 3/8"	10 3/4"	11 1/16"	11 3/8"	9 7/16"
TOTAL RUN OF STAIRWAY	16'1 1/2"	17'6 3/16"	18'11 1/2"	11'9 9/16"	13'3"	14'8 3/8"	16'1 1/2"	17'6 3/16"	18'11 1/2"	11'9 9/16"
WELL OPENING FOR 6'-8" HEADROOM	12'11 11/6"	13'5"	14'9 13/16"	8'10 3/32"	10'2 7/16"	11'7 1/16"	12'11 11/6"	13'5"	14'9 13/16"	8'10 3/32"
ANGLE OF INCLINE	32°7'	30°12'	28°18'	40°30'	37°16'	34°29'	32°7'	30°12'	28°18'	40°30'
LENGTH OF CARRIAGE	19' 15/32"	20'3 3/16"	21'6 3/8"	15'6 3/16"	16'7 13/16"	17'9 31/32"	19' 15/32"	20'3 3/16"	21'6 3/8"	15'6 3/16"
L. F. OF RISER PER INCH OF STAIRWAY WIDTH	1.583	1.666	1.75	1.333	1.416	1.5	1.583	1.666	1.75	1.333
L. F. OF TREAD PER INCH OF STAIRWAY WIDTH	1.5	1.583	1.666	1.25	1.333	1.416	1.5	1.583	1.666	1.25
FRAMING SQUARE SETTINGS FOR CARRIAGE CUTS — TONGUE	6 3/4"	6 7/16"	6 1/8"	8 1/16"	7 9/16"	7 1/8"	6 3/4"	6 7/16"	6 1/8"	8 1/16"
FRAMING SQUARE SETTINGS FOR CARRIAGE CUTS — BODY	10 3/4"	11 1/16"	11 3/8"	9 7/16"	9 15/16"	10 3/8"	10 3/4"	11 1/16"	11 3/8"	9 7/16"

FLOOR TO FLOOR RISE	10'8 3/4"	10'8 3/4"	10'8 3/4"	10'8 3/4"	10'8 3/4"	10'8 7/8"	10'8 7/8"	10'8 7/8"	10'8 7/8"	10'8 7/8"
NUMBER OF RISERS	17	18	19	20	21	16	17	18	19	20
HEIGHT OF EACH RISER	7 9/16"	7 1/8"	6 3/4"	6 7/16"	6 1/8"	8 1/16"	7 9/16"	7 3/16"	6 25/32"	6 7/16"
TOTAL HEIGHT OF RISERS	10'8 9/16"	10'8 1/4"	10'8 1/4"	10'8 3/4"	10'8 5/8"	10'9"	10'8 9/16"	10'9 3/8"	10'8 27/32"	10'8 3/4"
RISERS OVERAGE (+) OR UNDERAGE (—)	−3/16"	−1/2"	−1/2"	0	−1/8"	+1/8"	−5/16"	+1/2"	−1/32"	−1/8"
NUMBER OF TREADS	16	17	18	19	20	15	16	17	18	19
WIDTH OF EACH TREAD	9 15/16"	10 3/8"	10 3/4"	11 1/16"	11 3/8"	9 7/16"	9 15/16"	10 5/16"	10 23/32"	11 1/16"
TOTAL RUN OF STAIRWAY	13'3"	14'8 3/8"	16'1 1/2"	17'6 3/16"	18'11 1/2"	11'9 9/16"	13'3"	14'7 5/16"	16' 15/16"	17'6 3/16"
WELL OPENING FOR 6'-8" HEADROOM	10'2 7/16"	11'7 1/16"	12'11 11/6"	13'5"	14'9 13/16"	8'10 3/32"	10'2 7/16"	11'6 3/32"	12' 7/16"	13'5"
ANGLE OF INCLINE	37°16'	34°29'	32°7'	30°12'	28°18'	40°30'	37°16'	34°53'	32°19'	30°12'
LENGTH OF CARRIAGE	16'7 13/16"	17'9 31/32"	19' 15/32"	20'3 3/16"	21'6 3/8"	15'6 3/16"	16'7 13/16"	17'9 11/16"	19' 5/16"	20'3 3/16"
L. F. OF RISER PER INCH OF STAIRWAY WIDTH	1.416	1.5	1.583	1.666	1.75	1.333	1.416	1.5	1.583	1.666
L. F. OF TREAD PER INCH OF STAIRWAY WIDTH	1.333	1.416	1.5	1.583	1.666	1.25	1.333	1.416	1.5	1.583
FRAMING SQUARE SETTINGS FOR CARRIAGE CUTS — TONGUE	7 9/16"	7 1/8"	6 3/4"	6 7/16"	6 1/8"	8 1/16"	7 9/16"	7 3/16"	6 25/32"	6 7/16"
FRAMING SQUARE SETTINGS FOR CARRIAGE CUTS — BODY	9 15/16"	10 3/8"	10 3/4"	11 1/16"	11 3/8"	9 7/16"	9 15/16"	10 5/16"	10 23/32"	11 1/16"

FLOOR TO FLOOR RISE	10'8 7/8"	10'9"	10'9"	10'9"	10'9"	10'9"	10'9"	10'9 1/8"	10'9 1/8"	10'9 1/8"
NUMBER OF RISERS	21	16	17	18	19	20	21	16	17	18
HEIGHT OF EACH RISER	6 1/8"	8 1/16"	7 9/16"	7 3/16"	6 13/16"	6 7/16"	6 1/8"	8 1/16"	7 5/8"	7 3/16"
TOTAL HEIGHT OF RISERS	10'8 5/8"	10'9"	10'8 9/16"	10'9 3/8"	10'9 7/16"	10'8 3/4"	10'8 5/8"	10'9"	10'9 5/8"	10'9 3/8"
RISERS OVERAGE (+) OR UNDERAGE (—)	−1/4"	0	−7/16"	+3/8"	+7/16"	−1/4"	−3/8"	−1/8"	+1/2"	+1/4"
NUMBER OF TREADS	20	15	16	17	18	19	20	15	16	17
WIDTH OF EACH TREAD	11 3/8"	9 7/16"	9 15/16"	10 5/16"	10 11/16"	11 1/16"	11 3/8"	9 7/16"	9 7/8"	10 5/16"
TOTAL RUN OF STAIRWAY	18'11 1/2"	11'9 9/16"	13'3"	14'7 5/16"	16'3/8"	17'6 3/16"	18'11 1/2"	11'9 9/16"	13'2"	14'7 5/16"
WELL OPENING FOR 6'-8" HEADROOM	14'9 13/16"	8'10 3/32"	10'2 7/16"	11'6 3/32"	11'11 15/16"	13'5"	14'9 13/16"	8'10 3/32"	10'1 9/16"	11'6 3/32"
ANGLE OF INCLINE	28°18'	40°30'	37°16'	34°53'	32°31'	30°12'	28°18'	40°30'	37°40'	34°53'
LENGTH OF CARRIAGE	21'6 3/8"	15'6 3/16"	16'7 13/16"	17'9 11/16"	19'1/8"	20'3 3/16"	21'6 3/8"	15'6 3/16"	16'7 5/8"	17'9 11/16"
L. F. OF RISER PER INCH OF STAIRWAY WIDTH	1.75	1.333	1.416	1.5	1.583	1.666	1.75	1.333	1.416	1.5
L. F. OF TREAD PER INCH OF STAIRWAY WIDTH	1.666	1.25	1.333	1.416	1.5	1.583	1.666	1.25	1.333	1.416
FRAMING SQUARE SETTINGS FOR CARRIAGE CUTS — TONGUE	6 1/8"	8 1/16"	7 9/16"	7 3/16"	6 13/16"	6 7/16"	6 1/8"	8 1/16"	7 5/8"	7 3/16"
FRAMING SQUARE SETTINGS FOR CARRIAGE CUTS — BODY	11 3/8"	9 7/16"	9 15/16"	10 5/16"	10 11/16"	11 1/16"	11 3/8"	9 7/16"	9 7/8"	10 5/16"

FLOOR TO FLOOR RISE	$10'9\frac{1}{8}''$	$10'9\frac{1}{8}''$	$10'9\frac{1}{8}''$	$10'9\frac{1}{4}''$	$10'9\frac{1}{4}''$	$10'9\frac{1}{4}''$	$10'9\frac{1}{4}''$	$10'9\frac{1}{4}''$	$10'9\frac{1}{4}''$	$10'9\frac{3}{8}''$
NUMBER OF RISERS	19	20	21	16	17	18	19	20	21	16
HEIGHT OF EACH RISER	$6\frac{13}{16}''$	$6\frac{7}{16}''$	$6\frac{1}{8}''$	$8\frac{1}{16}''$	$7\frac{5}{8}''$	$7\frac{3}{16}''$	$6\frac{13}{16}''$	$6\frac{7}{16}''$	$6\frac{5}{32}''$	$8\frac{1}{16}''$
TOTAL HEIGHT OF RISERS	$10'9\frac{7}{16}''$	$10'8\frac{3}{4}''$	$10'8\frac{5}{8}''$	$10'9''$	$10'9\frac{5}{8}''$	$10'9\frac{3}{8}''$	$10'9\frac{7}{16}''$	$10'8\frac{3}{4}''$	$10'9\frac{9}{32}''$	$10'9''$
RISERS OVERAGE (+) OR UNDERAGE (—)	$+\frac{5}{16}''$	$-\frac{3}{8}''$	$-\frac{1}{2}''$	$-\frac{1}{4}''$	$+\frac{3}{8}''$	$+\frac{1}{8}''$	$+\frac{3}{16}''$	$-\frac{1}{2}''$	$+\frac{1}{32}''$	$-\frac{3}{8}''$
NUMBER OF TREADS	18	19	20	15	16	17	18	19	20	15
WIDTH OF EACH TREAD	$10\frac{11}{16}''$	$11\frac{1}{16}''$	$11\frac{3}{8}''$	$9\frac{7}{16}''$	$9\frac{7}{8}''$	$10\frac{5}{16}''$	$10\frac{11}{16}''$	$11\frac{1}{16}''$	$11\frac{11}{32}''$	$9\frac{7}{16}''$
TOTAL RUN OF STAIRWAY	$16'\frac{3}{8}''$	$17'6\frac{3}{16}''$	$18'11\frac{1}{2}''$	$11'9\frac{9}{16}''$	$13'2''$	$14'7\frac{5}{16}''$	$16'\frac{3}{8}''$	$17'6\frac{3}{16}''$	$18'10\frac{7}{8}''$	$11'9\frac{9}{16}''$
WELL OPENING FOR 6'-8" HEADROOM	$11'11\frac{15}{16}''$	$13'5''$	$14'9\frac{13}{16}''$	$8'10\frac{3}{32}''$	$10'1\frac{9}{16}''$	$11'6\frac{3}{32}''$	$11'11\frac{15}{16}''$	$13'5''$	$14'9\frac{1}{4}''$	$8'10\frac{3}{32}''$
ANGLE OF INCLINE	32°31'	30°12'	28°18'	40°30'	37°40'	34°53'	32°31'	30°12'	28°29'	40°30'
LENGTH OF CARRIAGE	$19'\frac{1}{8}''$	$20'3\frac{3}{16}''$	$21'6\frac{3}{8}''$	$15'6\frac{3}{16}''$	$16'7\frac{5}{8}''$	$17'9\frac{11}{16}''$	$19'\frac{1}{8}''$	$20'3\frac{3}{16}''$	$21'6\frac{1}{8}''$	$15'6\frac{3}{16}''$
L. F. OF RISER PER INCH OF STAIRWAY WIDTH	1.583	1.666	1.75	1.333	1.416	1.5	1.583	1.666	1.75	1.333
L. F. OF TREAD PER INCH OF STAIRWAY WIDTH	1.5	1.583	1.666	1.25	1.333	1.416	1.5	1.583	1.666	1.25
FRAMING SQUARE SETTINGS FOR CARRIAGE CUTS — TONGUE	$6\frac{13}{16}''$	$6\frac{7}{16}''$	$6\frac{1}{8}''$	$8\frac{1}{16}''$	$7\frac{5}{8}''$	$7\frac{3}{16}''$	$6\frac{13}{16}''$	$6\frac{7}{16}''$	$6\frac{5}{32}''$	$8\frac{1}{16}''$
FRAMING SQUARE SETTINGS FOR CARRIAGE CUTS — BODY	$10\frac{11}{16}''$	$11\frac{1}{16}''$	$11\frac{3}{8}''$	$9\frac{7}{16}''$	$9\frac{7}{8}''$	$10\frac{5}{16}''$	$10\frac{11}{16}''$	$11\frac{1}{16}''$	$11\frac{11}{32}''$	$9\frac{7}{16}''$

FLOOR TO FLOOR RISE	10'9 3/8"	10'9 3/8"	10'9 3/8"	10'9 3/8"	10'9 3/8"	10'9 1/2"	10'9 1/2"	10'9 1/2"	10'9 1/2"	10'9 1/2"
NUMBER OF RISERS	17	18	19	20	21	16	17	18	19	20
HEIGHT OF EACH RISER	7 5/8"	7 3/16"	6 13/16"	6 15/32"	6 5/32"	8 1/8"	7 5/8"	7 3/16"	6 13/16"	6 1/2"
TOTAL HEIGHT OF RISERS	10'9 5/8"	10'9 3/8"	10'9 7/16"	10'9 3/8"	10'9 9/32"	10'10"	10'9 5/8"	10'9 3/8"	10'9 7/16"	10'10"
RISERS OVERAGE (+) OR UNDERAGE (—)	+1/4"	0	+1/16"	0	−3/32"	+1/2"	+1/8"	−1/8"	−1/16"	+1/2"
NUMBER OF TREADS	16	17	18	19	20	15	16	17	18	19
WIDTH OF EACH TREAD	9 7/8"	10 5/16"	10 11/16"	11 1/32"	11 11/32"	9 3/8"	9 7/8"	10 5/16"	10 11/16"	11"
TOTAL RUN OF STAIRWAY	13'2"	14'7 5/16"	16' 3/8"	17'5 19/32"	18'10 7/8"	11'8 5/8"	13'2"	14'7 5/16"	16' 3/8"	17'5"
WELL OPENING FOR 6'-8" HEADROOM	10'1 9/16"	11'6 3/32"	11'11 15/16"	13'4 15/32'"	14'9 1/4"	8'9 9/32"	10'1 9/16"	11'6 3/32"	11'11 15/16"	13'3 15/16"
ANGLE OF INCLINE	37°40'	34°53'	32°31'	30°23'	28°29'	40°55'	37°40'	34°53'	32°31'	30°35'
LENGTH OF CARRIAGE	16'7 5/8"	17'9 11/16"	19' 1/8"	20'2 31/32"	21'6 1/8"	15'6 3/32"	16'7 5/8"	17'9 11/16"	19' 1/8"	20'2 3/4"
L. F. OF RISER PER INCH OF STAIRWAY WIDTH	1.416	1.5	1.583	1.666	1.75	1.333	1.416	1.5	1.583	1.666
L. F. OF TREAD PER INCH OF STAIRWAY WIDTH	1.333	1.416	1.5	1.583	1.666	1.25	1.333	1.416	1.5	1.583
FRAMING SQUARE SETTINGS FOR CARRIAGE CUTS — TONGUE	7 5/8"	7 3/16"	6 13/16"	6 15/32"	6 5/32"	8 1/8"	7 5/8"	7 3/16"	6 13/16"	6 1/2"
FRAMING SQUARE SETTINGS FOR CARRIAGE CUTS — BODY	9 7/8"	10 5/16"	10 11/16"	11 1/32"	11 11/32"	9 3/8"	9 7/8"	10 5/16"	10 11/16"	11"

FLOOR TO FLOOR RISE		10'9$\frac{1}{2}$"	10'9$\frac{5}{8}$"	10'9$\frac{5}{8}$"	10'9$\frac{5}{8}$"	10'9$\frac{5}{8}$"	10'9$\frac{5}{8}$"	10'9$\frac{5}{8}$"	10'9$\frac{3}{4}$"	10'9$\frac{3}{4}$"	10'9$\frac{3}{4}$"
NUMBER OF RISERS		21	16	17	18	19	20	21	16	17	18
HEIGHT OF EACH RISER		6$\frac{3}{16}$"	8$\frac{1}{8}$"	7$\frac{5}{8}$"	7$\frac{3}{16}$"	6$\frac{13}{16}$"	6$\frac{1}{2}$"	6$\frac{3}{16}$"	8$\frac{1}{8}$"	7$\frac{5}{8}$"	7$\frac{3}{16}$"
TOTAL HEIGHT OF RISERS		10'9$\frac{15}{16}$"	10'10"	10'9$\frac{5}{8}$"	10'9$\frac{3}{8}$"	10'9$\frac{7}{16}$"	10'10"	10'9$\frac{15}{16}$"	10'10"	10'9$\frac{5}{8}$"	10'9$\frac{3}{8}$"
RISERS OVERAGE (+) OR UNDERAGE (—)		+$\frac{7}{16}$"	+$\frac{3}{8}$"	0	−$\frac{1}{4}$"	−$\frac{3}{16}$"	+$\frac{3}{8}$"	+$\frac{5}{16}$"	+$\frac{1}{4}$"	−$\frac{1}{8}$"	−$\frac{3}{8}$"
NUMBER OF TREADS		20	15	16	17	18	19	20	15	16	17
WIDTH OF EACH TREAD		11$\frac{5}{16}$"	9$\frac{3}{8}$"	9$\frac{7}{8}$"	10$\frac{5}{16}$"	10$\frac{11}{16}$"	11"	11$\frac{5}{16}$"	9$\frac{3}{8}$"	9$\frac{7}{8}$"	10$\frac{5}{16}$"
TOTAL RUN OF STAIRWAY		18'10$\frac{1}{4}$"	11'8$\frac{5}{8}$"	13'2"	14'7$\frac{5}{16}$"	16'$\frac{3}{8}$"	17'5"	18'10$\frac{1}{4}$"	11'8$\frac{5}{8}$"	13'2"	14'7$\frac{5}{16}$"
WELL OPENING FOR 6'-8" HEADROOM		14'8$\frac{21}{32}$"	8'9$\frac{9}{32}$"	10'1$\frac{9}{16}$"	11'6$\frac{3}{32}$"	11'11$\frac{15}{16}$"	13'3$\frac{15}{16}$"	14'8$\frac{21}{32}$"	8'9$\frac{9}{32}$"	10'1$\frac{9}{16}$"	11'6$\frac{3}{32}$"
ANGLE OF INCLINE		28°41'	40°55'	37°40'	34°53'	32°31'	30°35'	28°41'	40°55'	37°40'	34°53'
LENGTH OF CARRIAGE		21'5$\frac{7}{8}$"	15'6$\frac{3}{32}$"	16'7$\frac{5}{8}$"	17'9$\frac{11}{16}$"	19'$\frac{1}{8}$"	20'2$\frac{3}{4}$"	21'5$\frac{7}{8}$"	15'6$\frac{3}{32}$"	16'7$\frac{5}{8}$"	17'9$\frac{11}{16}$"
L. F. OF RISER PER INCH OF STAIRWAY WIDTH		1.75	1.333	1.416	1.5	1.583	1.666	1.75	1.333	1.416	1.5
L. F. OF TREAD PER INCH OF STAIRWAY WIDTH		1.666	1.25	1.333	1.416	1.5	1.583	1.666	1.25	1.333	1.416
FRAMING SQUARE SETTINGS FOR CARRIAGE CUTS	TONGUE	6$\frac{3}{16}$"	8$\frac{1}{8}$"	7$\frac{5}{8}$"	7$\frac{3}{16}$"	6$\frac{13}{16}$"	6$\frac{1}{2}$"	6$\frac{3}{16}$"	8$\frac{1}{8}$"	7$\frac{5}{8}$"	7$\frac{3}{16}$"
	BODY	11$\frac{5}{16}$"	9$\frac{3}{8}$"	9$\frac{7}{8}$"	10$\frac{5}{16}$"	10$\frac{11}{16}$"	11"	11$\frac{5}{16}$"	9$\frac{3}{8}$"	9$\frac{7}{8}$"	10$\frac{5}{16}$"

FLOOR TO FLOOR RISE	10'9 3/4"	10'9 3/4"	10'9 3/4"	10'9 7/8"	10'9 7/8"	10'9 7/8"	10'9 7/8"	10'9 7/8"	10'9 7/8"	10'10"
NUMBER OF RISERS	19	20	21	16	17	18	19	20	21	16
HEIGHT OF EACH RISER	6 13/16"	6 1/2"	6 3/16"	8 1/8"	7 5/8"	7 3/16"	6 13/16"	6 1/2"	6 3/16"	8 1/8"
TOTAL HEIGHT OF RISERS	10'9 7/16"	10'10"	10'9 15/16"	10'10"	10'9 5/8"	10'9 3/8"	10'9 7/16"	10'10"	10'9 15/16"	10'10"
RISERS OVERAGE (+) OR UNDERAGE (—)	−5/16"	+1/4"	+3/16"	+1/8"	−1/4"	−1/2"	−7/16"	+1/8"	+1/16"	0
NUMBER OF TREADS	18	19	20	15	16	17	18	19	20	15
WIDTH OF EACH TREAD	10 11/16"	11"	11 5/16"	9 3/8"	9 7/8"	10 5/16"	10 11/16"	11"	11 5/16"	9 3/8"
TOTAL RUN OF STAIRWAY	16'3/8"	17'5"	18'10 1/4"	11'8 5/8"	13'2"	14'7 5/16"	16'3/8"	17'5"	18'10 1/4"	11'8 5/8"
WELL OPENING FOR 6'-8" HEADROOM	11'11 15/16"	13'3 15/16"	14'8 21/32"	8'9 9/32"	10'1 9/16"	11'6 3/32"	11'11 15/16"	13'3 15/16"	14'8 21/32"	8'9 9/32"
ANGLE OF INCLINE	32°31'	30°35'	28°41'	40°55'	37°40'	34°53'	32°31'	30°35'	28°41'	40°55'
LENGTH OF CARRIAGE	19'1/8"	20'2 3/4"	21'5 7/8"	15'6 3/32"	16'7 5/8"	17'9 11/16"	19'1/8"	20'2 3/4"	21'5 7/8"	15'6 3/32"
L. F. OF RISER PER INCH OF STAIRWAY WIDTH	1.583	1.666	1.75	1.333	1.416	1.5	1.583	1.666	1.75	1.333
L. F. OF TREAD PER INCH OF STAIRWAY WIDTH	1.5	1.583	1.666	1.25	1.333	1.416	1.5	1.583	1.666	1.25
FRAMING SQUARE SETTINGS FOR CARRIAGE CUTS — TONGUE	6 13/16"	6 1/2"	6 3/16"	8 1/8"	7 5/8"	7 3/16"	6 13/16"	6 1/2"	6 3/16"	8 1/8"
FRAMING SQUARE SETTINGS FOR CARRIAGE CUTS — BODY	10 11/16"	11"	11 5/16"	9 3/8"	9 7/8"	10 5/16"	10 11/16"	11"	11 5/16"	9 3/8"

FLOOR TO FLOOR RISE	10'10"	10'10"	10'10"	10'10"	10'10"	10'10 1/8"	10'10 1/8"	10'10 1/8"	10'10 1/8"	10'10 1/8"
NUMBER OF RISERS	17	18	19	20	21	16	17	18	19	20
HEIGHT OF EACH RISER	7 5/8"	7 1/4"	6 27/32"	6 1/2"	6 3/16"	8 1/8"	7 5/8"	7 1/4"	6 7/8"	6 1/2"
TOTAL HEIGHT OF RISERS	10'9 5/8"	10'10 1/2"	10'10 1/32"	10'10"	10'9 15/16"	10'10"	10'9 5/8"	10'10 1/2"	10'10 5/8"	10'10"
RISERS OVERAGE (+) OR UNDERAGE (—)	−3/8"	+1/2"	+1/32"	0	−1/16"	−1/8"	−1/2"	+3/8"	+1/2"	−1/8"
NUMBER OF TREADS	16	17	18	19	20	15	16	17	18	19
WIDTH OF EACH TREAD	9 7/8"	10 1/4"	10 21/32"	11"	11 5/16"	9 3/8"	9 7/8"	10 1/4"	10 5/8"	11"
TOTAL RUN OF STAIRWAY	13'2	14'6 1/4"	15'11 13/16"	17'5"	18'10 1/4"	11'8 5/8"	13'2"	14'6 1/4"	15'11 1/4"	17'5"
WELL OPENING FOR 6'-8" HEADROOM	10'1 9/16"	11'5 1/8"	11'11 15/32"	13'3 15/16"	14'8 21/32"	8'9 9/32"	10'1 9/16"	11'5 1/8"	11'10 31/32"	13'3 15/16"
ANGLE OF INCLINE	37°40'	35°16'	32°43'	30°35'	28°41'	40°55'	37°40'	35°16'	32°54'	30°35'
LENGTH OF CARRIAGE	16'7 5/8"	17'9 7/16"	18'11 31/32"	20'2 3/4"	21'5 7/8"	15'6 3/32"	16'7 5/8"	17'9 7/16"	18'11 13/16"	20'2 3/4"
L. F. OF RISER PER INCH OF STAIRWAY WIDTH	1.416	1.5	1.583	1.666	1.75	1.333	1.416	1.5	1.583	1.666
L. F. OF TREAD PER INCH OF STAIRWAY WIDTH	1.333	1.416	1.5	1.583	1.666	1.25	1.333	1.416	1.5	1.583
FRAMING SQUARE SETTINGS FOR CARRIAGE CUTS — TONGUE	7 5/8"	7 1/4"	6 27/32"	6 1/2"	6 3/16"	8 1/8"	7 5/8"	7 1/4"	6 7/8"	6 1/2"
FRAMING SQUARE SETTINGS FOR CARRIAGE CUTS — BODY	9 7/8"	10 1/4"	10 21/32"	11"	11 5/16"	9 3/8"	9 7/8"	10 1/4"	10 5/8"	11"

FLOOR TO FLOOR RISE		10'10$\frac{1}{8}$"	10'10$\frac{1}{4}$"	10'10$\frac{1}{4}$"	10'10$\frac{1}{4}$"	10'10$\frac{1}{4}$"	10'10$\frac{1}{4}$"	10'10$\frac{1}{4}$"	10'10$\frac{3}{8}$"	10'10$\frac{3}{8}$"	10'10$\frac{3}{8}$"
NUMBER OF RISERS		21	16	17	18	19	20	21	16	17	18
HEIGHT OF EACH RISER		6$\frac{3}{16}$"	8$\frac{1}{8}$"	7$\frac{11}{16}$"	7$\frac{1}{4}$"	6$\frac{7}{8}$"	6$\frac{1}{2}$"	6$\frac{3}{16}$"	8$\frac{1}{8}$"	7$\frac{11}{16}$"	7$\frac{1}{4}$"
TOTAL HEIGHT OF RISERS		10'9$\frac{15}{16}$"	10'10"	10'10$\frac{11}{16}$"	10'10$\frac{1}{2}$"	10'10$\frac{5}{8}$"	10'10"	10'9$\frac{15}{16}$"	10'10"	10'10$\frac{11}{16}$"	10'10$\frac{1}{2}$"
RISERS OVERAGE (+) OR UNDERAGE (—)		$-\frac{3}{16}$"	$-\frac{1}{4}$"	$+\frac{7}{16}$"	$+\frac{1}{4}$"	$+\frac{3}{8}$"	$-\frac{1}{4}$"	$-\frac{5}{16}$"	$-\frac{3}{8}$"	$+\frac{5}{16}$"	$+\frac{1}{8}$"
NUMBER OF TREADS		20	15	16	17	18	19	20	15	16	17
WIDTH OF EACH TREAD		11$\frac{5}{16}$"	9$\frac{3}{8}$"	9$\frac{13}{16}$"	10$\frac{1}{4}$"	10$\frac{5}{8}$"	11"	11$\frac{5}{16}$"	9$\frac{3}{8}$"	9$\frac{13}{16}$"	10$\frac{1}{4}$"
TOTAL RUN OF STAIRWAY		18'10$\frac{1}{4}$"	11'8$\frac{5}{8}$"	13'1"	14'6$\frac{1}{4}$"	15'11$\frac{1}{4}$"	17'5"	18'10$\frac{1}{4}$"	11'8$\frac{5}{8}$"	13'1"	14'6$\frac{1}{4}$"
WELL OPENING FOR 6'-8" HEADROOM		14'8$\frac{21}{32}$"	8'9$\frac{9}{32}$"	10'$\frac{23}{32}$"	11'5$\frac{1}{8}$"	11'10$\frac{31}{32}$"	13'3$\frac{15}{16}$"	14'8$\frac{21}{32}$"	8'9$\frac{9}{32}$"	10'$\frac{23}{32}$"	11'5$\frac{1}{8}$"
ANGLE OF INCLINE		28°41'	40°55'	38°5'	35°16'	32°54'	30°35'	28°41'	40°55'	38°5'	35°16'
LENGTH OF CARRIAGE		21'5$\frac{7}{8}$"	15'6$\frac{3}{32}$"	16'7$\frac{7}{16}$"	17'9$\frac{7}{16}$"	18'11$\frac{13}{16}$"	20'2$\frac{3}{4}$"	21'5$\frac{7}{8}$"	15'6$\frac{3}{32}$"	16'7$\frac{7}{16}$"	17'9$\frac{7}{16}$"
L. F. OF RISER PER INCH OF STAIRWAY WIDTH		1.75	1.333	1.416	1.5	1.583	1.666	1.75	1.333	1.416	1.5
L. F. OF TREAD PER INCH OF STAIRWAY WIDTH		1.666	1.25	1.333	1.416	1.5	1.583	1.666	1.25	1.333	1.416
FRAMING SQUARE SETTINGS FOR CARRIAGE CUTS	TONGUE	6$\frac{3}{16}$"	8$\frac{1}{8}$"	7$\frac{11}{16}$"	7$\frac{1}{4}$"	6$\frac{7}{8}$"	6$\frac{1}{2}$"	6$\frac{3}{16}$"	8$\frac{1}{8}$"	7$\frac{11}{16}$"	7$\frac{1}{4}$"
	BODY	11$\frac{5}{16}$"	9$\frac{3}{8}$"	9$\frac{13}{16}$"	10$\frac{1}{4}$"	10$\frac{5}{8}$"	11"	11$\frac{5}{16}$"	9$\frac{3}{8}$"	9$\frac{13}{16}$"	10$\frac{1}{4}$"

FLOOR TO FLOOR RISE	10'10 3/8"	10'10 3/8"	10'11 3/8"	10'10 1/2"	10'10 1/2"	10'10 1/2"	10'10 1/2"	10'10 1/2"	10'10 1/2"	10'10 5/8"
NUMBER OF RISERS	19	20	21	16	17	18	19	20	21	16
HEIGHT OF EACH RISER	6 7/8"	6 1/2"	6 3/16"	8 3/16"	7 11/16"	7 1/4"	6 7/8"	6 1/2"	6 7/32"	8 3/16"
TOTAL HEIGHT OF RISERS	10'10 5/8"	10'10"	10'9 15/16"	10'11"	10'10 11/16"	10'10 1/2"	10'10 5/8"	10'10"	10'10 19/32"	10'11"
RISERS OVERAGE (+) OR UNDERAGE (—)	+1/4"	−3/8"	−7/16"	+1/2"	+3/16"	0	+1/8"	−1/2"	+3/32"	+3/8"
NUMBER OF TREADS	18	19	20	15	16	17	18	19	20	15
WIDTH OF EACH TREAD	10 5/8"	11"	11 5/16"	9 5/16"	9 13/16"	10 1/4"	10 5/8"	11"	11 9/32"	9 5/16"
TOTAL RUN OF STAIRWAY	15'11 1/4"	17'5"	18'10 1/4"	11'7 11/16"	13'1"	14'6 1/4"	15'11 1/4"	17'5"	18'9 5/8"	11'7 11/16"
WELL OPENING FOR 6'-8" HEADROOM	11'10 31/32"	13'3 15/16"	14'8 21/32"	8'8 1/2"	10' 23/32"	11'5 1/8"	11'10 31/32"	13'3 15/16"	14'8 3/32"	8'8 1/2"
ANGLE OF INCLINE	32°54'	30°35'	28°41'	41°19'	38°5'	35°16'	32°54'	30°35'	28°52'	41°19'
LENGTH OF CARRIAGE	18'11 13/16"	20'2 3/4"	21'5 7/8"	15'6"	16'7 7/16"	17'9 7/16"	18'11 13/16"	20'2 3/4"	21'5 5/8"	15'6"
L. F. OF RISER PER INCH OF STAIRWAY WIDTH	1.583	1.666	1.75	1.333	1.416	1.5	1.583	1.666	1.75	1.333
L. F. OF TREAD PER INCH OF STAIRWAY WIDTH	1.5	1.583	1.666	1.25	1.333	1.416	1.5	1.583	1.666	1.25
FRAMING SQUARE SETTINGS FOR CARRIAGE CUTS — TONGUE	6 7/8"	6 1/2"	6 3/16"	8 3/16"	7 11/16"	7 1/4"	6 7/8"	6 1/2"	6 7/32"	8 3/16"
FRAMING SQUARE SETTINGS FOR CARRIAGE CUTS — BODY	10 5/8"	11"	11 5/16"	9 5/16"	9 13/16"	10 1/4"	10 5/8"	11"	11 9/32"	9 5/16"

FLOOR TO FLOOR RISE		10'10 5/8"	10'10 5/8"	10'10 5/8"	10'10 5/8"	10'10 5/8"	10'10 3/4"	10'10 3/4"	10'10 3/4"	10'10 3/4"	10'10 3/4"
NUMBER OF RISERS		17	18	19	20	21	16	17	18	19	20
HEIGHT OF EACH RISER		7 11/16"	7 1/4"	6 7/8"	6 17/32"	6 7/32"	8 3/16"	7 11/16"	7 1/4"	6 7/8"	6 9/16"
TOTAL HEIGHT OF RISERS		10'10 11/16"	10'10 1/2"	10'10 5/8"	10'10 5/8"	10'10 19/32"	10'11"	10'10 11/16"	10'10 1/2"	10'10 5/8"	10'11 1/4"
RISERS OVERAGE (+) OR UNDERAGE (—)		+1/16"	−1/8"	0	0	−1/32"	+1/4"	−1/16"	−1/4"	−1/8"	+1/2"
NUMBER OF TREADS		16	17	18	19	20	15	16	17	18	19
WIDTH OF EACH TREAD		9 13/16"	10 1/4"	10 5/8"	10 31/32"	11 9/32"	9 5/16"	9 13/16"	10 1/4"	10 5/8"	10 15/16"
TOTAL RUN OF STAIRWAY		13'1"	14'6 1/4"	15'11 1/4"	17'4 13/32"	18'9 5/8"	11'7 11/16"	13'1"	14'6 1/4"	15'11 1/4"	17'3 13/16"
WELL OPENING FOR 6'-8" HEADROOM		10' 23/32"	11'5 1/8"	11'10 31/32"	13'3 13/32"	14'8 3/32"	8'8 1/2"	10' 23/32"	11'5 1/8"	11'10 31/32"	13'2 27/32"
ANGLE OF INCLINE		38°5'	35°16'	32°54'	30°46'	28°52'	41°19'	38°5'	35°16'	32°54'	30°58'
LENGTH OF CARRIAGE		16'7 7/16"	17'9 7/16"	18'11 13/16"	20'2 9/16"	21'6 5/8"	15'6"	16'7 7/16"	17'9 7/16"	18'11 13/16"	20'2 11/32"
L. F. OF RISER PER INCH OF STAIRWAY WIDTH		1.416	1.5	1.583	1.666	1.75	1.333	1.416	1.5	1.583	1.666
L. F. OF TREAD PER INCH OF STAIRWAY WIDTH		1.333	1.416	1.5	1.583	1.666	1.25	1.333	1.416	1.5	1.583
FRAMING SQUARE SETTINGS FOR CARRIAGE CUTS	TONGUE	7 11/16"	7 1/4"	6 7/8"	6 17/32"	6 7/32"	8 3/16"	7 11/16"	7 1/4"	6 7/8"	6 9/16"
	BODY	9 13/16"	10 1/4"	10 5/8"	10 31/32"	11 9/32"	9 5/16"	9 13/16"	10 1/4"	10 5/8"	10 15/16"

FLOOR TO FLOOR RISE	10'10$\frac{3}{4}$"	10'10$\frac{7}{8}$"	10'10$\frac{7}{8}$"	10'10$\frac{7}{8}$"	10'10$\frac{7}{8}$"	10'10$\frac{7}{8}$"	10'10$\frac{7}{8}$"	10'11"	10'11"	10'11"
NUMBER OF RISERS	21	16	17	18	19	20	21	16	17	18
HEIGHT OF EACH RISER	6$\frac{1}{4}$"	8$\frac{3}{16}$"	7$\frac{11}{16}$"	7$\frac{1}{4}$"	6$\frac{7}{8}$"	6$\frac{9}{16}$"	6$\frac{1}{4}$"	8$\frac{3}{16}$"	7$\frac{11}{16}$"	7$\frac{1}{4}$"
TOTAL HEIGHT OF RISERS	10'11$\frac{1}{4}$"	10'11"	10'10$\frac{11}{16}$"	10'10$\frac{1}{2}$"	10'10$\frac{5}{8}$"	10'11$\frac{1}{4}$"	10'11$\frac{1}{4}$"	10'11"	10'10$\frac{11}{16}$"	10'10$\frac{1}{2}$"
RISERS OVERAGE (+) OR UNDERAGE (—)	+$\frac{1}{2}$"	+$\frac{1}{8}$"	−$\frac{3}{16}$"	−$\frac{3}{8}$"	−$\frac{1}{4}$"	+$\frac{3}{8}$"	+$\frac{3}{8}$"	0	−$\frac{5}{16}$"	−$\frac{1}{2}$"
NUMBER OF TREADS	20	15	16	17	18	19	20	15	16	17
WIDTH OF EACH TREAD	11$\frac{1}{4}$"	9$\frac{5}{16}$"	9$\frac{13}{16}$"	10$\frac{1}{4}$"	10$\frac{5}{8}$"	10$\frac{15}{16}$"	11$\frac{1}{4}$"	9$\frac{5}{16}$"	9$\frac{13}{16}$"	10$\frac{1}{4}$"
TOTAL RUN OF STAIRWAY	18'9"	11'7$\frac{11}{16}$"	13'1"	14'6$\frac{1}{4}$"	15'11$\frac{1}{4}$"	17'3$\frac{13}{16}$"	18'9"	11'7$\frac{11}{16}$"	13'1"	14'6$\frac{1}{4}$"
WELL OPENING FOR 6'-8" HEADROOM	14'7$\frac{1}{2}$"	8'8$\frac{1}{2}$"	10'$\frac{23}{32}$"	11'5$\frac{1}{8}$"	11'10$\frac{31}{32}$"	13'2$\frac{27}{32}$"	14'7$\frac{1}{2}$"	8'8$\frac{1}{2}$"	10'$\frac{23}{32}$"	11'5$\frac{1}{8}$"
ANGLE OF INCLINE	29°3'	41°19'	38°5'	35°16'	32°54'	30°58'	29°3'	41°19'	38°5'	35°16'
LENGTH OF CARRIAGE	21'5$\frac{13}{32}$"	15'6"	16'7$\frac{7}{16}$"	17'9$\frac{7}{16}$"	18'11$\frac{13}{16}$"	20'2$\frac{11}{32}$"	21'5$\frac{13}{32}$"	15'6"	16'7$\frac{7}{16}$"	17'9$\frac{7}{16}$"
L. F. OF RISER PER INCH OF STAIRWAY WIDTH	1.75	1.333	1.416	1.5	1.583	1.666	1.75	1.333	1.416	1.5
L. F. OF TREAD PER INCH OF STAIRWAY WIDTH	1.666	1.25	1.333	1.416	1.5	1.583	1.666	1.25	1.333	1.416
FRAMING SQUARE SETTINGS FOR CARRIAGE CUTS — TONGUE	6$\frac{1}{4}$"	8$\frac{3}{16}$"	7$\frac{11}{16}$"	7$\frac{1}{4}$"	6$\frac{7}{8}$"	6$\frac{9}{16}$"	6$\frac{1}{4}$"	8$\frac{3}{16}$"	7$\frac{11}{16}$"	7$\frac{1}{4}$"
FRAMING SQUARE SETTINGS FOR CARRIAGE CUTS — BODY	11$\frac{1}{4}$"	9$\frac{5}{16}$"	9$\frac{13}{16}$"	10$\frac{1}{4}$"	10$\frac{5}{8}$"	10$\frac{15}{16}$"	11$\frac{1}{4}$"	9$\frac{5}{16}$"	9$\frac{13}{16}$"	10$\frac{1}{4}$"

FLOOR TO FLOOR RISE		10'11"	10'11"	10'11"	10'11 1/8"	10'11 1/8"	10'11 1/8"	10'11 1/8"	10'11 1/8"	10'11 1/8"	10'11 1/4"
NUMBER OF RISERS		19	20	21	16	17	18	19	20	21	16
HEIGHT OF EACH RISER		6 7/8"	6 9/16"	6 1/4"	8 3/16"	7 11/16"	7 5/16"	6 7/8"	6 9/16"	6 1/4"	8 3/16"
TOTAL HEIGHT OF RISERS		10'10 5/8"	10'11 1/4"	10'11 1/4"	10'11"	10'10 11/16"	10'11 5/8"	10'10 5/8"	10'11 1/4"	10'11 1/4"	10'11"
RISERS OVERAGE (+) OR UNDERAGE (—)		−3/8"	+1/4"	+1/4"	−1/8"	−7/16"	+1/2"	−1/2"	+1/8"	+1/8"	−1/4"
NUMBER OF TREADS		18	19	20	15	16	17	18	19	20	15
WIDTH OF EACH TREAD		10 5/8"	10 15/16"	11 1/4"	9 5/16"	9 13/16"	10 3/16"	10 5/8"	10 15/16"	11 1/4"	9 5/16"
TOTAL RUN OF STAIRWAY		15'11 1/4"	17'3 13/16"	18'9"	11'7 11/16"	13'1"	14'5 3/16"	15'11 1/4"	17'3 13/16"	18'9"	11'7 11/16"
WELL OPENING FOR 6'-8" HEADROOM		11'10 31/32"	13'2 27/32"	14'7 1/2"	8'8 1/2"	10' 23/32"	11'4 3/16"	11'10 31/32"	13'2 27/32"	14'7 1/2"	8'8 1/2"
ANGLE OF INCLINE		32°54'	30°58'	29°3'	41°19'	38°5'	35°40'	32°54'	30°58'	29°3'	41°19'
LENGTH OF CARRIAGE		18'11 13/16"	20'2 11/32"	21'5 13/32"	15'6"	16'7 7/16"	17'9 3/16"	18'11 13/16"	20'2 11/32"	21'5 13/32"	15'6"
L. F. OF RISER PER INCH OF STAIRWAY WIDTH		1.583	1.666	1.75	1.333	1.416	1.5	1.583	1.666	1.75	1.333
L. F. OF TREAD PER INCH OF STAIRWAY WIDTH		1.5	1.583	1.666	1.25	1.333	1.416	1.5	1.583	1.666	1.25
FRAMING SQUARE SETTINGS FOR CARRIAGE CUTS	TONGUE	6 7/8"	6 9/16"	6 1/4"	8 3/16"	7 11/16"	7 5/16"	6 7/8"	6 9/16"	6 1/4"	8 3/16"
	BODY	10 5/8"	10 15/16"	11 1/4"	9 5/16"	9 13/16"	10 3/16"	10 5/8"	10 15/16"	11 1/4"	9 5/16"

FLOOR TO FLOOR RISE		10'11 1/4"	10'11 1/4"	10'11 1/4"	10'11 1/4"	10'11 1/4"	10'11 3/8"	10'11 3/8"	10'11 3/8"	10'11 3/8"	10'11 3/8"
NUMBER OF RISERS		17	18	19	20	21	16	17	18	19	20
HEIGHT OF EACH RISER		7 3/4"	7 5/16"	6 29/32"	6 9/16"	6 1/4"	8 3/16"	7 3/4"	7 5/16"	6 15/16"	6 9/16"
TOTAL HEIGHT OF RISERS		10'11 3/4"	10'11 5/8"	10'11 7/32"	10'11 1/4"	10'11 1/4"	10'11"	10'11 3/4"	10'11 5/8"	10'11 13/16"	10'11 1/4"
RISERS OVERAGE (+) OR UNDERAGE (—)		+1/2"	+3/8"	−1/32"	0	0	−3/8"	+3/8"	+1/4"	+7/16"	−1/8"
NUMBER OF TREADS		16	17	18	19	20	15	16	17	18	19
WIDTH OF EACH TREAD		9 3/4"	10 3/16"	10 19/32"	10 15/16"	11 1/4"	9 5/16"	9 3/4"	10 3/16"	10 9/16"	10 15/16"
TOTAL RUN OF STAIRWAY		13'	14'5 3/16"	15'10 11/16"	17'3 13/16"	18'9"	11'7 11/16"	13'	14'5 3/16"	15'10 1/8"	17'3 13/16"
WELL OPENING FOR 6'-8" HEADROOM		9'11 27/32"	11'4 3/16"	11'10 5/32"	13'2 27/32"	14'7 1/2"	8'8 1/2"	9'11 27/32"	11'4 3/16"	11'9 31/32"	13'2 27/32"
ANGLE OF INCLINE		38°29'	35°40'	33°6'	30°58'	29°3'	41°19'	38°29'	35°40'	33°18'	30°58'
LENGTH OF CARRIAGE		16'7 9/32"	17'9 3/16"	18'11 5/8"	20'2 11/32"	21'5 13/32"	15'6"	16'7 9/32"	17'9 3/16"	18'11 15/32"	20'2 11/32"
L. F. OF RISER PER INCH OF STAIRWAY WIDTH		1.416	1.5	1.583	1.666	1.75	1.333	1.416	1.5	1.583	1.666
L. F. OF TREAD PER INCH OF STAIRWAY WIDTH		1.333	1.416	1.5	1.583	1.666	1.25	1.333	1.416	1.5	1.583
FRAMING SQUARE SETTINGS FOR CARRIAGE CUTS	TONGUE	7 3/4"	7 5/16"	6 29/32"	6 9/16"	6 1/4"	8 3/16"	7 3/4"	7 5/16"	6 15/16"	6 9/16"
	BODY	9 3/4"	10 3/16"	10 19/32"	10 15/16"	11 1/4"	9 5/16"	9 3/4"	10 3/16"	10 9/16"	10 15/16"

FLOOR TO FLOOR RISE	10'11 3/8"	10'11 1/2"	10'11 1/2"	10'11 1/2"	10'11 1/2"	10'11 1/2"	10'11 1/2"	10'11 5/8"	10'11 5/8"	10'11 5/8"
NUMBER OF RISERS	21	16	17	18	19	20	21	16	17	18
HEIGHT OF EACH RISER	6 1/4"	8 1/4"	7 3/4"	7 5/16"	6 15/16"	6 9/16"	6 1/4"	8 1/4"	7 3/4"	7 5/16"
TOTAL HEIGHT OF RISERS	10'11 1/4"	11'	10'11 3/4"	10'11 5/8"	10'11 13/16"	10'11 1/4"	10'11 1/4"	11'	10'11 3/4"	10'11 5/8"
RISERS OVERAGE (+) OR UNDERAGE (—)	−1/8"	+1/2"	+1/4"	+1/8"	+5/16"	−1/4"	−1/4"	+3/8"	+1/8"	0
NUMBER OF TREADS	20	15	16	17	18	19	20	15	16	17
WIDTH OF EACH TREAD	11 1/4"	9 1/4"	9 3/4"	10 3/16"	10 9/16"	10 15/16"	11 1/4"	9 1/4"	9 3/4"	10 3/16"
TOTAL RUN OF STAIRWAY	18'9"	11'6 3/4"	13'	14'5 3/16"	15'10 1/8"	17'3 13/16"	18'9"	11'6 3/4"	13'	14'5 3/16"
WELL OPENING FOR 6'-8" HEADROOM	14'7 1/2"	8'7 23/32"	9'11 27/32"	11'4 3/16"	11'9 31/32"	13'2 27/32"	14'7 1/2"	8'7 23/32"	9'11 27/32"	11'4 3/16"
ANGLE OF INCLINE	29°3'	41°44'	38°29'	35°40'	33°18'	30°58'	29°3'	41°44'	38°29'	35°40'
LENGTH OF CARRIAGE	21'5 13/32"	15'5 15/16"	16'7 9/32"	17'9 3/16"	18'11 15/32"	20'2 11/32"	21'5 13/32"	15'5 15/16"	16'7 9/32"	17'9 3/16"
L. F. OF RISER PER INCH OF STAIRWAY WIDTH	1.75	1.333	1.416	1.5	1.583	1.666	1.75	1.333	1.416	1.5
L. F. OF TREAD PER INCH OF STAIRWAY WIDTH	1.666	1.25	1.333	1.416	1.5	1.583	1.666	1.25	1.333	1.416
FRAMING SQUARE SETTINGS FOR CARRIAGE CUTS — TONGUE	6 1/4"	8 1/4"	7 3/4"	7 5/16"	6 15/16"	6 9/16"	6 1/4"	8 1/4"	7 3/4"	7 5/16"
FRAMING SQUARE SETTINGS FOR CARRIAGE CUTS — BODY	11 1/4"	9 1/4"	9 3/4"	10 3/16"	10 9/16"	10 15/16"	11 1/4"	9 1/4"	9 3/4"	10 3/16"

FLOOR TO FLOOR RISE		10'11 5/8"	10'11 5/8"	10'11 5/8"	10'11 3/4"	10'11 3/4"	10'11 3/4"	10'11 3/4"	10'11 3/4"	10'11 3/4"	10'11 7/8"
NUMBER OF RISERS		19	20	21	16	17	18	19	20	21	16
HEIGHT OF EACH RISER		6 15/16"	6 9/16"	6 1/4"	8 1/4"	7 3/4"	7 5/16"	6 15/16"	6 9/16"	6 1/4"	8 1/4"
TOTAL HEIGHT OF RISERS		10'11 13/16"	10'11 1/4"	10'11 1/4"	11'	10'11 3/4"	10'11 5/8"	10'11 13/16"	10'11 1/4"	10'11 1/4"	11'
RISERS OVERAGE (+) OR UNDERAGE (—)		+3/16"	−3/8"	−3/8"	+1/4"	0	−1/8"	+1/16"	−1/2"	−1/2"	+1/8"
NUMBER OF TREADS		18	19	20	15	16	17	18	19	20	15
WIDTH OF EACH TREAD		10 9/16"	10 15/16"	11 1/4"	9 1/4"	9 3/4"	10 3/16"	10 9/16"	10 15/16"	11 1/4"	9 1/4"
TOTAL RUN OF STAIRWAY		15'10 1/8"	17'3 13/16"	18'9"	11'6 3/4"	13'	14'5 3/16"	15'10 1/8"	17'3 13/16"	18'9"	11'6 3/4"
WELL OPENING FOR 6'-8" HEADROOM		11'9 31/32"	13'2 27/32"	14'7 1/2"	8'7 23/32"	9'11 27/32"	11'4 3/16"	11'9 31/32"	13'2 27/32"	14'7 1/2"	8'7 23/32"
ANGLE OF INCLINE		33°18'	30°58'	29°3'	41°44'	38°29'	35°40'	33°18'	30°58'	29°3'	41°44'
LENGTH OF CARRIAGE		18'11 15/32"	20'2 11/32"	21'5 13/32"	15'5 15/16"	16'7 9/32"	17'9 3/16"	18'11 15/32"	20'2 11/32"	21'5 13/32"	15'5 15/16"
L. F. OF RISER PER INCH OF STAIRWAY WIDTH		1.583	1.666	1.75	1.333	1.416	1.5	1.583	1.666	1.75	1.333
L. F. OF TREAD PER INCH OF STAIRWAY WIDTH		1.5	1.583	1.666	1.25	1.333	1.416	1.5	1.583	1.666	1.25
FRAMING SQUARE SETTINGS FOR CARRIAGE CUTS	TONGUE	6 15/16"	6 9/16"	6 1/4"	8 1/4"	7 3/4"	7 5/16"	6 15/16"	6 9/16"	6 1/4"	8 1/4"
	BODY	10 9/16"	10 15/16"	11 1/4"	9 1/4"	9 3/4"	10 3/16"	10 9/16"	10 15/16"	11 1/4"	9 1/4"

FLOOR TO FLOOR RISE	10'11$\frac{7}{8}$"	10'11$\frac{7}{8}$"	10'11$\frac{7}{8}$"	10'11$\frac{7}{8}$"	10'11$\frac{7}{8}$"	11'	11'	11'	11'	11'
NUMBER OF RISERS	17	18	19	20	21	16	17	18	19	20
HEIGHT OF EACH RISER	7$\frac{3}{4}$"	7$\frac{5}{16}$"	6$\frac{15}{16}$"	6$\frac{19}{32}$"	6$\frac{9}{32}$"	8$\frac{1}{4}$"	7$\frac{3}{4}$"	7$\frac{5}{16}$"	6$\frac{15}{16}$"	6$\frac{5}{8}$"
TOTAL HEIGHT OF RISERS	10'11$\frac{3}{4}$"	10'11$\frac{5}{8}$"	10'11$\frac{13}{16}$"	10'11$\frac{7}{8}$"	10'11$\frac{29}{32}$"	11'	10'11$\frac{3}{4}$"	10'11$\frac{5}{8}$"	10'11$\frac{13}{16}$"	11'$\frac{1}{2}$"
RISERS OVERAGE (+) OR UNDERAGE (—)	−$\frac{1}{8}$"	−$\frac{1}{4}$"	−$\frac{1}{16}$"	0	+$\frac{1}{32}$"	0	−$\frac{1}{4}$"	−$\frac{3}{8}$"	−$\frac{3}{16}$"	+$\frac{1}{2}$"
NUMBER OF TREADS	16	17	18	19	20	15	16	17	18	19
WIDTH OF EACH TREAD	9$\frac{3}{4}$"	10$\frac{3}{16}$"	10$\frac{9}{16}$"	10$\frac{29}{32}$"	11$\frac{7}{32}$"	9$\frac{1}{4}$"	9$\frac{3}{4}$"	10$\frac{3}{16}$"	10$\frac{9}{16}$"	10$\frac{7}{8}$"
TOTAL RUN OF STAIRWAY	13'	14'5$\frac{3}{16}$"	15'10$\frac{1}{8}$"	17'3$\frac{7}{32}$"	18'8$\frac{3}{8}$"	11'6$\frac{3}{4}$"	13'	14'5$\frac{3}{16}$"	15'10$\frac{1}{8}$"	17'2$\frac{5}{8}$"
WELL OPENING FOR 6'-8" HEADROOM	9'11$\frac{27}{32}$"	11'4$\frac{3}{16}$"	11'9$\frac{31}{32}$"	13'2$\frac{5}{16}$"	14'6$\frac{29}{32}$"	8'7$\frac{23}{32}$"	9'11$\frac{27}{32}$"	11'4$\frac{3}{16}$"	11'9$\frac{31}{32}$"	13'1$\frac{25}{32}$"
ANGLE OF INCLINE	38°29'	35°40'	33°18'	31°9'	29°15'	41°44'	38°29'	35°40'	33°18'	31°21'
LENGTH OF CARRIAGE	16'7$\frac{9}{32}$"	17'9$\frac{3}{16}$"	18'11$\frac{15}{32}$"	20'2$\frac{5}{32}$"	21'5$\frac{5}{32}$"	15'5$\frac{15}{16}$"	16'7$\frac{9}{32}$"	17'9$\frac{3}{16}$"	18'11$\frac{15}{32}$"	20'1$\frac{15}{16}$"
L. F. OF RISER PER INCH OF STAIRWAY WIDTH	1.416	1.5	1.583	1.666	1.75	1.333	1.416	1.5	1.583	1.666
L. F. OF TREAD PER INCH OF STAIRWAY WIDTH	1.333	1.416	1.5	1.583	1.666	1.25	1.333	1.416	1.5	1.583
FRAMING SQUARE SETTINGS FOR CARRIAGE CUTS — TONGUE	7$\frac{3}{4}$"	7$\frac{5}{16}$"	6$\frac{15}{16}$"	6$\frac{19}{32}$"	6$\frac{9}{32}$"	8$\frac{1}{4}$"	7$\frac{3}{4}$"	7$\frac{5}{16}$"	6$\frac{15}{16}$"	6$\frac{5}{8}$"
FRAMING SQUARE SETTINGS FOR CARRIAGE CUTS — BODY	9$\frac{3}{4}$"	10$\frac{3}{16}$"	10$\frac{9}{16}$"	10$\frac{29}{32}$"	11$\frac{7}{32}$"	9$\frac{1}{4}$"	9$\frac{3}{4}$"	10$\frac{3}{16}$"	10$\frac{9}{16}$"	10$\frac{7}{8}$"

FLOOR TO FLOOR RISE		11'	11'	11' $\frac{1}{8}$"	11' $\frac{1}{8}$"	11' $\frac{1}{8}$"	11' $\frac{1}{8}$"	11' $\frac{1}{8}$"	11' $\frac{1}{8}$"	11' $\frac{1}{8}$"	11' $\frac{1}{4}$"
NUMBER OF RISERS		21	22	16	17	18	19	20	21	22	16
HEIGHT OF EACH RISER		$6\frac{9}{32}$"	6"	$8\frac{1}{4}$"	$7\frac{3}{4}$"	$7\frac{5}{16}$"	$6\frac{15}{16}$"	$6\frac{5}{8}$"	$6\frac{5}{16}$"	6"	$8\frac{1}{4}$"
TOTAL HEIGHT OF RISERS		10' $11\frac{29}{32}$"	11'	11'	10' $11\frac{3}{4}$"	10' $11\frac{5}{8}$"	10' $11\frac{13}{16}$"	11' $\frac{1}{2}$"	11' $\frac{9}{16}$"	11'	11'
RISERS OVERAGE (+) OR UNDERAGE (—)		$-\frac{3}{32}$"	0	$-\frac{1}{8}$"	$-\frac{3}{8}$"	$-\frac{1}{2}$"	$-\frac{5}{16}$"	$+\frac{3}{8}$"	$+\frac{7}{16}$"	$-\frac{1}{8}$"	$-\frac{1}{4}$"
NUMBER OF TREADS		20	21	15	16	17	18	19	20	21	15
WIDTH OF EACH TREAD		$11\frac{7}{32}$"	$11\frac{1}{2}$"	$9\frac{1}{4}$"	$9\frac{3}{4}$"	$10\frac{3}{16}$"	$10\frac{9}{16}$"	$10\frac{7}{8}$"	$11\frac{3}{16}$"	$11\frac{1}{2}$"	$9\frac{1}{4}$"
TOTAL RUN OF STAIRWAY		18' $8\frac{3}{8}$"	20' $1\frac{1}{2}$"	11' $6\frac{3}{4}$"	13'	14' $5\frac{3}{16}$"	15' $10\frac{1}{8}$"	17' $2\frac{5}{8}$"	18' $7\frac{3}{4}$"	20' $1\frac{1}{2}$"	11' $6\frac{3}{4}$"
WELL OPENING FOR 6'-8" HEADROOM		14' $6\frac{29}{32}$"	15' $\frac{5}{32}$"	8' $7\frac{23}{32}$"	9' $11\frac{27}{32}$"	11' $4\frac{3}{16}$"	11' $9\frac{31}{32}$"	13' $1\frac{25}{32}$"	13' $7\frac{5}{32}$"	15' $\frac{5}{32}$"	8' $7\frac{25}{32}$"
ANGLE OF INCLINE		29°15'	27°33'	41°44'	38°29'	35°40'	33°18'	31°21'	29°26'	27°33'	41°44'
LENGTH OF CARRIAGE		21' $5\frac{5}{32}$"	22' $8\frac{13}{32}$"	15' $5\frac{15}{16}$"	16' $7\frac{9}{32}$"	17' $9\frac{3}{16}$"	18' $11\frac{15}{32}$"	20' $1\frac{15}{16}$"	21' $4\frac{29}{32}$"	22' $8\frac{13}{32}$"	15' $5\frac{15}{16}$"
L. F. OF RISER PER INCH OF STAIRWAY WIDTH		1.75	1.833	1.333	1.416	1.5	1.583	1.666	1.75	1.833	1.333
L. F. OF TREAD PER INCH OF STAIRWAY WIDTH		1.666	1.75	1.25	1.333	1.416	1.5	1.583	1.666	1.75	1.25
FRAMING SQUARE SETTINGS FOR CARRIAGE CUTS	TONGUE	$6\frac{9}{32}$"	6"	$8\frac{1}{4}$"	$7\frac{3}{4}$"	$7\frac{5}{16}$"	$6\frac{15}{16}$"	$6\frac{5}{8}$"	$6\frac{5}{16}$"	6"	$8\frac{1}{4}$"
	BODY	$11\frac{7}{32}$"	$11\frac{1}{2}$"	$9\frac{1}{4}$"	$9\frac{3}{4}$"	$10\frac{3}{16}$"	$10\frac{9}{16}$"	$10\frac{7}{8}$"	$11\frac{3}{16}$"	$11\frac{1}{2}$"	$9\frac{1}{4}$"

FLOOR TO FLOOR RISE	11'$\frac{1}{4}$"	11'$\frac{1}{4}$"	11'$\frac{1}{4}$"	11'$\frac{1}{4}$"	11'$\frac{1}{4}$"	11'$\frac{1}{4}$"	11'$\frac{3}{8}$"	11'$\frac{3}{8}$"	11'$\frac{3}{8}$"	11'$\frac{3}{8}$"
NUMBER OF RISERS	17	18	19	20	21	22	16	17	18	19
HEIGHT OF EACH RISER	7$\frac{3}{4}$"	7$\frac{3}{8}$"	6$\frac{15}{16}$"	6$\frac{5}{8}$"	6$\frac{5}{16}$"	6"	8$\frac{1}{4}$"	7$\frac{13}{16}$"	7$\frac{3}{8}$"	6$\frac{31}{32}$"
TOTAL HEIGHT OF RISERS	10'11$\frac{3}{4}$"	11'$\frac{3}{4}$"	10'11$\frac{13}{16}$"	11'$\frac{1}{2}$"	11'$\frac{9}{16}$"	11'	11'	11'$\frac{13}{16}$"	11'$\frac{3}{4}$"	11'$\frac{13}{32}$"
RISERS OVERAGE (+) OR UNDERAGE (—)	$-\frac{1}{2}$"	$+\frac{1}{2}$"	$-\frac{7}{16}$"	$+\frac{1}{4}$"	$+\frac{5}{16}$"	$-\frac{1}{4}$"	$-\frac{3}{8}$"	$+\frac{7}{16}$"	$+\frac{3}{8}$"	$+\frac{1}{32}$"
NUMBER OF TREADS	16	17	18	19	20	21	15	16	17	18
WIDTH OF EACH TREAD	9$\frac{3}{4}$"	10$\frac{1}{8}$"	10$\frac{9}{16}$"	10$\frac{7}{8}$"	11$\frac{3}{16}$"	11$\frac{1}{2}$"	9$\frac{1}{4}$"	9$\frac{11}{16}$"	10$\frac{1}{8}$"	10$\frac{17}{32}$"
TOTAL RUN OF STAIRWAY	13'	14'4$\frac{1}{8}$"	15'10$\frac{1}{8}$"	17'2$\frac{5}{8}$"	18'7$\frac{3}{4}$"	20'1$\frac{1}{2}$"	11'6$\frac{3}{4}$"	12'11"	14'4$\frac{1}{8}$"	15'9$\frac{9}{16}$"
WELL OPENING FOR 6'-8" HEADROOM	9'11$\frac{27}{32}$"	10'5$\frac{3}{32}$"	11'9$\frac{31}{32}$"	13'1$\frac{25}{32}$"	13'7$\frac{5}{32}$"	15'$\frac{5}{32}$"	8'7$\frac{23}{32}$"	9'10$\frac{15}{16}$"	10'5$\frac{3}{32}$"	11'9$\frac{1}{2}$"
ANGLE OF INCLINE	38°29'	36°4'	33°18'	31°21'	29°26'	27°33'	41°44'	38°53'	36°4'	33°30'
LENGTH OF CARRIAGE	16'7$\frac{9}{32}$"	17'8$\frac{15}{16}$"	18'11$\frac{15}{32}$"	20'1$\frac{15}{16}$"	21'4$\frac{29}{32}$"	22'8$\frac{13}{32}$"	15'5$\frac{15}{16}$"	16'7$\frac{1}{8}$"	17'8$\frac{15}{16}$"	18'11$\frac{5}{16}$"
L. F. OF RISER PER INCH OF STAIRWAY WIDTH	1.416	1.5	1.583	1.666	1.75	1.833	1.333	1.416	1.5	1.583
L. F. OF TREAD PER INCH OF STAIRWAY WIDTH	1.333	1.416	1.5	1.583	1.666	1.75	1.25	1.333	1.416	1.5
FRAMING SQUARE SETTINGS FOR CARRIAGE CUTS — TONGUE	7$\frac{3}{4}$"	7$\frac{3}{8}$"	6$\frac{15}{16}$"	6$\frac{5}{8}$"	6$\frac{5}{16}$"	6"	8$\frac{1}{4}$"	7$\frac{13}{16}$"	7$\frac{3}{8}$"	6$\frac{31}{32}$"
FRAMING SQUARE SETTINGS FOR CARRIAGE CUTS — BODY	9$\frac{3}{4}$"	10$\frac{1}{8}$"	10$\frac{9}{16}$"	10$\frac{7}{8}$"	11$\frac{3}{16}$"	11$\frac{1}{2}$"	9$\frac{1}{4}$"	9$\frac{11}{16}$"	10$\frac{1}{8}$"	10$\frac{17}{32}$"

FLOOR TO FLOOR RISE		11' 3/8"	11' 3/8"	11' 3/8"	11' 1/2"	11' 1/2"	11' 1/2"	11' 1/2"	11' 1/2"	11' 1/2"	11' 5/8"
NUMBER OF RISERS		20	21	22	17	18	19	20	21	22	17
HEIGHT OF EACH RISER		6 5/8"	6 5/16"	6"	7 13/16"	7 3/8"	7"	6 5/8"	6 5/16"	6"	7 13/16"
TOTAL HEIGHT OF RISERS		11' 1/2"	11' 9/16"	11'	11' 13/16"	11' 3/4"	11' 1"	11' 1/2"	11' 9/16"	11'	11' 13/16"
RISERS OVERAGE (+) OR UNDERAGE (—)		+1/8"	+3/16"	−3/8"	+5/16"	+1/4"	+1/2"	0	+1/16"	−1/2"	+3/16"
NUMBER OF TREADS		19	20	21	16	17	18	19	20	21	16
WIDTH OF EACH TREAD		10 7/8"	11 3/16"	11 1/2"	9 11/16"	10 1/8"	10 1/2"	10 7/8"	11 3/16"	11 1/2"	9 11/16"
TOTAL RUN OF STAIRWAY		17' 2 5/8"	18' 7 3/4"	20' 1 1/2"	12' 11"	14' 4 1/8"	15' 9"	17' 2 5/8"	18' 7 3/4"	20' 1 1/2"	12' 11"
WELL OPENING FOR 6'-8" HEADROOM		13' 1 25/32"	13' 7 5/32"	15' 5/32"	9' 10 15/16"	10' 5 3/32"	11' 9"	13' 1 25/32"	13' 7 5/32"	15' 5/32"	9' 10 15/16"
ANGLE OF INCLINE		31°21'	29°26'	27°33'	38°53'	36°4'	33°41'	31°21'	29°26'	27°33'	38°53'
LENGTH OF CARRIAGE		20' 1 15/16"	21' 4 29/32"	22' 8 13/32"	16' 7 1/8"	17' 8 15/16"	18' 11 5/32"	20' 1 15/16"	21' 4 29/32"	22' 8 13/32"	16' 7 1/8"
L. F. OF RISER PER INCH OF STAIRWAY WIDTH		1.666	1.75	1.833	1.416	1.5	1.583	1.666	1.75	1.833	1.416
L. F. OF TREAD PER INCH OF STAIRWAY WIDTH		1.583	1.666	1.75	1.333	1.416	1.5	1.583	1.666	1.75	1.333
FRAMING SQUARE SETTINGS FOR CARRIAGE CUTS	TONGUE	6 5/8"	6 5/16"	6"	7 13/16"	7 3/8"	7"	6 5/8"	6 5/16"	6"	7 13/16"
	BODY	10 7/8"	11 3/16"	11 1/2"	9 11/6"	10 1/8"	10 1/2"	10 7/8"	11 3/16"	11 1/2"	9 11/16"

FLOOR TO FLOOR RISE		11' 5/8"	11' 5/8"	11' 5/8"	11' 5/8"	11' 5/8"	11' 3/4"	11' 3/4"	11' 3/4"	11' 3/4"	11' 3/4"
NUMBER OF RISERS		18	19	20	21	22	17	18	19	20	21
HEIGHT OF EACH RISER		7 3/8"	7"	6 5/8"	6 5/16"	6 1/32"	7 13/16"	7 3/8"	7"	6 5/8"	6 5/16"
TOTAL HEIGHT OF RISERS		11' 3/4"	11'1"	11' 1/2"	11' 9/16"	11' 11/16"	11' 13/16"	11' 3/4"	11'1"	11' 1/2"	11' 9/16"
RISERS OVERAGE (+) OR UNDERAGE (—)		+1/8"	+3/8"	−1/8"	−1/16"	+1/16"	+1/16"	0	+1/4"	−1/4"	−3/16"
NUMBER OF TREADS		17	18	19	20	21	16	17	18	19	20
WIDTH OF EACH TREAD		10 1/8"	10 1/2"	10 7/8"	11 3/16"	11 15/32"	9 11/16"	10 1/8"	10 1/2"	10 7/8"	11 3/16"
TOTAL RUN OF STAIRWAY		14'4 1/8"	15'9"	17'2 5/8"	18'7 3/4"	20' 27/32"	12'11"	14'4 1/8"	15'9"	17'2 5/8"	18'7 3/4"
WELL OPENING FOR 6'-8" HEADROOM		10'5 3/32"	11'9"	13'1 25/32"	13'7 5/32"	14'11 19/32"	9'10 15/16"	10'5 3/32"	11'9"	13'1 25/32"	13'7 5/32"
ANGLE OF INCLINE		36°4'	33°41'	31°21'	29°26'	27°44'	38°53'	36°4'	33°41'	31°21'	29°26'
LENGTH OF CARRIAGE		17'8 15/16"	18'11 5/32"	20'1 15/16"	21'4 29/32"	22'8 1/8"	16'7 1/8"	17'8 15/16"	18'11 5/32"	20'1 15/16"	21'4 29/32"
L. F. OF RISER PER INCH OF STAIRWAY WIDTH		1.5	1.583	1.666	1.75	1.833	1.416	1.5	1.583	1.666	1.75
L. F. OF TREAD PER INCH OF STAIRWAY WIDTH		1.416	1.5	1.583	1.666	1.75	1.333	1.416	1.5	1.583	1.666
FRAMING SQUARE SETTINGS FOR CARRIAGE CUTS	TONGUE	7 3/8"	7"	6 5/8"	6 5/16"	6 1/32"	7 13/16"	7 3/8"	7"	6 5/8"	6 5/16"
	BODY	10 1/8"	10 1/2"	10 7/8"	11 3/16"	11 15/32"	9 11/16"	10 1/8"	10 1/2"	10 7/8"	11 3/16"

FLOOR TO FLOOR RISE		11'3/4"	11'7/8"	11'7/8"	11'7/8"	11'7/8"	11'7/8"	11'7/8"	11'1"	11'1"	11'1"
NUMBER OF RISERS		22	17	18	19	20	21	22	17	18	19
HEIGHT OF EACH RISER		6 1/32"	7 13/16"	7 3/8"	7"	6 5/8"	6 5/16"	6 1/16"	7 13/16"	7 3/8"	7"
TOTAL HEIGHT OF RISERS		11'11/16"	11'13/16"	11'3/4"	11'1"	11'1/2"	11'9/16"	11'1 3/8"	11'13/16"	11'3/4"	11'1"
RISERS OVERAGE (+) OR UNDERAGE (—)		−1/16"	−1/16"	−1/8"	+1/8"	−3/8"	−5/16"	+1/2"	−3/16"	−1/4"	0
NUMBER OF TREADS		21	16	17	18	19	20	21	16	17	18
WIDTH OF EACH TREAD		11 15/32"	9 11/16"	10 1/8"	10 1/2"	10 7/8"	11 3/16"	11 7/16"	9 11/16"	10 1/8"	10 1/2"
TOTAL RUN OF STAIRWAY		20'27/32"	12'11"	14'4 1/8"	15'9"	17'2 5/8"	18'7 3/4"	20'3/16"	12'11"	14'4 1/8"	15'9"
WELL OPENING FOR 6'-8" HEADROOM		14'11 19/32"	9'10 15/16"	10'5 3/32"	11'9"	13'1 25/32"	13'7 5/32"	14'11"	9'10 15/16"	10'5 3/32"	11'9"
ANGLE OF INCLINE		27°44'	38°53'	36°4'	33°41'	31°21'	29°26'	27°56'	38°53'	36°4'	33°41'
LENGTH OF CARRIAGE		22'8 1/8"	16'7 1/8"	17'8 15/16"	18'11 5/32"	20'1 15/16"	21'4 29/32"	22'7 27/32"	16'7 1/8"	17'8 15/16"	18'11 5/32"
L. F. OF RISER PER INCH OF STAIRWAY WIDTH		1.833	1.416	1.5	1.583	1.666	1.75	1.833	1.416	1.5	1.583
L. F. OF TREAD PER INCH OF STAIRWAY WIDTH		1.75	1.333	1.416	1.5	1.583	1.666	1.75	1.333	1.416	1.5
FRAMING SQUARE SETTINGS FOR CARRIAGE CUTS	TONGUE	6 1/32"	7 13/16"	7 3/8"	7"	6 5/8"	6 5/16"	6 1/16"	7 13/16"	7 3/8"	7"
	BODY	11 15/32"	9 11/16"	10 1/8"	10 1/2"	10 7/8"	11 3/16"	11 7/16"	9 11/16"	10 1/8"	10 1/2"

FLOOR TO FLOOR RISE		11'1"	11'1"	11'1"	11'1$\frac{1}{8}$"	11'1$\frac{1}{8}$"	11'1$\frac{1}{8}$"	11'1$\frac{1}{8}$"	11'1$\frac{1}{8}$"	11'1$\frac{1}{8}$"	11'1$\frac{1}{4}$"
NUMBER OF RISERS		20	21	22	17	18	19	20	21	22	17
HEIGHT OF EACH RISER		6$\frac{5}{8}$"	6$\frac{5}{16}$"	6$\frac{1}{16}$"	7$\frac{13}{16}$"	7$\frac{3}{8}$"	7"	6$\frac{21}{32}$"	6$\frac{11}{32}$"	6$\frac{1}{16}$"	7$\frac{13}{16}$"
TOTAL HEIGHT OF RISERS		11'$\frac{1}{2}$"	11'$\frac{9}{16}$"	11'1$\frac{3}{8}$"	11'$\frac{13}{16}$"	11'$\frac{3}{4}$"	11'1"	11'1$\frac{1}{8}$"	11'1$\frac{7}{32}$"	11'1$\frac{3}{8}$"	11'$\frac{13}{16}$"
RISERS OVERAGE (+) OR UNDERAGE (—)		$-\frac{1}{2}$"	$-\frac{7}{16}$"	$+\frac{3}{8}$"	$-\frac{5}{16}$"	$-\frac{3}{8}$"	$-\frac{1}{8}$"	0	$+\frac{3}{32}$"	$+\frac{1}{4}$"	$-\frac{7}{16}$"
NUMBER OF TREADS		19	20	21	16	17	18	19	20	21	16
WIDTH OF EACH TREAD		10$\frac{7}{8}$"	11$\frac{3}{16}$"	11$\frac{7}{16}$"	9$\frac{11}{16}$"	10$\frac{1}{8}$"	10$\frac{1}{2}$"	10$\frac{27}{32}$"	11$\frac{5}{32}$"	11$\frac{7}{16}$"	9$\frac{11}{16}$"
TOTAL RUN OF STAIRWAY		17'2$\frac{5}{8}$"	18'7$\frac{3}{4}$"	20'$\frac{3}{16}$"	12'11"	14'4$\frac{1}{8}$"	15'9"	17'2$\frac{1}{32}$"	18'7$\frac{1}{8}$"	20'$\frac{3}{16}$"	12'11"
WELL OPENING FOR 6'-8" HEADROOM		13'1$\frac{25}{32}$"	13'7$\frac{5}{32}$"	14'11"	9'10$\frac{15}{16}$"	10'5$\frac{3}{32}$"	11'9"	13'1$\frac{9}{32}$"	13'6$\frac{5}{8}$"	14'11"	9'10$\frac{15}{16}$"
ANGLE OF INCLINE		31°21'	29°26'	27°56'	38°53'	36°4'	33°41'	31°33'	29°37'	27°56'	38°53'
LENGTH OF CARRIAGE		20'1$\frac{15}{16}$"	21'4$\frac{29}{32}$"	22'7$\frac{27}{32}$"	16'7$\frac{1}{8}$"	17'8$\frac{15}{16}$"	18'11$\frac{5}{32}$"	20'1$\frac{3}{4}$"	21'4$\frac{11}{16}$"	22'7$\frac{27}{32}$"	16'7$\frac{1}{8}$"
L. F. OF RISER PER INCH OF STAIRWAY WIDTH		1.666	1.75	1.833	1.416	1.5	1.583	1.666	1.75	1.833	1.416
L. F. OF TREAD PER INCH OF STAIRWAY WIDTH		1.583	1.666	1.75	1.333	1.416	1.5	1.583	1.666	1.75	1.333
FRAMING SQUARE SETTINGS FOR CARRIAGE CUTS	TONGUE	6$\frac{5}{8}$"	6$\frac{5}{16}$"	6$\frac{1}{16}$"	7$\frac{13}{16}$"	7$\frac{3}{8}$"	7"	6$\frac{21}{32}$"	6$\frac{11}{32}$"	6$\frac{1}{16}$"	7$\frac{13}{16}$"
	BODY	10$\frac{7}{8}$"	11$\frac{3}{16}$"	11$\frac{7}{16}$"	9$\frac{11}{16}$"	10$\frac{1}{8}$"	10$\frac{1}{2}$"	10$\frac{27}{32}$"	11$\frac{5}{32}$"	11$\frac{7}{16}$"	9$\frac{11}{16}$"

FLOOR TO FLOOR RISE		11'1 1/4"	11'1 1/4"	11'1 1/4"	11'1 1/4"	11'1 1/4"	11'1 3/8"	11'1 3/8"	11'1 3/8"	11'1 3/8"	11'1 3/8"
NUMBER OF RISERS		18	19	20	21	22	17	18	19	20	21
HEIGHT OF EACH RISER		7 3/8"	7"	6 11/16"	6 11/32"	6 1/16"	7 7/8"	7 7/16"	7"	6 11/16"	6 3/8"
TOTAL HEIGHT OF RISERS		11' 3/4"	11'1"	11'1 3/4"	11'1 7/32"	11'1 3/8"	11'1 7/8"	11'1 7/8"	11'1"	11'1 3/4"	11'1 7/8"
RISERS OVERAGE (+) OR UNDERAGE (—)		−1/2"	−1/4"	+1/2"	−1/32"	+1/8"	+1/2"	+1/2"	−3/8"	+3/8"	+1/2"
NUMBER OF TREADS		17	18	19	20	21	16	17	18	19	20
WIDTH OF EACH TREAD		10 1/8"	10 1/2"	10 13/16"	11 5/32"	11 7/16"	9 5/8"	10 1/16"	10 1/2"	10 13/16"	11 1/8"
TOTAL RUN OF STAIRWAY		14'4 1/8"	15'9"	17'1 7/16"	18'7 1/8"	20' 3/16"	12'10"	14'3 1/16"	15'9"	17'1 7/16"	18'6 1/2"
WELL OPENING FOR 6'-8" HEADROOM		10'5 3/32"	11'9"	13' 3/4"	13'6 5/8"	14'11"	9'10 3/32"	10'4 3/16"	11'9"	13' 3/4"	13'6 1/16"
ANGLE OF INCLINE		36°4'	33°41'	32°44'	29°37'	27°56'	39°17'	36°28'	33°41'	31°44'	29°49'
LENGTH OF CARRIAGE		17'8 15/16"	18'11 5/32"	20'1 9/16"	21'4 11/16"	22'7 27/32"	16'6 31/32"	17'8 23/32"	18'11 5/32"	20'1 9/16"	21'4 7/16"
L. F. OF RISER PER INCH OF STAIRWAY WIDTH		1.5	1.583	1.666	1.75	1.833	1.416	1.5	1.583	1.666	1.75
L. F. OF TREAD PER INCH OF STAIRWAY WIDTH		1.416	1.5	1.583	1.666	1.75	1.333	1.416	1.5	1.583	1.666
FRAMING SQUARE SETTINGS FOR CARRIAGE CUTS	TONGUE	7 3/8"	7"	6 11/16"	6 11/32"	6 1/16"	7 7/8"	7 7/16"	7"	6 11/16"	6 3/8"
	BODY	10 1/8"	10 1/2"	10 13/16"	11 5/32"	11 7/16"	9 5/8"	10 1/16"	10 1/2"	10 13/16"	11 1/8"

FLOOR TO FLOOR RISE	11'1 3/8"	11'1 1/2"	11'1 1/2"	11'1 1/2"	11'1 1/2"	11'1 1/2"	11'1 1/2"	11'1 5/8"	11'1 5/8"	11'1 5/8"
NUMBER OF RISERS	22	17	18	19	20	21	22	17	18	19
HEIGHT OF EACH RISER	6 1/16"	7 7/8"	7 7/16"	7"	6 11/16"	6 3/8"	6 1/16"	7 7/8"	7 7/16"	7 1/32"
TOTAL HEIGHT OF RISERS	11'1 3/8"	11'1 7/8"	11'1 7/8"	11'1"	11'1 3/4"	11'1 7/8"	11'1 3/8"	11'1 7/8"	11'1 7/8"	11'1 19/32"
RISERS OVERAGE (+) OR UNDERAGE (—)	0	+3/8"	+3/8"	−1/2"	+1/4"	+3/8"	−1/8"	+1/4"	+1/4"	−1/32"
NUMBER OF TREADS	21	16	17	18	19	20	21	16	17	18
WIDTH OF EACH TREAD	11 7/16"	9 5/8"	10 1/16"	10 1/2"	10 13/16"	11 1/8"	11 7/16"	9 5/8"	10 1/16"	10 15/32"
TOTAL RUN OF STAIRWAY	20' 3/16"	12'10"	14'3 1/16"	15'9"	17'1 7/16"	18'6 1/2"	20' 3/16"	12'10"	14'3 1/16"	15'8 7/16"
WELL OPENING FOR 6'-8" HEADROOM	14'11"	9'10 3/32"	10'4 7/32"	11'9"	13' 3/4"	13'6 1/16"	14'11"	9'10 3/32"	10'4 7/32"	11'8 17/32"
ANGLE OF INCLINE	27°56'	39°17'	36°28'	33°41'	31°44'	29°49'	27°56'	39°17'	36°28'	33°53'
LENGTH OF CARRIAGE	22'7 27/32"	16'6 31/32"	17'8 23/32"	18'11 5/32"	20'1 9/16"	21'4 7/16"	22'7 27/32"	16'6 31/32"	17'8 23/32"	18'11"
L. F. OF RISER PER INCH OF STAIRWAY WIDTH	1.833	1.416	1.5	1.583	1.666	1.75	1.833	1.416	1.5	1.583
L. F. OF TREAD PER INCH OF STAIRWAY WIDTH	1.75	1.333	1.416	1.5	1.583	1.666	1.75	1.333	1.416	1.5
FRAMING SQUARE SETTINGS FOR CARRIAGE CUTS — TONGUE	6 1/16"	7 7/8"	7 7/16"	7"	6 11/16"	6 3/8"	6 1/16"	7 7/8"	7 7/16"	7 1/32"
FRAMING SQUARE SETTINGS FOR CARRIAGE CUTS — BODY	11 7/16"	9 5/8"	10 1/16"	10 1/2"	10 13/16"	11 1/8"	11 7/16"	9 5/8"	10 1/16"	10 15/32"

FLOOR TO FLOOR RISE		11'1$\frac{5}{8}$"	11'1$\frac{5}{8}$"	11'1$\frac{5}{8}$"	11'1$\frac{3}{4}$"	11'1$\frac{3}{4}$"	11'1$\frac{3}{4}$"	11'1$\frac{3}{4}$"	11'1$\frac{3}{4}$"	11'1$\frac{3}{4}$"	11'1$\frac{7}{8}$"
NUMBER OF RISERS		20	21	22	17	18	19	20	21	22	17
HEIGHT OF EACH RISER		6$\frac{11}{16}$"	6$\frac{3}{8}$"	6$\frac{1}{16}$"	7$\frac{7}{8}$"	7$\frac{7}{16}$"	7$\frac{1}{16}$"	6$\frac{11}{16}$"	6$\frac{3}{8}$"	6$\frac{1}{16}$"	7$\frac{7}{8}$"
TOTAL HEIGHT OF RISERS		11'1$\frac{3}{4}$"	11'1$\frac{7}{8}$"	11'1$\frac{3}{8}$"	11'1$\frac{7}{8}$"	11'1$\frac{7}{8}$"	11'2$\frac{3}{16}$"	11'1$\frac{3}{4}$"	11'1$\frac{7}{8}$"	11'1$\frac{3}{8}$"	11'1$\frac{7}{8}$"
RISERS OVERAGE (+) OR UNDERAGE (—)		+$\frac{1}{8}$"	+$\frac{1}{4}$"	−$\frac{1}{4}$"	+$\frac{1}{8}$"	+$\frac{1}{8}$"	+$\frac{7}{16}$"	0	+$\frac{1}{8}$"	−$\frac{3}{8}$"	0
NUMBER OF TREADS		19	20	21	16	17	18	19	20	21	16
WIDTH OF EACH TREAD		10$\frac{13}{16}$"	11$\frac{1}{8}$"	11$\frac{7}{16}$"	9$\frac{5}{8}$"	10$\frac{1}{16}$"	10$\frac{7}{16}$"	10$\frac{13}{16}$"	11$\frac{1}{8}$"	11$\frac{7}{16}$"	9$\frac{5}{8}$"
TOTAL RUN OF STAIRWAY		17'1$\frac{7}{16}$"	18'6$\frac{1}{2}$"	20'$\frac{3}{16}$"	12'10"	14'3$\frac{1}{16}$"	15'7$\frac{7}{8}$"	17'1$\frac{7}{16}$"	18'6$\frac{1}{2}$"	20'$\frac{3}{16}$"	12'10"
WELL OPENING FOR 6'-8" HEADROOM		13'$\frac{3}{4}$"	13'6$\frac{1}{16}$"	14'11"	9'10$\frac{3}{32}$"	10'4$\frac{7}{32}$"	11'8$\frac{1}{32}$"	13'$\frac{3}{4}$"	13'6$\frac{1}{16}$"	14'11"	9'10$\frac{3}{32}$"
ANGLE OF INCLINE		31°44'	29°49'	27°56'	39°17'	36°28'	34°5'	31°44'	29°49'	27°56'	39°17'
LENGTH OF CARRIAGE		20'1$\frac{9}{16}$"	21'4$\frac{7}{16}$"	22'7$\frac{27}{32}$"	16'6$\frac{31}{32}$"	17'8$\frac{23}{32}$"	18'10$\frac{27}{32}$"	20'1$\frac{9}{16}$"	21'4$\frac{7}{16}$"	22'7$\frac{27}{32}$"	16'6$\frac{31}{32}$"
L. F. OF RISER PER INCH OF STAIRWAY WIDTH		1.666	1.75	1.833	1.416	1.5	1.583	1.666	1.75	1.833	1.416
L. F. OF TREAD PER INCH OF STAIRWAY WIDTH		1.583	1.666	1.75	1.333	1.416	1.5	1.583	1.666	1.75	1.333
FRAMING SQUARE SETTINGS FOR CARRIAGE CUTS	TONGUE	6$\frac{11}{16}$"	6$\frac{3}{8}$"	6$\frac{1}{16}$"	7$\frac{7}{8}$"	7$\frac{7}{16}$"	7$\frac{1}{16}$"	6$\frac{11}{16}$"	6$\frac{3}{8}$"	6$\frac{1}{16}$"	7$\frac{7}{8}$"
	BODY	10$\frac{13}{16}$"	11$\frac{1}{8}$"	11$\frac{7}{16}$"	9$\frac{5}{8}$"	10$\frac{1}{16}$"	10$\frac{7}{16}$"	10$\frac{13}{16}$"	11$\frac{1}{8}$"	11$\frac{7}{16}$"	9$\frac{5}{8}$"

FLOOR TO FLOOR RISE		11'1 7/8"	11'1 7/8"	11'1 7/8"	11'1 7/8"	11'1 7/8"	11'2"	11'2"	11'2"	11'2"	11'2"
NUMBER OF RISERS		18	19	20	21	22	17	18	19	20	21
HEIGHT OF EACH RISER		7 7/16"	7 1/16"	6 11/16"	6 3/8"	6 1/16"	7 7/8"	7 7/16"	7 1/16"	6 11/16"	6 3/8"
TOTAL HEIGHT OF RISERS		11'1 7/8"	11'2 3/16"	11'1 3/4"	11'1 7/8"	11'1 3/8"	11'1 7/8"	11'1 7/8"	11'2 3/16"	11'1 3/4"	11'1 7/8"
RISERS OVERAGE (+) OR UNDERAGE (—)		0	+5/16"	−1/8"	0	−1/2"	−1/8"	−1/8"	+3/16"	−1/4"	−1/8"
NUMBER OF TREADS		17	18	19	20	21	16	17	18	19	20
WIDTH OF EACH TREAD		10 1/16"	10 7/16"	10 13/16"	11 1/8"	11 7/16"	9 5/8"	10 1/16"	10 7/16"	10 13/16"	11 1/8"
TOTAL RUN OF STAIRWAY		14'3 1/16"	15'7 7/8"	17'1 7/16"	18'6 1/2"	20' 3/16"	12'10"	14'3 1/16"	15'7 7/8"	17'1 7/16"	18'6 1/2"
WELL OPENING FOR 6'-8" HEADROOM		10'4 7/32"	11'8 1/32"	13' 3/4"	13'6 1/16"	14'11"	9'10 3/32"	10'4 7/32"	11'8 1/32"	13' 3/4"	13'6 1/16"
ANGLE OF INCLINE		36°28'	34°5'	31°44'	29°49'	27°56'	39°17'	36°28'	34°5'	31°44'	29°49'
LENGTH OF CARRIAGE		17'8 23/32"	18'10 27/32"	20'1 9/16"	21'4 7/16"	22'7 27/32"	16'6 31/32"	17'8 23/32"	18'10 27/32"	20'1 9/16"	21'4 7/16"
L. F. OF RISER PER INCH OF STAIRWAY WIDTH		1.5	1.583	1.666	1.75	1.833	1.416	1.5	1.583	1.666	1.75
L. F. OF TREAD PER INCH OF STAIRWAY WIDTH		1.416	1.5	1.583	1.666	1.75	1.333	1.416	1.5	1.583	1.666
FRAMING SQUARE SETTINGS FOR CARRIAGE CUTS	TONGUE	7 7/16"	7 1/16"	6 11/16"	6 3/8"	6 1/16"	7 7/8"	7 7/16"	7 1/16"	6 11/16"	6 3/8"
	BODY	10 1/16"	10 7/16"	10 13/16"	11 1/8"	11 7/16"	9 5/8"	10 1/16"	10 7/16"	10 13/16"	11 1/8"

FLOOR TO FLOOR RISE	11'2"	11'2$\frac{1}{8}$"	11'2$\frac{1}{8}$"	11'2$\frac{1}{8}$"	11'2$\frac{1}{8}$"	11'2$\frac{1}{8}$"	11'2$\frac{1}{8}$"	11'2$\frac{1}{4}$"	11'2$\frac{1}{4}$"	11'2$\frac{1}{4}$"
NUMBER OF RISERS	22	17	18	19	20	21	22	17	18	19
HEIGHT OF EACH RISER	6$\frac{3}{32}$"	7$\frac{7}{8}$"	7$\frac{7}{16}$"	7$\frac{1}{16}$"	6$\frac{11}{16}$"	6$\frac{3}{8}$"	6$\frac{3}{32}$"	7$\frac{7}{8}$"	7$\frac{7}{16}$"	7$\frac{1}{16}$"
TOTAL HEIGHT OF RISERS	11'2$\frac{1}{16}$"	11'1$\frac{7}{8}$"	11'1$\frac{7}{8}$"	11'2$\frac{3}{16}$"	11'1$\frac{3}{4}$"	11'1$\frac{7}{8}$"	11'2$\frac{1}{16}$"	11'1$\frac{7}{8}$"	11'1$\frac{7}{8}$"	11'2$\frac{3}{16}$"
RISERS OVERAGE (+) OR UNDERAGE (—)	+$\frac{1}{16}$"	−$\frac{1}{4}$"	−$\frac{1}{4}$"	+$\frac{1}{16}$"	−$\frac{3}{8}$"	−$\frac{1}{4}$"	−$\frac{1}{16}$"	−$\frac{3}{8}$"	−$\frac{3}{8}$"	−$\frac{1}{16}$"
NUMBER OF TREADS	21	16	17	18	19	20	21	16	17	18
WIDTH OF EACH TREAD	11$\frac{13}{32}$"	9$\frac{5}{8}$"	10$\frac{1}{16}$"	10$\frac{7}{16}$"	10$\frac{13}{16}$"	11$\frac{1}{8}$"	11$\frac{13}{32}$"	9$\frac{5}{8}$"	10$\frac{1}{16}$"	10$\frac{7}{16}$"
TOTAL RUN OF STAIRWAY	19'11$\frac{17}{32}$"	12'10"	14'3$\frac{1}{16}$"	15'7$\frac{7}{8}$"	17'1$\frac{7}{16}$"	18'6$\frac{1}{2}$"	19'11$\frac{17}{32}$"	12'10"	14'3$\frac{1}{16}$"	15'7$\frac{7}{8}$"
WELL OPENING FOR 6'-8" HEADROOM	14'10$\frac{13}{32}$"	9'10$\frac{3}{32}$"	10'4$\frac{7}{32}$"	11'8$\frac{1}{32}$"	13' $\frac{3}{4}$"	13'6$\frac{1}{16}$"	14'10$\frac{13}{32}$"	9'10$\frac{3}{32}$"	10'4$\frac{7}{32}$"	11'8$\frac{1}{32}$"
ANGLE OF INCLINE	28°7'	39°17'	36°28'	34°5'	31°44'	29°49'	38°7'	39°17'	36°28'	34°5'
LENGTH OF CARRIAGE	22'7$\frac{9}{16}$"	16'6$\frac{31}{32}$"	17'8$\frac{23}{32}$"	18'10$\frac{27}{32}$"	20'1$\frac{9}{16}$"	21'4$\frac{7}{16}$"	22'7$\frac{9}{16}$"	16'6$\frac{31}{32}$"	17'8$\frac{23}{32}$"	18'10$\frac{27}{32}$"
L. F. OF RISER PER INCH OF STAIRWAY WIDTH	1.833	1.416	1.5	1.583	1.666	1.75	1.833	1.416	1.5	1.583
L. F. OF TREAD PER INCH OF STAIRWAY WIDTH	1.75	1.333	1.416	1.5	1.583	1.666	1.75	1.333	1.416	1.5
FRAMING SQUARE SETTINGS FOR CARRIAGE CUTS — TONGUE	6$\frac{3}{32}$"	7$\frac{7}{8}$"	7$\frac{7}{16}$"	7$\frac{1}{16}$"	6$\frac{11}{16}$"	6$\frac{3}{8}$"	6$\frac{3}{32}$"	7$\frac{7}{8}$"	7$\frac{7}{16}$"	7$\frac{1}{16}$"
FRAMING SQUARE SETTINGS FOR CARRIAGE CUTS — BODY	11$\frac{13}{32}$"	9$\frac{5}{8}$"	10$\frac{1}{16}$"	10$\frac{7}{16}$"	10$\frac{13}{16}$"	11$\frac{1}{8}$"	11$\frac{13}{32}$"	9$\frac{5}{8}$"	10$\frac{1}{16}$"	10$\frac{7}{16}$"

FLOOR TO FLOOR RISE		11'2 1/4"	11'2 1/4"	11'2 1/4"	11'2 3/8"	11'2 3/8"	11'2 3/8"	11'2 3/8"	11'2 3/8"	11'2 3/8"	11'2 1/2"
NUMBER OF RISERS		20	21	22	17	18	19	20	21	22	17
HEIGHT OF EACH RISER		6 11/16"	6 3/8"	6 1/8"	7 7/8"	7 7/16"	7 1/16"	6 23/32"	6 3/8"	6 1/8"	7 15/16"
TOTAL HEIGHT OF RISERS		11'1 3/4"	11'1 7/8"	11'2 3/4"	11'1 7/8"	11'1 7/8"	11'2 3/16"	11'2 3/8"	11'1 7/8"	11'2 3/4"	11'2 15/16"
RISERS OVERAGE (+) OR UNDERAGE (—)		−1/2"	−3/8"	+1/2"	−1/2"	−1/2"	−3/16"	0	−1/2"	+3/8"	+7/16"
NUMBER OF TREADS		19	20	21	16	17	18	19	20	21	16
WIDTH OF EACH TREAD		10 13/16"	11 1/8"	11 3/8"	9 5/8"	10 1/16"	10 7/16"	10 25/32"	11 1/8"	11 3/8"	9 9/16"
TOTAL RUN OF STAIRWAY		17'1 7/16"	18'6 1/2"	19'10 7/8"	12'10"	14'3 1/16"	15'7 7/8"	17' 27/32"	18'6 1/2"	19'10 7/8"	12'9"
WELL OPENING FOR 6'-8" HEADROOM		13' 3/4"	13'6 1/16"	14'9 13/16"	9'10 3/32"	10'4 7/32"	11'8 1/32"	13' 7/32"	13'6 1/16"	14'9 13/16"	9'9 1/4"
ANGLE OF INCLINE		31°44'	29°49'	28°18'	39°17'	36°28'	34°5'	31°56'	29°49'	28°18'	39°42'
LENGTH OF CARRIAGE		20'1 9/16"	21'4 7/16"	22'7 5/16"	16'6 31/32"	17'8 23/32"	18'10 27/32"	20'1 3/8"	21'4 7/16"	22'7 5/16"	16'6 27/32"
L. F. OF RISER PER INCH OF STAIRWAY WIDTH		1.666	1.75	1.833	1.416	1.5	1.583	1.666	1.75	1.833	1.416
L. F. OF TREAD PER INCH OF STAIRWAY WIDTH		1.583	1.666	1.75	1.333	1.416	1.5	1.583	1.666	1.75	1.333
FRAMING SQUARE SETTINGS FOR CARRIAGE CUTS	TONGUE	6 11/16"	6 3/8"	6 1/8"	7 7/8"	7 7/16"	7 1/16"	6 23/32"	6 3/8"	6 1/8"	7 15/16"
	BODY	10 13/16"	11 1/8"	11 3/8"	9 5/8"	10 1/16"	10 7/16"	10 25/32"	11 1/8"	11 3/8"	9 9/16"

FLOOR TO FLOOR RISE		11'2½"	11'2½"	11'2½"	11'2½"	11'2½"	11'2⅝"	11'2⅝"	11'2⅝"	11'2⅝"	11'2⅝"
NUMBER OF RISERS		18	19	20	21	22	17	18	19	20	21
HEIGHT OF EACH RISER		7½"	7 1/16"	6¾"	6 13/32"	6⅛"	7 15/16"	7½"	7 1/16"	6¾"	6 13/32"
TOTAL HEIGHT OF RISERS		11'3"	11'2 3/16"	11'3"	11'2 17/32"	11'2¾"	11'2 15/16"	11'3"	11'2 3/16"	11'3"	11'2 17/32"
RISERS OVERAGE (+) OR UNDERAGE (—)		+½"	−5/16"	+½"	+1/32"	+¼"	+5/16"	+⅜"	−7/16"	+⅜"	−3/32"
NUMBER OF TREADS		17	18	19	20	21	16	17	18	19	20
WIDTH OF EACH TREAD		10"	10 7/16"	10¾"	11 3/32"	11⅜"	9 9/16"	10"	10 7/16"	10¾"	11 3/32"
TOTAL RUN OF STAIRWAY		14'2"	15'7⅞"	17'¼"	18'5⅞"	19'10⅞"	12'9"	14'2"	15'7⅞"	17'¼"	18'5⅞"
WELL OPENING FOR 6'-8" HEADROOM		10'3 11/32"	11'8 1/32"	12'11 11/6"	13'5 17/32"	14'9 13/16"	9'9¼"	10'3 11/32"	11'8 1/32"	12'11 11/16"	13'5 17/32"
ANGLE OF INCLINE		36°52'	34°5'	32°7'	30°0'	28°18'	39°42'	36°52'	34°5'	32°7'	30°0'
LENGTH OF CARRIAGE		17'8½"	18'10 27/32"	20'1 3/16"	21'4 7/32"	22'7 5/16"	16'6 27/32"	17'8½"	18'10 27/32"	20'1 3/16"	21'4 7/32"
L. F. OF RISER PER INCH OF STAIRWAY WIDTH		1.5	1.583	1.666	1.75	1.833	1.416	1.5	1.583	1.666	1.75
L. F. OF TREAD PER INCH OF STAIRWAY WIDTH		1.416	1.5	1.583	1.666	1.75	1.333	1.416	1.5	1.583	1.666
FRAMING SQUARE SETTINGS FOR CARRIAGE CUTS	TONGUE	7½"	7 1/16"	6¾"	6 13/32"	6⅛"	7 15/16"	7½"	7 1/16"	6¾"	6 13/32"
	BODY	10"	10 7/16"	10¾"	11 3/32"	11⅜"	9 9/16"	10"	10 7/16"	10¾"	11 3/32"

FLOOR TO FLOOR RISE	11'2 5/8"	11'2 3/4"	11'2 3/4"	11'2 3/4"	11'2 3/4"	11'2 3/4"	11'2 3/4"	11'2 7/8"	11'2 7/8"	11'2 7/8"
NUMBER OF RISERS	22	17	18	19	20	21	22	17	18	19
HEIGHT OF EACH RISER	6 1/8"	7 15/16"	7 1/2"	7 3/32"	6 3/4"	6 7/16"	6 1/8"	7 15/16"	7 1/2"	7 1/8"
TOTAL HEIGHT OF RISERS	11'2 3/4"	11'2 15/16"	11'3"	11'2 25/32"	11'3"	11'3 3/16"	11'2 3/4"	11'2 15/16"	11'3"	11'3 3/8"
RISERS OVERAGE (+) OR UNDERAGE (—)	+1/8"	+3/16"	+1/4"	+1/32"	+1/4"	+7/16"	0	+1/16"	+1/8"	+1/2"
NUMBER OF TREADS	21	16	17	18	19	20	21	16	17	18
WIDTH OF EACH TREAD	11 3/8"	9 9/16"	10"	10 13/32"	10 3/4"	11 1/16"	11 3/8"	9 9/16"	10"	10 3/8"
TOTAL RUN OF STAIRWAY	19'10 7/8"	12'9"	14'2"	15'7 5/16"	17' 1/4"	18'5 1/4"	19'10 7/8"	12'9"	14'2"	15'6 3/4"
WELL OPENING FOR 6'-8" HEADROOM	14'9 13/16"	9'9 1/4"	10'3 11/32"	11'7"	12'11 11/16"	13'5"	14'9 13/16"	9'9 1/4"	10'3 11/32"	11'7 1/16"
ANGLE OF INCLINE	28°18'	39°42'	36°52'	34°17'	32°7'	30°12'	28°18'	39°42'	36°52'	34°29'
LENGTH OF CARRIAGE	22'7 5/16"	16'6 27/32"	17'8 1/2"	18'10 11/16"	20'1 3/16"	21'4"	22'7 5/16"	16'6 27/32"	17'8 1/2"	18'10 17/32"
L. F. OF RISER PER INCH OF STAIRWAY WIDTH	1.833	1.416	1.5	1.583	1.666	1.75	1.833	1.416	1.5	1.583
L. F. OF TREAD PER INCH OF STAIRWAY WIDTH	1.75	1.333	1.416	1.5	1.583	1.666	1.75	1.333	1.416	1.5
FRAMING SQUARE SETTINGS FOR CARRIAGE CUTS — TONGUE	6 1/8"	7 15/16"	7 1/2"	7 3/32"	6 3/4"	6 7/16"	6 1/8"	7 15/16"	7 1/2"	7 1/8"
FRAMING SQUARE SETTINGS FOR CARRIAGE CUTS — BODY	11 3/8"	9 9/16"	10"	10 13/32"	10 3/4"	11 1/16"	11 3/8"	9 9/16"	10"	10 3/8"

FLOOR TO FLOOR RISE		11'2 7/8"	11'2 7/8"	11'2 7/8"	11'3"	11'3"	11'3"	11'3"	11'3"	11'3"	11'3 1/8"
NUMBER OF RISERS		20	21	22	17	18	19	20	21	22	17
HEIGHT OF EACH RISER		6 3/4"	6 7/16"	6 1/8"	7 15/16"	7 1/2"	7 1/8"	6 3/4"	6 7/16"	6 1/8"	7 15/16"
TOTAL HEIGHT OF RISERS		11'3"	11'3 3/16"	11'2 3/4"	11'2 15/16"	11'3"	11'3 3/8"	11'3"	11'3 3/16"	11'2 3/4"	11'2 15/16"
RISERS OVERAGE (+) OR UNDERAGE (—)		+1/8"	+5/16"	−1/8"	−1/16"	0	+3/8"	0	+3/16"	−1/4"	−3/16"
NUMBER OF TREADS		19	20	21	16	17	18	19	20	21	16
WIDTH OF EACH TREAD		10 3/4"	11 1/16"	11 3/8"	9 9/16"	10"	10 3/8"	10 3/4"	11 1/16"	11 3/8"	9 9/16"
TOTAL RUN OF STAIRWAY		17'1/4"	18'5 1/4"	19'10 7/8"	12'9"	14'2"	15'6 3/4"	17'1/4"	18'5 1/4"	19'10 7/8"	12'9"
WELL OPENING FOR 6'-8" HEADROOM		12'11 11/16"	13'5"	14'9 13/16"	9'9 1/4"	10'3 11/32"	11'7 1/16"	12'11 11/16"	13'5"	14'9 13/16"	9'9 1/4"
ANGLE OF INCLINE		32°7'	30°12'	28°18'	39°42'	36°52'	34°29'	32°7'	30°12'	28°18'	39°42'
LENGTH OF CARRIAGE		20'1 3/16"	21'4"	22'7 5/16"	16'6 27/32"	17'8 1/2"	18'10 17/32"	20'1 3/16"	21'4"	22'7 5/16"	16'6 27/32"
L. F. OF RISER PER INCH OF STAIRWAY WIDTH		1.666	1.75	1.833	1.416	1.5	1.583	1.666	1.75	1.833	1.416
L. F. OF TREAD PER INCH OF STAIRWAY WIDTH		1.583	1.666	1.75	1.333	1.416	1.5	1.583	1.66	1.75	1.333
FRAMING SQUARE SETTINGS FOR CARRIAGE CUTS	TONGUE	6 3/4"	6 7/16"	6 1/8"	7 15/16"	7 1/2"	7 1/8"	6 3/4"	6 7/16"	6 1/8"	7 15/16"
	BODY	10 3/4"	11 1/16"	11 3/8"	9 9/16"	10"	10 3/8"	10 3/4"	11 1/16"	11 3/8"	9 9/16"

FLOOR TO FLOOR RISE		11'3 1/8"	11'3 1/8"	11'3 1/8"	11'3 1/8"	11'3 1/8"	11'3 1/4"	11'3 1/4"	11'3 1/4"	11'3 1/4"	11'3 1/4"
NUMBER OF RISERS		18	19	20	21	22	17	18	19	20	21
HEIGHT OF EACH RISER		7 1/2"	7 1/8"	6 3/4"	6 7/16"	6 1/8"	7 15/16"	7 1/2"	7 1/8"	6 3/4"	6 7/16"
TOTAL HEIGHT OF RISERS		11'3"	11'3 3/8"	11'3"	11'3 3/16"	11'2 3/4"	11'2 15/16"	11'3"	11'3 3/8"	11'3"	11'3 3/16"
RISERS OVERAGE (+) OR UNDERAGE (—)		−1/8"	+1/4"	−1/8"	+1/16"	−3/8"	−5/16"	−1/4"	+1/8"	−1/4"	−1/16"
NUMBER OF TREADS		17	18	19	20	21	16	17	18	19	20
WIDTH OF EACH TREAD		10"	10 3/8"	10 3/4"	11 1/16"	11 3/8"	9 9/16"	10"	10 3/8"	10 3/4"	11 1/16"
TOTAL RUN OF STAIRWAY		14'2"	15'6 3/4"	17' 1/4"	18'5 1/4"	19'10 7/8"	12'9"	14'2"	15'6 3/4"	17' 1/4"	18'5 1/4"
WELL OPENING FOR 6'-8" HEADROOM		10'3 11/32"	11'7 1/16"	12'11 11/16"	13'5"	14'9 13/16"	9'9 1/4"	10'3 11/32"	11'7 1/16"	12'11 11/16"	13'5"
ANGLE OF INCLINE		36°52'	34°29'	32°7'	30°12'	28°18'	39°42'	36°52'	34°29'	32°7'	30°12'
LENGTH OF CARRIAGE		17'8 1/2"	18'10 17/32"	20'1 3/16"	21'4"	22'7 5/16"	16'6 27/32"	17'8 1/2"	18'10 17/32"	20'1 3/16"	21'4"
L. F. OF RISER PER INCH OF STAIRWAY WIDTH		1.5	1.583	1.666	1.75	1.833	1.416	1.5	1.583	1.666	1.75
L. F. OF TREAD PER INCH OF STAIRWAY WIDTH		1.416	1.5	1.583	1.666	1.75	1.333	1.416	1.5	1.583	1.666
FRAMING SQUARE SETTINGS FOR CARRIAGE CUTS	TONGUE	7 1/2"	7 1/8"	6 3/4"	6 7/16"	6 1/8"	7 15/16"	7 1/2"	7 1/8"	6 3/4"	6 7/16"
	BODY	10"	10 3/8"	10 3/4"	11 1/16"	11 3/8"	9 9/16"	10"	10 3/8"	10 3/4"	11 1/16"

FLOOR TO FLOOR RISE	11'3 1/4"	11'3 3/8"	11'3 3/8"	11'3 3/8"	11'3 3/8"	11'3 3/8"	11'3 3/8"	11'3 1/2"	11'3 1/2"	11'3 1/2"
NUMBER OF RISERS	22	17	18	19	20	21	22	17	18	19
HEIGHT OF EACH RISER	6 1/8"	7 15/16"	7 1/2"	7 1/8"	6 3/4"	6 7/16"	6 5/32"	8"	7 1/2"	7 1/8"
TOTAL HEIGHT OF RISERS	11'2 3/4"	11'2 15/16"	11'3"	11'3 3/8"	11'3"	11'3 3/16"	11'3 7/16"	11'4"	11'3"	11'3 3/8"
RISERS OVERAGE (+) OR UNDERAGE (—)	−1/2"	−7/16"	−3/8"	0	−3/8"	−3/16"	+1/16"	+1/2"	−1/2"	−1/8"
NUMBER OF TREADS	21	16	17	18	19	20	21	16	17	18
WIDTH OF EACH TREAD	11 3/8"	9 9/16"	10"	10 3/8"	10 3/4"	11 1/16"	11 11/32"	9 1/2"	10"	10 3/8"
TOTAL RUN OF STAIRWAY	19'10 7/8"	12'9"	14'2"	15'6 3/4"	17' 1/4"	18'5 1/4"	19'10 7/32"	12'8"	14'2"	15'6 3/4"
WELL OPENING FOR 6'-8" HEADROOM	14'9 13/16"	9'9 1/4"	10'3 11/32"	11'7 1/16"	12'11 11/16"	13'5"	14'9 1/4"	8'10 7/8"	10'3 11/32"	11'7 1/16"
ANGLE OF INCLINE	28°18'	39°42'	36°52'	34°29'	32°7'	30°12'	28°29'	40°6'	36°52'	34°29'
LENGTH OF CARRIAGE	22'7 5/16"	16'6 27/32"	17'8 1/2"	18'10 17/32"	20'1 3/16"	21'4"	22'7 1/32"	16'6 23/32"	17'8 1/2"	18'10 17/32"
L. F. OF RISER PER INCH OF STAIRWAY WIDTH	1.833	1.416	1.5	1.583	1.666	1.75	1.833	1.416	1.5	1.583
L. F. OF TREAD PER INCH OF STAIRWAY WIDTH	1.75	1.333	1.416	1.4	1.583	1.666	1.75	1.333	1.416	1.5
FRAMING SQUARE SETTINGS FOR CARRIAGE CUTS — TONGUE	6 1/8"	7 15/16"	7 1/2"	7 1/8"	6 3/4"	6 7/16"	6 5/32"	8"	7 1/2"	7 1/8"
FRAMING SQUARE SETTINGS FOR CARRIAGE CUTS — BODY	11 3/8"	9 9/16"	10"	10 3/8"	10 3/4"	11 1/16"	11 11/32"	9 1/2"	10"	10 3/8"

FLOOR TO FLOOR RISE	11'3 1/2"	11'3 1/2"	11'3 1/2"	11'3 5/8"	11'3 5/8"	11'3 5/8"	11'3 5/8"	11'3 5/8"	11'3 5/8"	11'3 3/4"
NUMBER OF RISERS	20	21	22	17	18	19	20	21	22	17
HEIGHT OF EACH RISER	6 3/4"	6 7/16"	6 5/32"	8"	7 9/16"	7 1/8"	6 25/32"	6 7/16"	6 3/16"	8"
TOTAL HEIGHT OF RISERS	11'3"	11'3 3/16"	11'3 7/16"	11'4"	11'4 1/8"	11'3 3/8"	11'3 5/8"	11'3 3/16"	11'4 1/8"	11'4"
RISERS OVERAGE (+) OR UNDERAGE (—)	−1/2"	−5/16"	−1/16"	+3/8"	+1/2"	−1/4"	0	−7/16"	+1/2"	+1/4"
NUMBER OF TREADS	19	20	21	16	17	18	19	20	21	16
WIDTH OF EACH TREAD	10 3/4"	11 1/16"	11 11/32"	9 1/2"	9 15/16"	10 3/8"	10 23/32"	11 1/16"	11 5/16"	9 1/2"
TOTAL RUN OF STAIRWAY	17' 1/4"	18'5 1/4"	19'10 7/32"	12'8"	14' 15/16"	15'6 3/4"	16'11 21/32"	18'5 1/4"	19'9 9/16"	12'8"
WELL OPENING FOR 6'-8" HEADROOM	12'11 11/16"	13'5"	14'9 1/4"	8'10 7/8"	10'2 7/16"	11'7 1/16"	12' 7/16"	13'5"	14'8 21/32"	8'10 7/8"
ANGLE OF INCLINE	32°7'	30°12'	28°29'	40°6'	37°16'	34°29'	32°19'	30°12'	28°41'	40°6'
LENGTH OF CARRIAGE	20'1 3/16"	21'4"	22'7 1/32"	16'6 23/32"	17'8 9/32"	18'10 17/32"	20'1"	21'4"	22'6 25/32"	16'6 23/32"
L. F. OF RISER PER INCH OF STAIRWAY WIDTH	1.666	1.75	1.833	1.416	1.5	1.583	1.666	1.75	1.833	1.416
L. F. OF TREAD PER INCH OF STAIRWAY WIDTH	1.583	1.666	1.75	1.333	1.416	1.5	1.583	1.666	1.75	1.333
FRAMING SQUARE SETTINGS FOR CARRIAGE CUTS — TONGUE	6 3/4"	6 7/16"	6 5/32"	8"	7 9/16"	7 1/8"	6 25/32"	6 7/16"	6 3/16"	8"
FRAMING SQUARE SETTINGS FOR CARRIAGE CUTS — BODY	10 3/4"	11 1/16"	11 11/32"	9 1/2"	9 15/16"	10 3/8"	10 23/32"	11 1/16"	11 5/16"	9 1/2"

FLOOR TO FLOOR RISE		11'3 3/4"	11'3 3/4"	11'3 3/4"	11'3 3/4"	11'3 3/4"	11'3 7/8"	11'3 7/8"	11'3 7/8"	11'3 7/8"	11'3 7/8"
NUMBER OF RISERS		18	19	20	21	22	17	18	19	20	21
HEIGHT OF EACH RISER		7 9/16"	7 1/8"	6 13/16"	6 15/32"	6 3/16"	8"	7 9/16"	7 1/8"	6 13/16"	6 15/16"
TOTAL HEIGHT OF RISERS		11'4 1/8"	11'3 3/8"	11'4 1/4"	11'3 27/32"	11'4 1/8"	11'4"	11'4 1/8"	11'3 3/8"	11'4 1/4"	11'3 27/32"
RISERS OVERAGE (+) OR UNDERAGE (—)		+3/8"	−3/8"	+1/2"	+3/32"	+3/8"	+1/8"	+1/4"	−1/2"	+3/8"	−1/32"
NUMBER OF TREADS		17	18	19	20	21	16	17	18	19	20
WIDTH OF EACH TREAD		9 15/16"	10 3/8"	10 11/16"	11 1/32"	11 5/16"	9 1/2"	9 15/16"	10 3/8"	10 11/16"	11 1/32"
TOTAL RUN OF STAIRWAY		14'15/16"	15'6 3/4"	16'11 1/16"	18'4 5/8"	19'9 9/16"	12'8"	14'15/16"	15'6 3/4"	16'11 1/16"	18'4 5/8"
WELL OPENING FOR 6'-8" HEADROOM		10'2 7/16"	11'7 1/16"	11'11 15/16"	13'4 15/32"	14'8 21/32"	8'10'7/8"	10'2 7/16"	11'7 1/16"	11'11 15/16"	11'9 31/32"
ANGLE OF INCLINE		37°16'	34°29'	32°31'	30°23'	28°41'	40°6'	37°16'	34°29'	32°31'	30°23'
LENGTH OF CARRIAGE		17'8 9/32"	18'10 17/32"	20'13/16"	21'3 3/4"	22'6 25/32"	16'6 23/32"	17'8 9/32"	18'10 17/32"	20'13/16"	21'3 3/4"
L. F. OF RISER PER INCH OF STAIRWAY WIDTH		1.5	1.583	1.666	1.75	1.833	1.416	1.5	1.583	1.666	1.75
L. F. OF TREAD PER INCH OF STAIRWAY WIDTH		1.416	1.5	1.583	1.666	1.75	1.333	1.416	1.5	1.583	1.666
FRAMING SQUARE SETTINGS FOR CARRIAGE CUTS	TONGUE	7 9/16"	7 1/8"	6 13/16"	6 15/32"	6 3/16"	8"	7 9/16"	7 1/8"	6 13/16"	6 15/32"
	BODY	9 15/16"	10 3/8"	10 11/16"	11 1/32"	11 5/16"	9 1/2"	9 15/16"	10 3/8"	10 11/16"	11 1/32"

FLOOR TO FLOOR RISE	11'3 7/8"	11'4"	11'4"	11'4"	11'4"	11'4"	11'4"	11'4 1/8"	11'4 1/8"	11'4 1/8"
NUMBER OF RISERS	22	17	18	19	20	21	22	17	18	19
HEIGHT OF EACH RISER	6 3/16"	8"	7 9/16"	7 5/32"	6 13/16"	6 1/2"	6 3/16"	8"	7 9/16"	7 3/16"
TOTAL HEIGHT OF RISERS	11'4 1/8"	11'4"	11'4 1/8"	11'3 31/32"	11'4 1/4"	11'4 1/2"	11'4 1/8"	11'4"	11'4 1/8"	11'4 9/16"
RISERS OVERAGE (+) OR UNDERAGE (—)	+1/4"	0	+1/8"	−1/32"	+1/4"	+1/2"	+1/8"	−1/8"	0	+7/16"
NUMBER OF TREADS	21	16	17	18	19	20	21	16	17	18
WIDTH OF EACH TREAD	11 5/16"	9 1/2"	9 15/16"	10 11/32"	10 11/16"	11"	11 5/16"	9 1/2"	9 15/16"	10 5/16"
TOTAL RUN OF STAIRWAY	19'9 9/16"	12'8"	14'15/16"	15'6 3/16"	16'11 1/16"	18'4"	19'9 9/16"	12'8"	14'15/16"	15'5 5/8"
WELL OPENING FOR 6'-8" HEADROOM	14'8 21/32"	8'10 7/8"	10'2 7/16"	11'6 19/32"	11'11 15/16"	13'3 15/16"	14'8 21/32"	8'10 7/8"	10'2 7/16"	11'6 3/32"
ANGLE OF INCLINE	28°41'	40°6'	37°16'	34°41'	32°31'	30°35'	28°41'	40°6'	37°16'	34°53'
LENGTH OF CARRIAGE	22'6 25/32"	16'6 23/32"	17'8 9/32"	18'10 13/32"	20'13/16"	21'3 17/32"	22'6 25/32"	16'6 23/32"	17'8 9/32"	18'10 1/4"
L. F. OF RISER PER INCH OF STAIRWAY WIDTH	1.833	1.416	1.5	1.583	1.666	1.75	1.833	1.416	1.5	1.583
L. F. OF TREAD PER INCH OF STAIRWAY WIDTH	1.75	1.333	1.416	1.5	1.583	1.666	1.75	1.333	1.416	1.5
FRAMING SQUARE SETTINGS FOR CARRIAGE CUTS — TONGUE	6 3/16"	8"	7 9/16"	7 5/32"	6 13/16"	6 1/2"	6 3/16"	8"	7 9/16"	7 3/16"
FRAMING SQUARE SETTINGS FOR CARRIAGE CUTS — BODY	11 5/16"	9 1/2"	9 15/16"	10 11/32"	10 11/16"	11"	11 5/16"	9 1/2"	9 15/16"	10 5/16"

FLOOR TO FLOOR RISE	11'4 1/8"	11'4 1/8"	11'4 1/8"	11'4 1/4"	11'4 1/4"	11'4 1/4"	11'4 1/4"	11'4 1/4"	11'4 1/4"	11'4 3/8"
NUMBER OF RISERS	20	21	22	17	18	19	20	21	22	17
HEIGHT OF EACH RISER	6 13/16"	6 1/2"	6 3/16"	8"	7 9/16"	7 3/16"	6 13/16"	6 1/2"	6 3/16"	8"
TOTAL HEIGHT OF RISERS	11'4 1/4"	11'4 1/2"	11'4 1/8"	11'4"	11'4 1/8"	11'4 9/16"	11'4 1/4"	11'4 1/2"	11'4 1/8"	11'4"
RISERS OVERAGE (+) OR UNDERAGE (—)	+1/8"	+3/8"	0	−1/4"	−1/8"	+5/16"	0	+1/4"	−1/8"	−3/8"
NUMBER OF TREADS	19	20	21	16	17	18	19	20	21	16
WIDTH OF EACH TREAD	10 11/16"	11"	11 5/16"	9 1/2"	9 15/16"	10 5/16"	10 11/16"	11"	11 5/16"	9 1/2"
TOTAL RUN OF STAIRWAY	16'11 1/16"	18'4"	19'9 9/16"	12'8"	14' 15/16"	15'5 5/8"	16'11 1/16"	18'4"	19'9 9/16"	12'8"
WELL OPENING FOR 6'-8" HEADROOM	11'11 15/16"	13'3 15/16"	14'8 21/32"	8'10 7/8"	10'2 7/16"	11'6 3/32"	11'11 15/16"	13'3 15/16"	14'8 21/32"	8'10 7/8"
ANGLE OF INCLINE	32°31'	30°35'	28°41'	40°6'	37°16'	34°53'	32°31'	30°35'	28°41'	40°6'
LENGTH OF CARRIAGE	20' 13/16"	21'3 17/32"	22'6 25/32"	16'6 23/32"	17'8 9/32"	18'10 1/4"	20' 13/16"	21'3 17/32"	22'6 25/32"	16'6 23/32"
L. F. OF RISER PER INCH OF STAIRWAY WIDTH	1.666	1.75	1.833	1.416	1.5	1.583	1.666	1.75	1.833	1.416
L. F. OF TREAD PER INCH OF STAIRWAY WIDTH	1.583	1.666	1.75	1.333	1.416	1.5	1.583	1.666	1.75	1.333
FRAMING SQUARE SETTINGS FOR CARRIAGE CUTS — TONGUE	6 13/16"	6 1/2"	6 3/16"	8"	7 9/16"	7 3/16"	6 13/16"	6 1/2"	6 3/16"	8"
FRAMING SQUARE SETTINGS FOR CARRIAGE CUTS — BODY	10 11/16"	11"	11 5/16"	9 1/2"	9 15/16"	10 5/16"	10 11/16"	11"	11 5/16"	9 1/2"

FLOOR TO FLOOR RISE		11'4 3/8"	11'4 3/8"	11'4 3/8"	11'4 3/8"	11'4 3/8"	11'4 1/2"	11'4 1/2"	11'4 1/2"	11'4 1/2"	11'4 1/2"
NUMBER OF RISERS		18	19	20	21	22	17	18	19	20	21
HEIGHT OF EACH RISER		7 9/16"	7 3/16"	6 13/16"	6 1/2"	6 3/16"	8"	7 9/16"	7 3/16"	6 13/16"	6 1/2"
TOTAL HEIGHT OF RISERS		11'4 1/8"	11'4 9/16"	11'4 1/4"	11'4 1/2"	11'4 1/8"	11'4"	11'4 1/8"	11'4 9/16"	11'4 1/4"	11'4 1/2"
RISERS OVERAGE (+) OR UNDERAGE (—)		−1/4"	+3/16"	−1/8"	+1/8"	−1/4"	−1/2"	−3/8"	+1/16"	−1/4"	0
NUMBER OF TREADS		17	18	19	20	21	16	17	18	19	20
WIDTH OF EACH TREAD		9 15/16"	10 5/16"	10 11/16"	11"	11 5/16"	9 1/2"	9 15/16"	10 5/16"	10 11/16"	11"
TOTAL RUN OF STAIRWAY		14'15/16"	15'5 5/8"	16'11 1/16"	18'4"	19'9 9/16"	12'8"	14'15/16"	15'5 5/8"	16'11 1/16"	18'4"
WELL OPENING FOR 6'-8" HEADROOM		10'2 7/16"	11'6 3/32"	11'11 15/16"	13'3 15/16"	14'8 21/32"	8'10 7/8"	10'2 7/16"	11'6 3/32"	11'11 15/16"	13'3 15/16"
ANGLE OF INCLINE		37°16'	34°53'	32°31'	30°35'	28°41'	40°6'	37°16'	34°53'	32°31'	30°35'
LENGTH OF CARRIAGE		17'8 9/32"	18'10 1/4"	20'13/16"	21'3 17/32"	22'6 25/32"	16'6 23/32"	17'8 9/32"	18'10 1/4"	20'13/16"	21'3 17/32"
L. F. OF RISER PER INCH OF STAIRWAY WIDTH		1.5	1.583	1.666	1.75	1.833	1.416	1.5	1.583	1.666	1.75
L. F. OF TREAD PER INCH OF STAIRWAY WIDTH		1.416	1.5	1.583	1.666	1.75	1.333	1.416	1.5	1.583	1.666
FRAMING SQUARE SETTINGS FOR CARRIAGE CUTS	TONGUE	7 9/16"	7 3/16"	6 13/16"	6 1/2"	6 3/16"	8"	7 9/16"	7 3/16"	6 13/16"	6 1/2"
	BODY	9 15/16"	10 5/16"	10 11/16"	11"	11 5/16"	9 1/2"	9 15/16"	10 5/16"	10 11/16"	11"

FLOOR TO FLOOR RISE		11'4 1/2"	11'4 5/8"	11'4 5/8"	11'4 5/8"	11'4 5/8"	11'4 5/8"	11'4 5/8"	11'4 3/4"	11'4 3/4"	11'4 3/4"
NUMBER OF RISERS		22	17	18	19	20	21	22	17	18	19
HEIGHT OF EACH RISER		6 3/16"	8 1/16"	7 9/16"	7 3/16"	6 13/16"	6 1/2"	6 3/16"	8 1/16"	7 5/8"	7 3/16"
TOTAL HEIGHT OF RISERS		11'4 1/8"	11'5 1/16"	11'4 1/8"	11'4 9/16"	11'4 1/4"	11'4 1/2"	11'4 1/8"	11'5 1/16"	11'5 1/4"	11'4 9/16"
RISERS OVERAGE (+) OR UNDERAGE (—)		−3/8"	+7/16"	−1/2"	−1/16"	−3/8"	−1/8"	−1/2"	+5/16"	+1/2"	−3/16"
NUMBER OF TREADS		21	16	17	18	19	20	21	16	17	18
WIDTH OF EACH TREAD		11 5/16"	9 7/16"	9 15/16"	10 5/16"	10 11/16"	11"	11 5/16"	9 7/16"	9 7/8"	10 5/16"
TOTAL RUN OF STAIRWAY		19'9 9/16"	12'7"	14' 15/16"	15'5 5/8"	16'11 1/16"	18'4"	19'9 9/16"	12'7"	13'11 7/8"	15'5 5/8"
WELL OPENING FOR 6'-8" HEADROOM		14'8 21/32"	8'10 3/32"	10'2 7/16"	11'6 3/32"	11'11 15/16"	13'3 15/16"	14'8 21/32"	8'10 3/32"	10'1 9/16"	11'6 3/32"
ANGLE OF INCLINE		28°41'	40°30'	37°16'	34°53'	32°31'	30°35'	28°41'	40°30'	37°40'	34°53'
LENGTH OF CARRIAGE		22'6 25/32"	16'6 19/32"	17'8 9/32"	18'10 1/4"	20' 13/16"	21'3 17/32"	22'6 25/32"	16'6 19/32"	17'8 3/32"	18'10 1/4"
L. F. OF RISER PER INCH OF STAIRWAY WIDTH		1.833	1.416	1.5	1.583	1.666	1.75	1.833	1.416	1.5	1.583
L. F. OF TREAD PER INCH OF STAIRWAY WIDTH		1.75	1.333	1.416	1.5	1.583	1.666	1.75	1.333	1.416	1.5
FRAMING SQUARE SETTINGS FOR CARRIAGE CUTS	TONGUE	6 3/16"	8 1/16"	7 9/16"	7 3/16"	6 13/16"	6 1/2"	6 3/16"	8 1/16"	7 5/8"	7 3/16"
	BODY	11 5/16"	9 7/16"	9 15/16"	10 5/16"	10 11/16"	11"	11 5/16"	9 7/16"	9 7/8"	10 5/16"

FLOOR TO FLOOR RISE		11'4 3/4"	11'4 3/4"	11'4 3/4"	11'4 7/8"	11'4 7/8"	11'4 7/8"	11'4 7/8"	11'4 7/8"	11'4 7/8"	11'5"
NUMBER OF RISERS		20	21	22	17	18	19	20	21	22	17
HEIGHT OF EACH RISER		6 13/16"	6 1/2"	6 7/32"	8 1/16"	7 5/8"	7 3/16"	6 27/32"	6 1/2"	6 7/32"	8 1/16"
TOTAL HEIGHT OF RISERS		11'4 1/4"	11'4 1/2"	11'4 13/16"	11'5 1/16"	11'5 1/4"	11'4 9/16"	11'4 7/8"	11'4 1/2"	11'4 13/16"	11'5 1/16"
RISERS OVERAGE (+) OR UNDERAGE (—)		−1/2"	−1/4"	+1/16"	+3/16"	+3/8"	−5/16"	0	−3/8"	−1/16"	+1/16"
NUMBER OF TREADS		19	20	21	16	17	18	19	20	21	16
WIDTH OF EACH TREAD		10 11/16"	11"	11 9/32"	9 7/16"	9 7/8"	10 5/16"	10 21/32"	11"	11 9/32"	9 7/16"
TOTAL RUN OF STAIRWAY		16'11 1/16"	18'4"	19'8 29/32"	12'7"	13'11 7/8"	15'5 5/8"	16'10 15/32"	18'4"	19'8 29/32"	12'7"
WELL OPENING FOR 6'-8" HEADROOM		11'11 15/16"	13'3 15/16"	14'8 3/32"	8'10 3/32"	10'1 9/16"	11'6 3/32"	11'11 5/16"	13'3 15/16"	14'8 3/32"	8'10 3/32"
ANGLE OF INCLINE		32°31'	30°35'	28°52'	40°30'	37°40'	34°53'	32°43'	30°35'	28°52'	40°30'
LENGTH OF CARRIAGE		20' 13/16"	21'3 17/32"	22'6 17/32"	16'6 19/32"	17'8 3/32"	18'10 1/4"	20' 5/8"	21'3 17/32"	22'6 17/32"	16'6 19/32"
L. F. OF RISER PER INCH OF STAIRWAY WIDTH		1.666	1.75	1.833	1.416	1.5	1.583	1.666	1.75	1.833	1.416
L. F. OF TREAD PER INCH OF STAIRWAY WIDTH		1.583	1.666	1.75	1.333	1.416	1.5	1.583	1.666	1.75	1.333
FRAMING SQUARE SETTINGS FOR CARRIAGE CUTS	TONGUE	6 13/16"	6 1/2"	6 7/32"	8 1/16"	7 5/8"	7 3/16"	6 27/32"	6 1/2"	6 7/32"	8 1/16"
	BODY	10 11/16"	11"	11 9/32"	9 7/16"	9 7/8"	10 5/16"	10 21/32"	11"	11 9/32"	9 7/16"

FLOOR TO FLOOR RISE		11'5"	11'5"	11'5"	11'5"	11'5"	11'$5\frac{1}{8}$"	11'$5\frac{1}{8}$"	11'$5\frac{1}{8}$"	11'$5\frac{1}{8}$"	11'$5\frac{1}{8}$"
NUMBER OF RISERS		18	19	20	21	22	17	18	19	20	21
HEIGHT OF EACH RISER		$7\frac{5}{8}$"	$7\frac{3}{16}$"	$6\frac{7}{8}$"	$6\frac{1}{2}$"	$6\frac{1}{4}$"	$8\frac{1}{16}$"	$7\frac{5}{8}$"	$7\frac{7}{32}$"	$6\frac{7}{8}$"	$6\frac{17}{32}$"
TOTAL HEIGHT OF RISERS		11'$5\frac{1}{4}$"	11'$4\frac{9}{16}$"	11'$5\frac{1}{2}$"	11'$4\frac{1}{2}$"	11'$5\frac{1}{2}$"	11'$5\frac{1}{16}$"	11'$5\frac{1}{4}$"	11'$5\frac{5}{32}$"	11'$5\frac{1}{2}$"	11'$5\frac{5}{32}$"
RISERS OVERAGE (+) OR UNDERAGE (—)		$+\frac{1}{4}$"	$-\frac{7}{16}$"	$+\frac{1}{2}$"	$-\frac{1}{2}$"	$+\frac{1}{2}$"	$-\frac{1}{16}$"	$+\frac{1}{8}$"	$+\frac{1}{32}$"	$+\frac{3}{8}$"	$+\frac{1}{32}$"
NUMBER OF TREADS		17	18	19	20	21	16	17	18	19	20
WIDTH OF EACH TREAD		$9\frac{7}{8}$"	$10\frac{5}{16}$"	$10\frac{5}{8}$"	11"	$11\frac{1}{4}$"	$9\frac{7}{16}$"	$9\frac{7}{8}$"	$10\frac{9}{32}$"	$10\frac{5}{8}$"	$10\frac{31}{32}$"
TOTAL RUN OF STAIRWAY		13'$11\frac{7}{8}$"	15'$5\frac{5}{8}$"	16'$9\frac{7}{8}$"	18'4"	19'$8\frac{1}{4}$"	12'7"	13'$11\frac{7}{8}$"	15'$5\frac{1}{16}$"	16'$9\frac{7}{8}$"	18'$3\frac{3}{8}$"
WELL OPENING FOR 6'-8" HEADROOM		10'$1\frac{9}{16}$"	11'$6\frac{3}{32}$"	11'$10\frac{31}{32}$"	13'$3\frac{15}{16}$"	14'$7\frac{1}{2}$"	8'$10\frac{3}{32}$"	10'$1\frac{9}{16}$"	11'$5\frac{5}{8}$"	11'$10\frac{31}{32}$"	13'$3\frac{13}{32}$"
ANGLE OF INCLINE		37°40'	34°53'	32°54'	30°35'	29°3'	40°30'	37°40'	35°4'	32°54'	30°46'
LENGTH OF CARRIAGE		17'$8\frac{3}{32}$"	18'$10\frac{1}{4}$"	20'$\frac{15}{32}$"	21'$3\frac{17}{32}$"	22'$6\frac{1}{4}$"	16'$6\frac{19}{32}$"	17'$8\frac{3}{32}$"	18'$10\frac{1}{8}$"	20'$\frac{15}{32}$"	21'$3\frac{9}{16}$"
L. F. OF RISER PER INCH OF STAIRWAY WIDTH		1.5	1.583	1.666	1.75	1.833	1.416	1.5	1.583	1.666	1.75
L. F. OF TREAD PER INCH OF STAIRWAY WIDTH		1.416	1.5	1.583	1.666	1.75	1.333	1.416	1.5	1.583	1.666
FRAMING SQUARE SETTINGS FOR CARRIAGE CUTS	**TONGUE**	$7\frac{5}{8}$"	$7\frac{3}{16}$"	$6\frac{7}{8}$"	$6\frac{1}{2}$"	$6\frac{1}{4}$"	$8\frac{1}{16}$"	$7\frac{5}{8}$"	$7\frac{7}{32}$"	$6\frac{7}{8}$"	$6\frac{17}{32}$"
	BODY	$9\frac{7}{8}$"	$10\frac{5}{16}$"	$10\frac{5}{8}$"	11"	$11\frac{1}{4}$"	$9\frac{7}{16}$"	$9\frac{7}{8}$"	$10\frac{9}{32}$"	$10\frac{5}{8}$"	$10\frac{31}{32}$"

FLOOR TO FLOOR RISE	11'5 5/8"	11'5 1/4"	11'5 1/4"	11'5 1/4"	11'5 1/4"	11'5 1/4"	11'5 1/4"	11'5 3/8"	11'5 3/8"	11'5 3/8"
NUMBER OF RISERS	22	17	18	19	20	21	22	17	18	19
HEIGHT OF EACH RISER	6 1/4"	8 1/16"	7 5/8"	7 1/4"	6 7/8"	6 17/32"	6 1/4"	8 1/16"	7 5/8"	7 1/4"
TOTAL HEIGHT OF RISERS	11'5 1/2"	11'5 1/16"	11'5 1/4"	11'5 3/4"	11'5 1/2"	11'5 5/32"	11'5 1/2"	11'5 1/16"	11'5 1/4"	11'5 3/4"
RISERS OVERAGE (+) OR UNDERAGE (—)	+3/8"	−3/16"	0	+1/2"	+1/4"	−3/32"	+1/4"	−5/16"	−1/8"	+3/8"
NUMBER OF TREADS	21	16	17	18	19	20	21	16	17	18
WIDTH OF EACH TREAD	11 1/4"	9 7/16"	9 7/8"	10 1/4"	10 5/8"	10 31/32"	11 1/4"	9 7/16"	9 7/8"	10 1/4"
TOTAL RUN OF STAIRWAY	19'8 1/4"	12'7"	13'11 7/8"	15'4 1/2"	16'9 7/8"	18'3 3/8"	19'8 1/4"	12'7"	13'11 7/8"	15'4 1/2"
WELL OPENING FOR 6'-8" HEADROOM	14'7 1/2"	8'10 3/32"	10'1 9/16"	11'5 1/8"	11'10 31/32"	13'3 13/32"	14'7 1/2"	8'10 3/32"	10'1 9/16"	11'5 1/8"
ANGLE OF INCLINE	29°3'	40°30'	37°40'	35°16'	32°54'	30°46'	29°3'	40°30'	37°40'	35°16'
LENGTH OF CARRIAGE	22'6 1/4"	16'6 19/32"	17'8 3/32"	18'10"	20' 15/32"	21'3 9/16"	22'6 1/4"	16'6 19/32"	17'8 3/32"	18'10"
L. F. OF RISER PER INCH OF STAIRWAY WIDTH	1.833	1.416	1.5	1.583	1.666	1.75	1.833	1.416	1.5	1.583
L. F. OF TREAD PER INCH OF STAIRWAY WIDTH	1.75	1.333	1.416	1.5	1.583	1.666	1.75	1.333	1.416	1.5
FRAMING SQUARE SETTINGS FOR CARRIAGE CUTS — TONGUE	6 1/4"	8 1/16"	7 5/8"	7 1/4"	6 7/8"	6 17/32"	6 1/4"	8 1/16"	7 5/8"	7 1/4"
FRAMING SQUARE SETTINGS FOR CARRIAGE CUTS — BODY	11 1/4"	9 7/16"	9 7/8"	10 1/4"	10 5/8"	10 31/32"	11 1/4"	9 7/16"	9 7/8"	10 1/4"

FLOOR TO FLOOR RISE		11'5 3/8"	11'5 3/8"	11'5 3/8"	11'5 1/2"	11'5 1/2"	11'5 1/2"	11'5 1/2"	11'5 1/2"	11'5 1/2"	11'5 5/8"
NUMBER OF RISERS		20	21	22	17	18	19	20	21	22	17
HEIGHT OF EACH RISER		6 7/8"	6 9/16"	6 1/4"	8 1/16"	7 5/8"	7 1/4"	6 7/8"	6 9/16"	6 1/4"	8 1/8"
TOTAL HEIGHT OF RISERS		11'5 1/2"	11'5 13/16"	11'5 1/2"	11'5 1/16"	11'5 1/4"	11'5 3/4"	11'5 1/2"	11'5 13/16"	11'5 1/2"	11'6 1/8"
RISERS OVERAGE (+) OR UNDERAGE (—)		+1/8"	+7/16"	+1/8"	−7/16"	−1/4"	+1/4"	0	+5/16"	0	+1/2"
NUMBER OF TREADS		19	20	21	16	17	18	19	20	21	16
WIDTH OF EACH TREAD		10 5/8"	10 15/16"	11 1/4"	9 7/16"	9 7/8"	10 1/4"	10 5/8"	10 15/16"	11 1/4"	9 3/8"
TOTAL RUN OF STAIRWAY		16'9 7/8"	18'2 3/4"	19'8 1/4"	12'7"	13'11 7/8"	15'4 1/2"	16'9 7/8"	18'2 3/4"	19'8 1/4"	12'6"
WELL OPENING FOR 6'-8" HEADROOM		11'10 31/32"	13'2 27/32"	14'7 1/2"	8'10 3/32"	10'1 9/16"	11'5 1/8"	11'10 31/32"	13'2 27/32"	14'7 1/2"	8'9 9/32"
ANGLE OF INCLINE		32°54'	30°58'	29°3'	40°30'	37°40'	35°16'	32°54'	30°58'	29°3'	40°55'
LENGTH OF CARRIAGE		20' 15/32"	21'3 3/32"	22'6 1/4"	16'6 19/32"	17'8 3/32"	18'10"	20' 15/32"	21'3 3/32"	22'6 1/4"	16'6 1/2"
L. F. OF RISER PER INCH OF STAIRWAY WIDTH		1.666	1.75	1.833	1.416	1.5	1.583	1.666	1.75	1.833	1.416
L. F. OF TREAD PER INCH OF STAIRWAY WIDTH		1.583	1.666	1.75	1.333	1.416	1.5	1.583	1.666	1.75	1.333
FRAMING SQUARE SETTINGS FOR CARRIAGE CUTS	TONGUE	6 7/8"	6 9/16"	6 1/4"	8 1/16"	7 5/8"	7 1/4"	6 7/8"	6 9/16"	6 1/4"	8 1/8"
	BODY	10 5/8"	10 15/16"	11 1/4"	9 7/16"	9 7/8"	10 1/4"	10 5/8"	10 15/16"	11 1/4"	9 3/8"

FLOOR TO FLOOR RISE		11'5$\frac{5}{8}$"	11'5$\frac{5}{8}$"	11'5$\frac{5}{8}$"	11'5$\frac{5}{8}$"	11'5$\frac{5}{8}$"	11'5$\frac{3}{4}$"	11'5$\frac{3}{4}$"	11'5$\frac{3}{4}$"	11'5$\frac{3}{4}$"	11'5$\frac{3}{4}$"
NUMBER OF RISERS		18	19	20	21	22	17	18	19	20	21
HEIGHT OF EACH RISER		7$\frac{5}{8}$"	7$\frac{1}{4}$"	6$\frac{7}{8}$"	6$\frac{9}{16}$"	6$\frac{1}{4}$"	8$\frac{1}{8}$"	7$\frac{5}{8}$"	7$\frac{1}{4}$"	6$\frac{7}{8}$"	6$\frac{9}{16}$"
TOTAL HEIGHT OF RISERS		11'5$\frac{1}{4}$"	11'5$\frac{3}{4}$"	11'5$\frac{1}{2}$"	11'5$\frac{13}{16}$"	11'5$\frac{1}{2}$"	11'6$\frac{1}{8}$"	11'5$\frac{1}{4}$"	11'5$\frac{3}{4}$"	11'5$\frac{1}{2}$"	11'5$\frac{13}{16}$"
RISERS OVERAGE (+) OR UNDERAGE (—)		−$\frac{3}{8}$"	+$\frac{1}{8}$"	−$\frac{1}{8}$"	+$\frac{3}{16}$"	−$\frac{1}{8}$"	+$\frac{3}{8}$"	−$\frac{1}{2}$"	0	−$\frac{1}{4}$"	+$\frac{1}{16}$"
NUMBER OF TREADS		17	18	19	20	21	16	17	18	19	20
WIDTH OF EACH TREAD		9$\frac{7}{8}$"	10$\frac{1}{4}$"	10$\frac{5}{8}$"	10$\frac{15}{16}$"	11$\frac{1}{4}$"	9$\frac{3}{8}$"	9$\frac{7}{8}$"	10$\frac{1}{4}$"	10$\frac{5}{8}$"	10$\frac{15}{16}$"
TOTAL RUN OF STAIRWAY		13'11$\frac{7}{8}$"	15'4$\frac{1}{2}$"	16'9$\frac{7}{8}$"	18'2$\frac{3}{4}$"	19'8$\frac{1}{4}$"	12'6"	13'11$\frac{7}{8}$"	15'4$\frac{1}{2}$"	16'9$\frac{7}{8}$"	18'2$\frac{3}{4}$"
WELL OPENING FOR 6'-8" HEADROOM		10'1$\frac{9}{16}$"	11'5$\frac{1}{8}$"	11'10$\frac{31}{32}$"	13'2$\frac{27}{32}$"	14'7$\frac{1}{2}$"	8'9$\frac{9}{32}$"	10'1$\frac{9}{16}$"	11'5$\frac{1}{8}$"	11'10$\frac{31}{32}$"	13'2$\frac{27}{32}$"
ANGLE OF INCLINE		37°40'	35°16'	32°54'	30°58'	29°3'	40°55'	37°40'	35°16'	32°54'	30°58'
LENGTH OF CARRIAGE		17'8$\frac{3}{32}$"	18'10"	20'$\frac{15}{32}$"	21'3$\frac{3}{32}$"	22'6$\frac{1}{4}$"	16'6$\frac{1}{2}$"	17'8$\frac{3}{32}$"	18'10"	20'$\frac{15}{32}$"	21'3$\frac{3}{32}$"
L. F. OF RISER PER INCH OF STAIRWAY WIDTH		1.5	1.583	1.666	1.75	1.833	1.416	1.5	1.583	1.666	1.75
L. F. OF TREAD PER INCH OF STAIRWAY WIDTH		1.416	1.5	1.583	1.666	1.75	1.333	1.416	1.5	1.583	1.666
FRAMING SQUARE SETTINGS FOR CARRIAGE CUTS	TONGUE	7$\frac{5}{8}$"	7$\frac{1}{4}$"	6$\frac{7}{8}$"	6$\frac{9}{16}$"	6$\frac{1}{4}$"	8$\frac{1}{8}$"	7$\frac{5}{8}$"	7$\frac{1}{4}$"	6$\frac{7}{8}$"	6$\frac{9}{16}$"
	BODY	9$\frac{7}{8}$"	10$\frac{1}{4}$"	10$\frac{5}{8}$"	10$\frac{15}{16}$"	11$\frac{1}{4}$"	9$\frac{3}{8}$"	9$\frac{7}{8}$"	10$\frac{1}{4}$"	10$\frac{5}{8}$"	10$\frac{15}{16}$"

FLOOR TO FLOOR RISE	11'5 3/4"	11'5 7/8"	11'5 7/8"	11'5 7/8"	11'5 7/8"	11'5 7/8"	11'5 7/8"	11'6"	11'6"	11'6"
NUMBER OF RISERS	22	17	18	19	20	21	22	17	18	19
HEIGHT OF EACH RISER	6 1/4"	8 1/8"	7 11/16"	7 1/4"	6 7/8"	6 9/16"	6 1/4"	8 1/8"	7 11/16"	7 1/4"
TOTAL HEIGHT OF RISERS	11'5 1/2"	11'6 1/8"	11'6 3/8"	11'5 3/4"	11'5 1/2"	11'5 13/16"	11'5 1/2"	11'6 1/8"	11'6 3/8"	11'5 3/4"
RISERS OVERAGE (+) OR UNDERAGE (—)	−1/4"	+1/4"	+1/2"	−1/8"	−3/8"	−1/16"	−3/8"	+1/8"	+3/8"	−1/4"
NUMBER OF TREADS	21	16	17	18	19	20	21	16	17	18
WIDTH OF EACH TREAD	11 1/4"	9 3/8"	9 13/16"	10 1/4"	10 5/8"	10 15/16"	11 1/4"	9 3/8"	9 13/16"	10 1/4"
TOTAL RUN OF STAIRWAY	19'8 1/4"	12'6"	13'10 13/16"	15'4 1/2"	16'9 7/8"	18'2 3/4"	19'8 1/4"	12'6"	13'10 13/16"	15'4 1/2"
WELL OPENING FOR 6'-8" HEADROOM	14'7 1/2"	8'9 9/32"	10' 23/32"	11'5 1/8"	11'10 31/32"	13'2 27/32"	14'7 1/2"	8'9 9/32"	10' 23/32"	11'5 1/8"
ANGLE OF INCLINE	29°3'	40°55'	38°5'	35°16'	32°54'	30°58'	29°3'	40°55'	38°5'	35°16'
LENGTH OF CARRIAGE	22'6 1/4"	16'6 1/2"	17'7 29/32"	18'10"	20' 15/32"	21'3 3/32"	22'6 1/4"	16'6 1/2"	17'7 29/32"	18'10"
L. F. OF RISER PER INCH OF STAIRWAY WIDTH	1.833	1.416	1.5	1.583	1.666	1.75	1.833	1.416	1.5	1.583
L. F. OF TREAD PER INCH OF STAIRWAY WIDTH	1.75	1.333	1.416	1.5	1.583	1.666	1.75	1.333	1.416	1.5
FRAMING SQUARE SETTINGS FOR CARRIAGE CUTS — TONGUE	6 1/4"	8 1/8"	7 11/16"	7 1/4"	6 7/8"	6 9/16"	6 1/4"	8 1/8"	7 11/16"	7 1/4"
FRAMING SQUARE SETTINGS FOR CARRIAGE CUTS — BODY	11 1/4"	9 3/8"	9 13/16"	10 1/4"	10 5/8"	10 15/16"	11 1/4"	9 3/8"	9 13/16"	10 1/4"

FLOOR TO FLOOR RISE	11'6"	11'6"	11'6"	11'6"	11'$6\frac{1}{8}$"	11'$6\frac{1}{8}$"	11'$6\frac{1}{8}$"	11'$6\frac{1}{8}$"	11'$6\frac{1}{8}$"	11'$6\frac{1}{8}$"
NUMBER OF RISERS	20	21	22	23	17	18	19	20	21	22
HEIGHT OF EACH RISER	$6\frac{7}{8}$"	$6\frac{9}{16}$"	$6\frac{1}{4}$"	6"	$8\frac{1}{8}$"	$7\frac{11}{16}$"	$7\frac{1}{4}$"	$6\frac{29}{32}$"	$6\frac{9}{16}$"	$6\frac{9}{32}$"
TOTAL HEIGHT OF RISERS	11'$5\frac{1}{2}$"	11'$5\frac{13}{16}$"	11'$5\frac{1}{2}$"	11'6"	11'$6\frac{1}{8}$"	11'$6\frac{3}{8}$"	11'$5\frac{3}{4}$"	11'$6\frac{1}{8}$"	11'$5\frac{13}{16}$"	11'$6\frac{3}{16}$"
RISERS OVERAGE (+) OR UNDERAGE (—)	$-\frac{1}{2}$"	$-\frac{3}{16}$"	$-\frac{1}{2}$"	0	0	$+\frac{1}{4}$"	$-\frac{3}{8}$"	0	$-\frac{5}{16}$"	$+\frac{1}{16}$"
NUMBER OF TREADS	19	20	21	22	16	17	18	19	20	21
WIDTH OF EACH TREAD	$10\frac{5}{8}$"	$10\frac{15}{16}$"	$11\frac{1}{4}$"	$11\frac{1}{2}$"	$9\frac{3}{8}$"	$9\frac{13}{16}$"	$10\frac{1}{4}$"	$10\frac{19}{32}$"	$10\frac{15}{16}$"	$11\frac{7}{32}$"
TOTAL RUN OF STAIRWAY	16'$9\frac{7}{8}$"	18'$2\frac{3}{4}$"	19'$8\frac{1}{4}$"	20'1"	12'6"	13'$10\frac{13}{16}$"	15'$4\frac{1}{2}$"	16'$9\frac{9}{32}$"	18'$2\frac{3}{4}$"	19'$7\frac{19}{32}$"
WELL OPENING FOR 6'-8" HEADROOM	11'$10\frac{31}{32}$"	13'$2\frac{27}{32}$"	14'$7\frac{1}{2}$"	15'$\frac{5}{32}$"	8'$9\frac{9}{32}$"	10'$\frac{23}{32}$"	11'$5\frac{1}{8}$"	11'$10\frac{15}{32}$"	13'$2\frac{27}{32}$"	14'$6\frac{29}{32}$"
ANGLE OF INCLINE	32°54'	30°58'	29°3'	27°33'	40°55'	38°5'	35°16'	33°6'	30°58'	29°15'
LENGTH OF CARRIAGE	20'$\frac{15}{32}$"	21'$3\frac{3}{32}$"	22'$6\frac{1}{4}$"	23'$9\frac{3}{8}$"	16'$6\frac{1}{2}$"	17'$7\frac{29}{32}$"	18'10"	20'$\frac{9}{32}$"	21'$3\frac{3}{32}$"	22'6"
L. F. OF RISER PER INCH OF STAIRWAY WIDTH	1.666	1.75	1.833	1.916	1.416	1.5	1.583	1.666	1.75	1.833
L. F. OF TREAD PER INCH OF STAIRWAY WIDTH	1.583	1.666	1.75	1.833	1.333	1.416	1.5	1.583	1.666	1.75
FRAMING SQUARE SETTINGS FOR CARRIAGE CUTS — TONGUE	$6\frac{7}{8}$"	$6\frac{9}{16}$"	$6\frac{1}{4}$"	6"	$8\frac{1}{8}$"	$7\frac{11}{16}$"	$7\frac{1}{4}$"	$6\frac{29}{32}$"	$6\frac{9}{16}$"	$6\frac{9}{32}$"
FRAMING SQUARE SETTINGS FOR CARRIAGE CUTS — BODY	$10\frac{5}{8}$"	$10\frac{15}{16}$"	$11\frac{1}{4}$"	$11\frac{1}{2}$"	$9\frac{3}{8}$"	$9\frac{13}{16}$"	$10\frac{1}{4}$"	$10\frac{19}{32}$"	$10\frac{15}{16}$"	$11\frac{7}{32}$"

FLOOR TO FLOOR RISE		11'6 1/8"	11'6 1/4"	11'6 1/4"	11'6 1/4"	11'6 1/4"	11'6 1/4"	11'6 1/4"	11'6 1/4"	11'6 3/8"	11'6 3/8"
NUMBER OF RISERS		23	17	18	19	20	21	22	23	17	18
HEIGHT OF EACH RISER		6"	8 1/8"	7 11/16"	7 1/4"	6 15/16"	6 9/16"	6 9/32"	6"	8 1/8"	7 11/16"
TOTAL HEIGHT OF RISERS		11'6"	11'6 1/8"	11'6 3/8"	11'5 3/4"	11'6 3/4"	11'5 13/16"	11'6 3/16"	11'6"	11'6 1/8"	11'6 3/8"
RISERS OVERAGE (+) OR UNDERAGE (—)		−1/8"	−1/8"	+1/8"	−1/2"	+1/2"	−7/16"	−1/16"	−1/4"	−1/4"	0
NUMBER OF TREADS		22	16	17	18	19	20	21	22	16	17
WIDTH OF EACH TREAD		11 1/2"	9 3/8"	9 13/16"	10 1/4"	10 9/16"	10 15/16"	11 7/32"	11 1/2"	9 3/8"	9 13/16"
TOTAL RUN OF STAIRWAY		20'1"	12'6"	13'10 13/16"	15'4 1/2"	16'8 11/16"	18'2 3/4"	19'7 19/32"	20'1"	12'6"	13'10 13/16"
WELL OPENING FOR 6'-8" HEADROOM		15' 5/32"	8'9 9/32"	10' 23/32"	11'5 1/8"	11'9 31/32"	13'2 27/32"	14'6 29/32"	15' 5/32"	8'9 9/32"	10' 23/32"
ANGLE OF INCLINE		27°33'	40°55'	38°5'	35°16'	33°18'	30°58'	29°15'	27°33'	40°55'	38°5'
LENGTH OF CARRIAGE		23'9 3/8"	16'6 1/2"	17'7 29/32"	18'10"	20' 3/32"	21'3 3/32"	22'6"	23'9 3/8"	16'6 1/2"	17'7 29/32"
L. F. OF RISER PER INCH OF STAIRWAY WIDTH		1.916	1.416	1.5	1.583	1.666	1.75	1.833	1.916	1.416	1.5
L. F. OF TREAD PER INCH OF STAIRWAY WIDTH		1.833	1.333	1.416	1.5	1.583	1.666	1.75	1.833	1.333	1.416
FRAMING SQUARE SETTINGS FOR CARRIAGE CUTS	TONGUE	6"	8 1/8"	7 11/16"	7 1/4"	6 15/16"	6 9/16"	6 9/32"	6"	8 1/8"	7 11/16"
	BODY	11 1/2"	9 3/8"	9 13/16"	10 1/4"	10 9/16"	10 15/16"	11 7/32"	11 1/2"	9 3/8"	9 13/16"

FLOOR TO FLOOR RISE	11'6 3/8"	11'6 3/8"	11'6 3/8"	11'6 3/8"	11'6 3/8"	11'6 1/2"	11'6 1/2"	11'6 1/2"	11'6 1/2"	11'6 1/2"
NUMBER OF RISERS	19	20	21	22	23	17	18	19	20	21
HEIGHT OF EACH RISER	7 9/32"	6 15/16"	6 19/32"	6 5/16"	6"	8 1/8"	7 11/16"	7 5/16"	6 15/16"	6 19/32"
TOTAL HEIGHT OF RISERS	11'6 11/32"	11'6 3/4"	11'6 15/32"	11'6 7/8"	11'6"	11'6 1/8"	11'6 3/8"	11'6 15/16"	11'6 3/4"	11'6 15/32"
RISERS OVERAGE (+) OR UNDERAGE (—)	−1/32"	+3/8"	+3/32"	+1/2"	−3/8"	−3/8"	−1/8"	+7/16"	+1/4"	−1/32"
NUMBER OF TREADS	18	19	20	21	22	16	17	18	19	20
WIDTH OF EACH TREAD	10 7/32"	10 9/16"	10 29/32"	11 3/16"	11 1/2"	9 3/8"	9 13/16"	10 3/16"	10 9/16"	10 29/32"
TOTAL RUN OF STAIRWAY	15'3 15/16"	16'8 11/16"	18'2 1/8"	19'6 15/16"	20'1"	12'6"	13'10 13/16"	15'3 3/8"	16'8 11/16"	18'2 1/8"
WELL OPENING FOR 6'-8" HEADROOM	11'4 21/32"	11'9 31/32"	13'2 5/16"	13'7 5/32"	15' 5/32"	8'9 9/32"	10' 23/32"	11'4 3/16"	11'9 31/32"	13'2 5/16"
ANGLE OF INCLINE	35°28'	33°18'	31°9'	29°26'	27°33'	40°55'	38°5'	35°40'	33°18'	31°9'
LENGTH OF CARRIAGE	18'9 27/32"	20' 3/32"	21'2 29/32"	22'5 3/4"	23'9 3/8"	16'6 1/2"	17'7 29/32"	18'9 23/32"	20' 3/32"	21'2 29/32"
L. F. OF RISER PER INCH OF STAIRWAY WIDTH	1.583	1.666	1.75	1.833	1.916	1.416	1.5	1.583	1.666	1.75
L. F. OF TREAD PER INCH OF STAIRWAY WIDTH	1.5	1.583	1.666	1.75	1.833	1.333	1.416	1.5	1.583	1.666
FRAMING SQUARE SETTINGS FOR CARRIAGE CUTS — TONGUE	7 9/32"	6 15/16"	6 19/32"	6 5/16"	6"	8 1/8"	7 11/16"	7 5/16"	6 15/16"	6 19/32"
FRAMING SQUARE SETTINGS FOR CARRIAGE CUTS — BODY	10 7/32"	10 9/16"	10 29/32"	11 3/16"	11 1/2"	9 3/8"	9 13/16"	10 3/16"	10 9/16"	10 29/32"

FLOOR TO FLOOR RISE	11'6 1/2"	11'6 1/2"	11'6 5/8"	11'6 5/8"	11'6 5/8"	11'6 5/8"	11'6 5/8"	11'6 5/8"	11'6 5/8"	11'6 3/4"
NUMBER OF RISERS	22	23	17	18	19	20	21	22	23	17
HEIGHT OF EACH RISER	6 5/16"	6"	8 1/8"	7 11/16"	7 5/16"	6 15/16"	6 5/8"	6 5/16"	6 1/32"	8 3/16"
TOTAL HEIGHT OF RISERS	11'6 7/8"	11'6"	11'6 1/8"	11'6 3/8"	11'6 15/16"	11'6 3/4"	11'7 1/8"	11'6 7/8"	11'6 23/32"	11'7 3/16"
RISERS OVERAGE (+) OR UNDERAGE (—)	+3/8"	−1/2"	−1/2"	−1/4"	+5/16"	+1/8"	+1/2"	+1/4"	+3/32"	+7/16"
NUMBER OF TREADS	21	22	16	17	18	19	20	21	22	16
WIDTH OF EACH TREAD	11 3/16"	11 1/2"	9 3/8"	9 13/16"	10 3/16"	10 9/16"	10 7/8"	11 3/16"	11 15/32"	9 5/16"
TOTAL RUN OF STAIRWAY	19'6 15/16"	20'1"	12'6"	13'10 13/16"	15'3 3/8"	16'8 11/16"	18'1 1/2"	19'6 15/16"	21' 5/16"	12'5"
WELL OPENING FOR 6'-8" HEADROOM	13'7 5/32"	15' 5/32"	8'9 9/32"	10' 23/32"	11'4 3/16"	11'9 31/32"	13'1 25/32"	13'7 5/32"	14'11 19/32"	8'8 1/2"
ANGLE OF INCLINE	29°26'	27°33'	40°55'	38°5'	35°40'	33°18'	31°21'	29°26'	27°44'	41°19'
LENGTH OF CARRIAGE	22'5 3/4"	23'9 3/8"	16'6 1/2"	17'7 29/32"	18'9 23/32"	20' 3/32"	21'2 11/16"	22'5 3/4"	23'9 1/16"	16'6 13/32"
L. F. OF RISER PER INCH OF STAIRWAY WIDTH	1.833	1.916	1.416	1.5	1.583	1.666	1.75	1.833	1.916	1.416
L. F. OF TREAD PER INCH OF STAIRWAY WIDTH	1.75	1.833	1.333	1.416	1.5	1.583	1.666	1.75	1.833	1.333
FRAMING SQUARE SETTINGS FOR CARRIAGE CUTS — TONGUE	6 5/16"	6"	8 1/8"	7 11/16"	7 5/16"	6 15/16"	6 5/8"	6 5/16"	6 1/32"	8 3/16"
FRAMING SQUARE SETTINGS FOR CARRIAGE CUTS — BODY	11 3/16"	11 1/2"	9 3/8"	9 13/16"	10 3/16"	10 9/16"	10 7/8"	11 3/16"	11 15/32"	9 5/16"

FLOOR TO FLOOR RISE	11'6 3/4"	11'6 3/4"	11'6 3/4"	11'6 3/4"	11'6 3/4"	11'6 3/4"	11'6 7/8"	11'6 7/8"	11'6 7/8"	11'6 7/8"
NUMBER OF RISERS	18	19	20	21	22	23	17	18	19	20
HEIGHT OF EACH RISER	7 11/16"	7 5/16"	6 15/16"	6 5/8"	6 5/16"	6 1/32"	8 3/16"	7 11/16"	7 5/16"	6 15/16"
TOTAL HEIGHT OF RISERS	11'6 3/8"	11'6 15/16"	11'6 3/4"	11'7 1/8"	11'6 7/8"	11'6 23/32"	11'7 3/16"	11'6 3/8"	11'6 15/16"	11'6 3/4"
RISERS OVERAGE (+) OR UNDERAGE (—)	−3/8"	+3/16"	0	+3/8"	+1/8"	−1/32"	+5/16"	−1/2"	+1/16"	−1/8"
NUMBER OF TREADS	17	18	19	20	21	22	16	17	18	19
WIDTH OF EACH TREAD	9 13/16"	10 3/16"	10 9/16"	10 7/8"	11 3/16"	11 15/32"	9 5/16"	9 13/16"	10 3/16"	10 9/16"
TOTAL RUN OF STAIRWAY	13'10 13/16"	15'3 3/8"	16'8 11/16"	18'1 1/2"	19'6 15/16"	21' 5/16"	12'5"	13'10 13/16"	15'3 3/8"	16'8 11/16"
WELL OPENING FOR 6'-8" HEADROOM	10' 23/32"	11'4 3/16"	11'9 31/32"	13'1 25/32"	13'7 5/32"	14'11 19/32"	8'8 1/2"	10' 23/32"	11'4 3/16"	11'9 31/32"
ANGLE OF INCLINE	38°5'	35°40'	33°18'	31°21'	29°26'	27°44'	41°19'	38°5'	35°40'	33°18'
LENGTH OF CARRIAGE	17'7 29/32"	18'9 23/32"	20' 3/32"	21'2 11/16"	22'5 3/4"	23'9 1/16"	16'6 13/32"	17'7 29/32"	18'9 23/32"	20' 3/32"
L. F. OF RISER PER INCH OF STAIRWAY WIDTH	1.5	1.583	1.666	1.75	1.833	1.916	1.416	1.5	1.583	1.666
L. F. OF TREAD PER INCH OF STAIRWAY WIDTH	1.416	1.5	1.583	1.666	1.75	1.833	1.333	1.416	1.5	1.583
FRAMING SQUARE SETTINGS FOR CARRIAGE CUTS — TONGUE	7 11/16"	7 5/16"	6 15/16"	6 5/8"	6 5/16"	6 1/32"	8 3/16"	7 11/16"	7 5/16"	6 15/16"
FRAMING SQUARE SETTINGS FOR CARRIAGE CUTS — BODY	9 13/16"	10 3/16"	10 9/16"	10 7/8"	11 3/16"	11 15/32"	9 5/16"	9 13/16"	10 3/16"	10 9/16"

FLOOR TO FLOOR RISE	11'$6\frac{7}{8}$"	11'$6\frac{7}{8}$"	11'$6\frac{7}{8}$"	11'7"	11'7"	11'7"	11'7"	11'7"	11'7"	11'7"
NUMBER OF RISERS	21	22	23	17	18	19	20	21	22	23
HEIGHT OF EACH RISER	$6\frac{5}{8}$"	$6\frac{5}{16}$"	$6\frac{1}{32}$"	$8\frac{3}{16}$"	$7\frac{3}{4}$"	$7\frac{5}{16}$"	$6\frac{15}{16}$"	$6\frac{5}{8}$"	$6\frac{5}{16}$"	$6\frac{1}{16}$"
TOTAL HEIGHT OF RISERS	11'$7\frac{1}{8}$"	11'$6\frac{7}{8}$"	11'$6\frac{23}{32}$"	11'$7\frac{3}{16}$"	11'$7\frac{1}{2}$"	11'$6\frac{15}{16}$"	11'$6\frac{3}{4}$"	11'$7\frac{1}{8}$"	11'$6\frac{7}{8}$"	11'$7\frac{7}{16}$"
RISERS OVERAGE (+) OR UNDERAGE (—)	$+\frac{1}{4}$"	0	$-\frac{5}{32}$"	$+\frac{3}{16}$"	$+\frac{1}{2}$"	$-\frac{1}{16}$"	$-\frac{1}{4}$"	$+\frac{1}{8}$"	$-\frac{1}{8}$"	$+\frac{7}{16}$"
NUMBER OF TREADS	20	21	22	16	17	18	19	20	21	22
WIDTH OF EACH TREAD	$10\frac{7}{8}$"	$11\frac{3}{16}$"	$11\frac{15}{32}$"	$9\frac{5}{16}$"	$9\frac{3}{4}$"	$10\frac{3}{16}$"	$10\frac{9}{16}$"	$10\frac{7}{8}$"	$11\frac{3}{16}$"	$11\frac{7}{16}$"
TOTAL RUN OF STAIRWAY	18'$1\frac{1}{2}$"	19'$6\frac{15}{16}$"	21'$\frac{5}{16}$"	12'5"	13'$9\frac{3}{4}$"	15'$3\frac{3}{8}$"	16'$8\frac{11}{16}$"	18'$1\frac{1}{2}$"	19'$6\frac{15}{16}$"	20'$11\frac{5}{8}$"
WELL OPENING FOR 6'-8" HEADROOM	13'$1\frac{25}{32}$"	13'$7\frac{5}{32}$"	14'$11\frac{19}{32}$"	8'$8\frac{1}{2}$"	9'$11\frac{27}{32}$"	11'$4\frac{3}{16}$"	11'$9\frac{31}{32}$"	13'$1\frac{25}{32}$"	13'$7\frac{5}{32}$"	14'11"
ANGLE OF INCLINE	31°21'	29°26'	27°44'	41°19'	38°29'	35°40'	33°18'	31°21'	29°26'	27°56'
LENGTH OF CARRIAGE	21'$2\frac{11}{16}$"	22'$5\frac{3}{4}$"	23'$9\frac{1}{16}$"	16'$6\frac{13}{32}$"	17'$7\frac{23}{32}$"	18'$9\frac{23}{32}$"	20'$\frac{3}{32}$"	21'$2\frac{11}{16}$"	22'$5\frac{3}{4}$"	23'$8\frac{25}{32}$"
L. F. OF RISER PER INCH OF STAIRWAY WIDTH	1.75	1.833	1.916	1.416	1.5	1.583	1.666	1.75	1.833	1.916
L. F. OF TREAD PER INCH OF STAIRWAY WIDTH	1.666	1.75	1.833	1.333	1.416	1.5	1.583	1.666	1.75	1.833
FRAMING SQUARE SETTINGS FOR CARRIAGE CUTS — TONGUE	$6\frac{5}{8}$"	$6\frac{5}{16}$"	$6\frac{1}{32}$"	$8\frac{3}{16}$"	$7\frac{3}{4}$"	$7\frac{5}{16}$"	$6\frac{15}{16}$"	$6\frac{5}{8}$"	$6\frac{5}{16}$"	$6\frac{1}{16}$"
FRAMING SQUARE SETTINGS FOR CARRIAGE CUTS — BODY	$10\frac{7}{8}$"	$11\frac{3}{16}$"	$11\frac{15}{32}$"	$9\frac{5}{16}$"	$9\frac{3}{4}$"	$10\frac{3}{16}$"	$10\frac{9}{16}$"	$10\frac{7}{8}$"	$11\frac{3}{16}$"	$11\frac{7}{16}$"

FLOOR TO FLOOR RISE	11'7$\frac{1}{8}$"	11'7$\frac{1}{8}$"	11'7$\frac{1}{8}$"	11'7$\frac{1}{8}$"	11'7$\frac{1}{8}$"	11'7$\frac{1}{8}$"	11'7$\frac{1}{8}$"	11'7$\frac{1}{4}$"	11'7$\frac{1}{4}$"	11'7$\frac{1}{4}$"
NUMBER OF RISERS	17	18	19	20	21	22	23	17	18	19
HEIGHT OF EACH RISER	8$\frac{3}{16}$"	7$\frac{3}{4}$"	7$\frac{5}{16}$"	6$\frac{15}{16}$"	6$\frac{5}{8}$"	6$\frac{5}{16}$"	6$\frac{1}{16}$"	8$\frac{3}{16}$"	7$\frac{3}{4}$"	7$\frac{5}{16}$"
TOTAL HEIGHT OF RISERS	11'7$\frac{3}{16}$"	11'7$\frac{1}{2}$"	11'6$\frac{15}{16}$"	11'6$\frac{3}{4}$"	11'7$\frac{1}{8}$"	11'6$\frac{7}{8}$"	11'7$\frac{7}{16}$"	11'7$\frac{3}{16}$"	11'7$\frac{1}{2}$"	11'6$\frac{15}{16}$"
RISERS OVERAGE (+) OR UNDERAGE (—)	+$\frac{1}{16}$"	+$\frac{3}{8}$"	−$\frac{3}{16}$"	−$\frac{3}{8}$"	0	−$\frac{1}{4}$"	+$\frac{5}{16}$"	−$\frac{1}{16}$"	+$\frac{1}{4}$"	−$\frac{5}{16}$"
NUMBER OF TREADS	16	17	18	19	20	21	22	16	17	18
WIDTH OF EACH TREAD	9$\frac{5}{16}$"	9$\frac{3}{4}$"	10$\frac{3}{16}$"	10$\frac{9}{16}$"	10$\frac{7}{8}$"	11$\frac{3}{16}$"	11$\frac{7}{16}$"	9$\frac{5}{16}$"	9$\frac{3}{4}$"	10$\frac{3}{16}$"
TOTAL RUN OF STAIRWAY	12'5"	13'9$\frac{3}{4}$"	15'3$\frac{3}{8}$"	16'8$\frac{11}{16}$"	18'1$\frac{1}{2}$"	19'6$\frac{15}{16}$"	20'11$\frac{5}{8}$"	12'5"	13'9$\frac{3}{4}$"	15'3$\frac{3}{8}$"
WELL OPENING FOR 6'-8" HEADROOM	8'8$\frac{1}{2}$"	9'11$\frac{27}{32}$"	11'4$\frac{3}{16}$"	11'9$\frac{31}{32}$"	13'1$\frac{25}{32}$"	13'7$\frac{5}{32}$"	14'11"	8'8$\frac{1}{2}$"	9'11$\frac{27}{32}$"	11'4$\frac{3}{16}$"
ANGLE OF INCLINE	41°19'	38°29'	35°40'	33°18'	31°21'	29°26'	27°56'	41°19'	38°29'	35°40'
LENGTH OF CARRIAGE	16'6$\frac{13}{32}$"	17'7$\frac{23}{32}$"	18'9$\frac{23}{32}$"	20'$\frac{3}{32}$"	21'2$\frac{11}{16}$"	22'5$\frac{3}{4}$"	23'8$\frac{25}{32}$"	16'6$\frac{13}{32}$"	17'7$\frac{23}{32}$"	18'9$\frac{23}{32}$"
L. F. OF RISER PER INCH OF STAIRWAY WIDTH	1.416	1.5	1.583	1.666	1.75	1.833	1.916	1.416	1.5	1.583
L. F. OF TREAD PER INCH OF STAIRWAY WIDTH	1.333	1.416	1.5	1.583	1.666	1.75	1.833	1.333	1.416	1.5
FRAMING SQUARE SETTINGS FOR CARRIAGE CUTS — TONGUE	8$\frac{3}{16}$"	7$\frac{3}{4}$"	7$\frac{5}{16}$"	6$\frac{15}{16}$"	6$\frac{5}{8}$"	6$\frac{5}{16}$"	6$\frac{1}{16}$"	8$\frac{3}{16}$"	7$\frac{3}{4}$"	7$\frac{5}{16}$"
FRAMING SQUARE SETTINGS FOR CARRIAGE CUTS — BODY	9$\frac{5}{16}$"	9$\frac{3}{4}$"	10$\frac{3}{16}$"	10$\frac{9}{16}$"	10$\frac{7}{8}$"	11$\frac{3}{16}$"	11$\frac{7}{16}$"	9$\frac{5}{16}$"	9$\frac{3}{4}$"	10$\frac{3}{16}$"

FLOOR TO FLOOR RISE		11'7 1/4"	11'7 1/4"	11'7 1/4"	11'7 1/4"	11'7 3/8"	11'7 3/8"	11'7 3/8"	11'7 3/8"	11'7 3/8"	11'7 3/8"
NUMBER OF RISERS		20	21	22	23	17	18	19	20	21	22
HEIGHT OF EACH RISER		6 15/16"	6 5/8"	6 5/16"	6 1/16"	8 3/16"	7 3/4"	7 5/16"	6 31/32"	6 5/8"	6 5/16"
TOTAL HEIGHT OF RISERS		11'6 3/4"	11'7 1/8"	11'6 7/8"	11'7 7/16"	11'7 3/16"	11'7 1/2"	11'6 15/16"	11'7 3/8"	11'7 1/8"	11'6 7/8"
RISERS OVERAGE (+) OR UNDERAGE (—)		−1/2"	−1/8"	−3/8"	+3/16"	−3/16"	+1/8"	−7/16"	0	−1/4"	−1/2"
NUMBER OF TREADS		19	20	21	22	16	17	18	19	20	21
WIDTH OF EACH TREAD		10 9/16"	10 7/8"	11 3/16"	11 7/16"	9 5/16"	9 3/4"	10 3/16"	10 17/32"	10 7/8"	11 3/16"
TOTAL RUN OF STAIRWAY		16'8 11/16"	18'1 1/2"	19'6 15/16"	20'11 5/8"	12'5"	13'9 3/4"	15'3 3/8"	16'8 3/32"	18'1 1/2"	19'6 15/16"
WELL OPENING FOR 6'-8" HEADROOM		11'9 31/32"	13'1 25/32"	13'7 5/32"	14'11"	8'8 1/2"	9'11 27/32"	11'4 3/16"	11'9 1/2"	13'1 25/32"	13'7 5/32"
ANGLE OF INCLINE		33°18'	31°21'	29°26'	27°56'	41°19'	38°29'	35°40'	33°30'	31°21'	29°26'
LENGTH OF CARRIAGE		20' 3/32"	21'2 11/16"	22'5 3/4"	23'8 25/32"	16'6 13/32"	17'7 23/32"	18'9 23/32"	19'11 15/16"	21'2 11/16"	22'5 3/4"
L. F. OF RISER PER INCH OF STAIRWAY WIDTH		1.666	1.75	1.833	1.916	1.416	1.5	1.583	1.666	1.75	1.833
L. F. OF TREAD PER INCH OF STAIRWAY WIDTH		1.583	1.666	1.75	1.833	1.333	1.416	1.5	1.583	1.666	1.75
FRAMING SQUARE SETTINGS FOR CARRIAGE CUTS	TONGUE	6 15/16"	6 5/8"	6 5/16"	6 1/16"	8 3/16"	7 3/4"	7 5/16"	6 31/32"	6 5/8"	6 5/16"
	BODY	10 9/16"	10 7/8"	11 3/16"	11 7/16"	9 5/16"	9 3/4"	10 3/16"	10 17/32"	10 7/8"	11 3/16"

FLOOR TO FLOOR RISE		11'7 3/8"	11'7 1/2"	11'7 1/2"	11'7 1/2"	11'7 1/2"	11'7 1/2"	11'7 1/2"	11'7 1/2"	11'7 5/8"	11'7 5/8"
NUMBER OF RISERS		23	17	18	19	20	21	22	23	17	18
HEIGHT OF EACH RISER		6 1/16"	8 3/16"	7 3/4"	7 11/32"	7"	6 5/8"	6 11/32"	6 1/16"	8 3/16"	7 3/4"
TOTAL HEIGHT OF RISERS		11'7 7/16"	11'7 3/16"	11'7 1/2"	11'7 17/32"	11'8"	11'7 1/8"	11'7 9/16"	11'7 7/16"	11'7 3/16"	11'7 1/2"
RISERS OVERAGE (+) OR UNDERAGE (—)		+1/16"	−5/16"	0	+1/32"	+1/2"	−3/8"	+1/16"	−1/16"	−7/16"	−1/8"
NUMBER OF TREADS		22	16	17	18	19	20	21	22	16	17
WIDTH OF EACH TREAD		11 7/16"	9 5/16"	9 3/4"	10 5/32"	10 1/2"	10 7/8"	11 5/32"	11 7/16"	9 5/16"	9 3/4"
TOTAL RUN OF STAIRWAY		20'11 5/8"	12'5"	13'9 3/4"	15'2 13/16"	16'7 1/2"	18'1 1/2"	19'6 9/32"	20'11 5/8"	12'5"	13'9 3/4"
WELL OPENING FOR 6'-8" HEADROOM		14'11"	8'8 1/2"	9'11 27/32"	10'5 9/16"	11'9"	13'1 25/32"	13'6 5/8"	14'11"	8'8 1/2"	9'11 27/32"
ANGLE OF INCLINE		27°56'	41°19'	38°29'	35°52'	33°41'	31°21'	29°37'	27°56'	41°19'	38°29'
LENGTH OF CARRIAGE		23'8 25/32"	16'6 13/32"	17'7 23/32"	18'9 19/32"	19'11 25/32"	21'2 11/16"	22'5 1/2"	23'8 25/32"	16'6 13/32"	17'7 23/32"
L. F. OF RISER PER INCH OF STAIRWAY WIDTH		1.916	1.416	1.5	1.583	1.666	1.75	1.833	1.916	1.416	1.5
L. F. OF TREAD PER INCH OF STAIRWAY WIDTH		1.833	1.333	1.416	1.5	1.583	1.666	1.75	1.833	1.333	1.416
FRAMING SQUARE SETTINGS FOR CARRIAGE CUTS	TONGUE	6 1/16"	8 3/16"	7 3/4"	7 11/32"	7"	6 5/8"	6 11/32"	6 1/16"	8 3/16"	7 3/4"
	BODY	11 7/16"	9 5/16"	9 3/4"	10 5/32"	10 1/2"	10 7/8"	11 5/32"	11 7/16"	9 5/16"	9 3/4"

FLOOR TO FLOOR RISE		11'7 5/8"	11'7 5/8"	11'7 5/8"	11'7 5/8"	11'7 5/8"	11'7 3/4"	11'7 3/4"	11'7 3/4"	11'7 3/4"	11'7 3/4"
NUMBER OF RISERS		19	20	21	22	23	17	18	19	20	21
HEIGHT OF EACH RISER		7 3/8"	7"	6 5/8"	6 11/32"	6 1/16"	8 1/4"	7 3/4"	7 3/8"	7"	6 21/32"
TOTAL HEIGHT OF RISERS		11'8 1/8"	11'8"	11'7 1/8"	11'7 9/16"	11'7 7/16"	11'8 1/4"	11'7 1/2"	11'8 1/8"	11'8"	11'7 25/32"
RISERS OVERAGE (+) OR UNDERAGE (—)		+1/2"	+3/8"	−1/2"	−1/16"	−3/16"	+1/2"	−1/4"	+3/8"	+1/4"	+1/32"
NUMBER OF TREADS		18	19	20	21	22	16	17	18	19	20
WIDTH OF EACH TREAD		10 1/8"	10 1/2"	10 7/8"	11 5/32"	11 7/16"	9 1/4"	9 3/4"	10 1/8"	10 1/2"	10 27/32"
TOTAL RUN OF STAIRWAY		15'2 1/4"	16'7 1/2"	18'1 1/2"	19'6 9/32"	20'11 5/8"	12'4"	13'9 3/4"	15'2 1/4"	16'7 1/2"	18' 7/8"
WELL OPENING FOR 6'-8" HEADROOM		10'5 3/32"	11'9"	13'1 25/32"	13'6 5/8"	14'11"	8'7 23/32"	9'11 27/32"	10'5 3/32"	11'9"	13'1 9/32"
ANGLE OF INCLINE		36°4'	33°41'	31°21'	29°37'	27°56'	41°44'	38°29'	36°4'	33°41'	31°33'
LENGTH OF CARRIAGE		18'9 15/32"	19'11 25/32"	21'2 11/16"	22'5 1/2"	23'8 25/32"	16'6 5/16"	17'7 23/32"	18'9 15/32"	19'11 25/32"	21'2 15/32"
L. F. OF RISER PER INCH OF STAIRWAY WIDTH		1.583	1.666	1.75	1.833	1.916	1.416	1.5	1.583	1.666	1.75
L. F. OF TREAD PER INCH OF STAIRWAY WIDTH		1.5	1.583	1.666	1.75	1.833	1.333	1.416	1.5	1.583	1.666
FRAMING SQUARE SETTINGS FOR CARRIAGE CUTS	TONGUE	7 3/8"	7"	6 5/8"	6 11/32"	6 1/16"	8 1/4"	7 3/4"	7 3/8"	7"	6 21/32"
	BODY	10 1/8"	10 1/2"	10 7/8"	11 5/32"	11 7/16"	9 1/4"	9 3/4"	10 1/8"	10 1/2"	10 27/32"

FLOOR TO FLOOR RISE	11'7 3/4"	11'7 3/4"	11'7 7/8"	11'7 7/8"	11'7 7/8"	11'7 7/8"	11'7 7/8"	11'7 7/8"	11'7 7/8"	11'8"
NUMBER OF RISERS	22	23	17	18	19	20	21	22	23	17
HEIGHT OF EACH RISER	6 3/8"	6 1/16"	8 1/4"	7 3/4"	7 3/8"	7"	6 21/32"	6 3/8"	6 1/16"	8 1/4"
TOTAL HEIGHT OF RISERS	11'8 1/4"	11'7 7/16"	11'8 1/4"	11'7 1/2"	11'8 1/8"	11'8"	11'7 25/32"	11'8 1/4"	11'7 7/16"	11'8 1/4"
RISERS OVERAGE (+) OR UNDERAGE (—)	+1/2"	−5/16"	+3/8"	−3/8"	+1/4"	+1/8"	−3/32"	+3/8"	−7/16"	+1/4"
NUMBER OF TREADS	21	22	16	17	18	19	20	21	22	16
WIDTH OF EACH TREAD	11 1/8"	11 7/16"	9 1/4"	9 3/4"	10 1/8"	10 1/2"	10 27/32"	11 1/8"	11 7/16"	9 1/4"
TOTAL RUN OF STAIRWAY	19'5 5/8"	20'11 5/8"	12'4"	13'9 3/4"	15'2 1/4"	16'7 1/2"	18'7/8"	19'5 5/8"	20'11 5/8"	12'4"
WELL OPENING FOR 6'-8" HEADROOM	13'6 1/16"	14'11"	8'7 23/32"	9'11 27/32"	10'5 3/32"	11'9"	13'1 9/32"	13'6 1/16"	14'11"	8'7 23/32"
ANGLE OF INCLINE	29°49'	27°56'	41°44'	38°29'	36°4'	33°41'	31°33'	29°49'	27°56'	41°44'
LENGTH OF CARRIAGE	22'5 9/32"	23'8 25/32"	16'6 5/16"	17'7 23/32"	18'9 15/32"	19'11 25/32"	21'2 15/32"	22'5 9/32"	23'8 25/32"	16'6 5/16"
L. F. OF RISER PER INCH OF STAIRWAY WIDTH	1.833	1.916	1.416	1.5	1.583	1.666	1.75	1.833	1.916	1.416
L. F. OF TREAD PER INCH OF STAIRWAY WIDTH	1.75	1.833	1.333	1.416	1.5	1.583	1.666	1.75	1.833	1.333
FRAMING SQUARE SETTINGS FOR CARRIAGE CUTS — TONGUE	6 3/8"	6 1/16"	8 1/4"	7 3/4"	7 3/8"	7"	6 21/32"	6 3/8"	6 1/16"	8 1/4"
FRAMING SQUARE SETTINGS FOR CARRIAGE CUTS — BODY	11 1/8"	11 7/16"	9 1/4"	9 3/4"	10 1/8"	10 1/2"	10 27/32"	11 1/8"	11 7/16"	9 1/4"

FLOOR TO FLOOR RISE	11'8"	11'8"	11'8"	11'8"	11'8"	11'8"	11'8 1/8"	11'8 1/8"	11'8 1/8"	11'8 1/8"
NUMBER OF RISERS	18	19	20	21	22	23	17	18	19	20
HEIGHT OF EACH RISER	7 3/4"	7 3/8"	7"	6 11/16"	6 3/8"	6 3/32"	8 1/4"	7 13/16"	7 3/8"	7"
TOTAL HEIGHT OF RISERS	11'7 1/2"	11'8 1/8"	11'8"	11'8 7/16"	11'8 1/4"	11'8 5/32"	11'8 1/4"	11'8 5/8"	11'8 1/8"	11'8"
RISERS OVERAGE (+) OR UNDERAGE (—)	−1/2"	+1/8"	0	+7/16"	+1/4"	+5/32"	+1/8"	+1/2"	0	−1/8"
NUMBER OF TREADS	17	18	19	20	21	22	16	17	18	19
WIDTH OF EACH TREAD	9 3/4"	10 1/8"	10 1/2"	10 13/16"	11 1/8"	11 13/32"	9 1/4"	9 11/16"	10 1/8"	10 1/2"
TOTAL RUN OF STAIRWAY	13'9 3/4"	15'2 1/4"	16'7 1/2"	18' 1/4"	19'5 5/8"	20'10 15/16"	12'4"	13'8 11/16"	15'2 1/4"	16'7 1/2"
WELL OPENING FOR 6'-8" HEADROOM	9'11 27/32"	10'5 3/32"	11'9"	13' 3/4"	13'6 1/16"	14'10 13/32"	8'7 23/32"	9'10 15/16"	10'5 3/32"	11'9"
ANGLE OF INCLINE	38°29'	36°4'	33°41'	31°44'	29°49'	28°7'	41°44'	38°53'	36°4'	33°41'
LENGTH OF CARRIAGE	17'7 23/32"	18'9 15/32"	19'11 25/32"	21'2 9/32"	22'5 9/32"	23'8 1/2"	16'6 5/16"	17'7 9/16"	18'9 15/32"	19'11 25/32"
L. F. OF RISER PER INCH OF STAIRWAY WIDTH	1.5	1.583	1.666	1.75	1.833	1.916	1.416	1.5	1.583	1.666
L. F. OF TREAD PER INCH OF STAIRWAY WIDTH	1.416	1.5	1.583	1.666	1.75	1.833	1.333	1.416	1.5	1.583
FRAMING SQUARE SETTINGS FOR CARRIAGE CUTS — TONGUE	7 3/4"	7 3/8"	7"	6 11/16"	6 3/8"	6 3/32"	8 1/4"	7 13/16"	7 3/8"	7"
FRAMING SQUARE SETTINGS FOR CARRIAGE CUTS — BODY	9 3/4"	10 1/8"	10 1/2"	10 13/16"	11 1/8"	11 13/32"	9 1/4"	9 11/16"	10 1/8"	10 1/2"

FLOOR TO FLOOR RISE		11'8 1/8"	11'8 1/8"	11'8 1/8"	11'8 1/4"	11'8 1/4"	11'8 1/4"	11'8 1/4"	11'8 1/4"	11'8 1/4"	11'8 1/4"
NUMBER OF RISERS		21	22	23	17	18	19	20	21	22	23
HEIGHT OF EACH RISER		6 11/16"	6 3/8"	6 3/32"	8 1/4"	7 13/16"	7 3/8"	7"	6 11/16"	6 3/8"	6 3/32"
TOTAL HEIGHT OF RISERS		11'8 7/16"	11'8 1/4"	11'8 5/32"	11'8 1/4"	11'8 5/8"	11'8 1/8"	11'8"	11'8 7/16"	11'8 1/4"	11'8 5/32"
RISERS OVERAGE (+) OR UNDERAGE (—)		+5/16"	+1/8"	+1/32"	0	+3/8"	−1/8"	−1/4"	+3/16"	0	−3/32"
NUMBER OF TREADS		20	21	22	16	17	18	19	20	21	22
WIDTH OF EACH TREAD		10 13/16"	11 1/8"	11 13/32"	9 1/4"	9 11/16"	10 1/8"	10 1/2"	10 13/16"	11 1/8"	11 13/32"
TOTAL RUN OF STAIRWAY		18' 1/4"	19'5 5/8"	20'10 15/16"	12'4"	13'8 11/16"	15'2 1/4"	16'7 1/2"	18' 1/4"	19'5 5/8"	20'10 15/16"
WELL OPENING FOR 6'-8" HEADROOM		13' 3/4"	13'6 1/16"	14'10 13/32"	8'7 23/32"	9'10 15/16"	10'5 3/32"	11'9"	13' 3/4"	13'6 1/16"	14'10 13/32"
ANGLE OF INCLINE		31°44'	29°49'	28°7'	41°44'	38°53'	36°4'	33°41'	31°44'	29°49'	28°7'
LENGTH OF CARRIAGE		21'2 9/32"	22'5 9/32"	23'8 1/2"	16'6 5/16"	17'7 9/16"	18'9 15/32"	19'11 25/32"	21'2 9/32"	22'5 9/32"	23'8 1/2"
L. F. OF RISER PER INCH OF STAIRWAY WIDTH		1.75	1.833	1.916	1.416	1.5	1.583	1.666	1.75	1.833	1.916
L. F. OF TREAD PER INCH OF STAIRWAY WIDTH		1.666	1.75	1.833	1.333	1.416	1.5	1.583	1.666	1.75	1.833
FRAMING SQUARE SETTINGS FOR CARRIAGE CUTS	TONGUE	6 11/16"	6 3/8"	6 3/32"	8 1/4"	7 13/16"	7 3/8"	7"	6 11/16"	6 3/8"	6 3/32"
	BODY	10 13/16"	11 1/8"	11 13/32"	9 1/4"	9 11/16"	10 1/8"	10 1/2"	10 13/16"	11 1/8"	11 13/32"

FLOOR TO FLOOR RISE		11'8$\frac{3}{8}$"	11'8$\frac{3}{8}$"	11'8$\frac{3}{8}$"	11'8$\frac{3}{8}$"	11'8$\frac{3}{8}$"	11'8$\frac{3}{8}$"	11'8$\frac{3}{8}$"	11'8$\frac{1}{2}$"	11'8$\frac{1}{2}$"	11'8$\frac{1}{2}$"
NUMBER OF RISERS		17	18	19	20	21	22	23	17	18	19
HEIGHT OF EACH RISER		8$\frac{1}{4}$"	7$\frac{13}{16}$"	7$\frac{3}{8}$"	7"	6$\frac{11}{16}$"	6$\frac{3}{8}$"	6$\frac{1}{8}$"	8$\frac{1}{4}$"	7$\frac{13}{16}$"	7$\frac{3}{8}$"
TOTAL HEIGHT OF RISERS		11'8$\frac{1}{4}$"	11'8$\frac{5}{8}$"	11'8$\frac{1}{8}$"	11'8"	11'8$\frac{7}{16}$"	11'8$\frac{1}{4}$"	11'8$\frac{7}{8}$"	11'8$\frac{1}{4}$"	11'8$\frac{5}{8}$"	11'8$\frac{1}{8}$"
RISERS OVERAGE (+) OR UNDERAGE (—)		$-\frac{1}{8}$"	$+\frac{1}{4}$"	$-\frac{1}{4}$"	$-\frac{3}{8}$"	$+\frac{1}{16}$"	$-\frac{1}{8}$"	$+\frac{1}{2}$"	$-\frac{1}{4}$"	$+\frac{1}{8}$"	$-\frac{3}{8}$"
NUMBER OF TREADS		16	17	18	19	20	21	22	16	17	18
WIDTH OF EACH TREAD		9$\frac{1}{4}$"	9$\frac{11}{16}$"	10$\frac{1}{8}$"	10$\frac{1}{2}$"	10$\frac{13}{16}$"	11$\frac{1}{8}$"	11$\frac{3}{8}$"	9$\frac{1}{4}$"	9$\frac{11}{16}$"	10$\frac{1}{8}$"
TOTAL RUN OF STAIRWAY		12'4"	13'8$\frac{11}{16}$"	15'2$\frac{1}{4}$"	16'7$\frac{1}{2}$"	18'$\frac{1}{4}$"	19'5$\frac{5}{8}$"	20'10$\frac{1}{4}$"	12'4"	13'8$\frac{11}{16}$"	15'2$\frac{1}{4}$"
WELL OPENING FOR 6'-8" HEADROOM		8'7$\frac{23}{32}$"	9'10$\frac{15}{16}$"	10'5$\frac{3}{32}$"	11'9"	13'$\frac{3}{4}$"	13'6$\frac{1}{16}$"	14'9$\frac{13}{16}$"	8'7$\frac{23}{32}$"	9'10$\frac{15}{16}$"	10'5$\frac{3}{32}$"
ANGLE OF INCLINE		41°44'	38°53'	36°4'	33°41'	31°44'	29°49'	28°18'	41°44'	38°53'	36°4'
LENGTH OF CARRIAGE		16'6$\frac{5}{16}$"	17'7$\frac{9}{16}$"	18'9$\frac{15}{32}$"	19'11$\frac{25}{32}$"	21'2$\frac{9}{32}$"	22'5$\frac{9}{32}$"	23'8$\frac{7}{32}$"	16'6$\frac{5}{16}$"	17'7$\frac{9}{16}$"	18'9$\frac{15}{32}$"
L. F. OF RISER PER INCH OF STAIRWAY WIDTH		1.416	1.5	1.583	1.666	1.75	1.833	1.916	1.416	1.5	1.583
L. F. OF TREAD PER INCH OF STAIRWAY WIDTH		1.333	1.416	1.5	1.583	1.666	1.75	1.833	1.333	1.416	1.5
FRAMING SQUARE SETTINGS FOR CARRIAGE CUTS	TONGUE	8$\frac{1}{4}$"	7$\frac{13}{16}$"	7$\frac{3}{8}$"	7"	6$\frac{11}{16}$"	6$\frac{3}{8}$"	6$\frac{1}{8}$"	8$\frac{1}{4}$"	7$\frac{13}{16}$"	7$\frac{3}{8}$"
	BODY	9$\frac{1}{4}$"	9$\frac{11}{16}$"	10$\frac{1}{8}$"	10$\frac{1}{2}$"	10$\frac{13}{16}$"	11$\frac{1}{8}$"	11$\frac{3}{8}$"	9$\frac{1}{4}$"	9$\frac{11}{16}$"	10$\frac{1}{8}$"

FLOOR TO FLOOR RISE	11'8$\frac{1}{2}$"	11'8$\frac{1}{2}$"	11'8$\frac{1}{2}$"	11'8$\frac{1}{2}$"	11'8$\frac{5}{8}$"	11'8$\frac{5}{8}$"	11'8$\frac{5}{8}$"	11'8$\frac{5}{8}$"	11'8$\frac{5}{8}$"	11'8$\frac{5}{8}$"
NUMBER OF RISERS	20	21	22	23	17	18	19	20	21	22
HEIGHT OF EACH RISER	7"	6$\frac{11}{16}$"	6$\frac{3}{8}$"	6$\frac{1}{8}$"	8$\frac{1}{4}$"	7$\frac{13}{16}$"	7$\frac{3}{8}$"	7$\frac{1}{32}$"	6$\frac{11}{16}$"	6$\frac{3}{8}$"
TOTAL HEIGHT OF RISERS	11'8"	11'8$\frac{7}{16}$"	11'8$\frac{1}{4}$"	11'8$\frac{7}{8}$"	11'8$\frac{1}{4}$"	11'8$\frac{5}{8}$"	11'8$\frac{1}{8}$"	11'8$\frac{5}{8}$"	11'8$\frac{7}{16}$"	11'8$\frac{1}{4}$"
RISERS OVERAGE (+) OR UNDERAGE (—)	$-\frac{1}{2}$"	$-\frac{1}{16}$"	$-\frac{1}{4}$"	$+\frac{3}{8}$"	$-\frac{3}{8}$"	0	$-\frac{1}{2}$"	0	$-\frac{3}{16}$"	$-\frac{3}{8}$"
NUMBER OF TREADS	19	20	21	22	16	17	18	19	20	21
WIDTH OF EACH TREAD	10$\frac{1}{2}$"	10$\frac{13}{16}$"	11$\frac{1}{8}$"	11$\frac{3}{8}$"	9$\frac{1}{4}$"	9$\frac{11}{16}$"	10$\frac{1}{8}$"	10$\frac{15}{16}$"	10$\frac{13}{16}$"	11$\frac{1}{8}$"
TOTAL RUN OF STAIRWAY	16'7$\frac{1}{2}$"	18'$\frac{1}{4}$"	19'5$\frac{5}{8}$"	20'10$\frac{1}{4}$"	12'4"	13'8$\frac{11}{16}$"	15'2$\frac{1}{4}$"	16'6$\frac{29}{32}$"	18'$\frac{1}{4}$"	19'5$\frac{5}{8}$"
WELL OPENING FOR 6'-8" HEADROOM	11'9"	13'$\frac{3}{4}$"	13'6$\frac{1}{16}$"	14'9$\frac{13}{16}$"	8'7$\frac{23}{32}$"	9'10$\frac{15}{16}$"	10'5$\frac{3}{32}$"	11'8$\frac{19}{32}$"	13'$\frac{3}{4}$"	13'6$\frac{1}{16}$"
ANGLE OF INCLINE	33°41'	31°44'	29°49'	28°18'	41°44'	38°53'	36°4'	33°53'	31°44'	29°49'
LENGTH OF CARRIAGE	19'11$\frac{25}{32}$"	21'2$\frac{9}{32}$"	22'5$\frac{9}{32}$"	23'8$\frac{7}{32}$"	16'6$\frac{5}{16}$"	17'7$\frac{9}{16}$"	18'9$\frac{15}{32}$"	19'11$\frac{19}{32}$"	21'2$\frac{9}{32}$"	22'5$\frac{9}{32}$"
L. F. OF RISER PER INCH OF STAIRWAY WIDTH	1.666	1.75	1.833	1.916	1.416	1.5	1.583	1.666	1.75	1.833
L. F. OF TREAD PER INCH OF STAIRWAY WIDTH	1.583	1.666	1.75	1.833	1.333	1.416	1.5	1.583	1.666	1.75
FRAMING SQUARE SETTINGS FOR CARRIAGE CUTS — TONGUE	7"	6$\frac{11}{16}$"	6$\frac{3}{8}$"	6$\frac{1}{8}$"	8$\frac{1}{4}$"	7$\frac{13}{16}$"	7$\frac{3}{8}$"	7$\frac{1}{32}$"	6$\frac{11}{16}$"	6$\frac{3}{8}$"
BODY	10$\frac{1}{2}$"	10$\frac{13}{16}$"	11$\frac{1}{8}$"	11$\frac{3}{8}$"	9$\frac{1}{4}$"	9$\frac{11}{16}$"	10$\frac{1}{8}$"	10$\frac{15}{16}$"	10$\frac{13}{16}$"	11$\frac{1}{8}$"

FLOOR TO FLOOR RISE	11'8 5/8"	11'8 3/4"	11'8 3/4"	11'8 3/4"	11'8 3/4"	11'8 3/4"	11'8 3/4"	11'8 3/4"	11'8 7/8"	11'8 7/8"
NUMBER OF RISERS	23	17	18	19	20	21	22	23	18	19
HEIGHT OF EACH RISER	6 1/8"	8 1/4"	7 13/16"	7 13/32"	7 1/16"	6 11/16"	6 3/8"	6 1/8"	7 13/16"	7 7/16"
TOTAL HEIGHT OF RISERS	11'8 7/8"	11'8 1/4"	11'8 5/8"	11'8 23/32"	11'9 1/4"	11'8 7/16"	11'8 1/4"	11'8 7/8"	11'8 5/8"	11'9 5/16"
RISERS OVERAGE (+) OR UNDERAGE (—)	+1/4"	−1/2"	−1/8"	−1/32"	+1/2"	−5/16"	−1/2"	+1/8"	−1/4"	+7/16"
NUMBER OF TREADS	22	16	17	18	19	20	21	22	17	18
WIDTH OF EACH TREAD	11 3/8"	9 1/4"	9 11/16"	10 3/32"	10 7/16"	10 13/16"	11 1/8"	11 3/8"	9 11/16"	10 1/16"
TOTAL RUN OF STAIRWAY	20'10 1/4"	12'4"	13'8 11/16"	15'1 11/16"	16'6 5/16"	18' 1/4"	19'5 5/8"	20'10 1/4"	13'8 11/16"	15'1 1/8"
WELL OPENING FOR 6'-8" HEADROOM	14'9 13/16"	8'7 23/32"	9'10 15/16"	10'4 21/32"	11'8 1/32"	13' 3/4"	13'6 1/16"	14'9 13/16"	9'10 15/16"	10'4 7/32"
ANGLE OF INCLINE	28°18'	41°44'	38°53'	36°16'	34°5'	31°44'	29°49'	28°18'	38°53'	36°28'
LENGTH OF CARRIAGE	23'8 7/32"	16'6 5/16"	17'7 9/16"	18'9 11/32"	19'11 7/16"	21'2 9/32"	22'5 9/32"	23'8 7/32"	17'7 9/16"	18'9 7/32"
L. F. OF RISER PER INCH OF STAIRWAY WIDTH	1.916	1.416	1.5	1.583	1.666	1.75	1.833	1.916	1.5	1.583
L. F. OF TREAD PER INCH OF STAIRWAY WIDTH	1.833	1.333	1.416	1.5	1.583	1.666	1.75	1.833	1.416	1.5
FRAMING SQUARE SETTINGS FOR CARRIAGE CUTS — TONGUE	6 1/8"	8 1/4"	7 13/16"	7 13/32"	7 1/16"	6 11/16"	6 3/8"	6 1/8"	7 13/16"	7 7/16"
FRAMING SQUARE SETTINGS FOR CARRIAGE CUTS — BODY	11 3/8"	9 1/4"	9 11/16"	10 3/32"	10 7/16"	10 13/16"	11 1/8"	11 3/8"	9 11/16"	10 1/16"

FLOOR TO FLOOR RISE	11'8 7/8"	11'8 7/8"	11'8 7/8"	11'8 7/8"	11'9"	11'9"	11'9"	11'9"	11'9"	11'9"
NUMBER OF RISERS	20	21	22	23	18	19	20	21	22	23
HEIGHT OF EACH RISER	7 1/16"	6 11/16"	6 13/32"	6 1/8"	7 13/16"	7 7/16"	7 1/16"	6 23/32"	6 13/32"	6 1/8"
TOTAL HEIGHT OF RISERS	11'9 1/4"	11'8 7/16"	11'8 15/16"	11'8 7/8"	11'8 5/8"	11'9 5/16"	11'9 1/4"	11'9 3/32"	11'8 15/16"	11'8 7/8"
RISERS OVERAGE (+) OR UNDERAGE (—)	+3/8"	−7/16"	+1/16"	0	−3/8"	+5/16"	+1/4"	+3/32"	−1/16"	−1/8"
NUMBER OF TREADS	19	20	21	22	17	18	19	20	21	22
WIDTH OF EACH TREAD	10 7/16"	10 13/16"	11 3/32"	11 3/8"	9 11/16"	10 1/16"	10 7/16"	10 25/32"	11 3/32"	11 3/8"
TOTAL RUN OF STAIRWAY	16'6 5/16"	18'1/4"	19'4 31/32"	20'10 1/4"	13'8 11/16"	15'1 1/8"	16'6 5/16"	17'11 5/8"	19'4 31/32"	20'10 1/4"
WELL OPENING FOR 6'-8" HEADROOM	11'8 1/32"	13'3/4"	13'5 17/32"	14'9 13/16"	9'10 15/16"	10'4 7/32"	11'8 1/32"	13'7/32"	13'5 17/32"	14'9 13/16"
ANGLE OF INCLINE	34°5'	31°44'	30°0'	28°18'	38°53'	36°28'	34°5'	31°56'	30°0'	28°18'
LENGTH OF CARRIAGE	19'11 7/16"	21'2 9/32"	22'5 1/32"	23'8 7/32"	17'7 9/16"	18'9 7/32"	19'11 7/16"	21'2 1/16"	22'5 1/32"	23'8 7/32"
L. F. OF RISER PER INCH OF STAIRWAY WIDTH	1.666	1.75	1.833	1.916	1.5	1.583	1.666	1.75	1.833	1.916
L. F. OF TREAD PER INCH OF STAIRWAY WIDTH	1.583	1.666	1.75	1.833	1.416	1.5	1.583	1.666	1.75	1.833
FRAMING SQUARE SETTINGS FOR CARRIAGE CUTS — TONGUE	7 1/16"	6 11/16"	6 13/32"	6 1/8"	7 13/16"	7 7/16"	7 1/16"	6 23/32"	6 13/32"	6 1/8"
FRAMING SQUARE SETTINGS FOR CARRIAGE CUTS — BODY	10 7/16"	10 13/16"	11 3/32"	11 3/8"	9 11/16"	10 1/16"	10 7/16"	10 25/32"	11 3/32"	11 3/8"

FLOOR TO FLOOR RISE	11'9 1/8"	11'9 1/8"	11'9 1/8"	11'9 1/8"	11'9 1/8"	11'9 1/8"	11'9 1/4"	11'9 1/4"	11'9 1/4"	11'9 1/4"
NUMBER OF RISERS	18	19	20	21	22	23	18	19	20	21
HEIGHT OF EACH RISER	7 13/16"	7 7/16"	7 1/16"	6 23/32"	6 7/16"	6 1/8"	7 7/8"	7 7/16"	7 1/16"	6 3/4"
TOTAL HEIGHT OF RISERS	11'8 5/8"	11'9 5/16"	11'9 1/4"	11'9 3/32"	11'9 5/8"	11'8 7/8"	11'9 3/4"	11'9 5/16"	11'9 1/4"	11'9 3/4"
RISERS OVERAGE (+) OR UNDERAGE (—)	−1/2"	+3/16"	+1/8"	−1/32"	+1/2"	−1/4"	+1/2"	+1/16"	0	+1/2"
NUMBER OF TREADS	17	18	19	20	21	22	17	18	19	20
WIDTH OF EACH TREAD	9 11/16"	10 1/16"	10 7/16"	10 25/32"	11 1/16"	11 3/8"	9 5/8"	10 1/16"	10 7/16"	10 3/4"
TOTAL RUN OF STAIRWAY	13'8 11/16"	15'1 1/8"	16'6 5/16"	17'11 5/8"	19'4 5/16"	20'10 1/4"	13'7 5/8"	15'1 1/8"	16'6 5/16"	17'11"
WELL OPENING FOR 6'-8" HEADROOM	9'10 15/16"	10'4 7/32"	11'8 1/32"	13' 7/32"	13'5"	14'9 13/16"	9'10 3/32"	10'4 7/32"	11'8 1/32"	12'11 11/16"
ANGLE OF INCLINE	38°53'	36°28'	34°5'	31°56'	30°12'	28°18'	39°17'	36°28'	34°5'	32°7'
LENGTH OF CARRIAGE	17'7 9/16"	18'9 7/32"	19'11 7/16"	21'2 1/16"	22'4 25/32"	23'8 7/32"	17'7 13/32"	18'9 7/32"	19'11 7/16"	21'1 7/8"
L. F. OF RISER PER INCH OF STAIRWAY WIDTH	1.5	1.583	1.666	1.75	1.833	1.916	1.5	1.583	1.666	1.75
L. F. OF TREAD PER INCH OF STAIRWAY WIDTH	1.416	1.5	1.583	1.666	1.75	1.833	1.416	1.5	1.583	1.666
FRAMING SQUARE SETTINGS FOR CARRIAGE CUTS — TONGUE	7 13/16"	7 7/16"	7 1/16"	6 23/32"	6 7/16"	6 1/8"	7 7/8"	7 7/16"	7 1/16"	6 3/4"
FRAMING SQUARE SETTINGS FOR CARRIAGE CUTS — BODY	9 11/16"	10 1/16"	10 7/16"	10 25/32"	11 1/16"	11 3/8"	9 5/8"	10 1/16"	10 7/16"	10 3/4"

FLOOR TO FLOOR RISE	11'9$\frac{1}{4}$"	11'9$\frac{1}{4}$"	11'9$\frac{3}{8}$"	11'9$\frac{3}{8}$"	11'9$\frac{3}{8}$"	11'9$\frac{3}{8}$"	11'9$\frac{3}{8}$"	11'9$\frac{3}{8}$"	11'9$\frac{1}{2}$"	11'9$\frac{1}{2}$"
NUMBER OF RISERS	22	23	18	19	20	21	22	23	18	19
HEIGHT OF EACH RISER	6$\frac{7}{16}$"	6$\frac{1}{8}$"	7$\frac{7}{8}$"	7$\frac{7}{16}$"	7$\frac{1}{16}$"	6$\frac{3}{4}$"	6$\frac{7}{16}$"	6$\frac{1}{8}$"	7$\frac{7}{8}$"	7$\frac{7}{16}$"
TOTAL HEIGHT OF RISERS	11'9$\frac{5}{8}$"	11'8$\frac{7}{8}$"	11'9$\frac{3}{4}$"	11'9$\frac{5}{16}$"	11'9$\frac{1}{4}$"	11'9$\frac{3}{4}$"	11'9$\frac{5}{8}$"	11'8$\frac{7}{8}$"	11'9$\frac{3}{4}$"	11'9$\frac{5}{16}$"
RISERS OVERAGE (+) OR UNDERAGE (—)	+$\frac{3}{8}$"	−$\frac{3}{8}$"	+$\frac{3}{8}$"	−$\frac{1}{16}$"	−$\frac{1}{8}$"	+$\frac{3}{8}$"	+$\frac{1}{4}$"	−$\frac{1}{2}$"	+$\frac{1}{4}$"	−$\frac{3}{16}$"
NUMBER OF TREADS	21	22	17	18	19	20	21	22	17	18
WIDTH OF EACH TREAD	11$\frac{1}{16}$"	11$\frac{3}{8}$"	9$\frac{5}{8}$"	10$\frac{1}{16}$"	10$\frac{7}{16}$"	10$\frac{3}{4}$"	11$\frac{1}{16}$"	11$\frac{3}{8}$"	9$\frac{5}{8}$"	10$\frac{1}{16}$"
TOTAL RUN OF STAIRWAY	19'4$\frac{5}{16}$"	20'10$\frac{1}{4}$"	13'7$\frac{5}{8}$"	15'1$\frac{1}{8}$"	16'6$\frac{5}{16}$"	17'11"	19'4$\frac{5}{16}$"	20'10$\frac{1}{4}$"	13'7$\frac{5}{8}$"	15'1$\frac{1}{8}$"
WELL OPENING FOR 6'-8" HEADROOM	13'5"	14'9$\frac{13}{16}$"	9'10$\frac{3}{32}$"	10'4$\frac{7}{32}$"	11'8$\frac{1}{32}$"	12'11$\frac{11}{16}$"	13'5"	14'9$\frac{13}{16}$"	9'10$\frac{3}{32}$"	10'4$\frac{7}{32}$"
ANGLE OF INCLINE	30°12'	28°18'	39°17'	36°28'	34°5'	32°7'	30°12'	28°18'	39°17'	36°28'
LENGTH OF CARRIAGE	22'4$\frac{25}{32}$"	23'8$\frac{7}{32}$"	17'7$\frac{13}{32}$"	18'9$\frac{7}{32}$"	19'11$\frac{7}{16}$"	21'1$\frac{7}{8}$"	22'4$\frac{25}{32}$"	23'8$\frac{7}{32}$"	17'7$\frac{13}{32}$"	18'9$\frac{7}{32}$"
L. F. OF RISER PER INCH OF STAIRWAY WIDTH	1.833	1.916	1.5	1.583	1.666	1.75	1.833	1.916	1.5	1.583
L. F. OF TREAD PER INCH OF STAIRWAY WIDTH	1.75	1.833	1.416	1.5	1.583	1.666	1.75	1.833	1.416	1.5
FRAMING SQUARE SETTINGS FOR CARRIAGE CUTS — TONGUE	6$\frac{7}{16}$"	6$\frac{1}{8}$"	7$\frac{7}{8}$"	7$\frac{7}{16}$"	7$\frac{1}{16}$"	6$\frac{3}{4}$"	6$\frac{7}{16}$"	6$\frac{1}{8}$"	7$\frac{7}{8}$"	7$\frac{7}{16}$"
BODY	11$\frac{1}{16}$"	11$\frac{3}{8}$"	9$\frac{5}{8}$"	10$\frac{1}{16}$"	10$\frac{7}{16}$"	10$\frac{3}{4}$"	11$\frac{1}{16}$"	11$\frac{3}{8}$"	9$\frac{5}{8}$"	10$\frac{1}{16}$"

FLOOR TO FLOOR RISE	11'9$\frac{1}{2}$"	11'9$\frac{1}{2}$"	11'9$\frac{1}{2}$"	11'9$\frac{1}{2}$"	11'9$\frac{5}{8}$"	11'9$\frac{5}{8}$"	11'9$\frac{5}{8}$"	11'9$\frac{5}{8}$"	11'9$\frac{5}{8}$"	11'9$\frac{5}{8}$"
NUMBER OF RISERS	20	21	22	23	18	19	20	21	22	23
HEIGHT OF EACH RISER	7$\frac{1}{16}$"	6$\frac{3}{4}$"	6$\frac{7}{16}$"	6$\frac{5}{32}$"	7$\frac{7}{8}$"	7$\frac{7}{16}$"	7$\frac{1}{16}$"	6$\frac{3}{4}$"	6$\frac{7}{16}$"	6$\frac{5}{32}$"
TOTAL HEIGHT OF RISERS	11'9$\frac{1}{4}$"	11'9$\frac{3}{4}$"	11'9$\frac{5}{8}$"	11'9$\frac{19}{32}$"	11'9$\frac{3}{4}$"	11'9$\frac{5}{16}$"	11'9$\frac{1}{4}$"	11'9$\frac{3}{4}$"	11'9$\frac{5}{8}$"	11'9$\frac{19}{32}$"
RISERS OVERAGE (+) OR UNDERAGE (—)	$-\frac{1}{4}$"	$+\frac{1}{4}$"	$+\frac{1}{8}$"	$+\frac{3}{32}$"	$+\frac{1}{8}$"	$-\frac{5}{16}$"	$-\frac{3}{8}$"	$+\frac{1}{8}$"	0	$-\frac{1}{32}$"
NUMBER OF TREADS	19	20	21	22	17	18	19	20	21	22
WIDTH OF EACH TREAD	10$\frac{7}{16}$"	10$\frac{3}{4}$"	11$\frac{1}{16}$"	11$\frac{11}{32}$"	9$\frac{5}{8}$"	10$\frac{1}{16}$"	10$\frac{7}{16}$"	10$\frac{3}{4}$"	11$\frac{1}{16}$"	11$\frac{11}{32}$"
TOTAL RUN OF STAIRWAY	16'6$\frac{5}{16}$"	17'11"	19'4$\frac{5}{16}$"	20'9$\frac{9}{16}$"	13'7$\frac{5}{8}$"	15'1$\frac{1}{8}$"	16'6$\frac{5}{16}$"	17'11"	19'4$\frac{5}{16}$"	20'9$\frac{9}{16}$"
WELL OPENING FOR 6'-8" HEADROOM	11'8$\frac{1}{32}$"	12'11$\frac{11}{16}$"	13'5"	14'9$\frac{1}{4}$"	9'10$\frac{3}{32}$"	10'4$\frac{7}{32}$"	11'8$\frac{1}{32}$"	12'11$\frac{11}{16}$"	13'5"	14'9$\frac{1}{4}$"
ANGLE OF INCLINE	34°5'	32°7'	30°12'	28°29'	39°17'	36°28'	34°5'	32°7'	30°12'	28°29'
LENGTH OF CARRIAGE	19'11$\frac{7}{16}$"	21'1$\frac{7}{8}$"	22'4$\frac{25}{32}$"	23'7$\frac{15}{16}$"	17'7$\frac{13}{32}$"	18'9$\frac{7}{32}$"	19'11$\frac{7}{16}$"	21'1$\frac{7}{8}$"	22'4$\frac{25}{32}$"	23'7$\frac{15}{16}$"
L. F. OF RISER PER INCH OF STAIRWAY WIDTH	1.666	1.75	1.833	1.916	1.5	1.583	1.666	1.75	1.833	1.916
L. F. OF TREAD PER INCH OF STAIRWAY WIDTH	1.583	1.666	1.75	1.833	1.416	1.5	1.583	1.666	1.75	1.833
FRAMING SQUARE SETTINGS FOR CARRIAGE CUTS — TONGUE	7$\frac{1}{16}$"	6$\frac{3}{4}$"	6$\frac{7}{16}$"	6$\frac{5}{32}$"	7$\frac{7}{8}$"	7$\frac{7}{16}$"	7$\frac{1}{16}$"	6$\frac{3}{4}$"	6$\frac{7}{16}$"	6$\frac{5}{32}$"
FRAMING SQUARE SETTINGS FOR CARRIAGE CUTS — BODY	10$\frac{7}{16}$"	10$\frac{3}{4}$"	11$\frac{1}{16}$"	11$\frac{11}{32}$"	9$\frac{5}{8}$"	10$\frac{1}{16}$"	10$\frac{7}{16}$"	10$\frac{3}{4}$"	11$\frac{1}{16}$"	11$\frac{11}{32}$"

FLOOR TO FLOOR RISE	11'9 3/4"	11'9 3/4"	11'9 3/4"	11'9 3/4"	11'9 3/4"	11'9 3/4"	11'9 7/8"	11'9 7/8"	11'9 7/8"	11'9 7/8"
NUMBER OF RISERS	18	19	20	21	22	23	18	19	20	21
HEIGHT OF EACH RISER	7 7/8"	7 7/16"	7 1/16"	6 3/4"	6 7/16"	6 5/32"	7 7/8"	7 15/32"	7 3/32"	6 3/4"
TOTAL HEIGHT OF RISERS	11'9 3/4"	11'9 5/16"	11'9 1/4"	11'9 3/4"	11'9 5/8"	11'9 19/32"	11'9 3/4"	11'9 29/32"	11'9 7/8"	11'9 3/4"
RISERS OVERAGE (+) OR UNDERAGE (—)	0	−7/16"	−1/2"	0	−1/8"	−5/32"	−1/8"	+1/32"	0	−1/8"
NUMBER OF TREADS	17	18	19	20	21	22	17	18	19	20
WIDTH OF EACH TREAD	9 5/8"	10 1/16"	10 7/16"	10 3/4"	11 1/16"	11 11/32"	9 5/8"	10 1/32"	10 13/32"	10 3/4"
TOTAL RUN OF STAIRWAY	13'7 5/8"	15'1 1/8"	16'6 5/16"	17'11"	19'4 5/16"	20'9 9/16"	13'7 5/8"	15' 9/16"	16'5 23/32"	17'11"
WELL OPENING FOR 6'-8" HEADROOM	9'10 3/32"	10'4 7/32"	11'8 1/32"	12'11 11/16"	13'5"	14'9 1/4"	9'10 3/32"	10'3 25/32"	11'7"	12'11 11/16"
ANGLE OF INCLINE	39°17'	36°28'	34°5'	32°7'	30°12'	28°29'	39°17'	36°40'	34°17'	32°7'
LENGTH OF CARRIAGE	17'7 13/32"	18'9 7/32"	19'11 7/16"	21'1 7/8"	22'4 25/32"	23'7 15/16"	17'7 13/32"	18'9 1/8"	19'11 9/32"	21'1 7/8"
L. F. OF RISER PER INCH OF STAIRWAY WIDTH	1.5	1.583	1.666	1.75	1.833	1.916	1.5	1.583	1.666	1.75
L. F. OF TREAD PER INCH OF STAIRWAY WIDTH	1.416	1.5	1.583	1.666	1.75	1.833	1.416	1.5	1.583	1.666
FRAMING SQUARE SETTINGS FOR CARRIAGE CUTS — TONGUE	7 7/8"	7 7/16"	7 1/16"	6 3/4"	6 7/16"	6 5/32"	7 7/8"	7 15/32"	7 3/32"	6 3/4"
FRAMING SQUARE SETTINGS FOR CARRIAGE CUTS — BODY	9 5/8"	10 1/16"	10 7/16"	10 3/4"	11 1/16"	11 11/32"	9 5/8"	10 1/32"	10 13/32"	10 3/4"

FLOOR TO FLOOR RISE	11'9$\frac{7}{8}$"	11'9$\frac{7}{8}$"	11'10"	11'10"	11'10"	11'10"	11'10"	11'10"	11'10$\frac{1}{8}$"	11'10$\frac{1}{8}$"
NUMBER OF RISERS	22	23	18	19	20	21	22	23	18	19
HEIGHT OF EACH RISER	6$\frac{7}{16}$"	6$\frac{3}{16}$"	7$\frac{7}{8}$"	7$\frac{1}{2}$"	7$\frac{1}{8}$"	6$\frac{3}{4}$"	6$\frac{7}{16}$"	6$\frac{3}{16}$"	7$\frac{7}{8}$"	7$\frac{1}{2}$"
TOTAL HEIGHT OF RISERS	11'9$\frac{5}{8}$"	11'10$\frac{5}{16}$"	11'9$\frac{3}{4}$"	11'10$\frac{1}{2}$"	11'10$\frac{1}{2}$"	11'9$\frac{3}{4}$"	11'9$\frac{5}{8}$"	11'10$\frac{5}{16}$"	11'9$\frac{3}{4}$"	11'10$\frac{1}{2}$"
RISERS OVERAGE (+) OR UNDERAGE (—)	−$\frac{1}{4}$"	+$\frac{7}{16}$"	−$\frac{1}{4}$"	+$\frac{1}{2}$"	+$\frac{1}{2}$"	−$\frac{1}{4}$"	−$\frac{3}{8}$"	+$\frac{5}{16}$"	−$\frac{3}{8}$"	+$\frac{3}{8}$"
NUMBER OF TREADS	21	22	17	18	19	20	21	22	17	18
WIDTH OF EACH TREAD	11$\frac{1}{16}$"	11$\frac{5}{16}$"	9$\frac{5}{8}$"	10"	10$\frac{3}{8}$"	10$\frac{3}{4}$"	11$\frac{1}{16}$"	11$\frac{5}{16}$"	9$\frac{5}{8}$"	10"
TOTAL RUN OF STAIRWAY	19'4$\frac{5}{16}$"	20'8$\frac{7}{8}$"	13'7$\frac{5}{8}$"	15'	16'5$\frac{1}{8}$"	17'11"	19'4$\frac{5}{16}$"	20'8$\frac{7}{8}$"	13'7$\frac{5}{8}$"	15'
WELL OPENING FOR 6'-8" HEADROOM	13'5"	14'8$\frac{21}{32}$"	9'10$\frac{3}{32}$"	10'3$\frac{11}{32}$"	11'7$\frac{1}{16}$"	12'11$\frac{11}{16}$"	13'5"	14'8$\frac{21}{32}$"	9'10$\frac{3}{32}$"	10'3$\frac{11}{32}$"
ANGLE OF INCLINE	30°12'	28°41'	39°17'	36°52'	34°29'	32°7'	30°12'	28°41'	39°17'	36°52'
LENGTH OF CARRIAGE	22'4$\frac{25}{32}$"	23'7$\frac{11}{16}$"	17'7$\frac{13}{32}$"	18'9"	19'11$\frac{1}{8}$"	21'1$\frac{7}{8}$"	22'4$\frac{25}{32}$"	23'7$\frac{11}{16}$"	17'7$\frac{13}{32}$"	18'9"
L. F. OF RISER PER INCH OF STAIRWAY WIDTH	1.833	1.916	1.5	1.583	1.666	1.75	1.833	1.916	1.5	1.583
L. F. OF TREAD PER INCH OF STAIRWAY WIDTH	1.75	1.833	1.416	1.5	1.583	1.666	1.75	1.833	1.416	1.5
FRAMING SQUARE SETTINGS FOR CARRIAGE CUTS — TONGUE	6$\frac{7}{16}$"	6$\frac{3}{16}$"	7$\frac{7}{8}$"	7$\frac{1}{2}$"	7$\frac{1}{8}$"	6$\frac{3}{4}$"	6$\frac{7}{16}$"	6$\frac{3}{16}$"	7$\frac{7}{8}$"	7$\frac{1}{2}$"
FRAMING SQUARE SETTINGS FOR CARRIAGE CUTS — BODY	11$\frac{1}{16}$"	11$\frac{5}{16}$"	9$\frac{5}{8}$"	10"	10$\frac{3}{8}$"	10$\frac{3}{4}$"	11$\frac{1}{16}$"	11$\frac{5}{16}$"	9$\frac{5}{8}$"	10"

FLOOR TO FLOOR RISE	11'10 1/8"	11'10 1/8"	11'10 1/8"	11'10 1/8"	11'10 1/4"	11'10 1/4"	11'10 1/4"	11'10 1/4"	11'10 1/4"	11'10 1/4"
NUMBER OF RISERS	20	21	22	23	18	19	20	21	22	23
HEIGHT OF EACH RISER	7 1/8"	6 3/4"	6 7/16"	6 3/16"	7 7/8"	7 1/2"	7 1/8"	6 3/4"	6 15/32"	6 3/16"
TOTAL HEIGHT OF RISERS	11'10 1/2"	11'9 3/4"	11'9 5/8"	11'10 5/16"	11'9 3/4"	11'10 1/2"	11'10 1/2"	11'9 3/4"	11'10 5/16"	11'10 5/16"
RISERS OVERAGE (+) OR UNDERAGE (—)	+3/8"	−3/8"	−1/2"	+3/16"	−1/2"	+1/4"	+1/4"	−1/2"	+1/16"	+1/16"
NUMBER OF TREADS	19	20	21	22	17	18	19	20	21	22
WIDTH OF EACH TREAD	10 3/8"	10 3/4"	11 1/16"	11 5/16"	9 5/8"	10"	10 3/8"	10 3/4"	11 1/32"	11 5/16"
TOTAL RUN OF STAIRWAY	16'5 1/8"	17'11"	19'4 5/16"	20'8 7/8"	13'7 5/8"	15'	16'5 1/8"	17'11"	19'3 21/32"	20'8 7/8"
WELL OPENING FOR 6'-8" HEADROOM	11'7 1/16"	12'11 11/16"	13'5"	14'8 21/32"	9'10 3/32"	10'3 11/32"	11'7 1/16"	12'11 11/6"	13'4 15/32"	14'8 21/32"
ANGLE OF INCLINE	34°29'	32°7'	30°12'	28°41'	39°17'	36°52'	34°29'	32°7'	30°23'	28°41'
LENGTH OF CARRIAGE	19'11 1/8"	21'1 7/8"	22'4 25/32"	23'7 11/16"	17'7 13/32"	18'9"	19'11 1/8"	21'1 7/8"	22'4 9/16"	23'7 11/16"
L. F. OF RISER PER INCH OF STAIRWAY WIDTH	1.666	1.75	1.833	1.916	1.5	1.583	1.666	1.75	1.833	1.916
L. F. OF TREAD PER INCH OF STAIRWAY WIDTH	1.583	1.666	1.75	1.833	1.416	1.5	1.583	1.666	1.75	1.833
FRAMING SQUARE SETTINGS FOR CARRIAGE CUTS — TONGUE	7 1/8"	6 3/4"	6 7/16"	6 3/16"	7 7/8"	7 1/2"	7 1/8"	6 3/4"	6 15/32"	6 3/16"
FRAMING SQUARE SETTINGS FOR CARRIAGE CUTS — BODY	10 3/8"	10 3/4"	11 1/16"	11 5/16"	9 5/8"	10"	10 3/8"	10 3/4"	11 1/32"	11 5/16"

FLOOR TO FLOOR RISE	11'10$\frac{3}{8}$"	11'10$\frac{3}{8}$"	11'10$\frac{3}{8}$"	11'10$\frac{3}{8}$"	11'10$\frac{3}{8}$"	11'10$\frac{3}{8}$"	11'10$\frac{1}{2}$"	11'10$\frac{1}{2}$"	11'10$\frac{1}{2}$"	11'10$\frac{1}{2}$"
NUMBER OF RISERS	18	19	20	21	22	23	18	19	20	21
HEIGHT OF EACH RISER	7$\frac{15}{16}$"	7$\frac{1}{2}$"	7$\frac{1}{8}$"	6$\frac{25}{32}$"	6$\frac{15}{32}$"	6$\frac{3}{16}$"	7$\frac{15}{16}$"	7$\frac{1}{2}$"	7$\frac{1}{8}$"	6$\frac{25}{32}$"
TOTAL HEIGHT OF RISERS	11'10$\frac{7}{8}$"	11'10$\frac{1}{2}$"	11'10$\frac{1}{2}$"	11'10$\frac{13}{32}$"	11'10$\frac{5}{16}$"	11'10$\frac{5}{16}$"	11'10$\frac{7}{8}$"	11'10$\frac{1}{2}$"	11'10$\frac{1}{2}$"	11'10$\frac{13}{32}$"
RISERS OVERAGE (+) OR UNDERAGE (—)	+$\frac{1}{2}$"	+$\frac{1}{8}$"	+$\frac{1}{8}$"	+$\frac{1}{32}$"	−$\frac{1}{16}$"	−$\frac{1}{16}$"	+$\frac{3}{8}$"	0	0	−$\frac{3}{32}$"
NUMBER OF TREADS	17	18	19	20	21	22	17	18	19	20
WIDTH OF EACH TREAD	9$\frac{9}{16}$"	10"	10$\frac{3}{8}$"	10$\frac{23}{32}$"	11$\frac{1}{32}$"	11$\frac{5}{16}$"	9$\frac{9}{16}$"	10"	10$\frac{3}{8}$"	10$\frac{23}{32}$"
TOTAL RUN OF STAIRWAY	13'6$\frac{9}{16}$"	15'	16'5$\frac{1}{8}$"	17'10$\frac{3}{8}$"	19'3$\frac{21}{32}$"	20'8$\frac{7}{8}$"	13'6$\frac{9}{16}$"	15'	16'5$\frac{1}{8}$"	17'10$\frac{3}{8}$"
WELL OPENING FOR 6'-8" HEADROOM	9'9$\frac{1}{4}$"	10'3$\frac{11}{32}$"	11'7$\frac{1}{16}$"	12'$\frac{7}{16}$"	13'4$\frac{15}{32}$"	14'8$\frac{21}{32}$"	9'9$\frac{1}{4}$"	10'3$\frac{11}{32}$"	11'7$\frac{1}{16}$"	12'$\frac{7}{16}$"
ANGLE OF INCLINE	39°42'	36°52'	34°29'	32°19'	30°23'	28°41'	39°42'	36°52'	34°29'	32°19'
LENGTH OF CARRIAGE	17'7$\frac{9}{32}$"	18'9"	19'11$\frac{1}{8}$"	21'1$\frac{11}{16}$"	22'4$\frac{9}{16}$"	23'7$\frac{11}{16}$"	17'7$\frac{9}{32}$"	18'9"	19'11$\frac{1}{8}$"	21'1$\frac{11}{16}$"
L. F. OF RISER PER INCH OF STAIRWAY WIDTH	1.5	1.583	1.666	1.75	1.833	1.916	1.5	1.583	1.666	1.75
L. F. OF TREAD PER INCH OF STAIRWAY WIDTH	1.416	1.5	1.583	1.666	1.75	1.833	1.416	1.5	1.583	1.666
FRAMING SQUARE SETTINGS FOR CARRIAGE CUTS — TONGUE	7$\frac{15}{16}$"	7$\frac{1}{2}$"	7$\frac{1}{8}$"	6$\frac{25}{32}$"	6$\frac{15}{32}$"	6$\frac{3}{16}$"	7$\frac{15}{16}$"	7$\frac{1}{2}$"	7$\frac{1}{8}$"	6$\frac{25}{32}$"
FRAMING SQUARE SETTINGS FOR CARRIAGE CUTS — BODY	9$\frac{9}{16}$"	10"	10$\frac{3}{8}$"	10$\frac{23}{32}$"	11$\frac{1}{32}$"	11$\frac{5}{16}$"	9$\frac{9}{16}$"	10"	10$\frac{3}{8}$"	10$\frac{23}{32}$"

FLOOR TO FLOOR RISE	11'10 1/2"	11'10 1/2"	11'10 5/8"	11'10 5/8"	11'10 5/8"	11'10 5/8"	11'10 5/8"	11'10 5/8"	11'10 3/4"	11'10 3/4"
NUMBER OF RISERS	22	23	18	19	20	21	22	23	18	19
HEIGHT OF EACH RISER	6 1/2"	6 3/16"	7 15/16"	7 1/2"	7 1/8"	6 13/16"	6 1/2"	6 3/16"	7 15/16"	7 1/2"
TOTAL HEIGHT OF RISERS	11'11"	11'10 5/16"	11'10 7/8"	11'10 1/2"	11'10 1/2"	11'11 1/16"	11'11"	11'10 5/16"	11'10 7/8"	11'10 1/2"
RISERS OVERAGE (+) OR UNDERAGE (—)	+1/2"	−3/16"	+1/4"	−1/8"	−1/8"	+7/16"	+3/8"	−5/16"	+1/8"	−1/4"
NUMBER OF TREADS	21	22	17	18	19	20	21	22	17	18
WIDTH OF EACH TREAD	11"	11 5/16"	9 9/16"	10"	10 3/8"	10 11/16"	11"	11 5/16"	9 9/16"	10"
TOTAL RUN OF STAIRWAY	19'3"	20'8 7/8"	13'6 9/16"	15'	16'5 1/8"	17'9 3/4"	19'3"	20'8 7/8"	13'6 9/16"	15'
WELL OPENING FOR 6'-8" HEADROOM	13'3 15/16"	14'8 21/32"	9'9 1/4"	10'3 11/32"	11'7 1/16"	11'11 15/16"	13'3 15/16"	14'8 21/32"	9'9 1/4"	10'3 11/32"
ANGLE OF INCLINE	30°35'	28°41'	39°42'	36°52'	34°29'	32°31'	30°35'	28°41'	39°42'	36°52'
LENGTH OF CARRIAGE	22'4 5/16"	23'7 11/16"	17'7 9/32"	18'9"	19'11 1/8"	21'1 15/32"	22'4 5/16"	23'7 11/16"	17'7 9/32"	18'9"
L. F. OF RISER PER INCH OF STAIRWAY WIDTH	1.833	1.916	1.5	1.583	1.666	1.75	1.833	1.916	1.5	1.583
L. F. OF TREAD PER INCH OF STAIRWAY WIDTH	1.75	1.833	1.416	1.5	1.583	1.666	1.75	1.833	1.416	1.5
FRAMING SQUARE SETTINGS FOR CARRIAGE CUTS — TONGUE	6 1/2"	6 3/16"	7 15/16"	7 1/2"	7 1/8"	6 13/16"	6 1/2"	6 3/16"	7 15/16"	7 1/2"
BODY	11"	11 5/16"	9 9/16"	10"	10 3/8"	10 11/16"	11"	11 5/16"	9 9/16"	10"

FLOOR TO FLOOR RISE	11'10 3/4"	11'10 3/4"	11'10 3/4"	11'10 3/4"	11'10 7/8"	11'10 7/8"	11'10 7/8"	11'10 7/8"	11'10 7/8"	11'10 7/8"
NUMBER OF RISERS	20	21	22	23	18	19	20	21	22	23
HEIGHT OF EACH RISER	7 1/8"	6 13/16"	6 1/2"	6 3/16"	7 15/16"	7 1/2"	7 1/8"	6 13/16"	6 1/2"	6 7/32"
TOTAL HEIGHT OF RISERS	11'10 1/2"	11'11 1/16"	11'11"	11'10 5/16"	11'10 7/8"	11'10 1/2"	11'10 1/2"	11'11 1/16"	11'11"	11'11 1/32"
RISERS OVERAGE (+) OR UNDERAGE (—)	−1/4"	+5/16"	+1/4"	−7/16"	0	−3/8"	−3/8"	+3/16"	+1/8"	+5/32"
NUMBER OF TREADS	19	20	21	22	17	18	19	20	21	22
WIDTH OF EACH TREAD	10 3/8"	10 11/16"	11"	11 5/16"	9 9/16"	10"	10 3/8"	10 11/16"	11"	11 9/32"
TOTAL RUN OF STAIRWAY	16'5 1/8"	17'9 3/4"	19'3"	20'8 7/8"	13'6 9/16"	15'	16'5 1/8"	17'9 3/4"	19'3"	20'8 3/16"
WELL OPENING FOR 6'-8" HEADROOM	11'7 1/16"	11'11 15/16"	13'3 15/16"	14'8 21/32"	9'9 1/4"	10'3 11/32"	11'7 1/16"	11'11 15/16"	13'3 15/16"	14'8 3/32"
ANGLE OF INCLINE	34°29'	32°31'	30°35'	28°41'	39°42'	36°52'	34°29'	32°31'	30°35'	28°52'
LENGTH OF CARRIAGE	19'11 1/8"	21'1 15/32"	22'4 5/16"	23'7 11/16"	17'7 9/32"	18'9"	19'11 1/8"	21'1 15/32"	22'4 5/16"	23'7 13/32"
L. F. OF RISER PER INCH OF STAIRWAY WIDTH	1.666	1.75	1.833	1.916	1.5	1.583	1.666	1.75	1.833	1.916
L. F. OF TREAD PER INCH OF STAIRWAY WIDTH	1.583	1.666	1.75	1.833	1.416	1.5	1.583	1.666	1.75	1.833
FRAMING SQUARE SETTINGS FOR CARRIAGE CUTS — TONGUE	7 1/8"	6 13/16"	6 1/2"	6 3/16"	7 15/16"	7 1/2"	7 1/8"	6 13/16"	6 1/2"	6 7/32"
FRAMING SQUARE SETTINGS FOR CARRIAGE CUTS — BODY	10 3/8"	10 11/16"	11"	11 5/16"	9 9/16"	10"	10 3/8"	10 11/16"	11"	11 9/32"

FLOOR TO FLOOR RISE	11'11"	11'11"	11'11"	11'11"	11'11"	11'11"	11'11 1/8"	11'11 1/8"	11'11 1/8"	11'11 1/8"
NUMBER OF RISERS	18	19	20	21	22	23	18	19	20	21
HEIGHT OF EACH RISER	7 15/16"	7 1/2"	7 1/8"	6 13/16"	6 1/2"	6 7/32"	7 15/16"	7 17/32"	7 5/32"	6 13/16"
TOTAL HEIGHT OF RISERS	11'10 7/8"	11'10 1/2"	11'10 1/2"	11'11 1/16"	11'11"	11'11 1/32"	11'10 7/8"	11'11 3/32"	11'11 1/8"	11'11 1/16"
RISERS OVERAGE (+) OR UNDERAGE (—)	−1/8"	−1/2"	−1/2"	+1/16"	0	+1/32"	−1/4"	−1/32"	0	−1/16"
NUMBER OF TREADS	17	18	19	20	21	22	17	18	19	20
WIDTH OF EACH TREAD	9 9/16"	10"	10 3/8"	10 11/16"	11"	11 9/32"	9 9/16"	9 31/32"	10 11/32"	10 11/16"
TOTAL RUN OF STAIRWAY	13'6 9/16"	15'	16'5 1/8"	17'9 3/4"	19'3"	20'8 3/16"	13'6 9/16"	14'11 7/16"	16'4 17/32"	17'9 3/4"
WELL OPENING FOR 6'-8" HEADROOM	9'9 1/4"	10'3 11/32"	11'7 1/16"	11'11 15/16"	13'3 15/16"	14'8 3/32"	9'9 1/4"	10'2 29/32"	11'6 19/32"	11'11 15/16"
ANGLE OF INCLINE	39°42'	36°52'	34°29'	32°31'	30°35'	28°52'	39°42'	37°4'	34°41'	32°31'
LENGTH OF CARRIAGE	17'7 9/32"	18'9"	19'11 1/8"	21'1 15/32"	22'4 5/16"	23'7 13/32"	17'7 9/32"	18'8 7/8"	19'10 31/32"	21'1 15/32"
L. F. OF RISER PER INCH OF STAIRWAY WIDTH	1.5	1.583	1.666	1.75	1.833	1.916	1.5	1.583	1.666	1.75
L. F. OF TREAD PER INCH OF STAIRWAY WIDTH	1.416	1.5	1.583	1.666	1.75	1.833	1.416	1.5	1.583	1.666
FRAMING SQUARE SETTINGS FOR CARRIAGE CUTS — TONGUE	7 15/16"	7 1/2"	7 1/8"	6 13/16"	6 1/2"	6 7/32"	7 15/16"	7 17/32"	7 5/32"	6 13/16"
FRAMING SQUARE SETTINGS FOR CARRIAGE CUTS — BODY	9 9/16"	10"	10 3/8"	10 11/16"	11"	11 9/32"	9 9/16"	9 31/32"	10 11/32"	10 11/16"

FLOOR TO FLOOR RISE	11'11 1/8"	11'11 1/8"	11'11 1/4"	11'11 1/4"	11'11 1/4"	11'11 1/4"	11'11 1/4"	11'11 1/4"	11'11 3/8"	11'11 3/8"
NUMBER OF RISERS	22	23	18	19	20	21	22	23	18	19
HEIGHT OF EACH RISER	6 1/2"	6 7/32"	7 15/16"	7 9/16"	7 3/16"	6 13/16"	6 1/2"	6 1/4"	7 15/16"	7 9/16"
TOTAL HEIGHT OF RISERS	11'11"	11'11 1/32"	11'10 7/8"	11'11 11/16"	11'11 3/4"	11'11 1/16"	11'11"	11'11 3/4"	11'10 7/8"	11'11 11/16"
RISERS OVERAGE (+) OR UNDERAGE (—)	−1/8"	−3/32"	−3/8"	+7/16"	+1/2"	−3/16"	−1/4"	+1/2"	−1/2"	+5/16"
NUMBER OF TREADS	21	22	17	18	19	20	21	22	17	18
WIDTH OF EACH TREAD	11"	11 9/32"	9 9/16"	9 15/16"	10 5/16"	10 11/16"	11"	11 1/4"	9 9/16"	9 15/16"
TOTAL RUN OF STAIRWAY	19'3"	20'8 3/16"	13'6 9/16"	14'10 7/8"	16'3 15/16"	17'9 3/4"	19'3"	20'7 1/2"	13'6 9/16"	14'10 7/8"
WELL OPENING FOR 6'-8" HEADROOM	13'3 15/16"	14'8 3/32"	9'9 1/4"	10'2 7/16"	11'6 3/32"	11'11 15/16"	13'3 15/16"	14'7 1/2"	9'9 1/4"	10'2 7/16"
ANGLE OF INCLINE	30°35'	28°52'	39°42'	37°16'	34°53'	32°31'	30°35'	29°3'	39°42'	37°16'
LENGTH OF CARRIAGE	22'4 5/16"	23'7 13/32"	17'7 9/32"	18'8 25/32"	19'10 27/32"	21'1 15/32"	22'4 5/16"	23'7 1/8"	17'7 9/32"	18'8 25/32"
L. F. OF RISER PER INCH OF STAIRWAY WIDTH	1.833	1.916	1.5	1.583	1.666	1.75	1.833	1.916	1.5	1.583
L. F. OF TREAD PER INCH OF STAIRWAY WIDTH	1.75	1.833	1.416	1.5	1.583	1.666	1.75	1.833	1.416	1.5
FRAMING SQUARE SETTINGS FOR CARRIAGE CUTS — TONGUE	6 1/2"	6 7/32"	7 15/16"	7 9/16"	7 3/16"	6 13/16"	6 1/2"	6 1/4"	7 15/16"	7 9/16"
FRAMING SQUARE SETTINGS FOR CARRIAGE CUTS — BODY	11"	11 9/32"	9 9/16"	9 15/16"	10 5/16"	10 11/16"	11"	11 1/4"	9 9/16"	9 15/16"

FLOOR TO FLOOR RISE		11'11 3/8"	11'11 3/8"	11'11 3/8"	11'11 3/8"	11'11 1/2"	11'11 1/2"	11'11 1/2"	11'11 1/2"	11'11 1/2"	11'11 1/2"
NUMBER OF RISERS		20	21	22	23	18	19	20	21	22	23
HEIGHT OF EACH RISER		7 3/16"	6 13/16"	6 1/2"	6 1/4"	8"	7 9/16"	7 3/16"	6 13/16"	6 1/2"	6 1/4"
TOTAL HEIGHT OF RISERS		11'11 3/4"	11'11 1/16"	11'11"	11'11 3/4"	12'	11'11 11/16"	11'11 3/4"	11'11 1/16"	11'11"	11'11 3/4"
RISERS OVERAGE (+) OR UNDERAGE (—)		+3/8"	−5/16"	−3/8"	+3/8"	+1/2"	+3/16"	+1/4"	−7/16"	−1/2"	+1/4"
NUMBER OF TREADS		19	20	21	22	17	18	19	20	21	22
WIDTH OF EACH TREAD		10 5/16"	10 11/16"	11"	11 1/4"	9 1/2"	9 15/16"	10 5/16"	10 11/16"	11"	11 1/4"
TOTAL RUN OF STAIRWAY		16'3 15/16"	17'9 3/4"	19'3"	20'7 1/2"	13'5 1/2"	14'10 7/8"	16'3 15/16"	17'9 3/4"	19'3"	20'7 1/2"
WELL OPENING FOR 6'-8" HEADROOM		11'6 3/32"	11'11 15/16"	13'3 15/16"	14'7 1/2"	8'10 7/8"	10'2 7/16"	11'6 3/32"	11'11 15/16"	13'3 15/16"	14'7 1/2"
ANGLE OF INCLINE		34°53'	32°31'	30°35'	29°3'	40°6'	37°16'	34°53'	32°31'	30°35'	29°3'
LENGTH OF CARRIAGE		19'10 27/32"	21'1 15/32"	22'4 5/16"	23'7 1/8"	17'7 1/8"	18'8 25/32"	19'10 27/32"	21'1 15/32"	22'4 5/16"	23'7 1/8"
L. F. OF RISER PER INCH OF STAIRWAY WIDTH		1.666	1.75	1.833	1.916	1.5	1.583	1.666	1.75	1.833	1.916
L. F. OF TREAD PER INCH OF STAIRWAY WIDTH		1.583	1.666	1.75	1.833	1.416	1.5	1.583	1.666	1.75	1.833
FRAMING SQUARE SETTINGS FOR CARRIAGE CUTS	TONGUE	7 3/16"	6 13/16"	6 1/2"	6 1/4"	8"	7 9/16"	7 3/16"	6 13/16"	6 1/2"	6 1/4"
	BODY	10 5/16"	10 11/16"	11"	11 1/4"	9 1/2"	9 15/16"	10 5/16"	10 11/16"	11"	11 1/4"

FLOOR TO FLOOR RISE		11'11 5/8"	11'11 5/8"	11'11 5/8"	11'11 5/8"	11'11 5/8"	11'11 5/8"	11'11 3/4"	11'11 3/4"	11'11 3/4"	11'11 3/4"
NUMBER OF RISERS		18	19	20	21	22	23	18	19	20	21
HEIGHT OF EACH RISER		8"	7 9/16"	7 3/16"	6 27/32"	6 17/32"	6 1/4"	8"	7 9/16"	7 3/16"	6 27/32"
TOTAL HEIGHT OF RISERS		12'	11'11 11/16"	11'11 3/4"	11'11 23/32"	11'11 11/16"	11'11 3/4"	12'	11'11 11/16"	11'11 3/4"	11'11 23/32"
RISERS OVERAGE (+) OR UNDERAGE (—)		+3/8"	+1/16"	+1/8"	+3/32"	+1/16"	+1/8"	+1/4"	−1/16"	0	−1/32"
NUMBER OF TREADS		17	18	19	20	21	22	17	18	19	20
WIDTH OF EACH TREAD		9 1/2"	9 15/16"	10 5/16"	10 21/32"	10 31/32"	11 1/4"	9 1/2"	9 15/16"	10 5/16"	10 21/32"
TOTAL RUN OF STAIRWAY		13'5 1/2"	14'10 7/8"	16'3 15/16"	17'9 1/8"	19'2 11/32"	20'7 1/2"	13'5 1/2"	14'10 7/8"	16'3 15/16"	17'9 1/8"
WELL OPENING FOR 6'-8" HEADROOM		8'10 7/8"	10'2 7/16"	11'6 3/32"	11'11 5/16"	13'3 13/32"	14'7 1/2"	8'10 7/8"	10'2 7/16"	11'6 3/32"	11'11 15/32"
ANGLE OF INCLINE		40°6'	37°16'	34°53'	32°43'	30°46'	29°3'	40°6'	37°16'	34°53'	32°43'
LENGTH OF CARRIAGE		17'7 1/8"	18'8 25/32"	19'10 27/32"	21'1 9/32"	22'4 3/32"	23'7 1/8"	17'7 1/8"	18'8 25/32"	19'10 27/32"	21'1 9/32"
L. F. OF RISER PER INCH OF STAIRWAY WIDTH		1.5	1.583	1.666	1.75	1.833	1.916	1.5	1.583	1.666	1.75
L. F. OF TREAD PER INCH OF STAIRWAY WIDTH		1.416	1.5	1.583	1.666	1.75	1.833	1.416	1.5	1.583	1.666
FRAMING SQUARE SETTINGS FOR CARRIAGE CUTS	TONGUE	8"	7 9/16"	7 3/16"	6 27/32"	6 17/32"	6 1/4"	8"	7 9/16"	7 3/16"	6 27/32"
	BODY	9 1/2"	9 15/16"	10 5/16"	10 21/32"	10 31/32"	11 1/4"	9 1/2"	9 15/16"	10 5/16"	10 21/32"

FLOOR TO FLOOR RISE		$11'11\frac{3}{4}''$	$11'11\frac{3}{4}''$	$11'11\frac{7}{8}''$	$11'11\frac{7}{8}''$	$11'11\frac{7}{8}''$	$11'11\frac{7}{8}''$	$11'11\frac{7}{8}''$	$11'11\frac{7}{8}''$	12'	12'
NUMBER OF RISERS		22	23	18	19	20	21	22	23	18	19
HEIGHT OF EACH RISER		$6\frac{17}{32}''$	$6\frac{1}{4}''$	8"	$7\frac{9}{16}''$	$7\frac{3}{16}''$	$6\frac{7}{8}''$	$6\frac{9}{16}''$	$6\frac{1}{4}''$	8"	$7\frac{9}{16}''$
TOTAL HEIGHT OF RISERS		$11'11\frac{11}{16}''$	$11'11\frac{3}{4}''$	12'	$11'11\frac{11}{16}''$	$11'11\frac{3}{4}''$	$12'\frac{3}{8}''$	$12'\frac{3}{8}''$	$11'11\frac{3}{4}''$	12'	$11'11\frac{11}{16}''$
RISERS OVERAGE (+) OR UNDERAGE (—)		$-\frac{1}{16}''$	0	$+\frac{1}{8}''$	$-\frac{3}{16}''$	$-\frac{1}{8}''$	$+\frac{1}{2}''$	$+\frac{1}{2}''$	$-\frac{1}{8}''$	0	$-\frac{5}{16}''$
NUMBER OF TREADS		21	22	17	18	19	20	21	22	17	18
WIDTH OF EACH TREAD		$10\frac{31}{32}''$	$11\frac{1}{4}''$	$9\frac{1}{2}''$	$9\frac{15}{16}''$	$10\frac{5}{16}''$	$10\frac{5}{8}''$	$10\frac{15}{16}''$	$11\frac{1}{4}''$	$9\frac{1}{2}''$	$9\frac{15}{16}''$
TOTAL RUN OF STAIRWAY		$19'2\frac{11}{32}''$	$20'7\frac{1}{2}''$	$13'5\frac{1}{2}''$	$14'10\frac{7}{8}''$	$16'3\frac{15}{16}''$	$17'8\frac{1}{2}''$	$19'1\frac{11}{16}''$	$20'7\frac{1}{2}''$	$13'5\frac{1}{2}''$	$14'10\frac{7}{8}''$
WELL OPENING FOR 6'-8" HEADROOM		$13'3\frac{13}{32}''$	$14'7\frac{1}{2}''$	$8'10\frac{7}{8}''$	$10'2\frac{7}{16}''$	$11'6\frac{3}{32}''$	$11'10\frac{31}{32}''$	$13'2\frac{27}{32}''$	$14'7\frac{1}{2}''$	$8'10\frac{7}{8}''$	$10'2\frac{7}{16}''$
ANGLE OF INCLINE		30°46'	29°3'	40°6'	37°16'	34°53'	32°54'	30°58'	29°3'	40°6'	37°16'
LENGTH OF CARRIAGE		$22'4\frac{3}{32}''$	$23'7\frac{1}{8}''$	$17'7\frac{1}{8}''$	$18'8\frac{25}{32}''$	$19'10\frac{27}{32}''$	$21'1\frac{3}{32}''$	$22'3\frac{7}{8}''$	$23'7\frac{1}{8}''$	$17'7\frac{1}{8}''$	$18'8\frac{25}{32}''$
L. F. OF RISER PER INCH OF STAIRWAY WIDTH		1.833	1.916	1.5	1.583	1.666	1.75	1.833	1.916	1.5	1.583
L. F. OF TREAD PER INCH OF STAIRWAY WIDTH		1.75	1.833	1.416	1.5	1.583	1.666	1.75	1.833	1.416	1.5
FRAMING SQUARE SETTINGS FOR CARRIAGE CUTS	TONGUE	$6\frac{17}{32}''$	$6\frac{1}{4}''$	8"	$7\frac{9}{16}''$	$7\frac{3}{16}''$	$6\frac{7}{8}''$	$6\frac{9}{16}''$	$6\frac{1}{4}''$	8"	$7\frac{9}{16}''$
	BODY	$10\frac{31}{32}''$	$11\frac{1}{4}''$	$9\frac{1}{2}''$	$9\frac{15}{16}''$	$10\frac{5}{16}''$	$10\frac{5}{8}''$	$10\frac{15}{16}''$	$11\frac{1}{4}''$	$9\frac{1}{2}''$	$9\frac{15}{16}''$

FLOOR TO FLOOR RISE	12'	12'	12'	12'	12'
NUMBER OF RISERS	20	21	22	23	24
HEIGHT OF EACH RISER	$7\frac{3}{16}$"	$6\frac{7}{8}$"	$6\frac{9}{16}$"	$6\frac{1}{4}$"	6"
TOTAL HEIGHT OF RISERS	11'$11\frac{3}{4}$"	12'$\frac{3}{8}$"	12'$\frac{3}{8}$"	11'$11\frac{3}{4}$"	12'
RISERS OVERAGE (+) OR UNDERAGE (—)	$-\frac{1}{4}$"	$+\frac{3}{8}$"	$+\frac{3}{8}$"	$-\frac{1}{4}$"	0
NUMBER OF TREADS	19	20	21	22	23
WIDTH OF EACH TREAD	$10\frac{5}{16}$"	$10\frac{5}{8}$"	$10\frac{15}{16}$"	$11\frac{1}{4}$"	$11\frac{1}{2}$"
TOTAL RUN OF STAIRWAY	16'$3\frac{15}{16}$"	17'$8\frac{1}{2}$"	19'$1\frac{11}{16}$"	20'$7\frac{1}{2}$"	22'$\frac{1}{2}$"
WELL OPENING FOR 6'-8" HEADROOM	11'$6\frac{3}{32}$"	11'$10\frac{31}{32}$"	13'$2\frac{27}{32}$"	14'$7\frac{1}{2}$"	15'$\frac{5}{32}$"
ANGLE OF INCLINE	34°53'	32°54'	30°58'	29°3'	27°33'
LENGTH OF CARRIAGE	19'$10\frac{27}{32}$"	21'$1\frac{3}{32}$"	22'$3\frac{7}{8}$"	23'$7\frac{1}{8}$"	24'$10\frac{11}{32}$"
L. F. OF RISER PER INCH OF STAIRWAY WIDTH	1.666	1.75	1.833	1.916	2.0
L. F. OF TREAD PER INCH OF STAIRWAY WIDTH	1.583	1.666	1.75	1.833	1.916
FRAMING SQUARE SETTINGS FOR CARRIAGE CUTS — TONGUE	$7\frac{3}{16}$"	$6\frac{7}{8}$"	$6\frac{9}{16}$"	$6\frac{1}{4}$"	6"
FRAMING SQUARE SETTINGS FOR CARRIAGE CUTS — BODY	$10\frac{5}{16}$"	$10\frac{5}{8}$"	$10\frac{15}{16}$"	$11\frac{1}{4}$"	$11\frac{1}{2}$"

Other Practical References

Rough Carpentry

All rough carpentry is covered in detail: sills, girders, columns, joists, sheathing, ceiling, roof and wall framing, roof trusses, dormers, bay windows, furring and grounds, stairs and insulation. Many of the 24 chapters explain practical code approved methods for saving lumber and time without sacrificing quality. Chapters on columns, headers, rafters, joists and girders show how to use simple engineering principles to selected the right lumber dimension for whatever species and grade you are using. **288 pages, 8½ x 11, $17.00**

National Construction Estimator

Current building costs in dollars and cents for residential, commercial, and industrial construction. Estimated prices for every commonly used building material. The manhours, recommended crew and labor cost for installation. Includes Estimate Writer, an electronic version of the book on computer disk -- at no extra cost on 5-1/4" high density (1.2Mb) disk. The 1991 National Construction Estimator and Estimate Writer on 1.2Mb disk cost **$22.50** (Add $10 if you want Estimate Writer on 5-1/4" double density 360K disks or 3½" 720K disks.)

How to Succeed With Your Own Construction Business

Everything you need to know and do to start your own construction business: setting up the paperwork, finding the work, advertising, using contracts, dealing with lenders, estimating, scheduling, finding and keeping good employees, keeping the books, and coping with success. If you're tired of working for someone else and considering starting your own construction business, all the knowledge, tips, and blank forms you need are in this book. **336 pages, 8½ x 11, $19.50**

Carpentry for Residential Construction

How to do professional quality carpentry work in homes and apartments. Illustrated instructions show you everything from setting batterboards to framing floors and walls, installing floor, wall and roof sheathing, and applying roofing. Covers finish carpentry, also: how to install each type of cornice, frieze, lookout, ledger, fascia and soffit; how to hang windows and doors; how to install siding, drywall and trim. Each job description includes the tools and materials needed, the estimated manhours required, and a step-by-step guide to each part of the task. **400 pages, 5½ x 8½, $19.75**

Building Layout

Shows how to use a transit to locate the building on the lot correctly, plan proper grades with minimum excavation, find utility lines and easements, establish correct elevations, lay out accurate foundations and set correct floor heights. Explains planning sewer connections, leveling a foundation out of level, using a story pole and batterboards, working on steep sites, and minimizing excavation costs. **240 pages, 5½ x 8½, $11.75**

Carpentry Estimating

Simple, clear instructions show you how to take off quantities and figure costs for all rough and finish carpentry. Shows how much overhead and profit to include, how to convert piece prices to MBF prices or linear foot prices, and how to use the tables included to quickly estimate manhours. All carpentry is covered; floor joists, exterior and interior walls and finishes, ceiling joists and rafters, stairs, trim, windows, doors, and much more. Includes sample forms, checklists, and the author's factor worksheets to save you time and help prevent errors. **320 pages, 8½ x 11, $25.50**

Running Your Remodeling Business

Everything you need to know about operating a remodeling business, from making your first sale to insuring your profits: how to advertise, write up a contract, estimate, schedule your jobs, arrange financing (for both you and your customers), and when and how to expand your business. Explains what you need to know about insurance, bonds, and liens, and how to collect the moeny you've earned. Includes sample business forms for your use. **272 pages, 8½ x 11, $21.00**

Bookkeeping for Builders

This book will show you simple, practical instructions for setting up and keeping accurate records — with a minimum of effort and frustration. Shows how to set up the essentials of a record-keeping system: the payment journal, income journal, general journal, records for fixed assets, accounts receivable, payables and purchases, petty cash, and job costs. You'll be able to keep the records required by the I.R.S., as well as accurate and organized business records for your own use. **208 pages, 8½ x 11, $19.75**

Remodeling Kitchens & Baths

This book is your guide to succeeding in a very lucrative area of the remodeling market: how to repair and replace damaged floors; how to redo walls, ceilings, and plumbing; and how to modernize the home wiring system to accommodate today's heavy electrical demands. Show how to install new sinks and countertops, ceramic tile, sunken tubs, whirlpool baths, luminous ceilings, skylights, and even special lighting effects. Completely illustrated, with manhour tables for figuring your labor costs. **384 pages, 8½ x 11, $26.25**

Video: Roof Framing 1

A complete step-by-step training video on the basics of roof cutting by Marshall Gross, the author of the book *Roof Framing*. Shows and explains calculating rise, run, and pitch, and laying out and cutting common rafters. **90 minutes, VHS, $80.00**

Video: Roof Framing 2

A complete training video on the more advanced techniques of roof framing by Marshall Gross, the author of *Roof Framing*. Shows and explains layout and framing an irregular roof, and making tie-ins to an existing roof. **90 minutes, VHS, $80.00**

Video: Stair Framing

Shows how to use a calculator to figure the rise and run of each step, the height of each riser, the number of treads, and the tread depth. Then watch how to take these measurements to construct an actual set of stairs. You'll see how to mark and cut your cartridges, treads, and risers, and install a stairway that fits your calculations for the perfect set of stairs. **60 minutes, VHS, $24.75**

Profits in Buying & Renovating Homes

Step-by-step instructions for selecting, repairing, improving, and selling highly profitable "fixer-uppers." Shows which price ranges offer the highest profit-to-investment ratios, which neighborhoods offer the best return, practical directions for repairs, and tips on dealing with buyers, sellers, and real estate agents. Shows you how to determine your profit before you buy, what bargains to avoid, and simple, inexpensive upgrades that will charm your buyers and ensure your profits. **304 pages, 8½ x 11, $19.75**

Rafter Length Manual

Complete rafter length tables and the "how to" of roof framing. Shows how to use the tables to find the actual length of common, hip, valley and jack rafters. Shows how to measure, mark, cut and erect the rafters, find the drop of the hip, shorten jack rafters, mark the ridge and much more. Has the tables, explanations and illustrations every professional roof framer needs. **369 pages, 5½ x 8½, $14.25**

NO POSTAGE
NECESSARY
IF MAILED
IN THE
UNITED STATES

BUSINESS REPLY MAIL

FIRST CLASS PERMIT NO. 271 CARLSBAD, CA

POSTAGE WILL BE PAID BY ADDRESSEE

Craftsman Book Company
6058 Corte Del Cedro
P. O. Box 6500
Carlsbad, CA 92008-0992